11 – 19 PROGRESSION

KU-012-831

endorsed for edexcel

College of Richard Collyer
Department Copy

Edexcel AS and A level Further Mathematics

Core Pure Mathematics Book 1/AS

Series Editor: Harry Smith

Authors: Greg Attwood, Jack Barraclough, Ian Bettison, Lee Cope, Charles Garnet Cox, Daniel Goldberg, Alistair Macpherson, Bronwen Moran, Su Nicholson, Laurence Pateman, Joe Petran, Keith Pledger, Harry Smith, Geoff Staley, Dave Wilkins

Pearson

Published by Pearson Education Limited, 80 Strand, London WC2R 0RL.

www.pearsonschoolsandfecolleges.co.uk

Copies of official specifications for all Pearson qualifications may be found on the website: qualifications.pearson.com

Text © Pearson Education Limited 2017
Edited by Tech-Set Ltd, Gateshead
Typeset by Tech-Set Ltd, Gateshead
Original illustrations © Pearson Education Limited 2017
Cover illustration Marcus@kja-artists

The rights of Greg Attwood, Jack Barraclough, Ian Bettison, Lee Cope, Charles Garnet Cox, Daniel Goldberg, Alistair Macpherson, Bronwen Moran, Su Nicholson, Laurence Pateman, Joe Petran, Keith Pledger, Harry Smith, Geoff Staley, Dave Wilkins to be identified as authors of this work have been asserted by them in accordance with the Copyright, Designs and Patents Act 1988.

First published 2017

20 19 18 17
10 9 8 7 6 5 4 3 2

British Library Cataloguing in Publication Data
A catalogue record for this book is available from the British Library

ISBN 978 1 292 18333 6

Copyright notice
All rights reserved. No part of this publication may be reproduced in any form or by any means (including photocopying or storing it in any medium by electronic means and whether or not transiently or incidentally to some other use of this publication) without the written permission of the copyright owner, except in accordance with the provisions of the Copyright, Designs and Patents Act 1988 or under the terms of a licence issued by the Copyright Licensing Agency, Barnards Inn 86 Fetter Lane, London EC4A 1EN (www.cla.co.uk). Applications for the copyright owner's written permission should be addressed to the publisher.

Printed in the UK by Bell and Bain Ltd, Glasgow

Acknowledgements
The authors and publisher would like to thank the following for their kind permission to reproduce their photographs:

(Key: b-bottom; c-centre; l-left; r-right; t-top)

Alamy Stock Photo: Eric Robison 155, 209 (c), Paul Fleet 126, 209 (b), Photo12 94, 209 (a), Science History Images 17, 89 (b), Zoonar GmbH 43, 89 (c); **Getty Images:** Henrik Sorenson 1, 89 (a), John Foxx 167, 209 (d); **Paul Nylander:** 54, 89 (d); **Shutterstock.com:** Marc Sitkin 71, 89 (e)

All other images © Pearson Education

A note from the publisher
In order to ensure that this resource offers high-quality support for the associated Pearson qualification, it has been through a review process by the awarding body. This process confirms that this resource fully covers the teaching and learning content of the specification or part of a specification at which it is aimed. It also confirms that it demonstrates an appropriate balance between the development of subject skills, knowledge and understanding, in addition to preparation for assessment.

Endorsement does not cover any guidance on assessment activities or processes (e.g. practice questions or advice on how to answer assessment questions), included in the resource nor does it prescribe any particular approach to the teaching or delivery of a related course.

While the publishers have made every attempt to ensure that advice on the qualification and its assessment is accurate, the official specification and associated assessment guidance materials are the only authoritative source of information and should always be referred to for definitive guidance.

Pearson examiners have not contributed to any sections in this resource relevant to examination papers for which they have responsibility.

Examiners will not use endorsed resources as a source of material for any assessment set by Pearson.

Endorsement of a resource does not mean that the resource is required to achieve this Pearson qualification, nor does it mean that it is the only suitable material available to support the qualification, and any resource lists produced by the awarding body shall include this and other appropriate resources.

Pearson has robust editorial processes, including answer and fact checks, to ensure the accuracy of the content in this publication, and every effort is made to ensure this publication is free of errors. We are, however, only human, and occasionally errors do occur. Pearson is not liable for any misunderstandings that arise as a result of errors in this publication, but it is our priority to ensure that the content is accurate. If you spot an error, please do contact us at resourcescorrections@pearson.com so we can make sure it is corrected.

Contents

Overarching themes

The following three overarching themes have been fully integrated throughout the Pearson Edexcel AS and A level Mathematics series, so they can be applied alongside your learning and practice.

1. Mathematical argument, language and proof

- Rigorous and consistent approach throughout
- Notation boxes explain key mathematical language and symbols
- Dedicated sections on mathematical proof explain key principles and strategies
- Opportunities to critique arguments and justify methods

2. Mathematical problem solving

- Hundreds of problem-solving questions, fully integrated into the main exercises
- Problem-solving boxes provide tips and strategies
- Structured and unstructured questions to build confidence
- Challenge boxes provide extra stretch

The Mathematical Problem-solving cycle

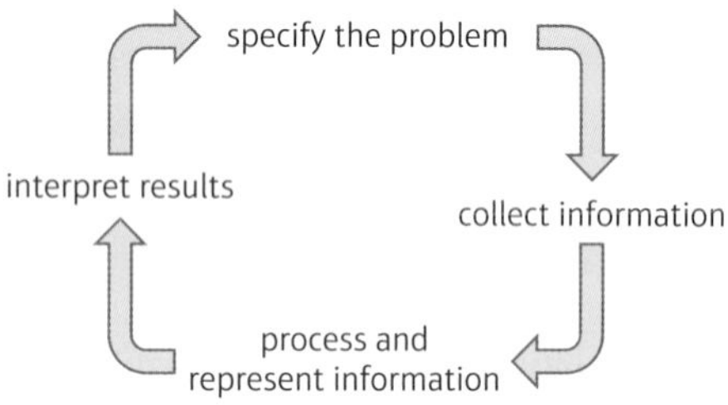

3. Mathematical modelling

- Dedicated modelling sections in relevant topics provide plenty of practice where you need it
- Examples and exercises include qualitative questions that allow you to interpret answers in the context of the model
- Dedicated chapter in Statistics & Mechanics Year 1/AS explains the principles of modelling in mechanics

Finding your way around the book

Access an online digital edition using the code at the front of the book.

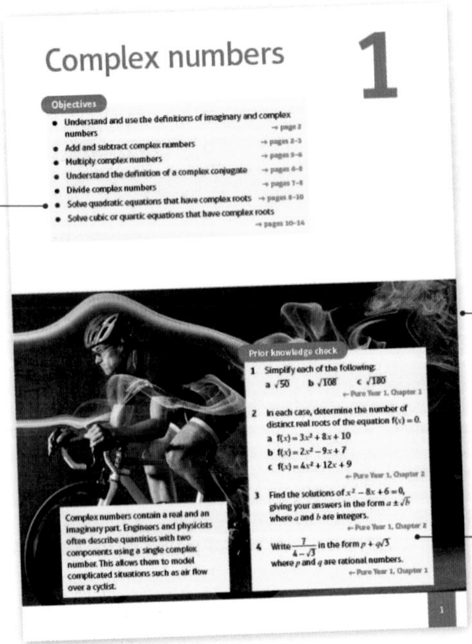

Each chapter starts with a list of objectives

The real world applications of the maths you are about to learn are highlighted at the start of the chapter with links to relevant questions in the chapter

The *Prior knowledge check* helps make sure you are ready to start the chapter

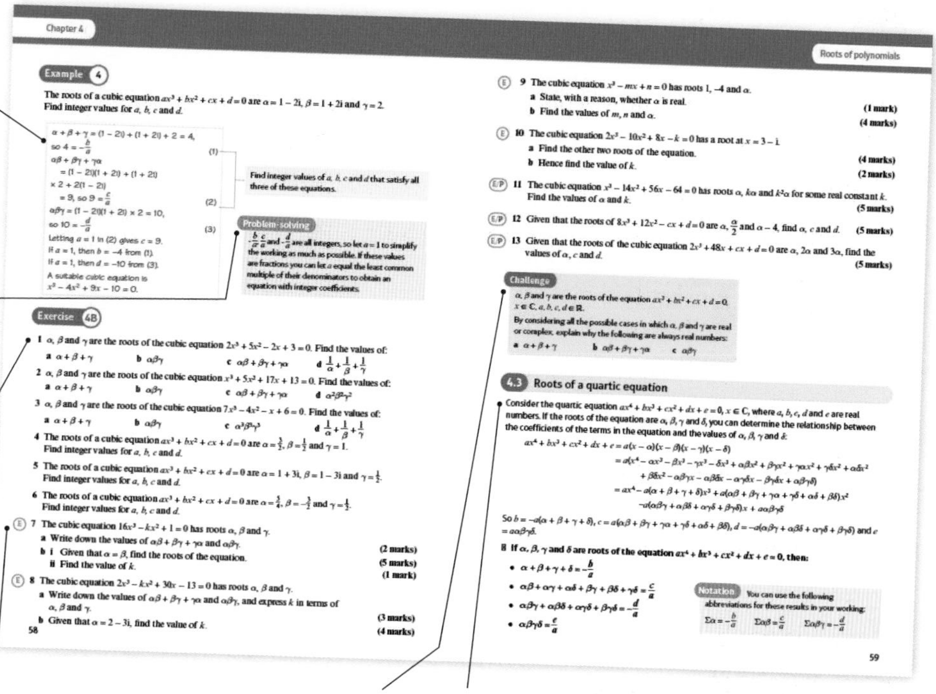
Chapter 4

Roots of polynomials

Example 4

Exercise 4B

Problem-solving

Challenge

4.3 Roots of a quartic equation

Notation

Step-by-step worked examples focus on the key types of questions you'll need to tackle

Problem-solving boxes provide hints, tips and strategies, and *Watch out* boxes highlight areas where students often lose marks in their exams

Exercise questions are carefully graded so they increase in difficulty and gradually bring you up to exam standard

Exam-style questions are flagged with Ⓔ

Problem-solving questions are flagged with Ⓟ

Exercises are packed with exam-style questions to ensure you are ready for the exams

Challenge boxes give you a chance to tackle some more difficult questions

Each section begins with explanation and key learning points

Each chapter ends with a *Mixed exercise* and a *Summary of key points*

Every few chapters a *Review exercise* helps you consolidate your learning with lots of exam-style questions

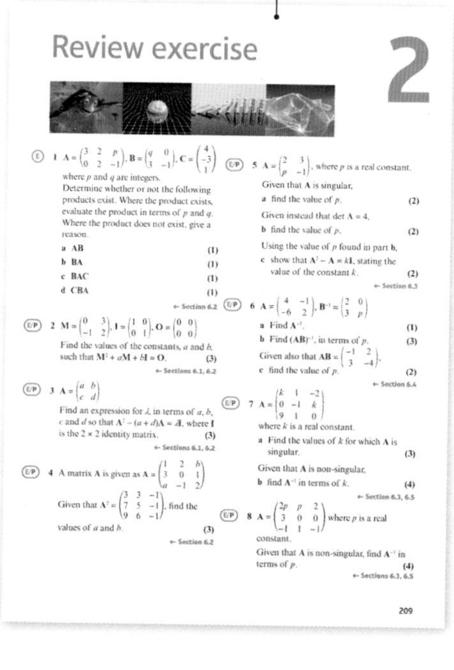
Review exercise 2

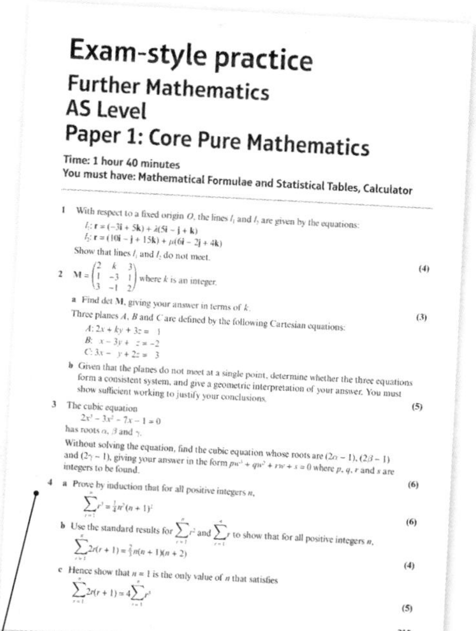
Exam-style practice

Further Mathematics

AS Level

Paper 1: Core Pure Mathematics

Time: 1 hour 40 minutes

You must have: Mathematical Formulae and Statistical Tables, Calculator

A full AS level practice paper at the back of the book helps you prepare for the real thing.

Extra online content

Whenever you see an *Online* box, it means that there is extra online content available to support you.

SolutionBank

SolutionBank provides a full worked solution for every question in the book.

Online Full worked solutions are available in SolutionBank.

Download all the solutions as a PDF or quickly find the solution you need online

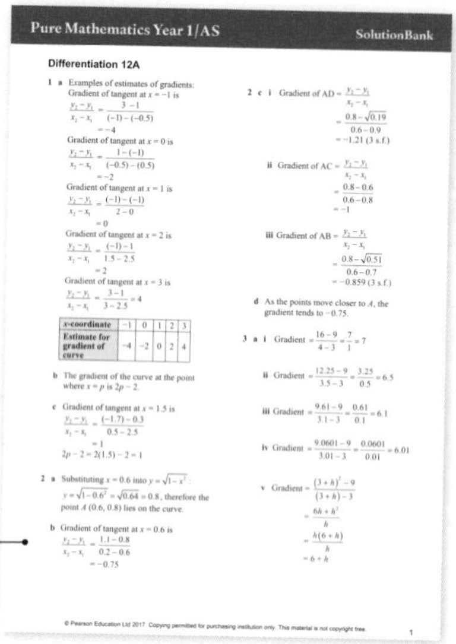

Pure Mathematics Year 1/AS — SolutionBank

Differentiation 12A

1 a Examples of estimates of gradients:

Gradient of tangent at $x = -1$ is

$\frac{y_2 - y_1}{x_2 - x_1} = \frac{3 - 1}{(-1) - (-0.5)} = -4$

Gradient of tangent at $x = 0$ is

$\frac{y_2 - y_1}{x_2 - x_1} = \frac{1 - (-1)}{(-0.5) - (0.5)} = -2$

Gradient of tangent at $x = 1$ is

$\frac{y_2 - y_1}{x_2 - x_1} = \frac{(-1) - (-1)}{2 - 0} = 0$

Gradient of tangent at $x = 2$ is

$\frac{y_2 - y_1}{x_2 - x_1} = \frac{(-1) - 1}{1.5 - 2.5} = 2$

Gradient of tangent at $x = 3$ is

$\frac{y_2 - y_1}{x_2 - x_1} = \frac{3 - 1}{3 - 2.5} = 4$

x-coordinate	−1	0	1	2	3
Estimate for gradient of curve	−4	−2	0	2	4

b The gradient of the curve at the point where $x = p$ is $2p - 2$.

c Gradient of tangent at $x = 1.5$ is

$\frac{y_2 - y_1}{x_2 - x_1} = \frac{(-1.7) - 0.3}{0.5 - 2.5} = 1$

$2p - 2 = 2(1.5) - 2 = 1$

2 a Substituting $x = 0.6$ into $y = \sqrt{1 - x^2}$:

$y = \sqrt{1 - 0.6^2} = \sqrt{0.64} = 0.8$, therefore the point A (0.6, 0.8) lies on the curve.

b Gradient of tangent at $x = 0.6$ is

$\frac{y_2 - y_1}{x_2 - x_1} = \frac{1.1 - 0.8}{0.2 - 0.6} = -0.75$

2 c i Gradient of AD $= \frac{y_2 - y_1}{x_2 - x_1} = \frac{0.8 - \sqrt{0.19}}{0.6 - 0.9} = -1.21$ (3 s.f.)

ii Gradient of AC $= \frac{y_2 - y_1}{x_2 - x_1} = \frac{0.8 - 0.6}{0.6 - 0.8} = -1$

iii Gradient of AB $= \frac{y_2 - y_1}{x_2 - x_1} = \frac{0.8 - \sqrt{0.51}}{0.6 - 0.7} = -0.859$ (3 s.f.)

d As the points move closer to A, the gradient tends to −0.75.

3 a i Gradient $= \frac{16 - 9}{4 - 3} = \frac{7}{1} = 7$

ii Gradient $= \frac{12.25 - 9}{3.5 - 3} = \frac{3.25}{0.5} = 6.5$

iii Gradient $= \frac{9.61 - 9}{3.1 - 3} = \frac{0.61}{0.1} = 6.1$

iv Gradient $= \frac{9.0601 - 9}{3.01 - 3} = \frac{0.0601}{0.01} = 6.01$

v Gradient $= \frac{(3 + h)^2 - 9}{(3 + h) - 3} = \frac{6h + h^2}{h} = \frac{h(6 + h)}{h} = 6 + h$

© Pearson Education Ltd 2017. Copying permitted for purchasing institution only. This material is not copyright free. 1

Use of technology

Explore topics in more detail, visualise problems and consolidate your understanding using pre-made GeoGebra activities.

Online Find the point of intersection graphically using technology.

GeoGebra

GeoGebra-powered interactives

Interact with the maths you are learning using GeoGebra's easy-to-use tools

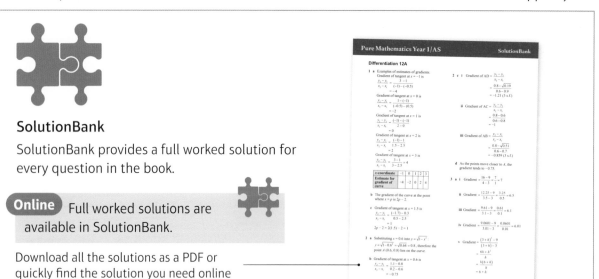

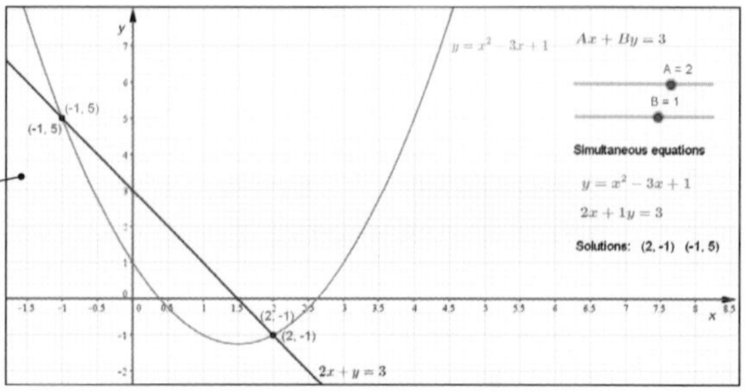

Access all the extra online content for free at:

www.pearsonschools.co.uk/cp1maths

You can also access the extra online content by scanning this QR code:

Complex numbers

1

Objectives

After completing this chapter you should be able to:

- Understand and use the definitions of imaginary and complex numbers → **page 2**
- Add and subtract complex numbers → **pages 2–3**
- Multiply complex numbers → **pages 5–6**
- Understand the definition of a complex conjugate → **pages 6–8**
- Divide complex numbers → **pages 7–8**
- Solve quadratic equations that have complex roots → **pages 8–10**
- Solve cubic or quartic equations that have complex roots → **pages 10–14**

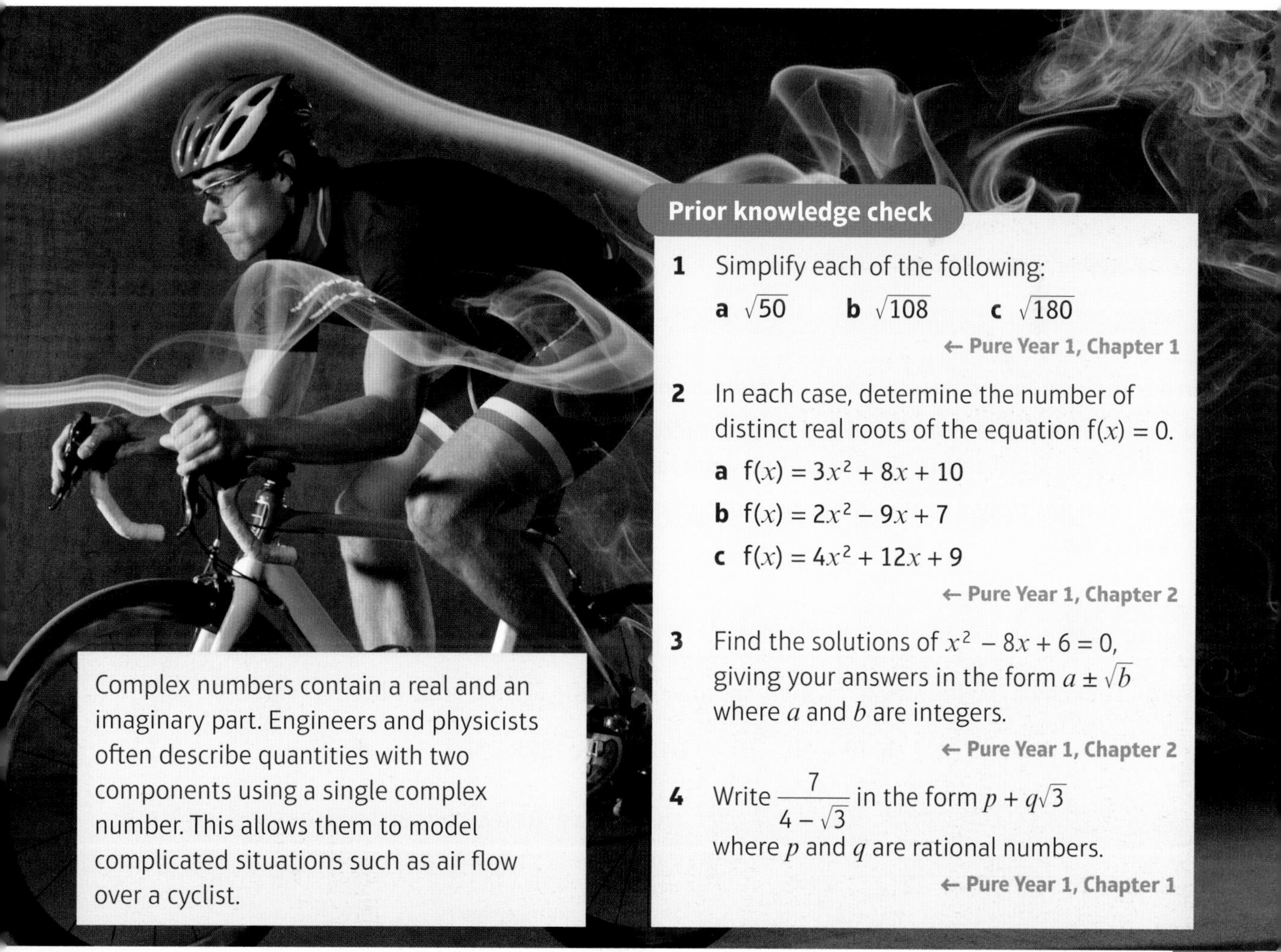

Complex numbers contain a real and an imaginary part. Engineers and physicists often describe quantities with two components using a single complex number. This allows them to model complicated situations such as air flow over a cyclist.

Prior knowledge check

1 Simplify each of the following:

a $\sqrt{50}$ **b** $\sqrt{108}$ **c** $\sqrt{180}$

← **Pure Year 1, Chapter 1**

2 In each case, determine the number of distinct real roots of the equation $f(x) = 0$.

a $f(x) = 3x^2 + 8x + 10$

b $f(x) = 2x^2 - 9x + 7$

c $f(x) = 4x^2 + 12x + 9$

← **Pure Year 1, Chapter 2**

3 Find the solutions of $x^2 - 8x + 6 = 0$, giving your answers in the form $a \pm \sqrt{b}$ where a and b are integers.

← **Pure Year 1, Chapter 2**

4 Write $\frac{7}{4-\sqrt{3}}$ in the form $p + q\sqrt{3}$ where p and q are rational numbers.

← **Pure Year 1, Chapter 1**

1.1 Imaginary and complex numbers

The quadratic equation $ax^2 + bx + c = 0$ has solutions given by

$$x = \frac{-b \pm \sqrt{b^2 - 4ac}}{2a}$$

If the expression under the square root is negative, there are no real solutions.

Links For the equation $ax^2 + bx + c = 0$, the **discriminant** is $b^2 - 4ac$.
- If $b^2 - 4ac > 0$, there are two distinct real roots.
- If $b^2 - 4ac = 0$, there are two equal real roots.
- If $b^2 - 4ac < 0$, there are no real roots.

← Pure Year 1, Section 2.5

You can find solutions to the equation in all cases by extending the number system to include $\sqrt{-1}$. Since there is no real number that squares to produce −1, the number $\sqrt{-1}$ is called an **imaginary number**, and is represented using the letter **i**. Sums of real and imaginary numbers, for example 3 + 2i, are known as **complex numbers**.

- $\mathbf{i = \sqrt{-1}}$
- **An imaginary number is a number of the form bi, where $b \in \mathbb{R}$.**
- **A complex number is written in the form $a + b$i, where $a, b \in \mathbb{R}$.**

Notation The set of all complex numbers is written as $\mathbb{C}$.
For the complex number $z = a + b\text{i}$:
- $\text{Re}(z) = a$ is the real part
- $\text{Im}(z) = b$ is the imaginary part

Example 1

Write each of the following in terms of i.

a $\sqrt{-36}$ **b** $\sqrt{-28}$

a $\sqrt{-36} = \sqrt{36 \times (-1)} = \sqrt{36}\sqrt{-1} = 6\text{i}$

b $\sqrt{-28} = \sqrt{28 \times (-1)} = \sqrt{28}\sqrt{-1}$
$= \sqrt{4}\sqrt{7}\sqrt{-1} = (2\sqrt{7})\text{i}$

You can use the rules of surds to manipulate imaginary numbers.

Watch out An alternative way of writing $(2\sqrt{7})\text{i}$ is $2\text{i}\sqrt{7}$. Avoid writing $2\sqrt{7}\text{i}$ as this can easily be confused with $2\sqrt{7\text{i}}$.

In a complex number, the real part and the imaginary part cannot be combined to form a single term.

- **Complex numbers can be added or subtracted by adding or subtracting their real parts and adding or subtracting their imaginary parts.**
- **You can multiply a real number by a complex number by multiplying out the brackets in the usual way.**

Example 2

Simplify each of the following, giving your answers in the form $a + b\text{i}$, where $a, b \in \mathbb{R}$.

a $(2 + 5\text{i}) + (7 + 3\text{i})$ **b** $(2 - 5\text{i}) - (5 - 11\text{i})$ **c** $2(5 - 8\text{i})$ **d** $\frac{10 + 6\text{i}}{2}$

a $(2 + 5\text{i}) + (7 + 3\text{i}) = (2 + 7) + (5 + 3)\text{i}$
$= 9 + 8\text{i}$

Add the real parts and add the imaginary parts.

b $(2 - 5\text{i}) - (5 - 11\text{i}) = (2 - 5) + (-5 - (-11))\text{i}$
$= -3 + 6\text{i}$

Subtract the real parts and subtract the imaginary parts.

c $2(5 - 8i) = (2 \times 5) - (2 \times 8)i = 10 - 16i$ — $2(5 - 8i)$ can also be written as $(5 - 8i) + (5 - 8i)$.

d $\frac{10 + 6i}{2} = \frac{10}{2} + \frac{6}{2}i = 5 + 3i$ — First separate into real and imaginary parts.

Exercise 1A

Do not use your calculator in this exercise.

1 Write each of the following in the form bi where b is a real number.

a $\sqrt{-9}$ **b** $\sqrt{-49}$ **c** $\sqrt{-121}$ **d** $\sqrt{-10000}$ **e** $\sqrt{-225}$

f $\sqrt{-5}$ **g** $\sqrt{-12}$ **h** $\sqrt{-45}$ **i** $\sqrt{-200}$ **j** $\sqrt{-147}$

2 Simplify, giving your answers in the form $a + bi$, where $a, b \in \mathbb{R}$.

a $(5 + 2i) + (8 + 9i)$ **b** $(4 + 10i) + (1 - 8i)$

c $(7 + 6i) + (-3 - 5i)$ **d** $(\frac{1}{2} + \frac{1}{3}i) + (\frac{5}{2} + \frac{5}{3}i)$

e $(20 + 12i) - (11 + 3i)$ **f** $(2 - i) - (-5 + 3i)$

g $(-4 - 6i) - (-8 - 8i)$ **h** $(3\sqrt{2} + i) - (\sqrt{2} - i)$

i $(-2 - 7i) + (1 + 3i) - (-12 + i)$ **j** $(18 + 5i) - (15 - 2i) - (3 + 7i)$

3 Simplify, giving your answers in the form $a + bi$, where $a, b \in \mathbb{R}$.

a $2(7 + 2i)$ **b** $3(8 - 4i)$

c $2(3 + i) + 3(2 + i)$ **d** $5(4 + 3i) - 4(-1 + 2i)$

e $\frac{6 - 4i}{2}$ **f** $\frac{15 + 25i}{5}$

g $\frac{9 + 11i}{3}$ **h** $\frac{-8 + 3i}{4} - \frac{7 - 2i}{2}$

(P) **4** Write in the form $a + bi$, where a and b are simplified surds.

a $\frac{4 - 2i}{\sqrt{2}}$ **b** $\frac{2 - 6i}{1 + \sqrt{3}}$

5 Given that $z = 7 - 6i$ and $w = 7 + 6i$, find, in the form $a + bi$, where $a, b \in \mathbb{R}$:

a $z - w$ **b** $w + z$

Notation Complex numbers are often represented by the letter z or the letter w.

(E) **6** Given that $z_1 = a + 9i$, $z_2 = -3 + bi$ and $z_2 - z_1 = 7 + 2i$, find a and b where $a, b \in \mathbb{R}$. **(2 marks)**

(P) **7** Given that $z_1 = 4 + i$ and $z_2 = 7 - 3i$, find, in the form $a + bi$, where $a, b \in \mathbb{R}$:

a $z_1 - z_2$ **b** $4z_2$ **c** $2z_1 + 5z_2$

(P) **8** Given that $z = a + bi$ and $w = a - bi$, $a, b \in \mathbb{R}$, show that:

a $z + w$ is always real **b** $z - w$ is always imaginary

You can use complex numbers to find solutions to any quadratic equation with real coefficients.

- **If $b^2 - 4ac < 0$ then the quadratic equation $ax^2 + bx + c = 0$ has two distinct complex roots, neither of which are real.**

Example 3

Solve the equation $z^2 + 9 = 0$.

$z^2 = -9$

$z = \pm\sqrt{-9} = \pm\sqrt{9 \times -1} = \pm\sqrt{9}\sqrt{-1} = \pm 3\text{i}$

$z = +3\text{i},\ z = -3\text{i}$

Note that just as $z^2 = 9$ has two roots $+3$ and -3, $z^2 = -9$ also has two roots $+3\text{i}$ and -3i.

Example 4

Solve the equation $z^2 + 6z + 25 = 0$.

Method 1 (Completing the square)

$z^2 + 6z = (z + 3)^2 - 9$

Because $(z + 3)^2 = (z + 3)(z + 3) = z^2 + 6z + 9$

$z^2 + 6z + 25 = (z + 3)^2 - 9 + 25 = (z + 3)^2 + 16$

$(z + 3)^2 + 16 = 0$

$(z + 3)^2 = -16$

$z + 3 = \pm\sqrt{-16} = \pm 4\text{i}$

$\sqrt{-16} = \sqrt{16 \times (-1)} = \sqrt{16}\sqrt{-1} = 4\text{i}$

$z = -3 \pm 4\text{i}$

$z = -3 + 4\text{i}, \quad z = -3 - 4\text{i}$

You can use your calculator to find the complex roots of a quadratic equation like this one.

Method 2 (Quadratic formula)

$z = \dfrac{-6 \pm \sqrt{6^2 - 4 \times 1 \times 25}}{2}$

Using $z = \dfrac{-b \pm \sqrt{b^2 - 4ac}}{2a}$

$= \dfrac{-6 \pm \sqrt{-64}}{2}$

$\sqrt{-64} = \sqrt{64 \times (-1)} = \sqrt{64}\sqrt{-1} = 8\text{i}$

$z = \dfrac{-6 \pm 8\text{i}}{2} = -3 \pm 4\text{i}$

$z = -3 + 4\text{i}, \quad z = -3 - 4\text{i}$

Exercise 1B

Do not use your calculator in this exercise.

1 Solve each of the following equations. Write your answers in the form $\pm b\text{i}$.

a $z^2 + 121 = 0$ **b** $z^2 + 40 = 0$ **c** $2z^2 + 120 = 0$

d $3z^2 + 150 = 38 - z^2$ **e** $z^2 + 30 = -3z^2 - 66$ **f** $6z^2 + 1 = 2z^2$

2 Solve each of the following equations. Write your answers in the form $a \pm b\text{i}$.

a $(z - 3)^2 - 9 = -16$

b $2(z - 7)^2 + 30 = 6$

c $16(z + 1)^2 + 11 = 2$

Hint The left-hand side of each equation is in completed square form already. Use inverse operations to find the values of z.

3 Solve each of the following equations. Write your answers in the form $a \pm bi$.

a $z^2 + 2z + 5 = 0$ **b** $z^2 - 2z + 10 = 0$ **c** $z^2 + 4z + 29 = 0$

d $z^2 + 10z + 26 = 0$ **e** $z^2 + 5z + 25 = 0$ **f** $z^2 + 3z + 5 = 0$

4 Solve each of the following equations. Write your answers in the form $a \pm bi$.

a $2z^2 + 5z + 4 = 0$ **b** $7z^2 - 3z + 3 = 0$ **c** $5z^2 - z + 3 = 0$

5 The solutions to the quadratic equation $z^2 - 8z + 21 = 0$ are z_1 and z_2.
Find z_1 and z_2, giving each in the form $a \pm \mathrm{i}\sqrt{b}$.

E/P **6** The equation $z^2 + bz + 11 = 0$, where $b \in \mathbb{R}$, has distinct non-real complex roots.
Find the range of possible values of b. **(3 marks)**

1.2 Multiplying complex numbers

You can multiply complex numbers using the same technique that you use for multiplying brackets in algebra. You can use the fact that $\mathrm{i} = \sqrt{-1}$ to simplify powers of i.

- $\mathbf{i^2 = -1}$

Example 5

Express each of the following in the form $a + bi$, where a and b are real numbers.

a $(2 + 3\mathrm{i})(4 + 5\mathrm{i})$ **b** $(7 - 4\mathrm{i})^2$

a $(2 + 3\mathrm{i})(4 + 5\mathrm{i}) = 2(4 + 5\mathrm{i}) + 3\mathrm{i}(4 + 5\mathrm{i})$ — Multiply the two brackets as you would with real numbers.
$= 8 + 10\mathrm{i} + 12\mathrm{i} + 15\mathrm{i}^2$
$= 8 + 10\mathrm{i} + 12\mathrm{i} - 15$ — Use the fact that $\mathrm{i}^2 = -1$.
$= (8 - 15) + (10\mathrm{i} + 12\mathrm{i})$ — Add real parts and add imaginary parts.
$= -7 + 22\mathrm{i}$

b $(7 - 4\mathrm{i})^2 = (7 - 4\mathrm{i})(7 - 4\mathrm{i})$
$= 7(7 - 4\mathrm{i}) - 4\mathrm{i}(7 - 4\mathrm{i})$ — Multiply out the two brackets as you would with real numbers.
$= 49 - 28\mathrm{i} - 28\mathrm{i} + 16\mathrm{i}^2$
$= 49 - 28\mathrm{i} - 28\mathrm{i} - 16$ — Use the fact that $\mathrm{i}^2 = -1$.
$= (49 - 16) + (-28\mathrm{i} - 28\mathrm{i})$ — Add real parts and add imaginary parts.
$= 33 - 56\mathrm{i}$

Example 6

Simplify: **a** i^3 **b** i^4 **c** $(2\mathrm{i})^5$

a $\mathrm{i}^3 = \mathrm{i} \times \mathrm{i} \times \mathrm{i} = \mathrm{i}^2 \times \mathrm{i} = -\mathrm{i}$ — $\mathrm{i}^2 = -1$
b $\mathrm{i}^4 = \mathrm{i} \times \mathrm{i} \times \mathrm{i} \times \mathrm{i} = \mathrm{i}^2 \times \mathrm{i}^2 = (-1) \times (-1) = 1$
c $(2\mathrm{i})^5 = 2\mathrm{i} \times 2\mathrm{i} \times 2\mathrm{i} \times 2\mathrm{i} \times 2\mathrm{i}$ — $(2\mathrm{i})^5 = 2^5 \times \mathrm{i}^5$. First work out $2^5 = 32$.
$= 32(\mathrm{i} \times \mathrm{i} \times \mathrm{i} \times \mathrm{i} \times \mathrm{i}) = 32(\mathrm{i}^2 \times \mathrm{i}^2 \times \mathrm{i})$
$= 32 \times (-1) \times (-1) \times \mathrm{i} = 32\mathrm{i}$

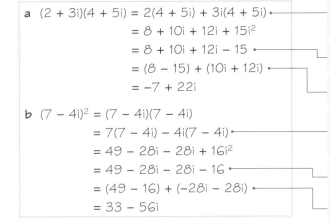

Exercise 1C

Do not use your calculator in this exercise.

1 Simplify each of the following, giving your answers in the form $a + b\text{i}$.

a $(5 + \text{i})(3 + 4\text{i})$ **b** $(6 + 3\text{i})(7 + 2\text{i})$ **c** $(5 - 2\text{i})(1 + 5\text{i})$

d $(13 - 3\text{i})(2 - 8\text{i})$ **e** $(-3 - \text{i})(4 + 7\text{i})$ **f** $(8 + 5\text{i})^2$

g $(2 - 9\text{i})^2$ **h** $(1 + \text{i})(2 + \text{i})(3 + \text{i})$

i $(3 - 2\text{i})(5 + \text{i})(4 - 2\text{i})$ **j** $(2 + 3\text{i})^3$

Hint For part **h**, begin by multiplying the first pair of brackets.

(P) **2 a** Simplify $(4 + 5\text{i})(4 - 5\text{i})$, giving your answer in the form $a + b\text{i}$.

b Simplify $(7 - 2\text{i})(7 + 2\text{i})$, giving your answer in the form $a + b\text{i}$.

c Comment on your answers to parts **a** and **b**.

d Prove that $(a + b\text{i})(a - b\text{i})$ is a real number for any real numbers a and b.

(P) **3** Given that $(a + 3\text{i})(1 + b\text{i}) = 25 - 39\text{i}$, find two possible pairs of values for a and b.

4 Write each of the following in its simplest form.

a i^6 **b** $(3\text{i})^4$ **c** $\text{i}^5 + \text{i}$ **d** $(4\text{i})^3 - 4\text{i}^3$

(P) **5** Express $(1 + \text{i})^6$ in the form $a - b\text{i}$, where a and b are integers to be found.

(P) **6** Find the value of the real part of $(3 - 2\text{i})^4$.

Problem-solving

You can use the binomial theorem to expand $(a + b)^n$. ← **Pure Year 1, Section 8.3**

(P) **7** $\text{f}(z) = 2z^2 - z + 8$

Find: **a** $\text{f}(2\text{i})$ **b** $\text{f}(3 - 6\text{i})$

(E/P) **8** $\text{f}(z) = z^2 - 2z + 17$

Show that $z = 1 - 4\text{i}$ is a solution to $\text{f}(z) = 0$. **(2 marks)**

9 a Given that $\text{i}^1 = \text{i}$ and $\text{i}^2 = -1$, write i^3 and i^4 in their simplest forms.

b Write i^5, i^6, i^7 and i^8 in their simplest forms.

c Write down the value of:

i i^{100} **ii** i^{253} **iii** i^{301}

Challenge

a Expand $(a + b\text{i})^2$.

b Hence, or otherwise, find $\sqrt{40 - 42\text{i}}$, giving your answer in the form $a - b\text{i}$, where a and b are positive integers.

Notation The **principal square root** of a complex number, $\sqrt{z}$, has a positive real part.

1.3 Complex conjugation

- **For any complex number $z = a + b\text{i}$, the complex conjugate of the number is defined as $z^* = a - b\text{i}$.**

Notation Together z and z^* are called a **complex conjugate pair**.

Example 7

Given that $z = 2 - 7\text{i}$,

a write down z^* **b** find the value of $z + z^*$ **c** find the value of zz^*

a $z^* = 2 + 7i$

b $z + z^* = (2 - 7i) + (2 + 7i)$
$= (2 + 2) + (-7 + 7)i = 4$

c $zz^* = (2 - 7i)(2 + 7i)$
$= 2(2 + 7i) - 7i(2 + 7i)$
$= 4 + 14i - 14i - 49i^2$
$= 4 + 49 = 53$

Change the sign of the imaginary part from – to +.

Note Notice that $z + z^*$ is real.

Remember $i^2 = -1$.

Note Notice that zz^* is real.

For any complex number z, the product of z and z^* is a real number. You can use this property to **divide two complex numbers**. To do this, you multiply both the numerator and the denominator by the complex conjugate of the denominator and then simplify the result.

Links The method used to divide complex numbers is similar to the method used to rationalise a denominator when simplifying surds.
← **Pure Year 1, Section 1.6**

Example 8

Write $\frac{5 + 4i}{2 - 3i}$ in the form $a + bi$.

$$\frac{5 + 4i}{2 - 3i} = \frac{5 + 4i}{2 - 3i} \times \frac{2 + 3i}{2 + 3i}$$

$$= \frac{(5 + 4i)(2 + 3i)}{(2 - 3i)(2 + 3i)}$$

$$(5 + 4i)(2 + 3i) = 5(2 + 3i) + 4i(2 + 3i)$$
$$= 10 + 15i + 8i + 12i^2$$
$$= -2 + 23i$$

$$(2 - 3i)(2 + 3i) = 2(2 + 3i) - 3i(2 + 3i)$$
$$= 4 + 6i - 6i - 9i^2 = 13$$

$$\frac{5 + 4i}{2 - 3i} = \frac{-2 + 23i}{13} = -\frac{2}{13} + \frac{23}{13}i$$

The complex conjugate of the denominator is $2 + 3i$. Multiply both the numerator and the denominator by the complex conjugate.

zz^* is real, so $(2 - 3i)(2 + 3i)$ will be a real number.

You can enter complex numbers directly into your calculator to multipy or divide them quickly.

Divide each term in the numerator by 13.

Exercise 1D

Do not use your calculator in this exercise.

1 Write down the complex conjugate z^* for:

a $z = 8 + 2i$ **b** $z = 6 - 5i$ **c** $z = \frac{2}{3} - \frac{1}{2}i$ **d** $z = \sqrt{5} + i\sqrt{10}$

2 Find $z + z^*$ and zz^* for:

a $z = 6 - 3i$ **b** $z = 10 + 5i$ **c** $z = \frac{3}{4} + \frac{1}{4}i$ **d** $z = \sqrt{5} - 3i\sqrt{5}$

3 Write each of the following in the form $a + bi$.

a $\frac{3 - 5i}{1 + 3i}$ **b** $\frac{3 + 5i}{6 - 8i}$ **c** $\frac{28 - 3i}{1 - i}$ **d** $\frac{2 + i}{1 + 4i}$

4 Write $\frac{(3-4i)^2}{1+i}$ in the form $x + iy$ where $x, y \in \mathbb{R}$.

5 Given that $z_1 = 1 + i$, $z_2 = 2 + i$ and $z_3 = 3 + i$, write each of the following in the form $a + bi$.

a $\frac{z_1 z_2}{z_3}$ **b** $\frac{(z_2)^2}{z_1}$ **c** $\frac{2z_1 + 5z_3}{z_2}$

(E) **6** Given that $\frac{5+2i}{z} = 2 - i$, find z in the form $a + bi$. **(2 marks)**

7 Simplify $\frac{6+8i}{1+i} + \frac{6+8i}{1-i}$, giving your answer in the form $a + bi$.

8 $w = \frac{4}{8 - i\sqrt{2}}$

Express w in the form $a + bi\sqrt{2}$, where a and b are rational numbers.

9 $w = 1 - 9i$

Express $\frac{1}{w}$ in the form $a + bi$, where a and b are rational numbers.

10 $z = 4 - i\sqrt{2}$

Use algebra to express $\frac{z+4}{z-3}$ in the form $p + qi\sqrt{2}$, where p and q are rational numbers.

(E/P) **11** The complex number z satisfies the equation $(4 + 2i)(z - 2i) = 6 - 4i$.
Find z, giving your answer in the form $a + bi$ where a and b are rational numbers. **(4 marks)**

(E/P) **12** The complex numbers z_1 and z_2 are given by $z_1 = p - 7i$ and $z_2 = 2 + 5i$ where p is an integer.
Find $\frac{z_1}{z_2}$ in the form $a + bi$ where a and b are rational, and are given in terms of p. **(4 marks)**

(E) **13** $z = \sqrt{5} + 4i$. z^* is the complex conjugate of z.
Show that $\frac{z}{z^*} = a + bi\sqrt{5}$, where a and b are rational numbers to be found. **(4 marks)**

(E/P) **14** The complex number z is defined by $z = \frac{p+5i}{p-2i}$, $p \in \mathbb{R}$, $p > 0$.
Given that the real part of z is $\frac{1}{2}$,

a find the value of p **(4 marks)**

b write z in the form $a + bi$, where a and b are real. **(1 mark)**

1.4 Roots of quadratic equations

- **For real numbers a, b and c, if the roots of the quadratic equation $az^2 + bz + c = 0$ are non-real complex numbers, then they occur as a conjugate pair.**

Another way of stating this is that for a real-valued quadratic function $f(z)$, if z_1 is a root of $f(z) = 0$ then z_1^* is also a root. You can use this fact to find one root if you know the other, or to find the original equation.

- **If the roots of a quadratic equation are α and β, then you can write the equation as $(z - \alpha)(z - \beta) = 0$**

  or $z^2 - (\alpha + \beta)z + \alpha\beta = 0$

Notation Roots of complex-valued polynomials are often written using Greek letters such as α (alpha), β (beta) and γ (gamma).

Example 9

Given that $\alpha = 7 + 2i$ is one of the roots of a quadratic equation with real coefficients,

a state the value of the other root, β

b find the quadratic equation

c find the values of $\alpha + \beta$ and $\alpha\beta$ and interpret the results.

a $\beta = 7 - 2i$

α and β will always be a complex conjugate pair.

b
$$(z - \alpha)(z - \beta) = 0$$
$$(z - (7 + 2i))(z - (7 - 2i)) = 0$$
$$z^2 - z(7 - 2i) - z(7 + 2i) + (7 + 2i)(7 - 2i) = 0$$
$$z^2 - 7z + 2iz - 7z - 2iz + 49 - 14i + 14i - 4i^2 = 0$$
$$z^2 - 14z + 49 + 4 = 0$$
$$z^2 - 14z + 53 = 0$$

The quadratic equation with roots α and β is $(z - \alpha)(z - \beta) = 0$

Collect like terms. Use the fact that $i^2 = -1$.

c $\alpha + \beta = (7 + 2i) + (7 - 2i)$
$= (7 + 7) + (2 + (-2))i = 14$

The coefficient of z in the above equation is $-(\alpha + \beta)$.

$\alpha\beta = (7 + 2i)(7 - 2i) = 49 - 14i + 14i - 4i^2$
$= 49 + 4 = 53$

The constant term in the above equation is $\alpha\beta$.

Problem-solving

For $z = a + bi$, you should learn the results:

$$z + z^* = 2a$$
$$zz^* = a^2 + b^2$$

You can use these to find the quadratic equation quickly.

Exercise 1E

1 The roots of the quadratic equation $z^2 + 2z + 26 = 0$ are α and β.

Find: **a** α and β **b** $\alpha + \beta$ **c** $\alpha\beta$

2 The roots of the quadratic equation $z^2 - 8z + 25 = 0$ are α and β.

Find: **a** α and β **b** $\alpha + \beta$ **c** $\alpha\beta$

(E) **3** Given that $2 + 3i$ is one of the roots of a quadratic equation with real coefficients,

a write down the other root of the equation **(1 mark)**

b find the quadratic equation, giving your answer in the form $z^2 + bz + c = 0$ where b and c are real constants. **(3 marks)**

(E) **4** Given that $5 - i$ is a root of the equation $z^2 + pz + q = 0$, where p and q are real constants,

a write down the other root of the equation **(1 mark)**

b find the value of p and the value of q. **(3 marks)**

(E/P) **5** Given that $z_1 = -5 + 4i$ is one of the roots of the quadratic equation $z^2 + bz + c = 0$, where b and c are real constants, find the values of b and c. **(4 marks)**

(E/P) **6** Given that $1 + 2i$ is one of the roots of a quadratic equation with real coefficients, find the equation giving your answer in the form $z^2 + bz + c = 0$ where b and c are integers to be found. **(4 marks)**

E/P **7** Given that $3 - 5\text{i}$ is one of the roots of a quadratic equation with real coefficients, find the equation giving your answer in the form $z^2 + bz + c = 0$ where b and c are real constants. **(4 marks)**

E/P **8** $z = \dfrac{5}{3 - \text{i}}$

a Find z in the form $a + b\text{i}$, where a and b are real constants. **(1 mark)**

Given that z is a complex root of the quadratic equation $x^2 + px + q = 0$, where p and q are real integers,

b find the value of p and the value of q. **(4 marks)**

E/P **9** Given that $z = 5 + q\text{i}$ is a root of the equation $z^2 - 4pz + 34 = 0$, where p and q are positive real constants, find the value of p and the value of q. **(4 marks)**

1.5 Solving cubic and quartic equations

You can generalise the rule for the roots of quadratic equations to any polynomial with real coefficients.

- **If f(z) is a polynomial with real coefficients, and z_1 is a root of f(z) = 0, then $z_1{}^*$ is also a root of f(z) = 0.**

Note If z_1 is real, then $z_1{}^* = z_1$.

You can use this property to find roots of cubic and quartic equations with real coefficients.

- **An equation of the form $az^3 + bz^2 + cz + d = 0$ is called a cubic equation, and has three roots.**
- **For a cubic equation with real coefficients, either:**
  - **all three roots are real, or**
  - **one root is real and the other two roots form a complex conjugate pair.**

Watch out A real-valued cubic equation might have two, or three, repeated real roots.

Example 10

Given that -1 is a root of the equation $z^3 - z^2 + 3z + k = 0$,

a find the value of k **b** find the other two roots of the equation.

a If -1 is a root,

$$(-1)^3 - (-1)^2 + 3(-1) + k = 0$$
$$-1 - 1 - 3 + k = 0$$
$$k = 5$$

b -1 is a root of the equation, so $z + 1$ is a factor of $z^3 - z^2 + 3z + 5$.

$$\begin{array}{r} z^2 - 2z + 5 \\ z + 1\overline{)z^3 - z^2 + 3z + 5} \\ \underline{z^3 + z^2} \qquad\qquad \\ -2z^2 + 3z \qquad \\ \underline{-2z^2 - 2z} \qquad \\ 5z + 5 \\ \underline{5z + 5} \\ 0 \end{array}$$

Problem-solving

Use the factor theorem to help: if f(α) = 0, then α is a root of the polynomial and $z - \alpha$ is a factor of the polynomial.

Use long division (or inspection) to find the quadratic factor.

$z^3 - z^2 + 3z + 5 = (z + 1)(z^2 - 2z + 5) = 0$

Solving $z^2 - 2z + 5 = 0$,

$z^2 - 2z = (z - 1)^2 - 1$

$z^2 - 2z + 5 = (z - 1)^2 - 1 + 5 = (z - 1)^2 + 4$

$(z - 1)^2 + 4 = 0$

$(z - 1)^2 = -4$

$z - 1 = \pm\sqrt{-4} = \pm 2i$

$z = 1 \pm 2i$

$z = 1 + 2i, z = 1 - 2i$

So the other two roots of the equation are $1 + 2i$ and $1 - 2i$.

The other two roots are found by solving the quadratic equation.

Solve by completing the square. Alternatively, you could use the quadratic formula.

The quadratic equation has complex roots, which must be a conjugate pair.

You could write the equation as $(z + 1)(z - (1 + 2i))(z - (1 - 2i)) = 0$

- **An equation of the form $az^4 + bz^3 + cz^2 + dz + e = 0$ is called a quartic equation, and has four roots.**
- **For a quartic equation with real coefficients, either:**
  - **all four roots are real, or**
  - **two roots are real and the other two roots form a complex conjugate pair, or**
  - **two roots form a complex conjugate pair and the other two roots also form a complex conjugate pair.**

Watch out A real-valued quartic equation might have repeated real roots or repeated complex roots.

Example 11

Given that $3 + i$ is a root of the quartic equation $2z^4 - 3z^3 - 39z^2 + 120z - 50 = 0$, solve the equation completely.

Another root is $3 - i$.

So $(z - (3 + i))(z - (3 - i))$ is a factor of $2z^4 - 3z^3 - 39z^2 + 120z - 50$

$(z - (3 + i))(z - (3 - i)) = z^2 - z(3 - i) - z(3 + i) + (3 + i)(3 - i)$

$= z^2 - 6z + 10$

So $z^2 - 6z + 10$ is a factor of $2z^4 - 3z^3 - 39z^2 + 120z - 50$.

$(z^2 - 6z + 10)(az^2 + bz + c) = 2z^4 - 3z^3 - 39z^2 + 120z - 50$

Consider $2z^4$:

The only z^4 term in the expansion is $z^2 \times az^2$, so $a = 2$.

$(z^2 - 6z + 10)(2z^2 + bz + c) = 2z^4 - 3z^3 - 39z^2 + 120z - 50$

Consider $-3z^3$:

The z^3 terms in the expansion are $z^2 \times bz$ and $-6z \times 2z^2$,

so $bz^3 - 12z^3 = -3z^3$

$b - 12 = -3$

$b = 9$

so $(z^2 - 6z + 10)(2z^2 + 9z + c) = 2z^4 - 3z^3 - 39z^2 + 120z - 50$

Complex roots occur in conjugate pairs.

If α and β are roots of $f(z) = 0$, then $(z - \alpha)(z - \beta)$ is a factor of $f(z)$.

You can work this out quickly by noting that $(z - (a + bi))(z - (a - bi)) = z^2 - 2az + a^2 + b^2$

Problem-solving

It is possible to factorise a polynomial without using a formal algebraic method. Here, the polynomial is factorised by 'inspection'. By considering each term of the quartic separately, it is possible to work out the missing coefficients.

Consider −50:
The only constant term in the expansion is $10 \times c$, so $c = -5$.
$2z^4 - 3z^3 - 39z^2 + 120z - 50 = (z^2 - 6z + 10)(2z^2 + 9z - 5)$

You can check this by considering the z and z^2 terms in the expansion.

Solving $2z^2 + 9z - 5 = 0$:
$(2z - 1)(z + 5) = 0$
$z = \frac{1}{2}$, $z = -5$

So the roots of $2z^4 - 3z^3 - 39z^2 + 120z - 50 = 0$ are $\frac{1}{2}$, -5, $3 + \text{i}$ and $3 - \text{i}$

Example 12

Show that $z^2 + 4$ is a factor of $z^4 - 2z^3 + 21z^2 - 8z + 68$.

Hence solve the equation $z^4 - 2z^3 + 21z^2 - 8z + 68 = 0$.

Using long division:

$$\begin{array}{r l}
 & z^2 - 2z + 17 \\
z^2 + 4 \,) & z^4 - 2z^3 + 21z^2 - 8z + 68 \\
 & z^4 \qquad\quad + 4z^2 \\
 & -2z^3 + 17z^2 - 8z \\
 & -2z^3 \qquad\quad - 8z \\
 & 17z^2 \qquad + 68 \\
 & 17z^2 \qquad + 68 \\
 & 0
\end{array}$$

So $z^4 - 2z^3 + 21z^2 - 8z + 68 = (z^2 + 4)(z^2 - 2z + 17) = 0$
Either $z^2 + 4 = 0$ or $z^2 - 2z + 17 = 0$
Solving $z^2 + 4 = 0$:
$z^2 = -4$
$z = \pm 2\text{i}$
Solving $z^2 - 2z + 17 = 0$:
$(z - 1)^2 + 16 = 0$
$(z - 1)^2 = -16$
$z - 1 = \pm 4\text{i}$
$z = 1 \pm 4\text{i}$
So the roots of $z^4 - 2z^3 + 21z^2 - 8z + 68 = 0$ are 2i, -2i, $1 + 4\text{i}$ and $1 - 4\text{i}$

Alternatively, the quartic can be factorised by inspection:
$z^4 - 2z^3 + 21z^2 - 8z + 68$
$= (z^2 + 4)(az^2 + bz + c)$
$a = 1$, as the leading coefficient is 1.
The only z^3 term is formed by $z^2 \times bz$ so $b = -2$.
The constant term is formed by $4 \times c$, so $4c = 68$, and $c = 17$.

Solve by completing the square. Alternatively, you could use the quadratic formula.

Watch out You could use your calculator to solve $z^2 - 2z + 17 = 0$. However, you should still write down the equation you are solving, and both roots.

Exercise 1F

(E) **1** $f(z) = z^3 - 6z^2 + 21z - 26$

a Show that $f(2) = 0$. **(1 mark)**

b Hence solve $f(z) = 0$ completely. **(3 marks)**

(E) **2** $f(z) = 2z^3 + 5z^2 + 9z - 6$

a Show that $f\left(\frac{1}{2}\right) = 0$. **(1 mark)**

b Hence write $f(z)$ in the form $(2z - 1)(z^2 + bz + c)$, where b and c are real constants to be found. **(2 marks)**

c Use algebra to solve $f(z) = 0$ completely. **(2 marks)**

(E/P) **3** $g(z) = 2z^3 - 4z^2 - 5z - 3$

Given that $z = 3$ is a root of the equation $g(z) = 0$, solve $g(z) = 0$ completely. **(4 marks)**

(E) **4** $p(z) = z^3 + 4z^2 - 15z - 68$

Given that $z = -4 + i$ is a solution to the equation $p(z) = 0$,

a show that $z^2 + 8z + 17$ is a factor of $p(z)$. **(2 marks)**

b Hence solve $p(z) = 0$ completely. **(2 marks)**

(E) **5** $f(z) = z^3 + 9z^2 + 33z + 25$

Given that $f(z) = (z + 1)(z^2 + az + b)$, where a and b are real constants,

a find the value of a and the value of b **(2 marks)**

b find the three roots of $f(z) = 0$ **(4 marks)**

c find the sum of the three roots of $f(z) = 0$. **(1 mark)**

(E/P) **6** $g(z) = z^3 - 12z^2 + cz + d = 0$, where $c, d \in \mathbb{R}$.

Given that 6 and $3 + i$ are roots of the equation $g(z) = 0$,

a write down the other complex root of the equation **(1 mark)**

b find the value of c and the value of d. **(4 marks)**

(E/P) **7** $h(z) = 2z^3 + 3z^2 + 3z + 1$

Given that $2z + 1$ is a factor of $h(z)$, find the three roots of $h(z) = 0$. **(4 marks)**

(E/P) **8** $f(z) = z^3 - 6z^2 + 28z + k$

Given that $f(2) = 0$,

a find the value of k **(1 mark)**

b find the other two roots of the equation. **(4 marks)**

9 Find the four roots of the equation $z^4 - 16 = 0$.

(E) **10** $f(z) = z^4 - 12z^3 + 31z^2 + 108z - 360$

a Write $f(z)$ in the form $(z^2 - 9)(z^2 + bz + c)$, where b and c are real constants to be found. **(2 marks)**

b Hence find all the solutions to $f(z) = 0$. **(3 marks)**

(P) **11** $g(z) = z^4 + 2z^3 - z^2 + 38z + 130$

Given that $g(2 + 3i) = 0$, find all the roots of $g(z) = 0$.

(E/P) **12** $f(z) = z^4 - 10z^3 + 71z^2 + Qz + 442$, where Q is a real constant.

Given that $z = 2 - 3i$ is a root of the equation $f(z) = 0$,

a show that $z^2 - 6z + 34$ is a factor of $f(z)$ **(4 marks)**

b find the value of Q **(1 mark)**

c solve completely the equation $f(z) = 0$. **(2 marks)**

Challenge

Three of the roots of the equation $z^5 + bz^4 + cz^3 + dz^2 + ez + f = 0$, where $b, c, d, e, f \in \mathbb{R}$, are -2, $2i$ and $1 + i$. Find the values of b, c, d, e and f.

Mixed exercise 1

1 Given that $z_1 = 8 - 3i$ and $z_2 = -2 + 4i$, find, in the form $a + bi$, where $a, b \in \mathbb{R}$:

a $z_1 + z_2$

b $3z_2$

c $6z_1 - z_2$

(E/P) **2** The equation $z^2 + bz + 14 = 0$, where $b \in \mathbb{R}$ has no real roots.

Find the range of possible values of b. **(3 marks)**

3 The solutions to the quadratic equation $z^2 - 6z + 12 = 0$ are z_1 and z_2.

Find z_1 and z_2, giving each answer in the form $a \pm i\sqrt{b}$.

(E/P) **4** By using the binomial expansion, or otherwise, show that $(1 + 2i)^5 = 41 - 38i$. **(3 marks)**

(E) **5** $f(z) = z^2 - 6z + 10$

Show that $z = 3 + i$ is a solution to $f(z) = 0$. **(2 marks)**

6 $z_1 = 4 + 2i$, $z_2 = -3 + i$

Express, in the form $a + bi$, where $a, b \in \mathbb{R}$:

a z_1^* **b** $z_1 z_2$ **c** $\frac{z_1}{z_2}$

7 Write $\frac{(7 - 2i)^2}{1 + i\sqrt{3}}$ in the form $x + iy$ where $x, y \in \mathbb{R}$.

(E/P) **8** Given that $\frac{4 - 7i}{z} = 3 + i$, find z in the form $a + bi$, where $a, b \in \mathbb{R}$. **(2 marks)**

9 $z = \frac{1}{2 + i}$

Express in the form $a + bi$, where $a, b \in \mathbb{R}$:

a z^2 **b** $z - \frac{1}{z}$

E/P **10** Given that $z = a + b\text{i}$, show that $\dfrac{z}{z^*} = \left(\dfrac{a^2 - b^2}{a^2 + b^2}\right) + \left(\dfrac{2ab}{a^2 + b^2}\right)\text{i}$ **(4 marks)**

E/P **11** The complex number z is defined by $z = \dfrac{3 + q\text{i}}{q - 5\text{i}}$, where $q \in \mathbb{R}$.

Given that the real part of z is $\frac{1}{13}$,

a find the possible values of q **(4 marks)**

b write the possible values of z in the form $a + b\text{i}$, where a and b are real constants. **(1 mark)**

E/P **12** Given that $z = x + \text{i}y$, find the value of x and the value of y such that $z + 4\text{i}z^* = -3 + 18\text{i}$ where z^* is the complex conjugate of z. **(5 marks)**

13 $z = 9 + 6\text{i}$, $w = 2 - 3\text{i}$

Express $\dfrac{z}{w}$ in the form $a + b\text{i}$, where a and b are real constants.

E/P **14** The complex number z is given by $z = \dfrac{q + 3\text{i}}{4 + q\text{i}}$ where q is an integer.

Express z in the form $a + b\text{i}$ where a and b are rational and are given in terms of q. **(4 marks)**

E **15** Given that $6 - 2\text{i}$ is one of the roots of a quadratic equation with real coefficients,

a write down the other root of the equation **(1 mark)**

b find the quadratic equation, giving your answer in the form $z^2 + bz + c = 0$ where b and c are real constants. **(2 marks)**

E/P **16** Given that $z = 4 - k\text{i}$ is a root of the equation $z^2 - 2mz + 52 = 0$, where k and m are positive real constants, find the value of k and the value of m. **(4 marks)**

E/P **17** $\text{h}(z) = z^3 - 11z + 20$

Given that $2 + \text{i}$ is a root of the equation $\text{h}(z) = 0$, solve $\text{h}(z) = 0$ completely. **(4 marks)**

E/P **18** $\text{f}(z) = z^3 + 6z + 20$

Given that $\text{f}(1 + 3\text{i}) = 0$, solve $\text{f}(z) = 0$ completely. **(4 marks)**

E/P **19** $\text{f}(z) = z^3 + 3z^2 + kz + 48$, $k \in \mathbb{R}$

Given that $\text{f}(4\text{i}) = 0$,

a find the value of k **(2 marks)**

b find the other two roots of the equation. **(3 marks)**

E **20** $\text{f}(z) = z^4 - z^3 - 16z^2 - 74z - 60$

a Write $\text{f}(z)$ in the form $(z^2 - 5z - 6)(z^2 + bz + c)$, where b and c are real constants to be found. **(2 marks)**

b Hence find all the solutions to $\text{f}(z) = 0$. **(3 marks)**

E/P **21** $\text{g}(z) = z^4 - 6z^3 + 19z^2 - 36z + 78$

Given that $\text{g}(3 - 2\text{i}) = 0$, find all the roots of $\text{g}(z) = 0$. **(4 marks)**

E/P **22** $\text{f}(z) = z^4 - 2z^3 - 5z^2 + pz + 24$

Given that $\text{f}(4) = 0$,

a find the value of p **(1 mark)**

b solve completely the equation $\text{f}(z) = 0$. **(5 marks)**

Challenge

a Explain why a cubic equation with real coefficients cannot have a repeated non-real root.

b By means of an example, show that a quartic equation with real coefficients can have a repeated non-real root.

Summary of key points

1 $\mathbf{i} = \sqrt{-1}$ and $\mathbf{i}^2 = -1$

2 An **imaginary number** is a number of the form bi, where $b \in \mathbb{R}$.

3 A **complex number** is written in the form $a + b$i, where $a, b \in \mathbb{R}$.

4 Complex numbers can be added or subtracted by adding or subtracting their real parts and adding or subtracting their imaginary parts.

5 You can multiply a real number by a complex number by multiplying out the brackets in the usual way.

6 If $b^2 - 4ac < 0$ then the quadratic equation $ax^2 + bx + c = 0$ has two distinct complex roots, neither of which is real.

7 For any complex number $z = a + b$i, the **complex conjugate** of the number is defined as $z^* = a - b$i.

8 For real numbers a, b and c, if the roots of the quadratic equation $az^2 + bz + c = 0$ are non-real complex numbers, then they occur as a conjugate pair.

9 If the roots of a quadratic equation are α and β, then you can write the equation as $(z - \alpha)(z - \beta) = 0$ or $z^2 - (\alpha + \beta)z + \alpha\beta = 0$.

10 If $\mathrm{f}(z)$ is a polynomial with real coefficients, and z_1 is a root of $\mathrm{f}(z) = 0$, then $z_1{}^*$ is also a root of $\mathrm{f}(z) = 0$.

11 An equation of the form $az^3 + bz^2 + cz + d = 0$ is called a cubic equation, and has three roots.

For a cubic equation with real coefficients, either:

- all three roots are real, or
- one root is real and the other two roots form a complex conjugate pair.

12 An equation of the form $az^4 + bz^3 + cz^2 + dz + e = 0$ is called a quartic equation, and has four roots.

For a quartic equation with real coefficients, either:

- all four roots are real, or
- two roots are real and the other two roots form a complex conjugate pair, or
- two roots form a complex conjugate pair and the other two roots also form a complex conjugate pair.

Argand diagrams

2

Objectives

After completing this chapter you should be able to:

- Show complex numbers on an Argand diagram → **pages 18–19**
- Find the modulus and argument of a complex number → **pages 20–23**
- Write a complex number in modulus-argument form → **pages 23–28**
- Represent loci on an Argand diagram → **pages 28–36**
- Represent regions on an Argand diagram → **pages 36–38**

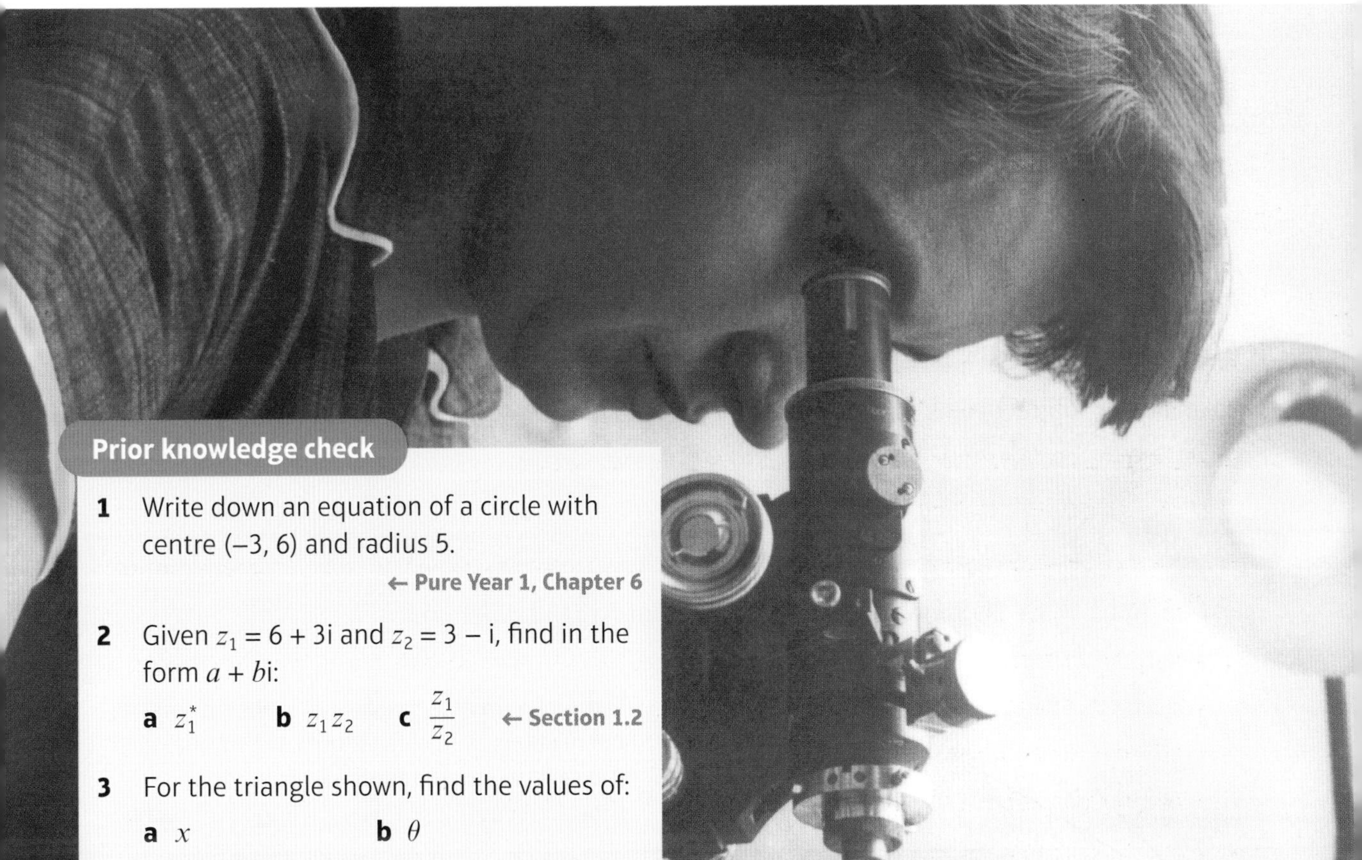

Prior knowledge check

1 Write down an equation of a circle with centre (–3, 6) and radius 5. ← **Pure Year 1, Chapter 6**

2 Given $z_1 = 6 + 3\text{i}$ and $z_2 = 3 - \text{i}$, find in the form $a + b\text{i}$:

a z_1^* **b** $z_1 z_2$ **c** $\frac{z_1}{z_2}$ ← **Section 1.2**

3 For the triangle shown, find the values of:

a x **b** θ

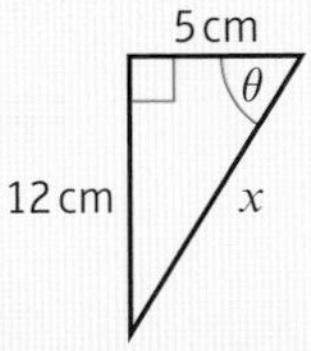

← **GCSE Mathematics**

4 Find the solutions of the quadratic equation $z^2 - 8z + 24 = 0$. ← **Section 1.4**

Complex numbers can be used to model electromagnetic waves. Rosalind Franklin helped discover DNA by using complex numbers to analyse the diffraction patterns of X-rays passing through crystals of DNA.

2.1 Argand diagrams

- **You can represent complex numbers on an Argand diagram. The x-axis on an Argand diagram is called the real axis and the y-axis is called the imaginary axis. The complex number $z = x + iy$ is represented on the diagram by the point $P(x, y)$, where x and y are Cartesian coordinates.**

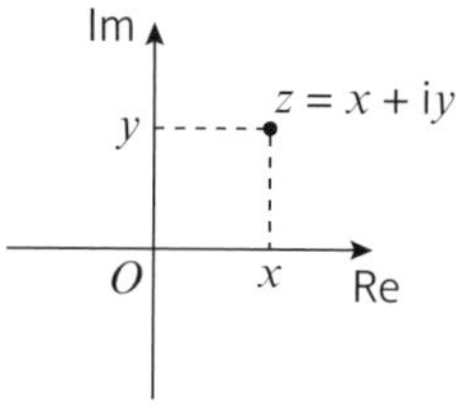

Example 1

Show the complex numbers $z_1 = -4 + i$, $z_2 = 2 + 3i$ and $z_3 = 2 - 3i$ on an Argand diagram.

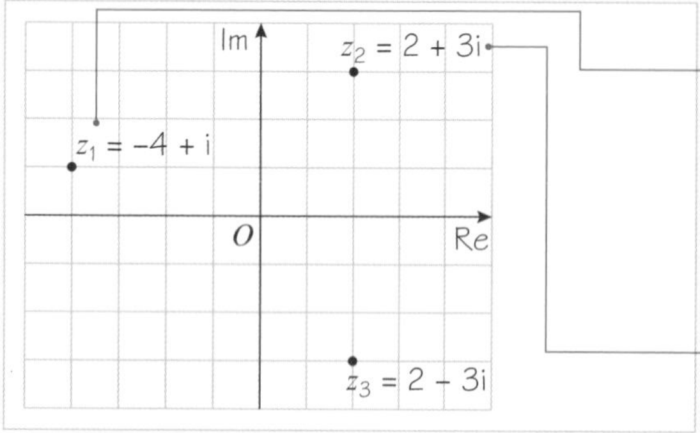

The real part of each number describes its horizontal position, and the imaginary part describes its vertical position. For example, $z_1 = -4 + i$ has real part -4 and imaginary part 1.

Note that z_2 and z_3 are complex conjugates. On an Argand diagram, complex conjugate pairs are symmetrical about the real axis. ← **Section 1.3**

Complex numbers can also be represented as vectors on the Argand diagram.

- **The complex number $z = x + iy$ can be represented as the vector $\begin{pmatrix} x \\ y \end{pmatrix}$ on an Argand diagram.**

You can add or subtract complex numbers on an Argand diagram by adding or subtracting their corresponding vectors.

Example 2

$z_1 = 4 + i$ and $z_2 = 3 + 3i$. Show z_1, z_2 and $z_1 + z_2$ on an Argand diagram.

$z_1 + z_2 = (4 + 3) + (1 + 3)i = 7 + 4i$

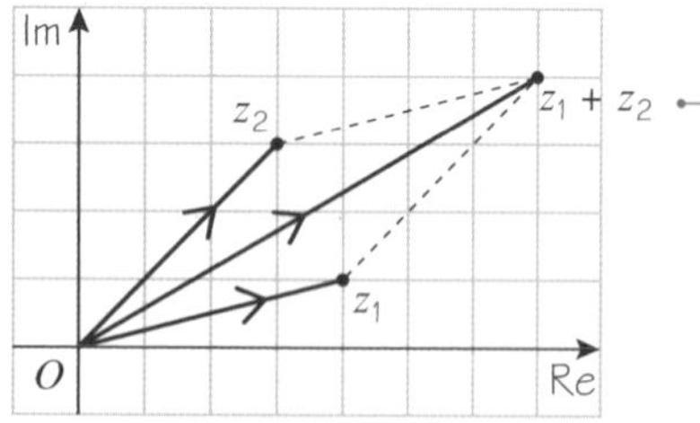

The vector representing $z_1 + z_2$ is the diagonal of the parallelogram with vertices at O, z_1 and z_2. You can use vector addition to find $z_1 + z_2$:

$$\begin{pmatrix} 4 \\ 1 \end{pmatrix} + \begin{pmatrix} 3 \\ 3 \end{pmatrix} = \begin{pmatrix} 7 \\ 4 \end{pmatrix}$$

Example 3

$z_1 = 2 + 5i$ and $z_2 = 4 + 2i$. Show z_1, z_2 and $z_1 - z_2$ on an Argand diagram.

$z_1 - z_2 = (2 - 4) + (5 - 2)\text{i} = -2 + 3\text{i}$

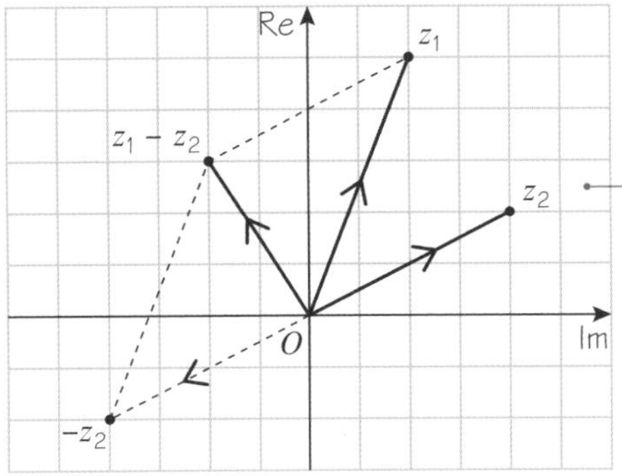

The vector corresponding to z_2 is $\begin{pmatrix}4\\2\end{pmatrix}$, so the vector corresponding to $-z_2$ is $\begin{pmatrix}-4\\-2\end{pmatrix}$.

The vector representing $z_1 - z_2$ is the diagonal of the parallelogram with vertices at O, z_1 and $-z_2$.

Online Explore adding and subtracting complex numbers on an Argand diagram using GeoGebra.

Exercise 2A

1 Show these numbers on an Argand diagram.

a $7 + 2\text{i}$ **b** $5 - 4\text{i}$ **c** $-6 - \text{i}$ **d** $-2 + 5\text{i}$

e 3i **f** $\sqrt{2} + 2\text{i}$ **g** $-\frac{1}{2} + \frac{5}{2}\text{i}$ **h** -4

2 $z_1 = 11 + 2\text{i}$ and $z_2 = 2 + 4\text{i}$. Show z_1, z_2 and $z_1 + z_2$ on an Argand diagram.

3 $z_1 = -3 + 6\text{i}$ and $z_2 = 8 - \text{i}$. Show z_1, z_2 and $z_1 + z_2$ on an Argand diagram.

4 $z_1 = 8 + 4\text{i}$ and $z_2 = 6 + 7\text{i}$. Show z_1, z_2 and $z_1 - z_2$ on an Argand diagram.

5 $z_1 = -6 - 5\text{i}$ and $z_2 = -4 + 4\text{i}$. Show z_1, z_2 and $z_1 - z_2$ on an Argand diagram.

(P) **6** $z_1 = 7 - 5\text{i}$, $z_2 = a + b\text{i}$ and $z_3 = -3 + 2\text{i}$ where $a, b \in \mathbb{Z}$. Given that $z_3 = z_1 + z_2$,

a find the values of a and b **b** show z_1, z_2 and z_3 on an Argand diagram.

(P) **7** $z_1 = p + q\text{i}$, $z_2 = 9 - 5\text{i}$ and $z_3 = -8 + 5\text{i}$ where $p, q \in \mathbb{Z}$. Given that $z_3 = z_1 + z_2$,

a find the values of p and q **b** show z_1, z_2 and z_3 on an Argand diagram.

(E) **8** The solutions to the quadratic equation $z^2 - 6z + 10 = 0$ are z_1 and z_2.

a Find z_1 and z_2, giving your answers in the form $p \pm q\text{i}$, where p and q are integers. **(3 marks)**

b Show, on an Argand diagram, the points representing the complex numbers z_1 and z_2. **(2 marks)**

(E/P) **9** $\text{f}(z) = 2z^3 - 19z^2 + 64z - 60$

a Show that $\text{f}\left(\frac{3}{2}\right) = 0$. **(1 mark)**

b Use algebra to solve $\text{f}(z) = 0$ completely. **(4 marks)**

c Show all three solutions on an Argand diagram. **(2 marks)**

Challenge

a Find all the solutions to the equation $z^6 = 1$.

b Show each solution on an Argand diagram.

c Show that each solution lies on a circle with centre (0, 0) and radius 1.

Hint There will be 6 distinct roots in total. Write $z^6 = 1$ as $(z^3 - 1)(z^3 + 1) = 0$, then find three distinct roots of $z^3 - 1 = 0$ and three distinct roots of $z^3 + 1 = 0$.

2.2 Modulus and argument

The **modulus** or absolute value of a complex number is the magnitude of its corresponding vector.

- **The modulus of a complex number, $|z|$, is the distance from the origin to that number on an Argand diagram. For a complex number $z = x + iy$, the modulus is given by $|z| = \sqrt{x^2 + y^2}$.**

Notation The modulus of the complex number z is written as r, $|z|$ or $|x + iy|$.

The **argument** of a complex number is the angle its corresponding vector makes with the positive real axis.

- **The argument of a complex number, arg z, is the angle between the positive real axis and the line joining that number to the origin on an Argand diagram, measured in an anticlockwise direction. For a complex number $z = x + iy$, the argument, θ, satisfies $\tan\theta = \frac{y}{x}$**

Notation The argument of the complex number z is written as arg z. It is usually given in radians, where
- 2π radians = 360°
- π radians = 180°

← Pure Year 2, Section 5.1

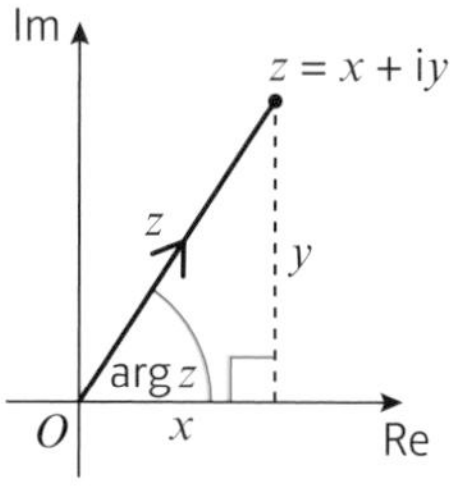

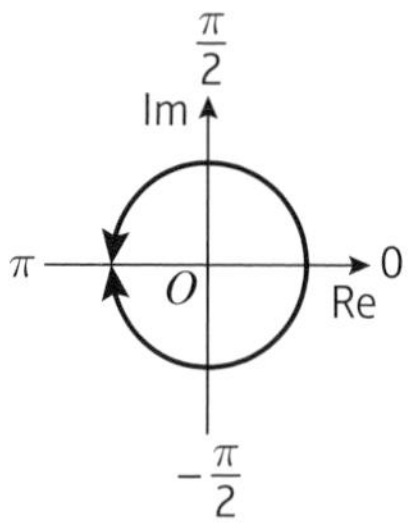

The argument θ of any complex number is usually given in the range $-\pi < \theta \leqslant \pi$. This is sometimes referred to as the **principal argument**.

Example 4

$z = 2 + 7i$, find:

a the modulus of z **b** the argument of z, giving your answer in radians to 2 decimal places.

a

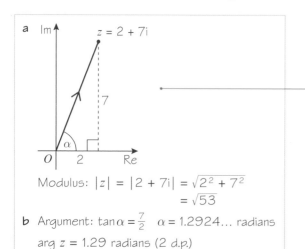

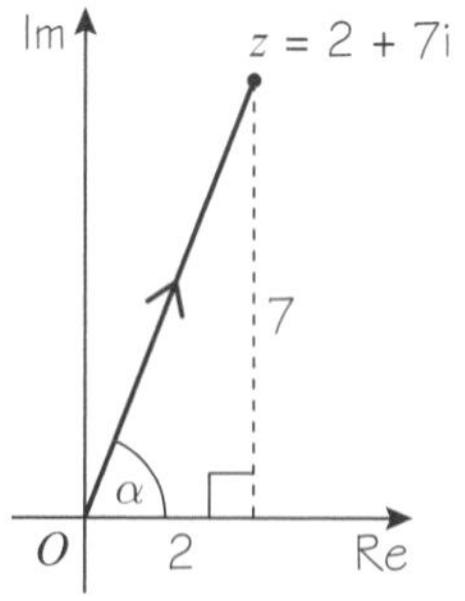

Sketch the Argand diagram, showing the position of the number.

Modulus: $|z| = |2 + 7i| = \sqrt{2^2 + 7^2}$
$= \sqrt{53}$

b Argument: $\tan\alpha = \frac{7}{2}$ $\alpha = 1.2924...$ radians

arg z = 1.29 radians (2 d.p.)

If z does not lie in the first quadrant, you can use the Argand diagram to help you find its argument.

- **Let α be the positive acute angle made with the real axis by the line joining the origin and z.**
  - **If z lies in the first quadrant then arg $z = \alpha$.**
  - **If z lies in the second quadrant then arg $z = \pi - \alpha$.**
  - **If z lies in the third quadrant then arg $z = -(\pi - \alpha)$.**
  - **If z lies in the fourth quadrant then arg $z = -\alpha$.**

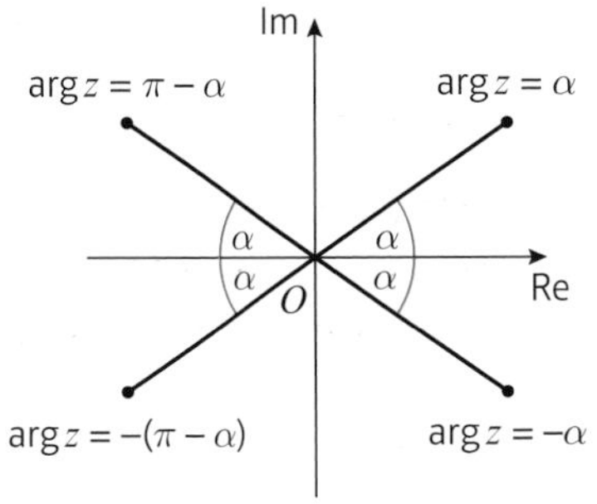

Example 5

$z = -4 - \text{i}$, find:

a the modulus of z **b** the argument of z, giving your answer in radians to 2 decimal places.

a

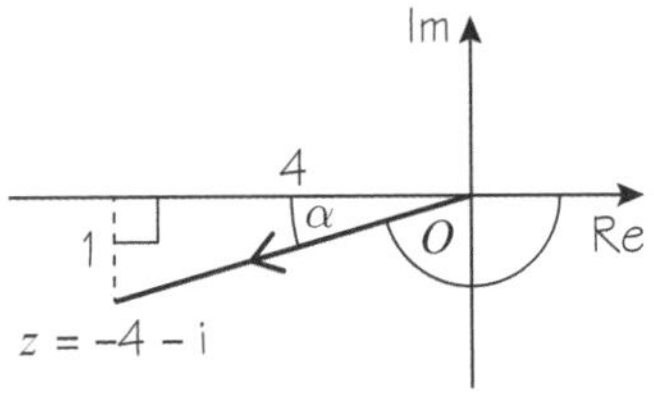

Sketch the Argand diagram, showing the position of the number.

Modulus: $|z| = |-4 - \text{i}| = \sqrt{(-4)^2 + (-1)^2}$
$= \sqrt{17}$

b Argument: $\tan\alpha = \frac{1}{4}$ $\alpha = 0.2449\ldots$ radians
$\arg z = -(\pi - 0.2449)$
$= -2.90$ radians (2 d.p.)

Here z is in the third quadrant, so the required argument is $-(\pi - \alpha)$.

Exercise 2B

1 For each of the following complex numbers,

i find the modulus, writing your answer in surd form if necessary

ii find the argument, writing your answer in radians to 2 decimal places.

a $z = 12 + 5\text{i}$ **b** $z = \sqrt{3} + \text{i}$ **c** $z = -3 + 6\text{i}$

d $z = 2 - 2\text{i}$ **e** $z = -8 - 7\text{i}$ **f** $z = -4 + 11\text{i}$

g $z = 2\sqrt{3} - \text{i}\sqrt{3}$ **h** $z = -8 - 15\text{i}$

Hint In part **c**, the complex number is in the second quadrant, so the argument will be $\pi - \alpha$. In part **d**, the complex number is in the fourth quadrant, so the argument will be $-\alpha$.

2 For each of the following complex numbers,

i find the modulus, writing your answer in surd form

ii find the argument, writing your answer in terms of π.

a $2 + 2\text{i}$ **b** $5 + 5\text{i}$ **c** $-6 + 6\text{i}$ **d** $-a - a\text{i}, a \in \mathbb{R}$

(E) **3** $z = -40 - 9i$

a Show z on an Argand diagram. **(1 mark)**

b Calculate $\arg z$, giving your answer in radians to 2 decimal places. **(2 marks)**

(E) **4** $z = 3 + 4i$

a Show that $z^2 = -7 + 24i$. **(2 marks)**

Find, showing your working:

b $|z^2|$ **(2 marks)**

c $\arg(z^2)$, giving your answer in radians to 2 decimal places. **(2 marks)**

d Show z and z^2 on an Argand diagram. **(1 mark)**

(E) **5** The complex numbers z_1 and z_2 are given by $z_1 = 4 + 6i$ and $z_2 = 1 + i$.

Find, showing your working:

a $\frac{z_1}{z_2}$ in the form $a + bi$, where a and b are real **(3 marks)**

b $\left|\frac{z_1}{z_2}\right|$ **(2 marks)**

c $\arg \frac{z_1}{z_2}$, giving your answer in radians to 2 decimal places. **(2 marks)**

(E/P) **6** The complex numbers z_1 and z_2 are such that $z_1 = 3 + 2pi$ and $\frac{z_1}{z_2} = 1 - i$ where p is a real constant.

a Find z_2 in the form $a + bi$, giving the real numbers a and b in terms of p. **(3 marks)**

Given that $\arg z_2 = \tan^{-1} 5$,

b find the value of p **(2 marks)**

c find the value of $|z_2|$ **(2 marks)**

d show z_1, z_2 and $\frac{z_1}{z_2}$ on a single Argand diagram. **(2 marks)**

(E) **7** $z = \frac{26}{2 - 3i}$, find:

a z in the form $a + ib$ where $a, b \in \mathbb{R}$ **(2 marks)**

b z^2 in the form $a + ib$ where $a, b \in \mathbb{R}$ **(2 marks)**

c $|z|$ **(2 marks)**

d $\arg(z^2)$, giving your answer in radians to 2 decimal places. **(2 marks)**

(E/P) **8** $z_1 = 4 + 2i$, $z_2 = 2 + 4i$, $z_3 = a + bi$ where $a, b \in \mathbb{R}$.

a Find the exact value of $|z_1 + z_2|$. **(2 marks)**

Given that $w = \frac{z_1 z_3}{z_2}$,

b find w in terms of a and b, giving your answer in the form $x + iy$, $x, y \in \mathbb{R}$. **(4 marks)**

Given also that $w = \frac{21}{5} - \frac{22}{5}i$, find:

c the values of a and b **(3 marks)**

d $\arg w$, giving your answer in radians to 2 decimal places. **(2 marks)**

E/P **9** The complex number w is given by $w = 6 + 3\text{i}$. Find:

a $|w|$ **(1 mark)**

b $\arg w$, giving your answer in radians to 2 decimal places. **(2 marks)**

Given that $\arg(\lambda + 5\text{i} + w) = \frac{\pi}{4}$ where λ is a real constant,

c find the value of λ. **(2 marks)**

E **10** $z = -1 - \text{i}\sqrt{3}$, find:

a $|z|$ **(1 mark)**

b $\left|\frac{z}{z^*}\right|$ **(4 marks)**

c $\arg z$, $\arg(z^*)$ and $\arg\frac{z}{z^*}$, giving your answers in terms of π. **(3 marks)**

E/P **11** The complex numbers w and z are given by $w = k + \text{i}$ and $z = -4 + 5k\text{i}$, where k is a real constant. Given that $\arg(w + z) = \frac{2\pi}{3}$, find the exact value of k. **(6 marks)**

E/P **12** The complex numbers w and z are defined such that $\arg w = \frac{\pi}{10}$, $|w| = 5$ and $\arg z = \frac{2\pi}{5}$

Given that $\arg(w + z) = \frac{\pi}{5}$, find the value of $|z|$. **(4 marks)**

2.3 Modulus–argument form of complex numbers

You can write any complex number in terms of its modulus and argument.

- **For a complex number z with $|z| = r$ and $\arg z = \theta$, the modulus–argument form of z is $z = r(\cos\theta + \text{i}\sin\theta)$**

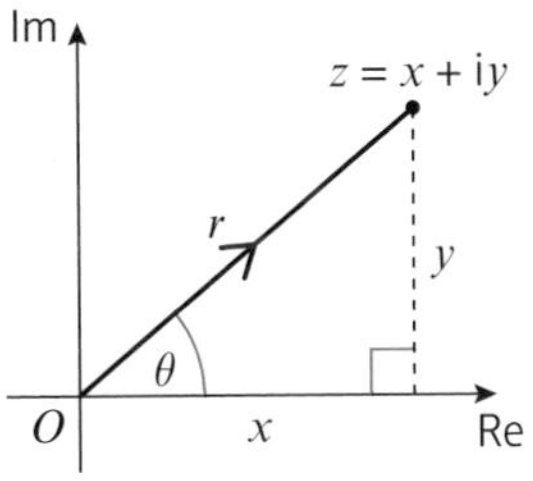

From the right-angled triangle, $x = r\cos\theta$ and $y = r\sin\theta$.

$z = x + \text{i}y = r\cos\theta + \text{i}r\sin\theta = r(\cos\theta + \text{i}\sin\theta)$

This formula works for a complex number in any quadrant of the Argand diagram. The argument, θ, is usually given in the range $-\pi < \theta \leqslant \pi$, although the formula works for any value of θ measured anticlockwise from the positive real axis.

Example 6

Express $z = -\sqrt{3} + \text{i}$ in the form $r(\cos\theta + \text{i}\sin\theta)$, where $-\pi < \theta \leqslant \pi$.

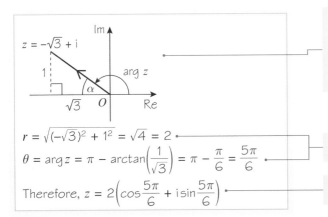

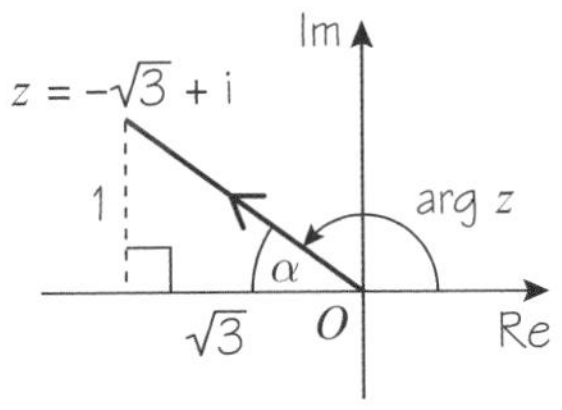

Sketch the Argand diagram, showing the position of the number.

Here z is in the second quadrant, so the required argument is $\pi - \alpha$.

$r = \sqrt{(-\sqrt{3})^2 + 1^2} = \sqrt{4} = 2$

$\theta = \arg z = \pi - \arctan\left(\frac{1}{\sqrt{3}}\right) = \pi - \frac{\pi}{6} = \frac{5\pi}{6}$

Find r and θ.

Therefore, $z = 2\left(\cos\frac{5\pi}{6} + \text{i}\sin\frac{5\pi}{6}\right)$

Apply $z = r(\cos\theta + \text{i}\sin\theta)$.

Example 7

Express $z = -1 - \mathrm{i}$ in the form $r(\cos\theta + \mathrm{i}\sin\theta)$, where $-\pi < \theta \leqslant \pi$.

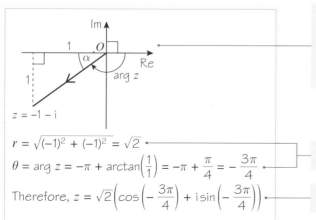

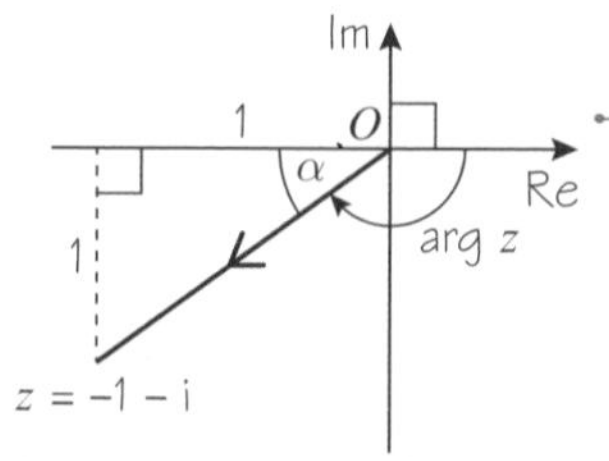

Sketch the Argand diagram, showing the position of the number.

Here z is in the third quadrant, so the required argument is $-(\pi - \alpha)$.

$r = \sqrt{(-1)^2 + (-1)^2} = \sqrt{2}$

$\theta = \arg z = -\pi + \arctan\left(\frac{1}{1}\right) = -\pi + \frac{\pi}{4} = -\frac{3\pi}{4}$

Find r and θ.

Therefore, $z = \sqrt{2}\left(\cos\left(-\frac{3\pi}{4}\right) + \mathrm{i}\sin\left(-\frac{3\pi}{4}\right)\right)$

Apply $z = r(\cos\theta + \mathrm{i}\sin\theta)$.

Exercise 2C

1 Express the following in the form $r(\cos\theta + \mathrm{i}\sin\theta)$, where $-\pi < \theta \leqslant \pi$. Give the exact values of r and θ where possible, or values to 2 decimal places otherwise.

a $2 + 2\mathrm{i}$ **b** $3\mathrm{i}$ **c** $-3 + 4\mathrm{i}$ **d** $1 - \sqrt{3}\mathrm{i}$

e $-2 - 5\mathrm{i}$ **f** -20 **g** $7 - 24\mathrm{i}$ **h** $-5 + 5\mathrm{i}$

2 Express these in the form $r(\cos\theta + \mathrm{i}\sin\theta)$, giving exact values of r and θ where possible, or values to two decimal places otherwise.

a $\dfrac{3}{1 + \mathrm{i}\sqrt{3}}$ **b** $\dfrac{1}{2 - \mathrm{i}}$ **c** $\dfrac{1 + \mathrm{i}}{1 - \mathrm{i}}$

3 Express the following in the form $x + \mathrm{i}y$, where $x, y \in \mathbb{R}$.

a $5\left(\cos\frac{\pi}{2} + \mathrm{i}\sin\frac{\pi}{2}\right)$ **b** $\frac{1}{2}\left(\cos\frac{\pi}{6} + \mathrm{i}\sin\frac{\pi}{6}\right)$ **c** $6\left(\cos\frac{5\pi}{6} + \mathrm{i}\sin\frac{5\pi}{6}\right)$

d $3\left(\cos\left(-\frac{2\pi}{3}\right) + \mathrm{i}\sin\left(-\frac{2\pi}{3}\right)\right)$ **e** $2\sqrt{2}\left(\cos\left(-\frac{\pi}{4}\right) + \mathrm{i}\sin\left(-\frac{\pi}{4}\right)\right)$ **f** $-4\left(\cos\frac{7\pi}{6} + \mathrm{i}\sin\frac{7\pi}{6}\right)$

(E) **4** **a** Express the complex number $z = 4\left(\cos\left(\frac{2\pi}{3}\right) + i\sin\left(\frac{2\pi}{3}\right)\right)$ in the form $x + \mathrm{i}y$, where $x, y \in \mathbb{R}$. **(2 marks)**

b Show the complex number z on an Argand diagram. **(1 mark)**

(E) **5** The complex number z is such that $|z| = 7$ and $\arg z = \frac{11\pi}{6}$. Find z in the form $p + q\mathrm{i}$, where p and q are exact real numbers to be found. **(3 marks)**

(E) **6** The complex number z is such that $|z| = 5$ and $\arg z = -\frac{4\pi}{3}$. Find z in the form $a + b\mathrm{i}$, where a and b are exact real numbers to be found. **(3 marks)**

You can use the following rules to multiply complex numbers quickly when they are given in modulus–argument form.

- **For any two complex numbers z_1 and z_2,**
  - $|z_1 z_2| = |z_1||z_2|$
  - **$\arg(z_1 z_2) = \arg z_1 + \arg z_2$**

Note You multiply the moduli and add the arguments.

To prove these results, consider z_1 and z_2 in modulus–argument form:

$$z_1 = r_1(\cos\theta_1 + \mathrm{i}\sin\theta_1) \text{ and } z_2 = r_2(\cos\theta_2 + \mathrm{i}\sin\theta_2)$$

Multiplying these numbers together, you get

$$\begin{aligned}
z_1z_2 &= r_1(\cos\theta_1 + \mathrm{i}\sin\theta_1) \times r_2(\cos\theta_2 + \mathrm{i}\sin\theta_2)\\
&= r_1r_2(\cos\theta_1 + \mathrm{i}\sin\theta_1)(\cos\theta_2 + \mathrm{i}\sin\theta_2)\\
&= r_1r_2(\cos\theta_1\cos\theta_2 + \mathrm{i}\cos\theta_1\sin\theta_2 + \mathrm{i}\sin\theta_1\cos\theta_2 + \mathrm{i}^2\sin\theta_1\sin\theta_2)\\
&= r_1r_2(\cos\theta_1\cos\theta_2 + \mathrm{i}\cos\theta_1\sin\theta_2 + \mathrm{i}\sin\theta_1\cos\theta_2 - \sin\theta_1\sin\theta_2)\\
&= r_1r_2((\cos\theta_1\cos\theta_2 - \sin\theta_1\sin\theta_2) + \mathrm{i}(\sin\theta_1\cos\theta_2 + \cos\theta_1\sin\theta_2))\\
&= r_1r_2(\cos(\theta_1 + \theta_2) + \mathrm{i}\sin(\theta_1 + \theta_2))
\end{aligned}$$

Links The last step of this working makes use of the trigonometric addition formulae:
$\sin(A \pm B) \equiv \sin A\cos B \pm \cos A\sin B$
$\cos(A \pm B) \equiv \cos A\cos B \mp \sin A\sin B$
← **Pure Year 2, Section 7.1**

This complex number is in modulus–argument form, with modulus r_1r_2 and argument $\theta_1 + \theta_2$, as required.

You can derive similar results for dividing two complex numbers given in modulus–argument form.

- **For any two complex numbers z_1 and z_2,**
  - $\left|\dfrac{z_1}{z_2}\right| = \dfrac{|z_1|}{|z_2|}$
  - **$\arg\left(\dfrac{z_1}{z_2}\right) = \arg z_1 - \arg z_2$**

Note You divide the moduli and subtract the arguments.

To prove these results, again consider z_1 and z_2 in modulus–argument form:

$$z_1 = r_1(\cos\theta_1 + \mathrm{i}\sin\theta_1) \text{ and } z_2 = r_2(\cos\theta_2 + \mathrm{i}\sin\theta_2)$$

Online Explore multiplying and dividing complex numbers on an Argand diagram using GeoGebra.

Dividing z_1 by z_2 you get

$$\begin{aligned}
\frac{z_1}{z_2} &= \frac{r_1(\cos\theta_1 + \mathrm{i}\sin\theta_1)}{r_2(\cos\theta_2 + \mathrm{i}\sin\theta_2)}\\
&= \frac{r_1(\cos\theta_1 + \mathrm{i}\sin\theta_1)}{r_2(\cos\theta_2 + \mathrm{i}\sin\theta_2)} \times \frac{(\cos\theta_2 - \mathrm{i}\sin\theta_2)}{(\cos\theta_2 - \mathrm{i}\sin\theta_2)}\\
&= \frac{r_1(\cos\theta_1\cos\theta_2 - \mathrm{i}\cos\theta_1\sin\theta_2 + \mathrm{i}\sin\theta_1\cos\theta_2 - \mathrm{i}^2\sin\theta_1\sin\theta_2)}{r_2(\cos\theta_2\cos\theta_2 - \mathrm{i}\cos\theta_2\sin\theta_2 + \mathrm{i}\sin\theta_2\cos\theta_2 - \mathrm{i}^2\sin\theta_2\sin\theta_2)}\\
&= \frac{r_1((\cos\theta_1\cos\theta_2 + \sin\theta_1\sin\theta_2) + \mathrm{i}(\sin\theta_1\cos\theta_2 - \cos\theta_1\sin\theta_2))}{r_2(\cos^2\theta_2 + \sin^2\theta_2)}\\
&= \frac{r_1}{r_2}(\cos(\theta_1 - \theta_2) + \mathrm{i}\sin(\theta_1 - \theta_2))
\end{aligned}$$

Links The last step of this working makes use of the trigonometric addition formulae together with the identity $\sin^2\theta + \cos^2\theta \equiv 1$
← **Pure Year 1, Section 10.3**

This complex number is in modulus–argument form, with modulus $\dfrac{r_1}{r_2}$ and argument $\theta_1 - \theta_2$, as required.

Example 8

$z_1 = 3\left(\cos\frac{5\pi}{12} + \mathrm{i}\sin\frac{5\pi}{12}\right)$ and $4\left(\cos\frac{\pi}{12} + \mathrm{i}\sin\frac{\pi}{12}\right)$

a Find: **i** $|z_1z_2|$ **ii** $\arg(z_1z_2)$

b Hence write z_1z_2 in the form: **i** $r(\cos\theta + \mathrm{i}\sin\theta)$ **ii** $x + \mathrm{i}y$

a i $|z_1z_2| = |z_1||z_2|$

$|z_1| = 3,\ |z_2| = 4$

$|z_1z_2| = 3 \times 4 = 12$

For a complex number in the form $z = r(\cos\theta + \mathrm{i}\sin\theta)$, $|z| = r$

ii $\arg(z_1z_2) = \arg z_1 + \arg z_2$

$\arg z_1 = \frac{5\pi}{12},\ \arg z_2 = \frac{\pi}{12}$

For a complex number in the form $z = r(\cos\theta + \mathrm{i}\sin\theta)$, $\arg z = \theta$

$\arg(z_1z_2) = \frac{5\pi}{12} + \frac{\pi}{12} = \frac{6\pi}{12} = \frac{\pi}{2}$

b i $z_1z_2 = r(\cos\theta + \mathrm{i}\sin\theta) = 12\left(\cos\frac{\pi}{2} + \mathrm{i}\sin\frac{\pi}{2}\right)$

$z_1z_2 = r_1r_2(\cos(\theta_1 + \theta_2) + \mathrm{i}\sin(\theta_1 + \theta_2))$

ii $\cos\frac{\pi}{2} = 0,\ \sin\frac{\pi}{2} = 1$

$z_1z_2 = 12(0 + \mathrm{i}(1)) = 12\mathrm{i}$

Evaluate $\cos\frac{\pi}{2}$ and $\sin\frac{\pi}{2}$

Example 9

$z_1 = 2\left(\cos\frac{\pi}{15} + \mathrm{i}\sin\frac{\pi}{15}\right)$ and $z_2 = 3\left(\cos\frac{2\pi}{5} - \mathrm{i}\sin\frac{2\pi}{5}\right)$

Express z_1z_2 in the form $x + \mathrm{i}y$.

Rewrite z_2 in the form $z_2 = r(\cos\theta + \mathrm{i}\sin\theta)$:

Watch out z_2 is not initially given in modulus–argument form.

$\cos\left(-\frac{2\pi}{5}\right) = \cos\frac{2\pi}{5}$ and $\sin\left(-\frac{2\pi}{5}\right) = -\sin\frac{2\pi}{5}$

Use $\cos(-\theta) = \cos\theta$ and $\sin(-\theta) = -\sin\theta$

$z_2 = 3\left(\cos\left(-\frac{2\pi}{5}\right) + \mathrm{i}\sin\left(-\frac{2\pi}{5}\right)\right)$

z_2 is now in modulus–argument form.

$z_1z_2 = 2\left(\cos\frac{\pi}{15} + \mathrm{i}\sin\frac{\pi}{15}\right) \times 3\left(\cos\left(-\frac{2\pi}{5}\right) + \mathrm{i}\sin\left(-\frac{2\pi}{5}\right)\right)$

$= 2 \times 3\left(\cos\left(\frac{\pi}{15} - \frac{2\pi}{5}\right) + \mathrm{i}\sin\left(\frac{\pi}{15} - \frac{2\pi}{5}\right)\right)$

Apply the result $z_1z_2 = r_1r_2(\cos(\theta_1 + \theta_2) + \mathrm{i}\sin(\theta_1 + \theta_2))$

$= 6\left(\cos\left(-\frac{\pi}{3}\right) + \mathrm{i}\sin\left(-\frac{\pi}{3}\right)\right)$

$= 6\left(\frac{1}{2} + \mathrm{i}\left(-\frac{\sqrt{3}}{2}\right)\right)$

Apply $\cos\left(-\frac{\pi}{3}\right) = \frac{1}{2}$ and $\sin\left(-\frac{\pi}{3}\right) = -\frac{\sqrt{3}}{2}$

$= 3 - 3\sqrt{3}\mathrm{i}$

Example 10

Express $\dfrac{\sqrt{2}\left(\cos\frac{\pi}{12} + \mathrm{i}\sin\frac{\pi}{12}\right)}{2\left(\cos\frac{5\pi}{6} + \mathrm{i}\sin\frac{5\pi}{6}\right)}$ in the form $x + \mathrm{i}y$.

$$\frac{\sqrt{2}\left(\cos\frac{\pi}{12} + \mathrm{i}\sin\frac{\pi}{12}\right)}{2\left(\cos\frac{5\pi}{6} + \mathrm{i}\sin\frac{5\pi}{6}\right)}$$

$$= \frac{\sqrt{2}}{2}\left(\cos\left(\frac{\pi}{12} - \frac{5\pi}{6}\right) + \mathrm{i}\sin\left(\frac{\pi}{12} - \frac{5\pi}{6}\right)\right)$$

$$= \frac{\sqrt{2}}{2}\left(\cos\left(-\frac{3\pi}{4}\right) + \mathrm{i}\sin\left(-\frac{3\pi}{4}\right)\right)$$

$$= \frac{\sqrt{2}}{2}\left(-\frac{1}{\sqrt{2}} + \mathrm{i}\left(-\frac{1}{\sqrt{2}}\right)\right)$$

$$= -\frac{1}{2} - \frac{1}{2}\mathrm{i}$$

Both numbers are in modulus–argument form, so you can divide the moduli and subtract the arguments.

Simplify.

Apply $\cos\left(-\frac{3\pi}{4}\right) = -\frac{1}{\sqrt{2}}$ and $\sin\left(-\frac{3\pi}{4}\right) = -\frac{1}{\sqrt{2}}$

Exercise 2D

1 For each given z_1 and z_2, find the following in the form $r(\cos\theta + \mathrm{i}\sin\theta)$:

i $|z_1z_2|$ **ii** $\arg(z_1z_2)$ **iii** z_1z_2

a $z_1 = 5\left(\cos\frac{3\pi}{8} + \mathrm{i}\sin\frac{3\pi}{8}\right)$, $z_2 = 6\left(\cos\frac{7\pi}{8} + \mathrm{i}\sin\frac{7\pi}{8}\right)$

b $z_1 = \sqrt{2}\left(\cos\frac{\pi}{3} + \mathrm{i}\sin\frac{\pi}{3}\right)$, $z_2 = 4\sqrt{2}\left(\cos\frac{3\pi}{4} + \mathrm{i}\sin\frac{3\pi}{4}\right)$

2 Given $z_1 = 8\left(\cos\frac{8\pi}{5} + \mathrm{i}\sin\frac{8\pi}{5}\right)$ and $z_2 = 4\left(\cos\frac{2\pi}{3} + \mathrm{i}\sin\frac{2\pi}{3}\right)$, write down the modulus and argument of:

a z_1z_2 **b** $\dfrac{z_1}{z_2}$ **c** z_1^2

3 Express the following in the form $x + \mathrm{i}y$:

a $(\cos 2\theta + \mathrm{i}\sin 2\theta)(\cos 3\theta + \mathrm{i}\sin 3\theta)$

b $\left(\cos\frac{3\pi}{11} + \mathrm{i}\sin\frac{3\pi}{11}\right)\left(\cos\frac{8\pi}{11} + \mathrm{i}\sin\frac{8\pi}{11}\right)$

c $3\left(\cos\frac{\pi}{4} + \mathrm{i}\sin\frac{\pi}{4}\right) \times 2\left(\cos\frac{\pi}{12} + \mathrm{i}\sin\frac{\pi}{12}\right)$

d $\sqrt{6}\left(\cos\frac{\pi}{3} - \mathrm{i}\sin\frac{\pi}{3}\right) \times \sqrt{3}\left(\cos\frac{\pi}{3} + \mathrm{i}\sin\frac{\pi}{3}\right)$

e $4\left(\cos\frac{5\pi}{9} - \mathrm{i}\sin\frac{5\pi}{9}\right) \times \frac{1}{2}\left(\cos\frac{5\pi}{18} - \mathrm{i}\sin\frac{5\pi}{18}\right)$

f $6\left(\cos\frac{\pi}{10} + \mathrm{i}\sin\frac{\pi}{10}\right) \times 5\left(\cos\frac{\pi}{3} + \mathrm{i}\sin\frac{\pi}{3}\right) \times \frac{1}{3}\left(\cos\frac{2\pi}{5} + \mathrm{i}\sin\frac{2\pi}{5}\right)$

g $(\cos 4\theta + \mathrm{i}\sin 4\theta)(\cos\theta - \mathrm{i}\sin\theta)$

h $3\left(\cos\frac{\pi}{12} + \mathrm{i}\sin\frac{\pi}{12}\right) \times \sqrt{2}\left(\cos\frac{\pi}{3} - \mathrm{i}\sin\frac{\pi}{3}\right)$

Hint First make sure both numbers are in modulus–argument form.

4 Express the following in the form $x + iy$:

a $\dfrac{\cos 5\theta + i\sin 5\theta}{\cos 2\theta + i\sin 2\theta}$ **b** $\dfrac{\sqrt{2}\left(\cos\frac{\pi}{2} + i\sin\frac{\pi}{2}\right)}{\frac{1}{2}\left(\cos\frac{\pi}{4} + i\sin\frac{\pi}{4}\right)}$ **c** $\dfrac{3\left(\cos\frac{\pi}{3} + i\sin\frac{\pi}{3}\right)}{4\left(\cos\frac{5\pi}{6} + i\sin\frac{5\pi}{6}\right)}$ **d** $\dfrac{\cos 2\theta - i\sin 2\theta}{\cos 3\theta + i\sin 3\theta}$

(E) **5** $z = -9 + 3i\sqrt{3}$

a Express z in the form $r(\cos\theta + i\sin\theta)$, $-\pi < \theta \leqslant \pi$ **(2 marks)**

b Given that $|w| = \sqrt{3}$ and $\arg w = \dfrac{7\pi}{12}$, express in the form $r(\cos\theta + i\sin\theta)$, $-\pi < \theta \leqslant \pi$:

i w **ii** zw **iii** $\dfrac{z}{w}$ **(4 marks)**

Challenge

By writing $z = 1 + i\sqrt{3}$ in modulus-argument form, show that:

a $z^7 = kz$ **b** $z^4 = pz$

where k and p are real constants to be found.

c Show z, z^7 and z^4 on an Argand diagram and describe the geometrical relationship between them.

2.4 Loci in the Argand diagram

Complex numbers can be used to represent a locus of points on an Argand diagram.

- **For two complex numbers $z_1 = x_1 + iy_1$ and $z_2 = x_2 + iy_2$, $|z_2 - z_1|$ represents the distance between the points z_1 and z_2 on an Argand diagram.**

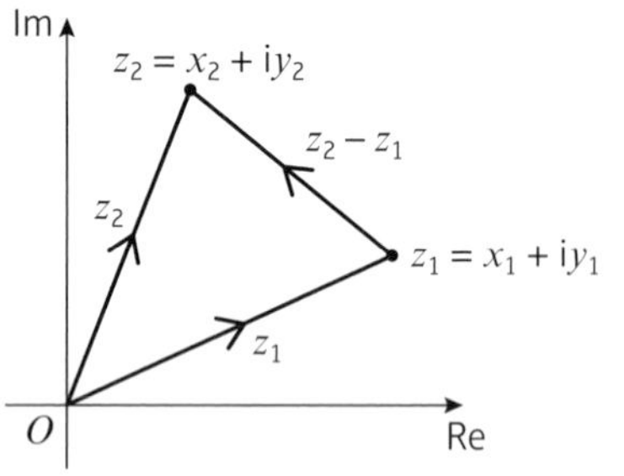

Using the above result, you can replace z_2 with the general point z. The locus of points described by $|z - z_1| = r$ is a circle with centre (x_1, y_1) and radius r.

Online Explore the locus of z, when $|z - z_1| = r$, using GeoGebra.

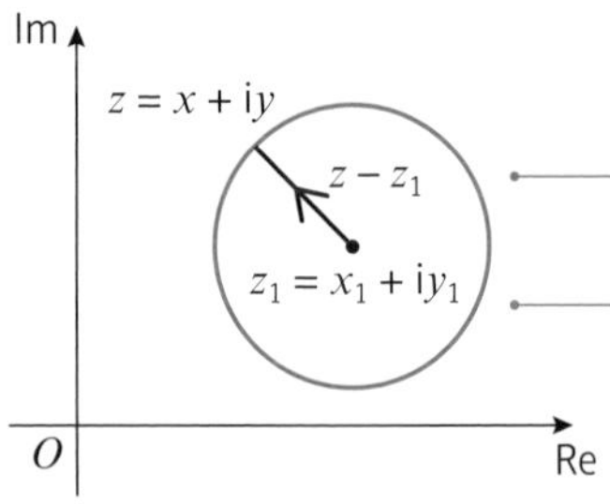
Im
$z = x + iy$
$z - z_1$
$z_1 = x_1 + iy_1$
O
Re
Locus of points.
Every point z, on the circumference of the circle, is a distance of r from the centre of the circle.

- **Given $z_1 = x_1 + iy_1$, the locus of point z on an Argand diagram such that $|z - z_1| = r$, or $|z - (x_1 + iy_1)| = r$, is a circle with centre (x_1, y_1) and radius r.**

You can derive a Cartesian form of the equation of a circle from this form by squaring both sides:

$$|z - z_1| = r$$
$$|(x - x_1) + i(y - y_1)| = r$$
$$(x - x_1)^2 + (y - y_1)^2 = r^2$$

Since $|p + qi| = \sqrt{p^2 + q^2}$

Links The Cartesian equation of a circle with centre (a, b) and radius r is $(x - a)^2 + (y - b)^2 = r^2$

← Pure Year 1, Section 6.2

The locus of points that are an equal distance from two different points z_1 and z_2 is the perpendicular bisector of the line segment joining the two points.

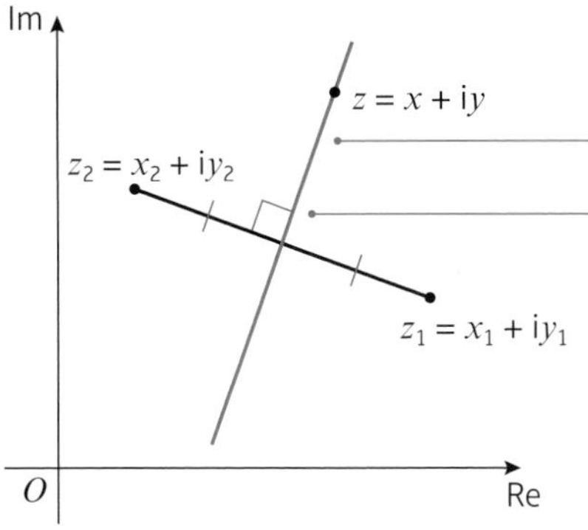

Locus of points.

Every point z on the line is an equal distance from points z_1 and z_2.

Online Explore the locus of z, when $|z - z_1| = |z - z_2|$, using GeoGebra.

- **Given $z_1 = x_1 + \mathrm{i}y_1$ and $z_2 = x_2 + \mathrm{i}y_2$, the locus of points z on an Argand diagram such that $|z - z_1| = |z - z_2|$ is the perpendicular bisector of the line segment joining z_1 and z_2.**

Example 11

Given that z satisfies $|z - 4| = 5$,

a sketch the locus of z on an Argand diagram.

b Find the values of z that satisfy:

i both $|z - 4| = 5$ and $\mathrm{Im}(z) = 0$ **ii** both $|z - 4| = 5$ and $\mathrm{Re}(z) = 0$

a $|z - 4| = 5$ is a circle with centre (4, 0) and radius 5.

$|z - (x_1 + \mathrm{i}y_1)| = r$ is represented by a circle with centre (x_1, y_1) and radius r.

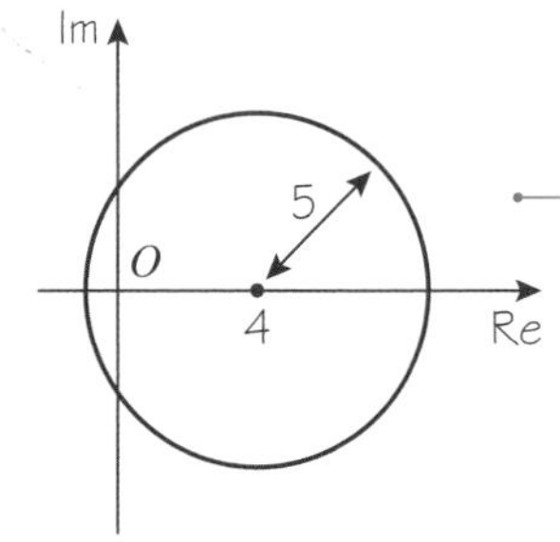

Sketch a circle with centre (4, 0) and radius 5 on an Argand diagram.

b i $\mathrm{Im}(z) = 0$ represents the real axis. The points where the circle cuts the real axis are (−1, 0) and (9, 0).

The values of z at these points are $z = -1$ and $z = 9$.

Centre of circle is (4, 0) and radius is 5. So consider $4 + 5 = 9$ and $4 - 5 = -1$.

Watch out Give your answers as complex numbers, not as coordinates.

ii $|z - 4| = 5 \Rightarrow (x - 4)^2 + y^2 = 5^2$

$(0 - 4)^2 + y^2 = 5^2$

$16 + y^2 = 25$

$y^2 = 9$

$y = \pm 3$

The points where the circle cuts the real axis are (0, 3) and (0, −3).

The values of z are $z = 3i$ and $z = -3i$.

This is the Cartesian equation of a circle with centre (4, 0) and radius 5.

Re(z) = 0 for all points on the imaginary axis, so set $x = 0$.

Example 12

A complex number z is represented by the point P in the Argand diagram.

Given that $|z - 5 - 3i| = 3$,

a sketch the locus of P **b** find the Cartesian equation of this locus

c find the maximum value of arg z in the interval $(-\pi, \pi)$.

a

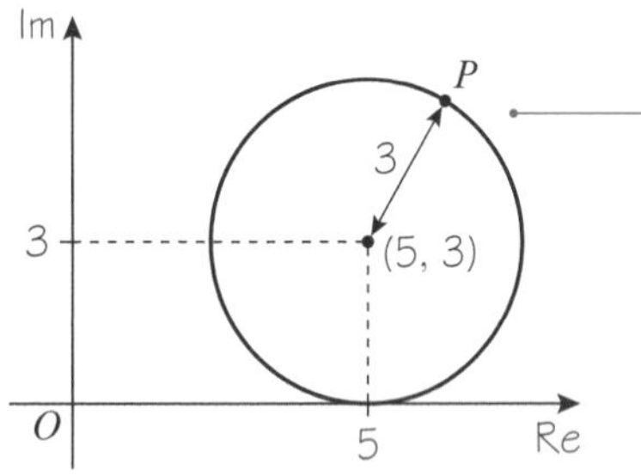

$|z - 5 - 3i|$ can be written as $|z - (5 + 3i)|$. As this distance is always equal to 3, the locus of P is a circle centre (5, 3), radius 3.

b The Cartesian equation of the locus is

$(x - 5)^2 + (y - 3)^2 = 9$

The standard Cartesian equation of a circle is $(x - a)^2 + (y - b)^2 = r^2$

c

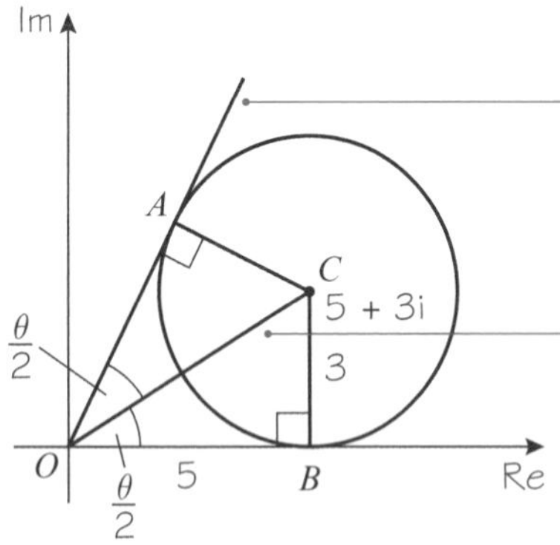

The maximum value of arg z is the angle OA makes with the positive real axis.

The line OC bisects the angle AOB.

Problem-solving

When solving geometrical problems like this one, it is helpful to draw an Argand diagram. The maximum value of arg(z) occurs when the line between the origin and P is a tangent to the circle.

Using triangle OBC:

$\tan\left(\frac{\theta}{2}\right) = \frac{3}{5}$

$\theta = 2\arctan\left(\frac{3}{5}\right) = 1.08$ rad (3 s.f.)

Use circle properties. OB is perpendicular to BC, and triangles OBC and OAC are congruent.

Example 13

Given that the complex number $z = x + iy$ satisfies the equation $|z - 12 - 5i| = 3$, find the minimum value of $|z|$ and maximum value of $|z|$.

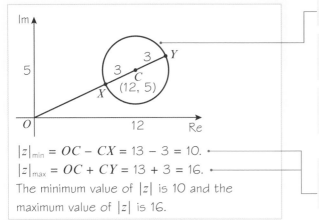

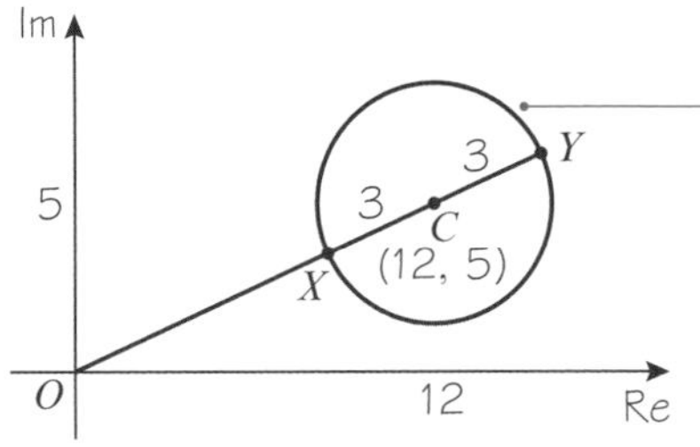

$|z|_{min} = OC - CX = 13 - 3 = 10.$
$|z|_{max} = OC + CY = 13 + 3 = 16.$
The minimum value of $|z|$ is 10 and the maximum value of $|z|$ is 16.

The locus of z is a circle centre $C(12, 5)$, radius 3.

$|z|$ represents the distance from the origin to any point on this locus.

$|z|_{min}$ and $|z|_{max}$ are represented by the distances OX and OY respectively.

The distance $OC = \sqrt{12^2 + 5^2} = 13$.
The radius $r = CX = CY = 3$.

Example 14

Given that $|z - 3| = |z + i|$,

a sketch the locus of z and find the Cartesian equation of this locus

b find the least possible value of $|z|$.

a $|z - 3| = |z + i|$ is the perpendicular bisector of the line segment joining the points (3, 0) and (0, −1).
The gradient of the line joining (0, −1) and (3, 0) is $\frac{1}{3}$
So, the gradient of the perpendicular bisector is −3.
The midpoint of the line joining (0, −1) and (3, 0) is $\left(\frac{3}{2}, -\frac{1}{2}\right)$.

$y - y_1 = m(x - x_1)$

$y + \frac{1}{2} = -3\left(x - \frac{3}{2}\right)$

$y + \frac{1}{2} = -3x + \frac{9}{2}$

$y = -3x + 4$

The locus of points z satisfying $|z - z_1| = |z - z_2|$ is the perpendicular bisector of the line segment joining z_1 to z_2.

The perpendicular bisector will pass through the midpoint.

Substitute $(x_1, y_1) = \left(\frac{3}{2}, -\frac{1}{2}\right)$ and $m = -3$ into the equation of a straight line.

Problem-solving

You could also square both sides of $|z - 3| = |z + i|$:

$$|x + iy - 3| = |x + iy + i|$$
$$|(x - 3) + iy| = |x + i(y + 1)|$$
$$(x - 3)^2 + y^2 = x^2 + (y + 1)^2$$
$$x^2 - 6x + 9 + y^2 = x^2 + y^2 + 2y + 1$$
$$y = -3x + 4$$

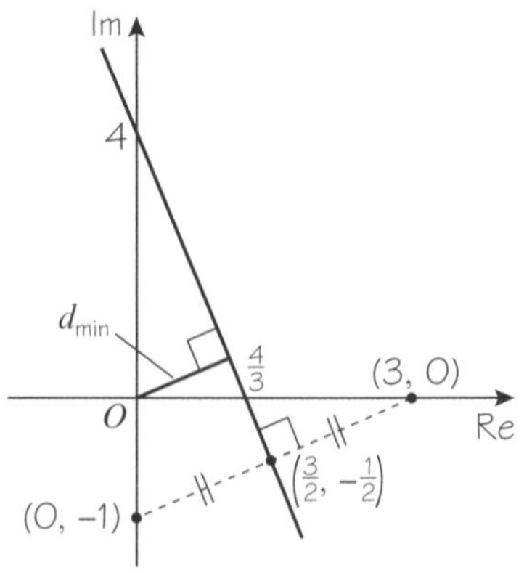

b The gradient of the line labelled d_{min} is $\frac{1}{3}$

The equation of this line is $y = \frac{1}{3}x$

$\frac{1}{3}x = -3x + 4$

$\frac{10}{3}x = 4$

$x = \frac{6}{5} \Rightarrow y = \frac{2}{5}$

$d_{min} = \sqrt{\left(\frac{6}{5}\right)^2 + \left(\frac{2}{5}\right)^2}$

$= \frac{2\sqrt{10}}{5}$

Problem-solving

The minimum distance is the perpendicular distance from O to the perpendicular bisector.

The line is parallel to the line joining (0, −1) and (3, 0).

The line passes through the origin.

Find the point where this line intersects the perpendicular bisector.

Solve to find x and substitute into $y = \frac{1}{3}x$ to find y.

Use Pythagoras' theorem.

Locus questions can also make use of the geometric property of the argument.

- **Given $z_1 = x_1 + \mathrm{i}y_1$, the locus of points z on an Argand diagram such that $\arg(z - z_1) = \theta$ is a half-line from, but not including, the fixed point z_1 making an angle θ with a line from the fixed point z_1 parallel to the real axis.**

Notation A **half-line** is a straight line extending from a point infinitely in one direction only.

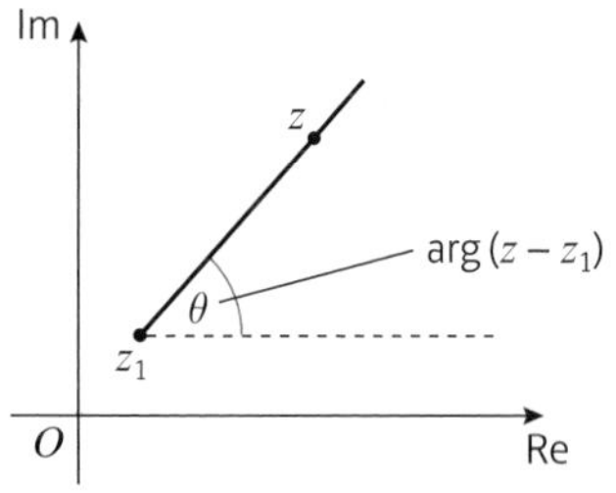

Online Explore the locus of z, when $\arg(z - z_1) = \theta$, using GeoGebra.

You can find the Cartesian equation of the half-line corresponding to $\arg(z - z_1) = \theta$ by considering how the argument is calculated:

$$\arg(z - z_1) = \theta$$

$$\arg((x - x_1) + \mathrm{i}(y - y_1)) = \theta$$

$$\frac{y - y_1}{x - x_1} = \tan\theta$$

$$y - y_1 = \tan\theta\,(x - x_1)$$

θ is a fixed angle so $\tan\theta$ is a constant.

This is the equation of a straight line with gradient $\tan\theta$ passing through the point (x_1, y_1).

Example 15

Given that $\arg(z + 3 + 2\text{i}) = \frac{3\pi}{4}$,

a sketch the locus of z on an Argand diagram

b find the Cartesian equation of the locus

c find the complex number z that satisfies both $|z + 3 + 2\text{i}| = 10$ and $\arg(z + 3 + 2\text{i}) = \frac{3\pi}{4}$

a

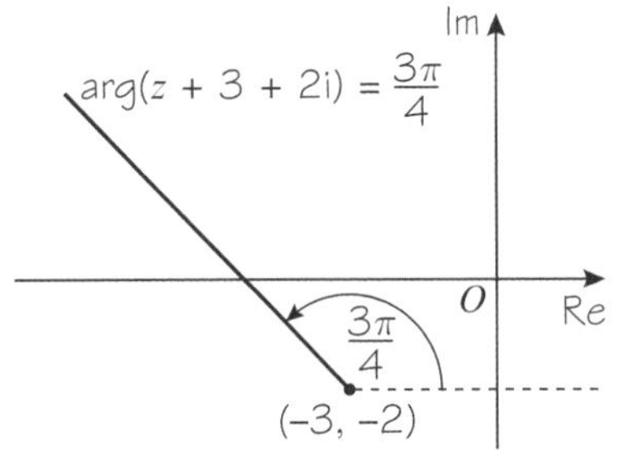

$z + 3 + 2\text{i}$ can be written as $z - (-3 - 2\text{i})$. As $\arg(z + 3 + 2\text{i}) = \frac{3\pi}{4}$, the locus of z is the half-line from (−3, −2) making an angle of $\frac{3\pi}{4}$ in an anticlockwise sense from a line in the same direction as the positive real axis.

b

$$\arg(z + 3 + 2\text{i}) = \frac{3\pi}{4}$$

z can be rewritten as $z = x + \text{i}y$.

$$\arg(x + \text{i}y + 3 + 2\text{i}) = \frac{3\pi}{4}$$

Group the real and imaginary parts.

$$\arg((x + 3) + \text{i}(y + 2)) = \frac{3\pi}{4}$$

Remove the argument.

$$\frac{y + 2}{x + 3} = \tan\frac{3\pi}{4}$$

$\tan\frac{3\pi}{4} = -1$

$$y + 2 = -(x + 3)$$

Hence the Cartesian equation of the locus is $y = -x - 5,\ x < -3$

Watch out The locus is the half-line so you need to give a suitable range of values for x.

c $|z + 3 + 2\text{i}| = 10$ is a circle with centre (−3, −2) and radius 10.

Use a geometric approach to find z.

Draw part of a circle with centre (−3, −2) and radius 10.

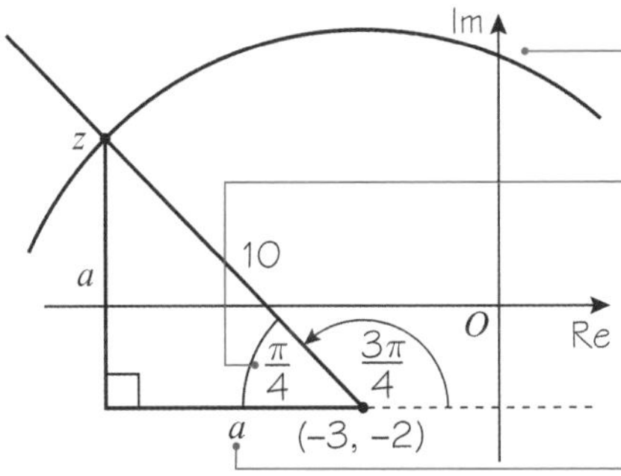

Angle inside the new triangle is $\pi - \frac{3\pi}{4} = \frac{\pi}{4}$

As the angle is $\frac{\pi}{4}$, the triangle is isosceles.
So the two shorter sides have the same length.

$$a^2 + a^2 = 10^2 \Rightarrow 2a^2 = 100$$

$$a = \sqrt{50} = \pm 5\sqrt{2}$$

$$z = (-3 - 5\sqrt{2}) + \text{i}(-2 + 5\sqrt{2})$$

Problem-solving An alternative algebraic approach would be to substitute the equation for the half-line, $y = -x - 5$, into the equation of the circle, $(x + 3)^2 + (y + 2)^2 = 10^2$, and then solve for x and y. You would need to choose the solution which lies on the correct half line.

Exercise 2E

(P) **1** Sketch the locus of z and give the Cartesian equation of the locus of z when:

a $|z| = 6$ **b** $|z| = 10$ **c** $|z - 3| = 2$

d $|z + 3i| = 3$ **e** $|z - 4i| = 5$ **f** $|z + 1| = 1$

g $|z - 1 - i| = 5$ **h** $|z + 3 + 4i| = 4$ **i** $|z - 5 + 6i| = 5$

Hint You may choose a geometric or an algebraic approach to answer these questions.

2 Given that z satisfies $|z - 5 - 4i| = 8$,

a sketch the locus of z on an Argand diagram

b find the exact values of z that satisfy:

i both $|z - 5 - 4i| = 8$ and $\text{Re}(z) = 0$ **ii** both $|z - 5 - 4i| = 8$ and $\text{Im}(z) = 0$

(P) **3** A complex number z is represented by the point P on the Argand diagram.
Given that $|z - 5 + 7i| = 5$,

a sketch the locus of P

b find the Cartesian equation of this locus

c find the maximum value of $\arg z$ in the interval $(-\pi, \pi)$.

(E/P) **4** On an Argand diagram the point P represents the complex number z.
Given that $|z - 4 - 3i| = 8$,

a find the Cartesian equation for the locus of P **(2 marks)**

b sketch the locus of P **(2 marks)**

c find the maximum and minimum values of $|z|$ for points on this locus. **(2 marks)**

(E/P) **5** The point P represents a complex number z on an Argand diagram.
Given that $|z + 2 - 2\sqrt{3}i| = 2$,

a sketch the locus of P on an Argand diagram **(2 marks)**

b write down the minimum value of $\arg z$ **(2 marks)**

c find the maximum value of $\arg z$. **(2 marks)**

6 Sketch the locus of z and give the Cartesian equation of the locus of z when:

a $|z - 6| = |z - 2|$ **b** $|z + 8| = |z - 4|$

c $|z| = |z + 6i|$ **d** $|z + 3i| = |z - 8i|$

e $|z - 2 - 2i| = |z + 2 + 2i|$ **f** $|z + 4 + i| = |z + 4 + 6i|$

g $|z + 3 - 5i| = |z - 7 - 5i|$ **h** $|z + 4 - 2i| = |z - 8 + 2i|$

i $\dfrac{|z + 3|}{|z - 6i|} = 1$ **j** $\dfrac{|z + 6 - i|}{|z - 10 - 5i|} = 1$

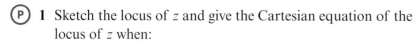

E/P **7** Given that $|z - 3| = |z - 6\text{i}|$,

a sketch the locus of z **(3 marks)**

b find the exact least possible value of $|z|$. **(4 marks)**

E/P **8** Given that $|z + 3 + 3\text{i}| = |z - 9 - 5\text{i}|$,

a sketch the locus of z **(3 marks)**

b find the Cartesian equation of this locus **(3 marks)**

c find the exact least possible value of $|z|$. **(3 marks)**

9 Sketch the locus of z and give the Cartesian equation of the locus of z when:

a $|2 - z| = 3$ **b** $|5\text{i} - z| = 4$ **c** $|3 - 2\text{i} - z| = 3$

10 Sketch the locus of z when:

a $\arg z = \frac{\pi}{3}$ **b** $\arg(z + 3) = \frac{\pi}{4}$ **c** $\arg(z - 2) = \frac{\pi}{2}$

d $\arg(z + 2 + 2\text{i}) = -\frac{\pi}{4}$ **e** $\arg(z - 1 - \text{i}) = \frac{3\pi}{4}$ **f** $\arg(z + 3\text{i}) = \pi$

g $\arg(z - 1 + 3\text{i}) = \frac{2\pi}{3}$ **h** $\arg(z - 3 + 4\text{i}) = -\frac{\pi}{2}$ **i** $\arg(z - 4\text{i}) = -\frac{3\pi}{4}$

P **11** Given that z satisfies $|z + 2\text{i}| = 3$,

a sketch the locus of z on an Argand diagram

b find $|z|$ that satisfies both $|z + 2\text{i}| = 3$ and $\arg z = \frac{\pi}{6}$

E/P **12** Given that the complex number z satisfies the equation $|z + 6 + 6\text{i}| = 4$,

a find the exact maximum and minimum value of $|z|$ **(3 marks)**

b find the range of values for θ, $-\pi < \theta < \pi$, for which $\arg(z - 4 + 2\text{i}) = \theta$ and $|z + 6 + 6\text{i}| = 4$ have no common solutions. **(4 marks)**

E/P **13** The point P represents a complex number z on an Argand diagram such that $|z| = 5$.
The point Q represents a complex number z on an Argand diagram such that $\arg(z + 4) = \frac{\pi}{2}$

a Sketch, on the same Argand diagram, the locus of P and the locus of Q as z varies. **(2 marks)**

b Find the complex number for which both $|z| = 5$ and $\arg(z + 4) = \frac{\pi}{2}$ **(2 marks)**

E/P **14** Given that the complex number z satisfies $|z - 2 - 2\text{i}| = 2$,

a sketch, on an Argand diagram, the locus of z **(2 marks)**

Given further that $\arg(z - 2 - 2\text{i}) = \frac{\pi}{6}$,

b find the value of z in the form $a + \text{i}b$, where $a, b \in \mathbb{R}$. **(4 marks)**

E/P **15** Sketch on the same Argand diagram the locus of points satisfying:

a $|z - 2\text{i}| = |z - 8\text{i}|$ **(2 marks)**

b $\arg(z - 2 - \text{i}) = \frac{\pi}{4}$ **(3 marks)**

The complex number z satisfies both $|z - 2\text{i}| = |z - 8\text{i}|$ and $\arg(z - 2 - \text{i}) = \frac{\pi}{4}$ **(2 marks)**

c Use your answers to parts **a** and **b** to find the value of z.

E/P **16** Sketch on the same Argand diagram the locus of points satisfying:

a $|z - 3 + 2\text{i}| = 4$ **(2 marks)**

b $\arg(z - 1) = -\frac{\pi}{4}$ **(3 marks)**

The complex number z satisfies both $|z - 3 + 2\text{i}| = 4$ and $\arg(z - 1) = -\frac{\pi}{4}$

Given that $z = a + \text{i}b$, where $a, b \in \mathbb{R}$,

c find the exact value of a and the exact value of b. **(3 marks)**

E/P **17** If the complex number z satisfies both $\arg z = \frac{\pi}{3}$ and $\arg(z - 4) = \frac{\pi}{2}$,

a find the value of z in the form $a + \text{i}b$, where $a, b \in \mathbb{R}$. **(3 marks)**

b Hence, find $\arg(z - 8)$. **(2 marks)**

E/P **18** Given that $\arg(z + 4) = \frac{\pi}{3}$,

a sketch the locus of z on an Argand diagram **(3 marks)**

b find the minimum value of $|z|$ for points on this locus. **(2 marks)**

E/P **19** A complex number z is represented by the point P on the Argand diagram.
Given $|z + 8 - 4\text{i}| = 2$,

a sketch the locus of P **(2 marks)**

b show that the maximum value of $\arg(z + 15 - 2\text{i})$ in the interval $(-\pi, \pi)$ is $2\arcsin\left(\frac{2}{\sqrt{53}}\right)$ **(3 marks)**

c find the exact values of the complex numbers that satisfy both $|z + 8 - 4\text{i}| = 2$ and $\arg(z + 4\text{i}) = \frac{3\pi}{4}$ **(3 marks)**

Challenge

The complex number z satisfies both $|z + \text{i}| = 5$ and $\arg(z - 2\text{i}) = \theta$, where θ is a real constant such that $-\pi < \theta \leqslant \pi$.
Given that $|z - 4\text{i}| < 3$, find the range of possible values of θ.

2.5 Regions in the Argand diagram

You can use complex numbers to represent regions on an Argand diagram.

Example 16

a On separate Argand diagrams, shade in the regions represented by:

i $|z - 4 - 2\text{i}| \leqslant 2$ **ii** $|z - 4| < |z - 6|$ **iii** $0 \leqslant \arg(z - 2 - 2\text{i}) \leqslant \frac{\pi}{4}$

b Hence, on the same Argand diagram, shade the region which satisfies
$\{z \in \mathbb{C} : |z - 4 - 2\text{i}| \leqslant 2\} \cap \{z \in \mathbb{C} : |z - 4| < |z - 6|\} \cap \left\{z \in \mathbb{C} : 0 \leqslant \arg(z - 2 - 2\text{i}) \leqslant \frac{\pi}{4}\right\}$

a i $|z - 4 - 2i| \leqslant 2$

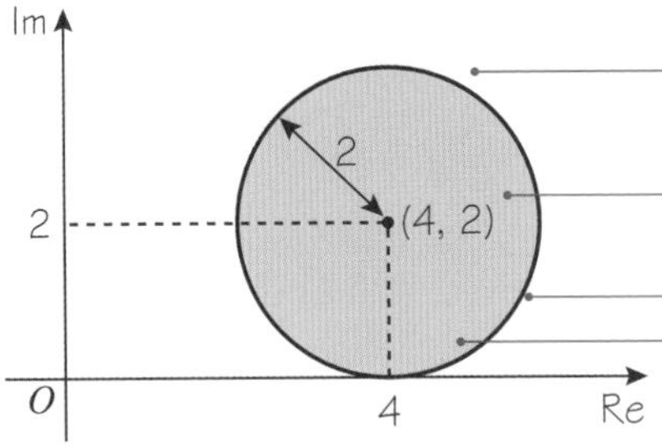

$|z - 4 - 2i| = 2$ represents a circle centre (4, 2), radius 2.

$|z - 4 - 2i| < 2$ represents the region on the inside of this circle.

$|z - 4 - 2i| \leqslant 2$ represents the boundary inside of this circle.

ii $|z - 4| < |z - 6|$

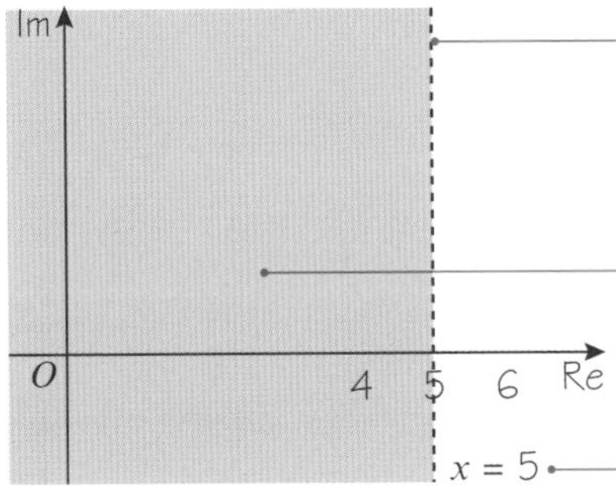

$|z - 4| = |z - 6|$ is represented by the line $x = 5$. This line is the perpendicular bisector of the line segment joining (4, 0) to (6, 0).

$|z - 4| < |z - 6|$ represents the region $x < 5$. All points in this region are closer to (4, 0) than to (6, 0).

Note this region does not include the line $x = 5$. So $x = 5$ is represented by a dashed line.

iii $0 \leqslant \arg(z - 2 - 2i) \leqslant \frac{\pi}{4}$

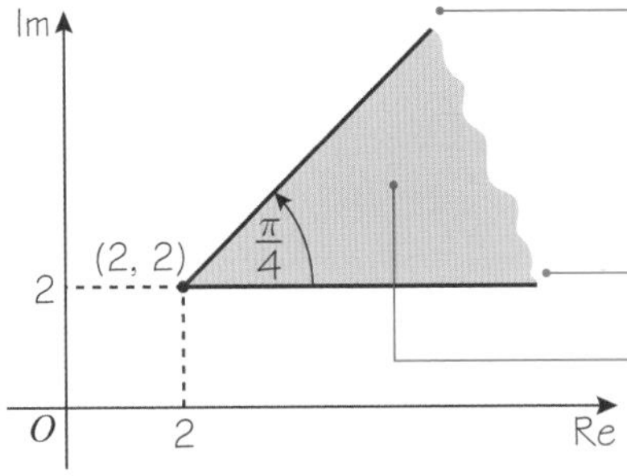

$\arg(z - 2 - 2i) = \frac{\pi}{4}$ is the half-line from the point (2, 2) at angle $\frac{\pi}{4}$ to the horizontal.

$\arg(z - 2 - 2i) = 0$ is the other half-line shown from the point (2, 2).

$0 \leqslant \arg(z - 2 - 2i) \leqslant \frac{\pi}{4}$ is represented by the region in between and including these two half-lines.

Notation The symbol $\cap$ is the symbol for the **intersection** of two sets. You need to find the region of points that lie in all three sets.

b $|z - 4 - 2i| \leqslant 2$, $|z - 4| < |z - 6|$ and $0 \leqslant \arg(z - 2 - 2i) \leqslant \frac{\pi}{4}$

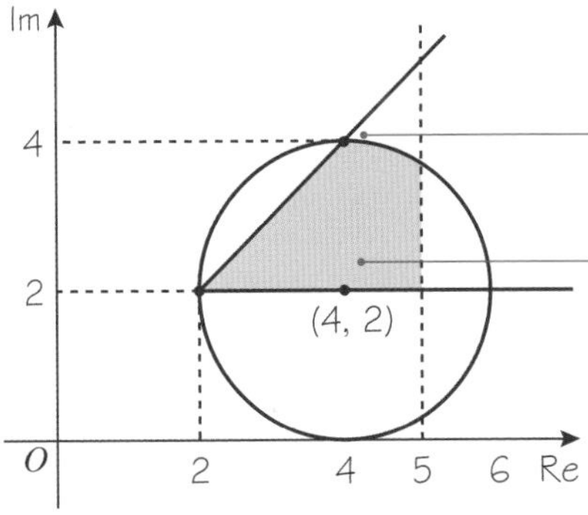

The line $\arg(z - 2 - 2i) = \frac{\pi}{4}$ and the circle $|z - 4 - 2i| = 2$ both go through the point (4, 4).

The region shaded is satisfied by all three of
$|z - 4 - 2i| \leqslant 2$
$|z - 4| < |z - 6|$
$0 \leqslant \arg(z - 2 - 2i) \leqslant \frac{\pi}{4}$

Online Explore this region using GeoGebra.

Exercise 2F

1 On an Argand diagram, shade in the regions represented by the following inequalities:

a $|z| < 3$
b $|z - 2i| > 2$
c $|z + 7| \geqslant |z - 1|$
d $|z + 6| > |z + 2 + 8i|$
e $2 \leqslant |z| \leqslant 3$
f $1 \leqslant |z + 4i| \leqslant 4$
g $3 \leqslant |z - 3 + 5i| \leqslant 5$

(E/P) 2 The region R in an Argand diagram is satisfied by the inequalities $|z| \leqslant 5$ and $|z| \leqslant |z - 6i|$. Draw an Argand diagram and shade in the region R. **(6 marks)**

(E/P) 3 The complex number z is represented by a point P on an Argand diagram.

Given that $|z + 1 - i| \leqslant 1$ and $0 \leqslant \arg z \leqslant \frac{3\pi}{4}$, shade the locus of P. **(6 marks)**

(E/P) 4 Shade on an Argand diagram the region satisfied by

$\{z \in \mathbb{C} : |z| \leqslant 3\} \cap \left\{z \in \mathbb{C} : \frac{\pi}{4} \leqslant \arg(z + 3) \leqslant \pi\right\}$ **(6 marks)**

(E/P) 5 a Sketch on the same Argand diagram:

i the locus of points representing $|z - 2| = |z - 6 - 8i|$ **(2 marks)**

ii the locus of points representing $\arg(z - 4 - 2i) = 0$ **(2 marks)**

iii the locus of points representing $\arg(z - 4 - 2i) = \frac{\pi}{2}$ **(2 marks)**

b Shade on an Argand diagram the set of points

$\{z \in \mathbb{C} : |z - 2| \leqslant |z - 6 - 8i|\} \cap \left\{z \in \mathbb{C} : 0 \leqslant \arg(z - 4 - 2i) \leqslant \frac{\pi}{2}\right\}$ **(2 marks)**

(E/P) 6 a Find the Cartesian equations of:

i the locus of points representing $|z + 10| = |z - 6 - 4i\sqrt{2}|$

ii the locus of points representing $|z + 1| = 3$. **(6 marks)**

b Find the two values of z that satisfy both $|z + 10| = |z - 6 - 4i\sqrt{2}|$ and $|z + 1| = 3$. **(2 marks)**

c Hence shade in the region R on an Argand diagram which satisfies both $|z + 10| \leqslant |z - 6 - 4i\sqrt{2}|$ and $|z + 1| \leqslant 3$. **(4 marks)**

Challenge

The sets A, B and C are defined as:

$A = \{z \in \mathbb{C} : |z + 5 + 8i| \leqslant 5\}$

$B = \{z \in \mathbb{C} : |z + 8 + 4i| \leqslant |z + 2 + 12i|\}$

$C = \left\{z \in \mathbb{C} : 0 \leqslant \arg(z + 10 + 8i) \leqslant \frac{\pi}{4}\right\}$

Shade the set of points $A \cap B \cap C'$, that are in set A and in set B, but not in set C.

Mixed exercise 2

(E) **1** $f(z) = z^2 + 5z + 10$

a Find the roots of the equation $f(z) = 0$, giving your answers in the form $a \pm ib$, where a and b are real numbers. **(3 marks)**

b Show these roots on an Argand diagram. **(1 mark)**

(E/P) **2** $f(z) = z^3 + z^2 + 3z - 5$

Given that $f(-1 + 2i) = 0$,

a find all the solutions to the equation $f(z) = 0$ **(4 marks)**

b show all the roots of $f(z) = 0$ on a single Argand diagram **(2 marks)**

c prove that these three points are the vertices of a right-angled triangle. **(2 marks)**

(E/P) **3** $f(z) = z^4 - z^3 + 13z^2 - 47z + 34$

Given that $z = -1 + 4i$ is a solution to the equation,

a find all the solutions to the equation $f(z) = 0$ **(4 marks)**

b show all the roots on a single Argand diagram. **(2 marks)**

(E) **4** The real and imaginary parts of the complex number $z = x + iy$ satisfy the equation

$$(4 - 3i)x - (1 + 6i)y - 3 = 0$$

a Find the value of x and the value of y. **(3 marks)**

b Show z on an Argand diagram. **(1 mark)**

Find the values of:

c $|z|$ **(2 marks)**

d $\arg z$ **(2 marks)**

(E) **5** $z_1 = 4 + 2i$, $z_2 = -3 + i$

a Draw points representing z_1 and z_2 on the same Argand diagram. **(1 mark)**

b Find the exact value of $|z_1 - z_2|$. **(2 marks)**

Given that $w = \frac{z_1}{z_2}$,

c express w in the form $a + ib$, where $a, b \in \mathbb{R}$ **(2 marks)**

d find $\arg w$, giving your answer in radians. **(2 marks)**

(E) **6** A complex number z is given by $z = a + 4i$ where a is a non-zero real number.

a Find $z^2 + 2z$ in the form $x + iy$, where x and y are real expressions in terms of a. **(4 marks)**

Given that $z^2 + 2z$ is real,

b find the value of a. **(1 mark)**

Using this value for a,

c find the values of the modulus and argument of z, giving the argument in radians and giving your answers correct to 3 significant figures. **(3 marks)**

d Show the complex numbers z, z^2 and $z^2 + 2z$ on a single Argand diagram. **(3 marks)**

(E) **7** The complex number z is defined by $z = \dfrac{3 + 5\text{i}}{2 - \text{i}}$

Find:

a $|z|$ **(4 marks)**

b $\arg z$ **(2 marks)**

(E) **8** $z = 1 + 2\text{i}$

a Show that $|z^2 - z| = 2\sqrt{5}$. **(4 marks)**

b Find $\arg(z^2 - z)$, giving your answer in radians to 2 decimal places. **(2 marks)**

c Show z and $z^2 - z$ on a single Argand diagram. **(2 marks)**

(E) **9** $z = \dfrac{1}{2 + \text{i}}$

a Express in the form $a + b\text{i}$, where $a, b \in \mathbb{R}$,

i z^2 **ii** $z - \dfrac{1}{z}$ **(4 marks)**

b Find $|z^2|$. **(2 marks)**

c Find $\arg\left(z - \dfrac{1}{z}\right)$, giving your answer in radians to two decimal places. **(2 marks)**

(E/P) **10** $z = \dfrac{a + 3\text{i}}{2 + a\text{i}}$, $a \in \mathbb{R}$

a Given that $a = 4$, find $|z|$.

b Show that there is only one value of a for which $\arg z = \dfrac{\pi}{4}$, and find this value.

(E) **11** $z_1 = -1 - \text{i}$, $z_2 = 1 + \text{i}\sqrt{3}$

a Express z_1 and z_2 in the form $r(\cos\theta + \text{i}\sin\theta)$, where $-\pi < \theta \leqslant \pi$. **(2 marks)**

b Find the modulus of:

i z_1z_2 **ii** $\dfrac{z_1}{z_2}$ **(2 marks)**

c Find the argument of:

i z_1z_2 **ii** $\dfrac{z_1}{z_2}$ **(2 marks)**

(E) **12** $z = 2 - 2\text{i}\sqrt{3}$

Find:

a $|z|$ **(1 mark)**

b $\arg z$, in terms of π. **(2 marks)**

$w = 4\left(\cos\left(-\frac{\pi}{4}\right) + \text{i}\sin\left(-\frac{\pi}{4}\right)\right)$

Find:

c $\left|\dfrac{w}{z}\right|$ **(1 mark)**

d $\arg\left(\dfrac{w}{z}\right)$, in terms of π. **(2 marks)**

(E) **13** Express $4 - 4\text{i}$ in the form $r(\cos\theta + \text{i}\sin\theta)$, where $r > 0$, $-\pi < \theta \leqslant \pi$, giving r and θ as exact values. **(3 marks)**

E/P **14** The point P represents a complex number z in an Argand diagram.
Given that $|z + 1 - i| = 1$,

a find a Cartesian equation for the locus of P **(2 marks)**

b sketch the locus of P on an Argand diagram **(2 marks)**

c find the greatest and least possible values of $|z|$ **(2 marks)**

d find the greatest and least possible values of $|z - 1|$. **(2 marks)**

P **15** Given that $\arg(z - 2 + 4i) = \frac{\pi}{4}$,

a sketch the locus of $P(x, y)$ which represents z on an Argand diagram

b find the minimum value of $|z|$ for points on this locus.

E/P **16** The complex number z satisfies $|z + 3 - 6i| = 3$. Show that the exact maximum value of $\arg z$ in the interval $(-\pi, \pi)$ is $\frac{\pi}{2} + 2\arcsin\left(\frac{1}{\sqrt{5}}\right)$. **(4 marks)**

E/P **17** A complex number z is represented by the point P on the Argand diagram.
Given that $|z - 5| = 4$,

a sketch the locus of P. **(2 marks)**

b Find the complex numbers that satisfy both $|z - 5| = 4$ and $\arg(z + 3i) = \frac{\pi}{3}$, giving your answers in radians to 2 decimal places. **(6 marks)**

c Given that $\arg(z + 5) = \theta$ and $|z - 5| = 4$ have no common solutions, find the range of possible values of θ, $-\pi < \theta < \pi$. **(3 marks)**

E/P **18** Given that $|z + 5 - 5i| = |z - 6 - 3i|$,

a sketch the locus of z **(3 marks)**

b find the Cartesian equation of this locus **(3 marks)**

c find the least possible value of $|z|$. **(3 marks)**

E/P **19** **a** Find the Cartesian equation of the locus of points that satisfies $|z - 4| = |z - 8i|$. **(3 marks)**

b Find the value of z that satisfies both $|z - 2| = |z - 4i|$ and $\arg z = \frac{\pi}{4}$ **(3 marks)**

c Shade on an Argand diagram the set of points
$\{z \in \mathbb{C} : |z - 4| \leqslant |z - 8i|\} \cap \left\{z \in \mathbb{C} : \frac{\pi}{4} \leqslant \arg z \leqslant \pi\right\}$ **(3 marks)**

E/P **20** **a** Find the Cartesian equations of:

i the locus of points representing $|z - 3 + i| = |z - 1 - i|$

ii the locus of points representing $|z - 2| = 2\sqrt{2}$. **(6 marks)**

b Find the two values of z that satisfy both $|z - 3 + i| = |z - 1 - i|$ and $|z - 2| = 2\sqrt{2}$. **(2 marks)**

The region R is defined by the inequalities $|z - 3 + i| \geqslant |z - 1 - i|$ and $|z - 2| \leqslant 2\sqrt{2}$.

c Show the region R on an Argand diagram. **(4 marks)**

Challenge

The complex number z satisfies $\arg(z - 3 + 3i) = -\frac{\pi}{4}$

The complex number w is such that $|w - z| = 3$.

a Sketch the locus of w.

b State the exact minimum value of $|w|$.

Summary of key points

1 You can represent complex numbers on an **Argand diagram**. The x-axis on an Argand diagram is called the **real axis** and the y-axis is called the **imaginary axis**. The complex number $z = x + iy$ is represented on the diagram by the point $P(x, y)$, where x and y are Cartesian coordinates.

2 The complex number $z = x + iy$ can be represented as the vector $\begin{pmatrix} x \\ y \end{pmatrix}$ on an Argand diagram.

3 The **modulus** of a complex number, $|z|$, is the distance from the origin to that number on an Argand diagram. For a complex number $z = x + iy$, the modulus is given by $|z| = \sqrt{x^2 + y^2}$.

4 The **argument** of a complex number, $\arg z$, is the angle between the positive real axis and the line joining that number to the origin on an Argand diagram. For a complex number $z = x + iy$, the argument, θ, satisfies $\tan\theta = \frac{y}{x}$

5 Let α be the positive acute angle made with the real axis by the line joining the origin and z.

- If z lies in the first quadrant then $\arg z = \alpha$.
- If z lies in the second quadrant then $\arg z = \pi - \alpha$.
- If z lies in the third quadrant then $\arg z = -(\pi - \alpha)$.
- If z lies in the fourth quadrant then $\arg z = -\alpha$.

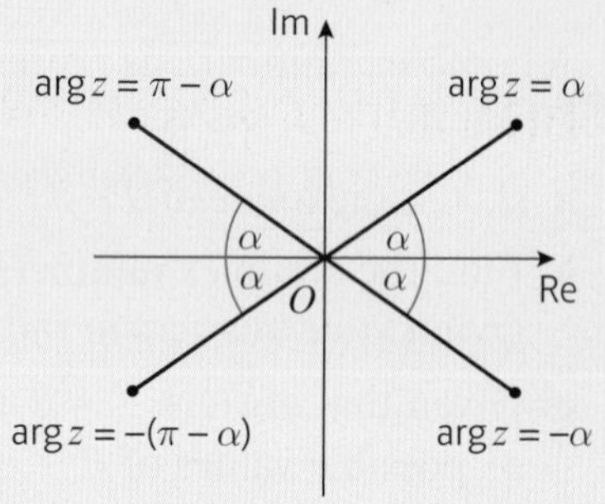

6 For a complex number z with $|z| = r$ and $\arg z = \theta$, the modulus–argument form of z is $z = r(\cos\theta + i\sin\theta)$

7 For any two complex numbers z_1 and z_2,

- $|z_1 z_2| = |z_1||z_2|$
- $\arg(z_1 z_2) = \arg z_1 + \arg z_2$
- $\left|\frac{z_1}{z_2}\right| = \frac{|z_1|}{|z_2|}$
- $\arg\left(\frac{z_1}{z_2}\right) = \arg z_1 - \arg z_2$

8 For two complex numbers $z_1 = x_1 + iy_1$ and $z_2 = x_2 + iy_2$, $|z_2 - z_1|$ represents the distance between the points z_1 and z_2 on an Argand diagram.

9 Given $z_1 = x_1 + iy_1$, the locus of points z on an Argand diagram such that $|z - z_1| = r$, or $|z - (x_1 + iy_1)| = r$, is a circle with centre (x_1, y_1) and radius r.

10 Given $z_1 = x_1 + iy_1$ and $z_2 = x_2 + iy_2$, the locus of points z on an Argand diagram such that $|z - z_1| = |z - z_2|$ is the perpendicular bisector of the line segment joining z_1 and z_2.

11 Given $z_1 = x_1 + iy_1$, the locus of points z on an Argand diagram such that $\arg(z - z_1) = \theta$ is a half-line from, but not including, the fixed point z_1 making an angle θ with a line from the fixed point z_1 parallel to the real axis.

Series

3

Objectives

After completing this chapter you should be able to:

- Use standard results for $\sum_{r=1}^{n} 1$ and $\sum_{r=1}^{n} r$ → pages 44–47
- Use standard results for $\sum_{r=1}^{n} r^2$ and $\sum_{r=1}^{n} r^3$ → pages 47–51
- Evaluate and simplify series of the form $\sum_{r=m}^{n} \mathrm{f}(r)$, where $\mathrm{f}(r)$ is linear, quadratic or cubic → pages 44–51

Prior knowledge check

1 Factorise:

a $x^2 + 5x + 6$ **b** $x^2 + 3x - 4$

c $2x^2 + 7x + 6$ ← Pure Year 1, Chapter 1

2 Simplify each expression by writing it as the product of two factors:

a $(k + 1) + (k + 1)(k + 2)$

b $\frac{1}{2}(k + 1)^2 + k^2(k + 1)^2$

c $k^2(2k - 1) + 10k - 5$ ← Pure Year 1, Chapter 1

The Greek letter $\sum$ is used in mathematics to represent a sum. For example, the infinite series $\frac{1}{1^2} + \frac{1}{2^2} + \frac{1}{3^2} + \ldots$ can be written as $\sum_{r=1}^{\infty} \frac{1}{r^2}$

This notation was first introduced by the Swiss mathematician Leonhard Euler, who also proved that this infinite sum is equal to $\frac{\pi^2}{6}$

3.1 Sums of natural numbers

You can use **sigma notation** to write series clearly and concisely. For example:

$$\sum_{r=1}^{3}(10r - 1) = (10 \times 1 - 1) + (10 \times 2 - 1) + (10 \times 3 - 1)$$
$$= 9 + 19 + 29 = 57$$

$$\sum_{r=1}^{n} r^2 = 1^2 + 2^2 + 3^2 + \dots + n^2$$

Links A series is the **sum** of the terms in a sequence.

← **Pure Year 2, Section 3.2**

Notation The numbers below and above the $\sum$ tell you which value of r to begin at, and which value to end at. You go up in increments of 1 each time.

- **To find the sum of a series of constant terms you can use the formula $\sum_{r=1}^{n} 1 = n$.**
- **The formula for the sum of the first n natural numbers is $\sum_{r=1}^{n} r = \frac{1}{2}n(n + 1)$.**

Notation $\sum_{r=1}^{n} r = 1 + 2 + 3 + \dots + n$

Example 1

Evaluate: **a** $\sum_{r=1}^{4}(2r - 1)$ **b** $\sum_{r=1}^{50} r$ **c** $\sum_{r=21}^{50} r$

a $\sum_{r=1}^{4}(2r - 1) = (2 \times 1 - 1) + (2 \times 2 - 1) + (2 \times 3 - 1) + (2 \times 4 - 1)$
$= 1 + 3 + 5 + 7$
$= 16$

There are only 4 terms in this series. Write out each one then find the sum.

b $\sum_{r=1}^{50} r = \frac{1}{2} \times 50 \times 51 = 1275$

Substitute $n = 50$ in $\sum_{r=1}^{n} r = \frac{1}{2}n(n + 1)$.

c $\sum_{r=21}^{50} r = \sum_{r=1}^{50} r - \sum_{r=1}^{20} r = 1275 - \frac{1}{2} \times 20 \times 21$
$= 1275 - 210 = 1065$

Problem-solving

$\sum_{r=21}^{50} r = 21 + 22 + 23 + \dots + 49 + 50$

Find the sum of the natural numbers up to 50, then subtract the sum of the natural numbers up to 20.

- **To find the sum of a series that does not start at $r = 1$, use $\sum_{r=k}^{n} f(r) = \sum_{r=1}^{n} f(r) - \sum_{r=1}^{k-1} f(r)$.**

Watch out You need to subtract the sum up to $k - 1$, not k.

Example 2

Show that $\sum_{r=5}^{2N-1} r = 2N^2 - N - 10, N \geqslant 3$.

$$\sum_{r=5}^{2N-1} r = \sum_{r=1}^{2N-1} r - \sum_{r=1}^{4} r$$
$$= \tfrac{1}{2} \times (2N-1)((2N-1)+1) - \tfrac{1}{2} \times 4 \times 5$$
$$= \tfrac{1}{2} \times (2N-1)(2N) - 10$$
$$= \tfrac{1}{2} \times (4N^2 - 2N) - 10$$
$$= 2N^2 - N - 10$$

Substitute $n = 2N - 1$ and $n = 4$ in $\frac{1}{2}n(n+1)$.

You can rearrange expressions involving sigma notation. This allows you to evaluate the sums of more complicated series.

- $\sum_{r=1}^{n} k\mathbf{f}(r) = k\sum_{r=1}^{n} \mathbf{f}(r)$
- $\sum_{r=1}^{n} (\mathbf{f}(r) + \mathbf{g}(r)) = \sum_{r=1}^{n} \mathbf{f}(r) + \sum_{r=1}^{n} \mathbf{g}(r)$

Example 3

Evaluate $\sum_{r=1}^{25} (3r + 1)$

$$\sum_{r=1}^{25} (3r+1) = 3\sum_{r=1}^{25} r + \sum_{r=1}^{25} 1$$
$$= 3 \times \tfrac{1}{2} \times 25 \times 26 + 25$$
$$= 975 + 25 = 1000$$

Use the rules given above to write the expression in terms of $\sum_{r=1}^{25} r$ and $\sum_{r=1}^{25} 1$.

Example 4

a Show that $\sum_{r=1}^{n} (7r - 4) = \frac{1}{2}n(7n - 1)$.

b Hence evaluate $\sum_{r=20}^{50} (7r - 4)$.

a $\sum_{r=1}^{n} (7r-4) = 7\sum_{r=1}^{n} r - 4\sum_{r=1}^{n} 1$
$$= \tfrac{7}{2}n(n+1) - 4n$$
$$= \tfrac{1}{2}n(7n + 7 - 8)$$
$$= \tfrac{1}{2}n(7n - 1)$$

$7\sum_{r=1}^{n} r = 7 \times \frac{1}{2}n(n+1)$ and $4\sum_{r=1}^{n} 1 = 4 \times n$

b $\sum_{r=20}^{50} (7r-4) = \sum_{r=1}^{50} (7r-4) - \sum_{r=1}^{19} (7r-4)$
$$= \tfrac{1}{2} \times 50 \times 349 - \tfrac{1}{2} \times 19 \times 132$$
$$= 8725 - 1254$$
$$= 7471$$

Use the result from part **a** to find each sum quickly.
$7 \times 50 - 1 = 349$ and $7 \times 19 - 1 = 132$

Exercise 3A

1 Evaluate:

a $\sum_{r=0}^{3}(2r+1)$ **b** $\sum_{r=1}^{40} r$ **c** $\sum_{r=1}^{20} r$ **d** $\sum_{r=1}^{99} r$

e $\sum_{r=10}^{40} r$ **f** $\sum_{r=100}^{200} r$ **g** $\sum_{r=21}^{40} r$ **h** $\sum_{r=1}^{k} r + \sum_{r=k+1}^{80} r,\ 0 < k < 80$

E/P **2** Given that $\sum_{r=1}^{n} r = 528$, find the value of n. **(4 marks)**

E/P **3** Given that $\sum_{r=1}^{k} r = \frac{1}{2}\sum_{r=1}^{20} r$, find the value of k. **(4 marks)**

E **4 a** Find an expression for $\sum_{r=1}^{2n-1} r$. **(3 marks)**

b Hence show that $\sum_{r=n+1}^{2n-1} r = \frac{3}{2}n(n-1),\ n \geqslant 2$. **(3 marks)**

E **5** Show that $\sum_{r=n-1}^{2n} r = \frac{1}{2}(n+2)(3n-1),\ n \geqslant 1$. **(5 marks)**

6 a Show that $\sum_{r=1}^{n^2} r - \sum_{r=1}^{n} r = \frac{1}{2}n(n^3-1)$.

b Hence evaluate $\sum_{r=10}^{81} r$.

Hint Use your result from part **a**.

7 Calculate the sum of each series:

a $\sum_{r=1}^{55}(3r-1)$ **b** $\sum_{r=1}^{90}(2-7r)$ **c** $\sum_{r=1}^{46}(9+2r)$

8 Show that:

a $\sum_{r=1}^{n}(3r+2) = \frac{1}{2}n(3n+7)$ **b** $\sum_{r=1}^{2n}(5r-4) = n(10n-3)$

c $\sum_{r=1}^{n+2}(2r+3) = (n+2)(n+6)$ **d** $\sum_{r=3}^{n}(4r+5) = (2n+11)(n-2)$

E/P **9 a** Show that $\sum_{r=1}^{k}(4r-5) = 2k^2 - 3k$. **(5 marks)**

b Find the smallest value of k for which $\sum_{r=1}^{k}(4r-5) > 4850$. **(4 marks)**

E/P **10** Given that $\mathrm{f}(r) = ar + b$ and $\sum_{r=1}^{n}\mathrm{f}(r) = \frac{1}{2}n(7n+1)$, find the constants a and b. **(5 marks)**

E **11 a** Show that $\sum_{r=1}^{4n-1}(3r+1) = 24n^2 - 2n - 1,\ n \geqslant 1$. **(5 marks)**

b Hence calculate $\sum_{r=1}^{99}(3r+1)$. **(2 marks)**

E **12 a** Show that $\sum_{r=1}^{2k+1}(4-5r) = -(2k+1)(5k+1),\ k \geqslant 0$. **(5 marks)**

b Hence evaluate $\sum_{r=1}^{25}(4-5r)$. **(2 marks)**

c Find the value of $\sum_{r=1}^{15}(5r-4)$. **(1 mark)**

(P) **13** Given that $\sum_{r=1}^{n} f(r) = n^2 + 4n$, deduce an expression for $f(r)$ in terms of r.

(E/P) **14** $f(r) = ar + b$, where a and b are rational constants.

Given that $\sum_{r=1}^{4} f(r) = 36$ and $\sum_{r=1}^{6} f(r) = 78$,

a find an expression for $\sum_{r=1}^{n} f(r)$ **(6 marks)**

b hence calculate $\sum_{r=1}^{10} f(r)$. **(2 marks)**

Challenge

Given that $\sum_{r=n}^{2n} (12 - 2r) = 0$, find the value of n.

3.2 Sums of squares and cubes

The expression for $\sum_{r=1}^{n} 1$ is linear, and the expression for $\sum_{r=1}^{n} r$ is quadratic. Similarly, you can find a cubic expression for the sum of the **squares** of the first n natural numbers, and a quartic expression for the sum of the **cubes** of the first n natural numbers.

- **The formula for the sum of the squares of the first n natural numbers is $\sum_{r=1}^{n} r^2 = \frac{1}{6}n(n + 1)(2n + 1)$.**
- **The formula for the sum of the cubes of the first n natural numbers is $\sum_{r=1}^{n} r^3 = \frac{1}{4}n^2 (n + 1)^2$.**

Links You can prove both of these results using **mathematical induction**. → Section 8.1

Example 5

Evaluate: **a** $\sum_{r=20}^{40} r^2$ **b** $\sum_{r=1}^{25} r^3$

a $\sum_{r=20}^{40} r^2 = \sum_{r=1}^{40} r^2 - \sum_{r=1}^{19} r^2$

$= \frac{1}{6} \times 40(40 + 1)(80 + 1)$

$- \frac{1}{6} \times 19(19 + 1)(38 + 1)$

$= 22\,140 - 2470 = 19\,670$

Substitute $n = 40$ and $n = 19$ in $\sum_{r=1}^{n} r^2 = \frac{1}{6}n(n + 1)(2n + 1)$

b $\sum_{r=1}^{25} r^3 = \frac{1}{4} \times 25^2 \times 26^2 = 105\,625$

Substitute $n = 25$ in $\frac{1}{4}n^2(n + 1)^2$.

Example 6

a Show that $\sum_{r=n+1}^{2n} r^2 = \frac{1}{6}n(2n + 1)(7n + 1)$.

b Verify that the result is true for $n = 1$ and $n = 2$.

a $\sum_{r=n+1}^{2n} r^2 = \sum_{r=1}^{2n} r^2 - \sum_{r=1}^{n} r^2$

$= \frac{1}{6} \times 2n(2n + 1)(4n + 1) - \frac{1}{6}n(n + 1)(2n + 1)$

$= \frac{1}{6}n(2n + 1)(2(4n + 1) - (n + 1))$

$= \frac{1}{6}n(2n + 1)(7n + 1)$

b When $n = 1$:

$\sum_{r=n+1}^{2n} r^2 = \sum_{r=2}^{2} r^2 = 2^2 = 4$

$\frac{1}{6}n(2n + 1)(7n + 1) = \frac{1}{6} \times 3 \times 8 = 4$ ✓

When $n = 2$:

$\sum_{r=n+1}^{2n} r^2 = \sum_{r=3}^{4} r^2 = 3^2 + 4^2 = 25$

$\frac{1}{6}n(2n + 1)(7n + 1) = \frac{2}{6} \times 5 \times 15 = 25$ ✓

Replace n by $2n$ in $\frac{1}{6}n(n + 1)(2n + 1)$.

Problem-solving

Look for common factors in each part of the expression. Here you can take out a factor of $\frac{1}{6}n(2n + 1)$.

Watch out When you have been asked to find a general result for a sum it is good practice to test it for small values of n. It will not prove that you are correct, but if one value of n does not work, you know that your result is incorrect.

Example 7

a Show that $\sum_{r=1}^{n}(r^2 + r - 2) = \frac{1}{3}n(n + 4)(n - 1)$.

b Hence find the sum of the series 4 + 10 + 18 + 28 + 40 + ... + 418.

a $\sum_{r=1}^{n}(r^2 + r - 2)$

$= \sum_{r=1}^{n} r^2 + \sum_{r=1}^{n} r - 2\sum_{r=1}^{n} 1$

$= \frac{1}{6}n(n + 1)(2n + 1) + \frac{1}{2}n(n + 1) - 2n$

$= \frac{1}{6}n((n + 1)(2n + 1) + 3(n + 1) - 12)$

$= \frac{1}{6}n(2n^2 + 3n + 1 + 3n + 3 - 12)$

$= \frac{1}{6}n(2n^2 + 6n - 8)$

$= \frac{1}{3}n(n^2 + 3n - 4)$

$= \frac{1}{3}n(n + 4)(n - 1)$

b 0 + 4 + 10 + 18 + 28 + 40 + ... + 418

$= \sum_{r=1}^{20}(r^2 + r - 2)$

$= \frac{1}{3} \times 20(20 + 4)(20 - 1)$

$= 3040$

Use the results for $\sum_{r=1}^{n} r^2$, $\sum_{r=1}^{n} r$ and $\sum_{r=1}^{n} 1$.

Problem-solving

The question says 'hence' so use your answer to part **a**. When $r = 1$, $r^2 + r - 2 = 0$, and when $r = 20$, $r^2 + r - 2 = 418$, so you can write the sum as $\sum_{r=1}^{20}(r^2 + r - 2)$.

Example 8

a Show that $\sum_{r=1}^{n} r(r+3)(2r-1) = \frac{1}{6}n(n+1)(3n^2+an+b)$, where a and b are integers to be found.

b Hence calculate $\sum_{r=11}^{40} r(r+3)(2r-1)$.

a $\sum_{r=1}^{n} r(r+3)(2r-1)$

$= \sum_{r=1}^{n}(2r^3+5r^2-3r)$

$= 2\sum_{r=1}^{n} r^3 + 5\sum_{r=1}^{n} r^2 - 3\sum_{r=1}^{n} r$

$= \frac{2}{4}n^2(n+1)^2 + \frac{5}{6}n(n+1)(2n+1) - \frac{3}{2}n(n+1)$

$= \frac{1}{6}n(n+1)(3n(n+1)+5(2n+1)-9)$

$= \frac{1}{6}n(n+1)(3n^2+13n-4)$

b $\sum_{r=11}^{40} r(r+3)(2r-1)$

$= \sum_{r=1}^{40} r(r+3)(2r-1) - \sum_{r=1}^{10} r(r+3)(2r-1)$

$= \frac{1}{6}(40 \times 41 \times 5316) - \frac{1}{6}(10 \times 11 \times 426)$

$= 1\,453\,040 - 7810$

$= 1\,445\,230$

First multiply out the brackets.

Use these rules:

$\sum_{r=1}^{n} k\mathrm{f}(r) = k\sum_{r=1}^{n}\mathrm{f}(r)$

$\sum_{r=1}^{n}(\mathrm{f}(r)+\mathrm{g}(r)) = \sum_{r=1}^{n}\mathrm{f}(r) + \sum_{r=1}^{n}\mathrm{g}(r)$

Use the results for $\sum_{r=1}^{n} r^3$, $\sum_{r=1}^{n} r^2$ and $\sum_{r=1}^{n} r$.

Substitute $n = 40$ and $n = 10$ in the result for **a**.

Exercise 3B

1 Evaluate:

a $\sum_{r=1}^{4} r^2$ **b** $\sum_{r=1}^{40} r^2$ **c** $\sum_{r=21}^{40} r^2$ **d** $\sum_{r=1}^{99} r^3$

e $\sum_{r=1}^{100} r^3$ **f** $\sum_{r=100}^{200} r^3$ **g** $\sum_{r=1}^{k} r^2 + \sum_{r=k+1}^{80} r^2$, $0 < k < 80$.

2 Show that:

a $\sum_{r=1}^{2n} r^2 = \frac{1}{3}n(2n+1)(4n+1)$ **b** $\sum_{r=1}^{2n-1} r^2 = \frac{1}{3}n(2n-1)(4n-1)$

c $\sum_{r=n}^{2n} r^2 = \frac{1}{6}n(n+1)(14n+1)$

(P) **3** Show that, for any $k \in \mathbb{N}$, $\sum_{r=1}^{n+k} r^3 = \frac{1}{4}(n+k)^2(n+k+1)^2$

(E) **4 a** Show that $\sum_{r=n+1}^{3n} r^3 = n^2(4n+1)(5n+2)$ **(3 marks)**

b Hence evaluate $\sum_{r=11}^{30} r^3$. **(2 marks)**

(E) **5** **a** Show that $\sum_{r=n}^{2n} r^3 = \frac{3}{4}n^2(n+1)(5n+1)$. **(3 marks)**

b Hence evaluate $\sum_{r=30}^{60} r^3$. **(2 marks)**

6 Evaluate:

a $\sum_{m=1}^{30}(m^2 - 1)$ **b** $\sum_{r=1}^{40} r(r+4)$ **c** $\sum_{r=1}^{80} r(r^2+3)$ **d** $\sum_{r=11}^{35}(r^3 - 2)$

(E) **7** **a** Show that $\sum_{r=1}^{n}(r+2)(r+5) = \frac{1}{3}n(n^2+12n+41)$ **(4 marks)**

b Hence calculate $\sum_{r=10}^{50}(r+2)(r+5)$. **(3 marks)**

(E) **8** **a** Show that $\sum_{r=1}^{n}(r^2+3r+1) = \frac{1}{3}n(n+a)(n+b)$, where a and b are integers to be found. **(4 marks)**

b Hence evaluate $\sum_{r=19}^{40}(r^2+3r+1)$. **(3 marks)**

(E) **9** **a** Show that $\sum_{r=1}^{n} r^2(r-1) = \frac{1}{12}n(n+1)(3n^2-n-2)$. **(4 marks)**

b Hence show that $\sum_{r=1}^{2n-1} r^2(r-1) = \frac{1}{3}n(2n-1)(6n^2-7n+1)$ **(4 marks)**

(E) **10** **a** Show that $\sum_{r=1}^{n}(r+1)(r+3) = \frac{1}{6}n(2n^2+an+b)$, where a and b are integers to be found. **(4 marks)**

b Hence find an expression, only in terms of n, for $\sum_{r=n+1}^{2n}(r+1)(r+3)$. **(3 marks)**

(E) **11** **a** Show that $\sum_{r=1}^{n}(r+3)(r+4) = \frac{1}{3}n(n^2+an+b)$, where a and b are integers to be found. **(4 marks)**

b Hence find an expression, only in terms of n, for $\sum_{r=n+1}^{3n}(r+3)(r+4)$. **(3 marks)**

(E) **12** **a** Show that $\sum_{r=1}^{n} r(r+3)^2 = \frac{1}{4}n(n+1)(n^2+an+b)$, where a and b are integers to be found. **(5 marks)**

b Hence evaluate $\sum_{r=10}^{20} r(r+3)^2$. **(3 marks)**

(P) **13** **a** Show that, for any $k \in \mathbb{N}$, $\sum_{r=1}^{kn}(2r-1) = k^2n^2$.

b Hence find a value of n such that $\sum_{r=1}^{5n}(2r-1) = \sum_{r=1}^{n} r^3$.

(P) **14 a** Show that $\sum_{r=1}^{n}(r^3 - r^2) = \frac{1}{12}n(n+1)(n-1)(3n+2)$. **(4 marks)**

b Hence find the value of n that satisfies $\sum_{r=1}^{n}(r^3 - r^2) = \sum_{r=1}^{n} 7r$. **(5 marks)**

Challenge

a Find polynomials $f_2(x)$, $f_3(x)$, $f_4(x)$ such that for every $n \in \mathbb{N}$:

$$\sum_{r=1}^{n} f_2(r) = n^2, \quad \sum_{r=1}^{n} f_3(r) = n^3, \quad \sum_{r=1}^{n} f_4(r) = n^4$$

b Hence show that for any linear, quadratic, or cubic polynomial $h(x)$ there exists a polynomial $g(x)$ such that $\sum_{r=1}^{n} g(r) = n(h(n))$.

Hint The polynomial $f_1(x) = 1$ satisfies $\sum_{r=1}^{n} f_1(r) = n$

Mixed exercise 3

Throughout this exercise you may assume the standard results for $\sum_{r=1}^{n} r$, $\sum_{r=1}^{n} r^2$ and $\sum_{r=1}^{n} r^3$.

1 Evaluate:

a $\sum_{r=1}^{10} r$ **b** $\sum_{r=10}^{50} r$ **c** $\sum_{r=1}^{10} r^2$ **d** $\sum_{r=1}^{10} r^3$

e $\sum_{r=26}^{50} r^2$ **f** $\sum_{r=50}^{100} r^3$ **g** $\sum_{r=1}^{60} r + \sum_{r=1}^{60} r^2$

2 Write each of the following as an expression in terms of n.

a $\sum_{r=1}^{n}(3r - 5)$ **b** $\sum_{r=1}^{n}(r^2 + r)$ **c** $\sum_{r=1}^{n}(3r^2 + 7r)$

d $\sum_{r=1}^{n}(4r^3 + 6r^2)$ **e** $\sum_{r=1}^{n}(r^2 - 2r)$ **f** $\sum_{r=1}^{n}(r^2 - 3r)$

g $\sum_{r=1}^{n}(r^2 - 5)$ **h** $\sum_{r=1}^{n}(2r^3 + 3r^2 + r + 4)$

(E) **3** Evaluate $\sum_{r=1}^{30} r(3r - 1)$. **(5 marks)**

(E) **4 a** Show that $\sum_{r=1}^{n} r^2(r - 3) = \frac{1}{4}n(n+1)(n^2 + an + b)$, where a and b are integers to be found. **(4 marks)**

b Hence evaluate $\sum_{r=1}^{20} r^2(r - 3)$. **(2 marks)**

(E) **5 a** Show that $\sum_{r=1}^{n}(2r - 1)^2 = \frac{1}{3}n(an + b)(an - b)$, where a and b are integers to be found. **(5 marks)**

b Hence find $\sum_{r=1}^{2n}(2r - 1)^2$. **(2 marks)**

(E) **6** **a** Show that $\sum_{r=1}^{n} r(r+2) = \frac{1}{6}n(n+1)(an+b)$, where a and b are integers to be found. **(4 marks)**

b Hence evaluate $\sum_{r=15}^{30} r(r+2)$. **(3 marks)**

(E/P) **7** **a** Show that $\sum_{r=n+1}^{2n} r^2 = \frac{1}{6}n(2n+1)(an+b)$, where a and b are integers to be found. **(4 marks)**

b Hence evaluate $\sum_{r=16}^{30} r^2$. **(2 marks)**

(E/P) **8** **a** Show that $\sum_{r=1}^{n} (r^2 - r - 1) = \frac{1}{3}n(n^2 - 4)$. **(4 marks)**

b Hence evaluate $\sum_{r=10}^{40} (r^2 - r - 1)$. **(3 marks)**

c Find the value of n such that $\sum_{r=1}^{n} (r^2 - r - 1) = \sum_{r=1}^{2n} r$. **(5 marks)**

(E/P) **9** **a** Show that $\sum_{r=1}^{n} r(2r^2+1) = \frac{1}{2}n(n+1)(n^2+n+1)$. **(4 marks)**

b Hence show that there are no values of n that satisfy $\sum_{r=1}^{n} r(2r^2+1) = \sum_{r=1}^{n} (100r^2 - r)$. **(6 marks)**

(E/P) **10** **a** Show that $\sum_{r=1}^{n} r(r+1)^2 = \frac{1}{12}n(n+1)(n+2)(an+b)$, where a and b are integers to be found. **(5 marks)**

b Hence find the value of n that satisfies $\sum_{r=1}^{n} r(r+1)^2 = \sum_{r=1}^{n} 70r$. **(6 marks)**

(E/P) **11** Find the value of n that satisfies $\sum_{r=1}^{n} r^2 = \sum_{r=1}^{n+1} (9r+1)$. **(7 marks)**

Challenge

Show that:

a $\sum_{i=1}^{n}\left(\sum_{r=1}^{i} r^2\right) = \frac{1}{12}n(n+1)^2(n+2)$

b $\sum_{j=1}^{n}\left(\sum_{i=1}^{j}\left(\sum_{r=1}^{i} r\right)\right) = \frac{1}{24}n(n+1)(n+2)(n+3)$

Summary of key points

1 To find the sum of a series of constant terms you can use the formula $\sum_{r=1}^{n} 1 = n$.

2 The formula for the sum of the first n natural numbers is $\sum_{r=1}^{n} r = \frac{1}{2}n(n+1)$.

3 To find the sum of a series that does not start at $r = 1$, use $\sum_{r=k}^{n} f(r) = \sum_{r=1}^{n} f(r) - \sum_{r=1}^{k-1} f(r)$

4 You can rearrange expressions involving sigma notation.

- $\sum_{r=1}^{n} kf(r) = k\sum_{r=1}^{n} f(r)$
- $\sum_{r=1}^{n} (f(r) + g(r)) = \sum_{r=1}^{n} f(r) + \sum_{r=1}^{n} g(r)$

5 The formula for the sum of the squares of the first n natural numbers is

$$\sum_{r=1}^{n} r^2 = \frac{1}{6}n(n+1)(2n+1)$$

6 The formula for the sum of the cubes of the first n natural numbers is

$$\sum_{r=1}^{n} r^3 = \frac{1}{4}n^2(n+1)^2$$

4 Roots of polynomials

Objectives

After completing this chapter you should be able to:

- Derive and use the relationships between the roots of a quadratic equation → pages 55–57
- Derive and use the relationships between the roots of a cubic equation → pages 57–59
- Derive and use the relationships between the roots of a quartic equation → pages 59–61
- Evaluate expressions relating to the roots of polynomials → pages 62–64
- Find the equation of a polynomial whose roots are a linear transformation of the roots of a given polynomial → pages 65–67

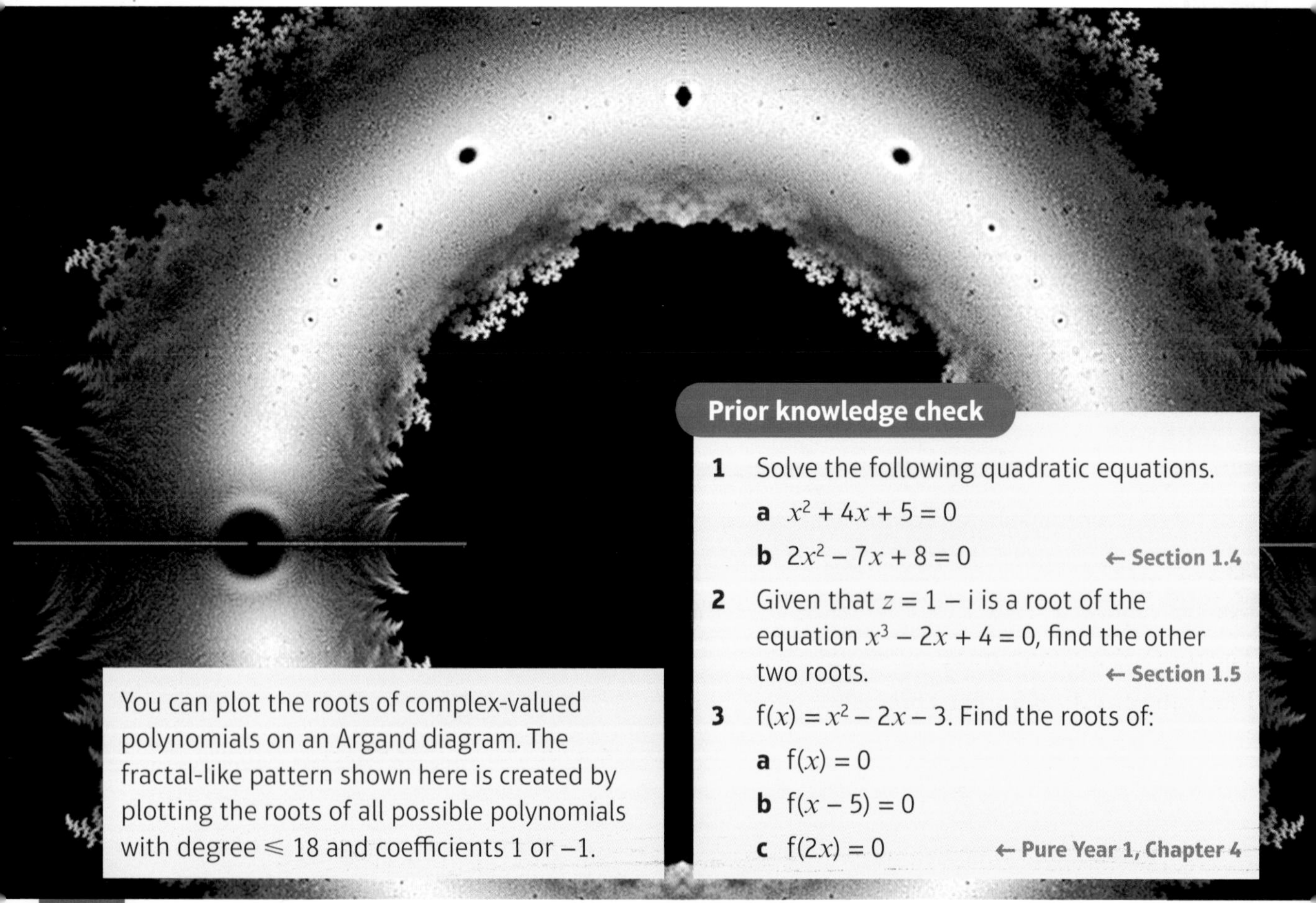

You can plot the roots of complex-valued polynomials on an Argand diagram. The fractal-like pattern shown here is created by plotting the roots of all possible polynomials with degree $\leqslant 18$ and coefficients 1 or -1.

Prior knowledge check

1 Solve the following quadratic equations.

a $x^2 + 4x + 5 = 0$

b $2x^2 - 7x + 8 = 0$ ← Section 1.4

2 Given that $z = 1 - i$ is a root of the equation $x^3 - 2x + 4 = 0$, find the other two roots. ← Section 1.5

3 $f(x) = x^2 - 2x - 3$. Find the roots of:

a $f(x) = 0$

b $f(x - 5) = 0$

c $f(2x) = 0$ ← Pure Year 1, Chapter 4

4.1 Roots of a quadratic equation

A quadratic equation of the form $ax^2 + bx + c = 0$, $x \in \mathbb{C}$, where a, b and c are real constants, can have two real roots, one repeated (real) root or two complex roots.

Links If the roots of a quadratic equation with real coefficients are complex, then they occur as a conjugate pair. ← Section 1.4

If the roots of this equation are α and β, you can determine the relationship between the **coefficients** of the terms in the quadratic equation and the values of α and β:

$$ax^2 + bx + c = a(x - \alpha)(x - \beta)$$
$$= a(x^2 - \alpha x - \beta x + \alpha\beta)$$
$$= ax^2 - a(\alpha + \beta)x + a\alpha\beta$$

Write the quadratic expression in factorised form, then rearrange into the form $ax^2 + bx + c$.

So $b = -a(\alpha + \beta)$ and $c = a\alpha\beta$.

- **If α and β are roots of the equation $ax^2 + bx + c = 0$, then:**
  - $\alpha + \beta = -\dfrac{b}{a}$
  - $\alpha\beta = \dfrac{c}{a}$

Note The sum of the roots is $-\frac{b}{a}$ and the product of the roots is $\frac{c}{a}$. Note that these values are real even if the roots are complex, because the sum or product of a conjugate pair is real.

Example 1

The roots of the quadratic equation $2x^2 - 5x - 4 = 0$ are α and β. Without solving the equation, find the values of:

a $\alpha + \beta$ **b** $\alpha\beta$ **c** $\dfrac{1}{\alpha} + \dfrac{1}{\beta}$ **d** $\alpha^2 + \beta^2$

a $\alpha + \beta = \frac{5}{2}$

Use the result $\alpha + \beta = -\frac{b}{a}$

b $\alpha\beta = -2$

Use the result $\alpha\beta = \frac{c}{a}$

c $\dfrac{1}{\alpha} + \dfrac{1}{\beta} = \dfrac{\alpha + \beta}{\alpha\beta} = \dfrac{\frac{5}{2}}{-2} = -\dfrac{5}{4}$

d $\alpha^2 + \beta^2 = (\alpha + \beta)^2 - 2\alpha\beta$

$= \left(\frac{5}{2}\right)^2 - 2(-2) = \frac{41}{4}$

Problem-solving

Write each expression in terms of $\alpha + \beta$ and $\alpha\beta$:
$(\alpha + \beta)^2 = \alpha^2 + \beta^2 + 2\alpha\beta \Rightarrow \alpha^2 + \beta^2 = (\alpha + \beta)^2 - 2\alpha\beta$

Example 2

The roots of a quadratic equation $ax^2 + bx + c = 0$ are $\alpha = -\frac{3}{2}$ and $\beta = \frac{5}{4}$

Find integer values for a, b and c.

$\alpha + \beta = -\frac{3}{2} + \frac{5}{4} = -\frac{1}{4}$, so $-\frac{1}{4} = -\frac{b}{a}$

$\alpha\beta = -\frac{3}{2} \times \frac{5}{4} = -\frac{15}{8}$, so $-\frac{15}{8} = \frac{c}{a}$

$ax^2 + bx + c = 0$ can be written as $x^2 + \frac{b}{a}x + \frac{c}{a} = 0$.
Use the values of $\alpha + \beta$ and $\alpha\beta$ to write down a quadratic equation with roots α and β.

$x^2 + \frac{1}{4}x - \frac{15}{8} = 0$ — Any constant multiple of this equation will have roots α and β.

$8x^2 + 2x - 15 = 0$

$a = 8, b = 2, c = -15$ — You could also set $a = 8$ to find integer solutions to the equations $-\frac{1}{4} = -\frac{b}{a}$ and $-\frac{15}{8} = \frac{c}{a}$

Exercise 4A

1 α and β are the roots of the quadratic equation $3x^2 + 7x - 4 = 0$. Without solving the equation, find the values of:

a $\alpha + \beta$ **b** $\alpha\beta$ **c** $\frac{1}{\alpha} + \frac{1}{\beta}$ **d** $\alpha^2 + \beta^2$

2 α and β are the roots of the quadratic equation $7x^2 - 3x + 1 = 0$. Without solving the equation, find the values of:

a $\alpha + \beta$ **b** $\alpha\beta$ **c** $\frac{1}{\alpha} + \frac{1}{\beta}$ **d** $\alpha^2 + \beta^2$

3 α and β are the roots of the quadratic equation $6x^2 - 9x + 2 = 0$. Without solving the equation, find the values of:

a $\alpha + \beta$ **b** $\alpha^2 \times \beta^2$

c $\frac{1}{\alpha} + \frac{1}{\beta}$ **d** $\alpha^3 + \beta^3$

Hint Try expanding $(\alpha + \beta)^3$.

4 The roots of a quadratic equation $ax^2 + bx + c = 0$ are $\alpha = 2$ and $\beta = -3$. Find integer values for a, b and c.

5 The roots of a quadratic equation $ax^2 + bx + c = 0$ are $\alpha = -\frac{1}{2}$ and $\beta = -\frac{1}{3}$. Find integer values for a, b and c.

6 The roots of a quadratic equation $ax^2 + bx + c = 0$ are $\alpha = \frac{-1 + \mathrm{i}}{2}$ and $\beta = \frac{-1 - \mathrm{i}}{2}$. Find integer values for a, b and c.

7 One of the roots of the quadratic equation $ax^2 + bx + c = 0$ is $\alpha = -1 - 4\mathrm{i}$.

a Write down the other root, β.

b Given that $a = 1$, find the values of b and c.

(P) **8** Given that $kx^2 + (k - 3)x - 2 = 0$, find the value of k if the sum of the roots is 4.

(P) **9** The equation $nx^2 - (16 + n)x + 256 = 0$ has real roots α and $-\alpha$. Find the value of n.

(P) **10** The roots of the equation $6x^2 + 36x + k = 0$ are reciprocals of each other. Find the value of k.

(P) **11** The equation $mx^2 + 4x + 4m = 0$ has roots of the form k and $2k$. Find the values of m and k.

(P) **12** The equation $ax^2 + 8x + c = 0$, where a and c are real constants, has roots α and α^*.

a Given that $\mathrm{Re}(\alpha) = 2$, find the value of a.

b Given that $\mathrm{Im}(\alpha) = 3\mathrm{i}$, find the value of c.

(P) **13** The equation $4x^2 + px + q = 0$, where p and q are real constants, has roots α and α^*.

a Given that $\text{Re}(\alpha) = -3$, find the value of p.

b Given that $\text{Im}(\alpha) \neq 0$, find the range of possible values of q.

4.2 Roots of a cubic equation

A cubic equation of the form $ax^3 + bx^2 + cx + d = 0$, $x \in \mathbb{C}$, where a, b, c and d are real constants, will always have at least one real root. It will also have either two further real roots, one further repeated (real) root or two complex roots.

Links If a cubic equation with real coefficients has two complex roots, then they will occur as a conjugate pair. ← Section 1.5

If the roots of this equation are α, β and γ, you can determine the relationship between the coefficients of the terms in the cubic equation and the values of α, β and γ:

$$ax^3 + bx^2 + cx + d = a(x - \alpha)(x - \beta)(x - \gamma)$$
$$= a(x^3 - \alpha x^2 - \beta x^2 - \gamma x^2 + \alpha\beta x + \beta\gamma x + \gamma\alpha x - \alpha\beta\gamma)$$
$$= ax^3 - a(\alpha + \beta + \gamma)x^2 + a(\alpha\beta + \beta\gamma + \gamma\alpha)x - a\alpha\beta\gamma$$

So $b = -a(\alpha + \beta + \gamma)$, $c = a(\alpha\beta + \beta\gamma + \gamma\alpha)$ and $d = -a\alpha\beta\gamma$.

- **If α, β and γ are roots of the equation $ax^3 + bx^2 + cx + d = 0$, then:**
  - $\alpha + \beta + \gamma = -\frac{b}{a}$
  - $\alpha\beta + \beta\gamma + \gamma\alpha = \frac{c}{a}$
  - $\alpha\beta\gamma = -\frac{d}{a}$

Note As with the rule for quadratic equations, the sum of the roots is $-\frac{b}{a}$, and the sum of the products of all possible pairs of roots is $\frac{c}{a}$

Example 3

α, β and γ are the roots of the cubic equation $2x^3 + 3x^2 - 4x + 2 = 0$. Without solving the equation, find the values of:

a $\alpha + \beta + \gamma$ **b** $\alpha\beta + \beta\gamma + \gamma\alpha$ **c** $\alpha\beta\gamma$ **d** $\frac{1}{\alpha} + \frac{1}{\beta} + \frac{1}{\gamma}$

a $\alpha + \beta + \gamma = -\frac{3}{2}$

Use the result $\alpha + \beta + \gamma = -\frac{b}{a}$

b $\alpha\beta + \beta\gamma + \gamma\alpha = -2$

Use the result $\alpha\beta + \beta\gamma + \gamma\alpha = \frac{c}{a}$

c $\alpha\beta\gamma = -1$

Use the result $\alpha\beta\gamma = -\frac{d}{a}$

d $\frac{1}{\alpha} + \frac{1}{\beta} + \frac{1}{\gamma} = \frac{\beta\gamma + \alpha\gamma + \alpha\beta}{\alpha\beta\gamma} = \frac{\frac{c}{a}}{-\frac{d}{a}}$

$= \frac{-2}{-1} = 2$

Notice that the numerator is just the sum of the products of pairs of roots with the terms in a different order.

Example 4

The roots of a cubic equation $ax^3 + bx^2 + cx + d = 0$ are $\alpha = 1 - 2i$, $\beta = 1 + 2i$ and $\gamma = 2$. Find integer values for a, b, c and d.

$\alpha + \beta + \gamma = (1 - 2i) + (1 + 2i) + 2 = 4$,

so $4 = -\frac{b}{a}$

$\alpha\beta + \beta\gamma + \gamma\alpha$

$= (1 - 2i)(1 + 2i) + (1 + 2i) \times 2 + 2(1 - 2i)$

$= 9$

so $9 = \frac{c}{a}$

$\alpha\beta\gamma = (1 - 2i)(1 + 2i) \times 2 = 10$,

so $10 = -\frac{d}{a}$

$x^3 - 4x^2 + 9x - 10 = 0$

$a = 1, b = -4, c = 9, d = -10$

$ax^3 + bx^2 + cx + d = 0$ can be written as $x^3 + \frac{b}{a}x^2 + \frac{c}{a}x + \frac{d}{a} = 0$. Use the values of $\alpha + \beta + \gamma$, $\alpha\beta + \beta\gamma + \gamma\alpha$ and $\alpha\beta\gamma$ to write down a cubic equation with roots α, β and γ.

Watch out Be careful with negative signs when writing out the equation.

You could also set $a = 1$ to find integer solutions to the equations $4 = -\frac{b}{a}$, $9 = \frac{c}{a}$ and $10 = -\frac{d}{a}$

Exercise 4B

1 α, β and γ are the roots of the cubic equation $2x^3 + 5x^2 - 2x + 3 = 0$. Find the values of:

a $\alpha + \beta + \gamma$ **b** $\alpha\beta\gamma$ **c** $\alpha\beta + \beta\gamma + \gamma\alpha$ **d** $\frac{1}{\alpha} + \frac{1}{\beta} + \frac{1}{\gamma}$

2 α, β and γ are the roots of the cubic equation $x^3 + 5x^2 + 17x + 13 = 0$. Find the values of:

a $\alpha + \beta + \gamma$ **b** $\alpha\beta\gamma$ **c** $\alpha\beta + \beta\gamma + \gamma\alpha$ **d** $\alpha^2\beta^2\gamma^2$

3 α, β and γ are the roots of the cubic equation $7x^3 - 4x^2 - x + 6 = 0$. Find the values of:

a $\alpha + \beta + \gamma$ **b** $\alpha\beta\gamma$ **c** $\alpha^3\beta^3\gamma^3$ **d** $\frac{1}{\alpha} + \frac{1}{\beta} + \frac{1}{\gamma}$

4 The roots of a cubic equation $ax^3 + bx^2 + cx + d = 0$ are $\alpha = \frac{3}{2}$, $\beta = \frac{1}{2}$ and $\gamma = 1$. Find integer values for a, b, c and d.

5 The roots of a cubic equation $ax^3 + bx^2 + cx + d = 0$ are $\alpha = 1 + 3i$, $\beta = 1 - 3i$ and $\gamma = \frac{1}{2}$ Find integer values for a, b, c and d.

6 The roots of a cubic equation $ax^3 + bx^2 + cx + d = 0$ are $\alpha = \frac{5}{4}$, $\beta = -\frac{3}{2}$ and $\gamma = \frac{1}{2}$ Find integer values for a, b, c and d.

(E) 7 The cubic equation $16x^3 - kx^2 + 1 = 0$ has roots α, β and γ.

a Write down the values of $\alpha\beta + \beta\gamma + \gamma\alpha$ and $\alpha\beta\gamma$. **(2 marks)**

b i Given that $\alpha = \beta$, find the roots of the equation. **(5 marks)**

ii Find the value of k. **(1 mark)**

(E) 8 The cubic equation $2x^3 - kx^2 + 30x - 13 = 0$ has roots α, β and γ.

a Write down the values of $\alpha\beta + \beta\gamma + \gamma\alpha$ and $\alpha\beta\gamma$, and express k in terms of α, β and γ. **(3 marks)**

b Given that $\alpha = 2 - 3i$, find the value of k. **(4 marks)**

E **9** The cubic equation $x^3 - mx + n = 0$ has roots 1, −4 and α.

a State, with a reason, whether α is real. **(1 mark)**

b Find the values of m, n and α. **(4 marks)**

E **10** The cubic equation $2x^3 - 10x^2 + 8x - k = 0$ has a root at $x = 3 - \mathrm{i}$.

a Find the other two roots of the equation. **(4 marks)**

b Hence find the value of k. **(2 marks)**

P **11** The cubic equation $x^3 - 14x^2 + 56x - 64 = 0$ has roots α, $k\alpha$ and $k^2\alpha$ for some real constant k. Find the values of α and k. **(5 marks)**

P **12** Given that the roots of $8x^3 + 12x^2 - cx + d = 0$ are α, $\frac{\alpha}{2}$ and $\alpha - 4$, find α, c and d. **(5 marks)**

P **13** Given that the roots of the cubic equation $2x^3 + 48x^2 + cx + d = 0$ are α, 2α and 3α, find the values of α, c and d. **(5 marks)**

Challenge

α, β and γ are the roots of the equation $ax^3 + bx^2 + cx + d = 0$, $x \in \mathbb{C}$, $a, b, c, d \in \mathbb{R}$.

By considering all the possible cases in which α, β and γ are real or complex, explain why the following are always real numbers:

a $\alpha + \beta + \gamma$ **b** $\alpha\beta + \beta\gamma + \gamma\alpha$ **c** $\alpha\beta\gamma$

4.3 Roots of a quartic equation

Consider the quartic equation $ax^4 + bx^3 + cx^2 + dx + e = 0$, $x \in \mathbb{C}$, where a, b, c, d and e are real numbers. If the roots of the equation are α, β, γ and δ, you can determine the relationship between the coefficients of the terms in the equation and the values of α, β, γ and δ:

$$\begin{aligned} ax^4 + bx^3 + cx^2 + dx + e &= a(x - \alpha)(x - \beta)(x - \gamma)(x - \delta) \\ &= a(x^4 - \alpha x^3 - \beta x^3 - \gamma x^3 - \delta x^3 + \alpha\beta x^2 + \beta\gamma x^2 + \gamma\alpha x^2 + \gamma\delta x^2 + \alpha\delta x^2 \\ &\quad + \beta\delta x^2 - \alpha\beta\gamma x - \alpha\beta\delta x - \alpha\gamma\delta x - \beta\gamma\delta x + \alpha\beta\gamma\delta) \\ &= ax^4 - a(\alpha + \beta + \gamma + \delta)x^3 + a(\alpha\beta + \beta\gamma + \gamma\alpha + \gamma\delta + \alpha\delta + \beta\delta)x^2 \\ &\quad - a(\alpha\beta\gamma + \alpha\beta\delta + \alpha\gamma\delta + \beta\gamma\delta)x + a\alpha\beta\gamma\delta \end{aligned}$$

So $b = -a(\alpha + \beta + \gamma + \delta)$, $c = a(\alpha\beta + \beta\gamma + \gamma\alpha + \gamma\delta + \alpha\delta + \beta\delta)$, $d = -a(\alpha\beta\gamma + \alpha\beta\delta + \alpha\gamma\delta + \beta\gamma\delta)$ and $e = a\alpha\beta\gamma\delta$.

- **If α, β, γ and δ are roots of the equation $ax^4 + bx^3 + cx^2 + dx + e = 0$, then:**
  - $\alpha + \beta + \gamma + \delta = -\frac{b}{a}$
  - $\alpha\beta + \alpha\gamma + \alpha\delta + \beta\gamma + \beta\delta + \gamma\delta = \frac{c}{a}$
  - $\alpha\beta\gamma + \alpha\beta\delta + \alpha\gamma\delta + \beta\gamma\delta = -\frac{d}{a}$
  - $\alpha\beta\gamma\delta = \frac{e}{a}$

Notation You can use the following abbreviations for these results in your working:

$\Sigma\alpha = -\frac{b}{a}$ $\Sigma\alpha\beta = \frac{c}{a}$ $\Sigma\alpha\beta\gamma = -\frac{d}{a}$

Example 5

The equation $x^4 + 2x^3 + px^2 + qx - 60 = 0$, $x \in \mathbb{C}$, $p, q \in \mathbb{R}$, has roots α, β, γ and δ.
Given that $\gamma = -2 + 4i$ and $\delta = \gamma^*$,

a show that $\alpha + \beta - 2 = 0$ and that $\alpha\beta + 3 = 0$.

b Hence find all the roots of the quartic equation and find the values of p and q.

a $\alpha + \beta + \gamma + \delta = -2$

Use $\Sigma\alpha = -\frac{b}{a}$ with $b = 2$ and $a = 1$.

$\alpha + \beta + (-2 + 4i) + (-2 - 4i) = -2$

$\alpha + \beta - 4 = -2$

Hence $\alpha + \beta - 2 = 0$ (1)

$\alpha\beta\gamma\delta = -60$

Use $\alpha\beta\gamma\delta = \frac{e}{a}$ with $e = -60$ and $a = 1$.

$\alpha\beta(-2 + 4i)(-2 - 4i) = -60$

You can find $(-2 + 4i)(-2 - 4i)$ quickly by remembering that $(a + bi)(a - bi) = a^2 + b^2$.

$20\alpha\beta = -60$

Hence $\alpha\beta + 3 = 0$ (2)

b Solve equations (1) and (2) simultaneously.
From (1), $\beta = 2 - \alpha$ so substitute into (2):

$\alpha(2 - \alpha) + 3 = 0$

$\alpha^2 - 2\alpha - 3 = 0$

$(\alpha - 3)(\alpha + 1) = 0$

$\alpha = 3$ or -1

If $\alpha = 3$, $\beta = -1$ and if $\alpha = -1$, $\beta = 3$

So the roots of the quartic equation are 3, -1, $-2 + 4i$ and $-2 - 4i$.

$\Sigma\alpha\beta = \frac{p}{a}$

Use $\Sigma\alpha\beta = \frac{c}{a}$ with $c = p$ and $a = 1$.

$\Rightarrow p = 3(-1) + 3(-2 + 4i) + 3(-2 - 4i) - (-2 + 4i) - (-2 - 4i) + (-2 + 4i)(-2 - 4i)$

So $p = 9$

$\Sigma\alpha\beta\gamma = -\frac{q}{a}$

$\Sigma\alpha\beta\gamma = -\frac{d}{a}$ with $d = q$ and $a = 1$.

$\Rightarrow -q = 3(-1)(-2 + 4i) + 3(-1)(-2 - 4i) + 3(-2 + 4i)(-2 - 4i) - (-2 + 4i)(-2 - 4i)$

So $q = -52$

Exercise 4C

1 α, β, γ and δ are the roots of the quartic equation $4x^4 + 3x^3 + 2x^2 - 5x - 4 = 0$. Without solving the equation, find the values of:

a $\alpha + \beta + \gamma + \delta$

b $\alpha\beta + \alpha\gamma + \alpha\delta + \beta\gamma + \beta\delta + \gamma\delta$

c $\alpha\beta\gamma + \alpha\beta\delta + \alpha\gamma\delta + \beta\gamma\delta$

d $\frac{1}{\alpha} + \frac{1}{\beta} + \frac{1}{\gamma} + \frac{1}{\delta}$

Hint $\frac{1}{\alpha} + \frac{1}{\beta} + \frac{1}{\gamma} + \frac{1}{\delta} = \frac{\beta\gamma\delta + \alpha\gamma\delta + \alpha\beta\delta + \alpha\beta\gamma}{\alpha\beta\gamma\delta}$

2 α, β, γ and δ are the roots of the quartic equation $2x^4 + 4x^3 - 3x^2 - x + 2 = 0$. Find the values of:

a $\alpha + \beta + \gamma + \delta$

b $\alpha\beta + \alpha\gamma + \alpha\delta + \beta\gamma + \beta\delta + \gamma\delta$

c $\alpha\beta\gamma + \alpha\beta\delta + \alpha\gamma\delta + \beta\gamma\delta$

d $\alpha\beta\gamma\delta$

e $\frac{1}{\alpha} + \frac{1}{\beta} + \frac{1}{\gamma} + \frac{1}{\delta}$

3 α, β, γ and δ are the roots of the quartic equation $x^4 + 3x^3 + 2x^2 - x + 4 = 0$. Find the values of:

a $\alpha + \beta + \gamma + \delta$
b $\alpha\beta + \alpha\gamma + \alpha\delta + \beta\gamma + \beta\delta + \gamma\delta$
c $\alpha\beta\gamma + \alpha\beta\delta + \alpha\gamma\delta + \beta\gamma\delta$
d $\alpha\beta\gamma\delta$
e $\alpha^2\beta^2\gamma^2\delta^2$

4 α, β, γ and δ are the roots of the quartic equation $7x^4 + 6x^3 - 5x^2 + 4x + 3 = 0$. Find the values of:

a $\alpha + \beta + \gamma + \delta$
b $\alpha\beta + \alpha\gamma + \alpha\delta + \beta\gamma + \beta\delta + \gamma\delta$
c $\alpha\beta\gamma + \alpha\beta\delta + \alpha\gamma\delta + \beta\gamma\delta$
d $\frac{1}{\alpha} + \frac{1}{\beta} + \frac{1}{\gamma} + \frac{1}{\delta}$
e $\alpha^3\beta^3\gamma^3\delta^3$

5 The roots of a quartic equation $ax^4 + bx^3 + cx^2 + dx + e = 0$ are $\alpha = -\frac{3}{2}$, $\beta = -\frac{1}{2}$, $\gamma = -2$ and $\delta = \frac{2}{3}$ Find integer values for a, b, c, d and e.

6 The roots of a quartic equation $ax^4 + bx^3 + cx^2 + dx + e = 0$ are $\alpha = -\frac{1}{2}$, $\beta = \frac{1}{3}$, $\gamma = 1 + \text{i}$ and $\delta = 1 - \text{i}$. Find integer values for a, b, c, d and e.

7 The roots of a quartic equation $ax^4 + bx^3 + cx^2 + dx + e = 0$ are such that $\Sigma\alpha = \frac{17}{12}$, $\Sigma\alpha\beta = -\frac{25}{72}$, $\Sigma\alpha\beta\gamma = -\frac{53}{72}$ and $\alpha\beta\gamma\delta = -\frac{1}{6}$. Find integer values for a, b, c, d and e.

E/P 8 The quartic equation $x^4 - 16x^3 + 86x^2 - 176x + 105 = 0$ has roots α, $\alpha + k$, $\alpha + 2k$ and $\alpha + 3k$ for some real constant k. Solve the equation. **(7 marks)**

E/P 9 The quartic equation $3072x^4 - 2880x^3 + 840x^2 - 90x + 3 = 0$ has roots α, $r\alpha$, $r^2\alpha$ and $r^3\alpha$ for some real constant r. Solve the equation. **(7 marks)**

E 10 Three of the roots of the quartic equation $40x^4 + 90x^3 - 115x^2 + mx + n = 0$ are 1, -3 and $\frac{1}{2}$

a Find the fourth root. **(2 marks)**
b Find the values of m and n. **(4 marks)**

E/P 11 The quartic equation $2x^4 - 34x^3 + 202x^2 + dx + e = 0$ has roots α, $\alpha + 1$, $2\alpha + 1$ and $3\alpha + 1$.

a Find α. **(2 marks)**
b Find the values of d and e. **(4 marks)**

E/P 12 The equation $4x^4 - 19x^3 + px^2 + qx + 10 = 0$, $x \in \mathbb{C}$, $p, q \in \mathbb{R}$, has roots α, β, γ and δ. Given that $\gamma = 3 + \text{i}$ and $\delta = \gamma^*$,

a show that $4\alpha + 4\beta + 5 = 0$ and that $4\alpha\beta - 1 = 0$. **(2 marks)**
b Hence find all the roots of the quartic equation and find the values of p and q. **(5 marks)**
c Show these roots on an Argand diagram. **(3 marks)**

E/P 13 A quartic equation $6x^4 - 10x^3 + 3x^2 + 6x - 40 = 0$ has roots α, β, γ and δ.

a Show that $\frac{1 - 3\text{i}}{2}$ is one root of the equation. **(3 marks)**
b Without solving the equation, find the other roots. **(5 marks)**
c Show these roots on an Argand diagram. **(3 marks)**

4.4 Expressions relating to the roots of a polynomial

You have already seen several results for finding the values of expressions relating to the roots of a polynomial.

- **The rules for reciprocals:**
  - **Quadratic:** $\frac{1}{\alpha} + \frac{1}{\beta} = \frac{\alpha + \beta}{\alpha\beta}$
  - **Cubic:** $\frac{1}{\alpha} + \frac{1}{\beta} + \frac{1}{\gamma} = \frac{\alpha\beta + \beta\gamma + \gamma\alpha}{\alpha\beta\gamma}$
  - **Quartic:** $\frac{1}{\alpha} + \frac{1}{\beta} + \frac{1}{\gamma} + \frac{1}{\delta} = \frac{\alpha\beta\gamma + \beta\gamma\delta + \gamma\delta\alpha + \delta\alpha\beta}{\alpha\beta\gamma\delta}$
- **The rules for products of powers:**
  - **Quadratic:** $\alpha^n \times \beta^n = (\alpha\beta)^n$
  - **Cubic:** $\alpha^n \times \beta^n \times \gamma^n = (\alpha\beta\gamma)^n$
  - **Quartic:** $\alpha^n \times \beta^n \times \gamma^n \times \delta^n = (\alpha\beta\gamma\delta)^n$

In addition to these you have also used the following results for the roots of quadratic equations:

- $\alpha^2 + \beta^2 = (\alpha + \beta)^2 - 2\alpha\beta$
- $\alpha^3 + \beta^3 = (\alpha + \beta)^3 - 3\alpha\beta(\alpha + \beta)$

There are equivalent results to these for the roots of cubic and quartic equations.

Example 6

a Expand $(\alpha + \beta + \gamma)^2$.

b A cubic equation has roots α, β, γ such that $\alpha\beta + \beta\gamma + \gamma\alpha = 7$ and $\alpha + \beta + \gamma = -3$.
Find the value of $\alpha^2 + \beta^2 + \gamma^2$.

a $(\alpha + \beta + \gamma)^2 = (\alpha + \beta + \gamma)(\alpha + \beta + \gamma)$
$= \alpha^2 + \alpha\beta + \alpha\gamma + \beta\alpha + \beta^2 + \beta\gamma + \gamma\alpha + \gamma\beta + \gamma^2$
$= \alpha^2 + \beta^2 + \gamma^2 + 2(\alpha\beta + \beta\gamma + \gamma\alpha)$

b $\alpha^2 + \beta^2 + \gamma^2 = (\alpha + \beta + \gamma)^2 - 2(\alpha\beta + \beta\gamma + \gamma\alpha)$
$= (-3)^2 - 2(7) = -5$

Rearrange the result from part **a**. You know the value of $\alpha + \beta + \gamma$ and the value of $\alpha\beta + \beta\gamma + \gamma\alpha$, so you have enough information to find $\alpha^2 + \beta^2 + \gamma^2$.

You can find an expression for the sum of the squares of a quartic equation in a similar way, by multiplying out $(\alpha + \beta + \gamma + \delta)^2$.

- **The rules for sums of squares:**
  - **Quadratic:** $\alpha^2 + \beta^2 = (\alpha + \beta)^2 - 2\alpha\beta$
  - **Cubic:** $\alpha^2 + \beta^2 + \gamma^2 = (\alpha + \beta + \gamma)^2 - 2(\alpha\beta + \beta\gamma + \gamma\alpha)$
  - **Quartic:** $\alpha^2 + \beta^2 + \gamma^2 + \delta^2 = (\alpha + \beta + \gamma + \delta)^2 - 2(\alpha\beta + \alpha\gamma + \alpha\delta + \beta\gamma + \beta\delta + \gamma\delta)$

Note If you learn these you can use them without proof in your exam.

You can find a similar result for the sum of the cubes of a cubic equation by multiplying out $(\alpha + \beta + \gamma)^3$.

- **The rules for sums of cubes:**
  - **Quadratic:** $\alpha^3 + \beta^3 = (\alpha + \beta)^3 - 3\alpha\beta(\alpha + \beta)$
  - **Cubic:** $\alpha^3 + \beta^3 + \gamma^3 = (\alpha + \beta + \gamma)^3 - 3(\alpha + \beta + \gamma)(\alpha\beta + \beta\gamma + \gamma\alpha) + 3\alpha\beta\gamma$

Note The result for the sum of cubes for a quartic equation is not required.

Example 7

The three roots of a cubic equation are α, β and γ. Given that $\alpha\beta\gamma = 4$, $\alpha\beta + \beta\gamma + \gamma\alpha = -5$ and $\alpha + \beta + \gamma = 3$, find the value of $(\alpha + 3)(\beta + 3)(\gamma + 3)$.

$(\alpha + 3)(\beta + 3)(\gamma + 3)$ — Expand the brackets.

$= \alpha\beta\gamma + 3\alpha\beta + 3\alpha\gamma + 9\alpha + 3\beta\gamma + 9\beta + 9\gamma + 27$

$= \alpha\beta\gamma + 3(\alpha\beta + \beta\gamma + \alpha\gamma) + 9(\alpha + \beta + \gamma) + 27$ — Group the terms and factorise to write the expression in terms of the expressions given in the question.

$= 4 + 3(-5) + 9(3) + 27$

$= 43$

Exercise 4D

1 A quadratic equation has roots α and β. Given that $\alpha + \beta = 4$ and $\alpha\beta = 3$, find:

a $\frac{1}{\alpha} + \frac{1}{\beta}$ **b** $\alpha^2\beta^2$ **c** $\alpha^2 + \beta^2$ **d** $\alpha^3 + \beta^3$

2 A quadratic equation has roots α and β. Given that $\alpha + \beta = -\frac{2}{3}$ and $\alpha\beta = \frac{3}{4}$, find:

a $\frac{1}{\alpha} + \frac{1}{\beta}$ **b** $\alpha^2\beta^2$ **c** $\alpha^2 + \beta^2$ **d** $\alpha^3 + \beta^3$

(P) 3 A quadratic equation has roots α and β. Given that $\alpha + \beta = \frac{5}{4}$ and $\alpha\beta = -\frac{1}{3}$, find:

a $(\alpha + 2)(\beta + 2)$ **b** $(\alpha - 4)(\beta - 4)$ **c** $(\alpha^2 + 1)(\beta^2 + 1)$

4 A cubic equation has roots α, β and γ. Given that $\alpha + \beta + \gamma = 2$, $\alpha\beta + \beta\gamma + \gamma\alpha = -3$ and $\alpha\beta\gamma = 4$, find:

a $\frac{1}{\alpha} + \frac{1}{\beta} + \frac{1}{\gamma}$ **b** $\alpha^2 + \beta^2 + \gamma^2$ **c** $\alpha^3 + \beta^3 + \gamma^3$ **d** $(\alpha\beta)^2 + (\beta\gamma)^2 + (\gamma\alpha)^2$

5 A cubic equation has roots α, β and γ. Given that $\Sigma\alpha = \frac{3}{2}$, $\Sigma\alpha\beta = -\frac{4}{3}$ and $\alpha\beta\gamma = \frac{1}{2}$, find:

a $\frac{1}{\alpha} + \frac{1}{\beta} + \frac{1}{\gamma}$ **b** $\alpha^2 + \beta^2 + \gamma^2$ **c** $\alpha^3 + \beta^3 + \gamma^3$ **d** $\alpha^3\beta^3\gamma^3$

(P) 6 A cubic equation has roots α, β and γ. Given that $\alpha + \beta + \gamma = -\frac{1}{2}$, $\alpha\beta + \beta\gamma + \gamma\alpha = \frac{3}{4}$ and $\alpha\beta\gamma = -\frac{2}{5}$, find:

a $(\alpha + 2)(\beta + 2)(\gamma + 2)$ **b** $(\alpha - 3)(\beta - 3)(\gamma - 3)$ **c** $(1 - \alpha)(1 - \beta)(1 - \gamma)$

d $(\alpha\beta)^2 + (\beta\gamma)^2 + (\gamma\alpha)^2$ **e** $(\alpha\beta)^3 + (\beta\gamma)^3 + (\gamma\alpha)^3$

7 A quartic equation has roots α, β, γ and δ. Given that $\alpha + \beta + \gamma + \delta = 3$, $\alpha\beta + \alpha\gamma + \alpha\delta + \beta\gamma + \beta\delta + \gamma\delta = 5$, $\alpha\beta\gamma + \alpha\beta\delta + \alpha\gamma\delta + \beta\gamma\delta = -4$ and $\alpha\beta\gamma\delta = -2$, find:

a $\frac{1}{\alpha} + \frac{1}{\beta} + \frac{1}{\gamma} + \frac{1}{\delta}$ **b** $\alpha^2 + \beta^2 + \gamma^2 + \delta^2$ **c** $\alpha^4\beta^4\gamma^4\delta^4$

8 A quartic equation has roots α, β, γ and δ. Given that $\Sigma\alpha = \frac{1}{2}$, $\Sigma\alpha\beta = -\frac{3}{4}$, $\Sigma\alpha\beta\gamma = -\frac{1}{5}$ and $\alpha\beta\gamma\delta = \frac{4}{3}$, find:

a $\frac{1}{\alpha} + \frac{1}{\beta} + \frac{1}{\gamma} + \frac{1}{\delta}$ **b** $\alpha^2 + \beta^2 + \gamma^2 + \delta^2$ **c** $\alpha^3\beta^3\gamma^3\delta^3$

d $(\alpha\beta)^2 + (\beta\gamma)^2 + (\gamma\alpha)^2 + (\gamma\delta)^2 + (\alpha\delta)^2 + (\beta\delta)^2$

e $(\alpha\beta\gamma)^2 + (\alpha\beta\delta)^2 + (\alpha\gamma\delta)^2 + (\beta\gamma\delta)^2$

(P) **9** A quartic equation has roots α, β, γ and δ. Given that $\Sigma\alpha = -\frac{1}{2}$, $\Sigma\alpha\beta = -\frac{1}{3}$, $\Sigma\alpha\beta\gamma = \frac{1}{4}$ and $\alpha\beta\gamma\delta = \frac{3}{2}$, find:

a $(\alpha + 1)(\beta + 1)(\gamma + 1)(\delta + 1)$ **b** $(2 - \alpha)(2 - \beta)(2 - \gamma)(2 - \delta)$

(E/P) **10** The roots of the equation $x^3 - 6x^2 + 9x - 15 = 0$ are α, β and γ.

a Write down the values of $\alpha + \beta + \gamma$, $\alpha\beta + \beta\gamma + \gamma\alpha$ and $\alpha\beta\gamma$. **(1 mark)**

b Hence find the values of:

i $\frac{1}{\alpha} + \frac{1}{\beta} + \frac{1}{\gamma}$ **(2 marks)**

ii $\alpha^2 + \beta^2 + \gamma^2$ **(2 marks)**

iii $(\alpha - 1)(\beta - 1)(\gamma - 1)$ **(3 marks)**

(E/P) **11** The roots of the equation $2x^3 + 4x^2 + 7 = 0$ are α, β and γ.

a Write down the values of $\alpha + \beta + \gamma$, $\alpha\beta + \beta\gamma + \gamma\alpha$ and $\alpha\beta\gamma$. **(1 mark)**

b Hence find the values of:

i $\alpha^2 + \beta^2 + \gamma^2$ **(2 marks)**

ii $\alpha^3\beta^3\gamma^3$ **(2 marks)**

iii $(\alpha + 2)(\beta + 2)(\gamma + 2)$ **(3 marks)**

(P) **12** Show that $\alpha^3 + \beta^3 + \gamma^3 \equiv (\alpha + \beta + \gamma)^3 - 3(\alpha + \beta + \gamma)(\alpha\beta + \beta\gamma + \gamma\alpha) + 3\alpha\beta\gamma$.

(E/P) **13** The roots of the equation $3x^3 - px + 11 = 0$ are α, β and γ.

a Given that $\alpha\beta + \beta\gamma + \gamma\alpha = 4$, write down the value of p. **(1 mark)**

b Write down the values of $\alpha + \beta + \gamma$ and $\alpha\beta\gamma$. **(1 mark)**

c Hence find the value of $(3 - \alpha)(3 - \beta)(3 - \gamma)$. **(3 marks)**

(E/P) **14** The roots of the equation $x^4 + 2x^2 - x + 3 = 0$ are α, β, γ and δ.

a Write down the values of $\Sigma\alpha$, $\Sigma\alpha\beta$, $\Sigma\alpha\beta\gamma$ and $\alpha\beta\gamma\delta$. **(1 mark)**

b Hence find the values of:

i $\frac{1}{\alpha} + \frac{1}{\beta} + \frac{1}{\gamma} + \frac{1}{\delta}$ **(3 marks)**

ii $\alpha^2 + \beta^2 + \gamma^2 + \delta^2$ **(3 marks)**

iii $(\alpha + 1)(\beta + 1)(\gamma + 1)(\delta + 1)$ **(3 marks)**

(E) **15** The roots of the equation $ax^4 + 3x^3 + 2x^2 + x - 6 = 0$ are α, β, γ and δ.

a Given that $\alpha\beta\gamma\delta = -3$, write down the value of a. **(1 mark)**

b Write down the values of $\Sigma\alpha$, $\Sigma\alpha\beta$ and $\Sigma\alpha\beta\gamma$. **(1 mark)**

c Hence find the value of $\frac{1}{\alpha} + \frac{1}{\beta} + \frac{1}{\gamma} + \frac{1}{\delta}$ **(3 marks)**

(P) **16** Prove that if a quartic equation has roots α, β, γ and δ then $\alpha^2 + \beta^2 + \gamma^2 + \delta^2 \equiv (\Sigma\alpha)^2 - 2\Sigma\alpha\beta$.

4.5 Linear transformations of roots

Given the sums and products of the roots of a polynomial, it is possible to find the equation of a second polynomial whose roots are a linear transformation of the roots of the first.

For example, if the roots of a cubic equation are α, β and γ, you need to be able to find the equation of a polynomial with roots $(\alpha + 2)$, $(\beta + 2)$ and $(\gamma + 2)$, or 3α, 3β and 3γ.

Example 8

The cubic equation $x^3 - 2x^2 + 3x - 4 = 0$ has roots α, β and γ. Find the equations of the polynomials with roots:

a 2α, 2β and 2γ **b** $(\alpha + 3)$, $(\beta + 3)$ and $(\gamma + 3)$

a Method 1

$\alpha + \beta + \gamma = 2$, $\alpha\beta + \beta\gamma + \gamma\alpha = 3$ and $\alpha\beta\gamma = 4$

Sum $= 2\alpha + 2\beta + 2\gamma = 2(\alpha + \beta + \gamma) = 4$

Pair sum $= (2\alpha)(2\beta) + (2\beta)(2\gamma) + (2\gamma)(2\alpha)$
$= 4(\alpha\beta + \beta\gamma + \gamma\alpha) = 12$

Product $= (2\alpha)(2\beta)(2\gamma) = 8\alpha\beta\gamma = 32$

Hence the new equation is $w^3 - 4w^2 + 12w - 32 = 0$

Method 2

Let $w = 2x$, hence $x = \frac{w}{2}$

Substituting: $\left(\frac{w}{2}\right)^3 - 2\left(\frac{w}{2}\right)^2 + 3\left(\frac{w}{2}\right) - 4 = 0$

Multiply through by 8: $w^3 - 4w^2 + 12w - 32 = 0$

b Method 1

Sum $= (\alpha + 3) + (\beta + 3) + (\gamma + 3) = \alpha + \beta + \gamma + 9 = 11$

Pair sum $= (\alpha + 3)(\beta + 3) + (\beta + 3)(\gamma + 3) + (\gamma + 3)(\alpha + 3)$
$= \alpha\beta + \beta\gamma + \gamma\alpha + 6(\alpha + \beta + \gamma) + 27$
$= 3 + 12 + 27$
$= 42$

Product $= (\alpha + 3)(\beta + 3)(\gamma + 3)$
$= \alpha\beta\gamma + 3(\alpha\beta + \beta\gamma + \gamma\alpha) + 9(\alpha + \beta + \gamma) + 27$
$= 4 + 9 + 18 + 27 = 58$

Hence the new equation is $w^3 - 11w^2 + 42w - 58 = 0$

Method 2

Let $w = x + 3$ hence $x = w - 3$.

Substituting: $(w - 3)^3 - 2(w - 3)^2 + 3(w - 3) - 4 = 0$
$w^3 - 9w^2 + 27w - 27 - 2(w^2 - 6w + 9) + 3w - 9 - 4 = 0$
$w^3 - 11w^2 + 42w - 58 = 0$

Problem-solving

Find the sum $\Sigma\alpha$, the pair sum $\Sigma\alpha\beta$ and the product $\alpha\beta\gamma$ for the original equation. Then use these values to find the equivalent sums and products for an equation with roots 2α, 2β and 2γ.

Set $a = 1$ in the new equation.

Watch out It's a good idea to choose a different letter such as w for the variable in your new equation.

This result could have been derived by direct substitution. Each root in the new equation is twice the corresponding root in the original equation, so set $w = 2x$.

You could leave this in the above form, or multiply through by 8 to get an equation with integer coefficients.

Follow the same steps. Use the results $\Sigma\alpha = 2$, $\Sigma\alpha\beta = 3$ and $\alpha\beta\gamma = 4$ from part **a**.

Each root in the new equation is 3 more than the equivalent root in the old equation.

Multiply out and simplify.

Example 9

The quartic equation $x^4 - 3x^3 + 15x + 1 = 0$ has roots α, β, γ and δ. Find the equation with roots $(2\alpha + 1)$, $(2\beta + 1)$, $(2\gamma + 1)$ and $(2\delta + 1)$.

Method 1

$\alpha + \beta + \gamma + \delta = 3$

$\alpha\beta + \alpha\gamma + \alpha\delta + \beta\gamma + \beta\delta + \gamma\delta = 0$

$\alpha\beta\gamma + \alpha\beta\delta + \alpha\gamma\delta + \beta\gamma\delta = -15$

$\alpha\beta\gamma\delta = 1$

Sum of roots:

$(2\alpha + 1) + (2\beta + 1) + (2\gamma + 1) + (2\delta + 1)$

$= 2(a + \beta + \gamma + \delta) + 4$

$= 10$

Pair sum:

$4(\alpha\beta + \alpha\gamma + \alpha\delta + \beta\gamma + \beta\delta + \gamma\delta) + 6(\alpha + \beta + \gamma + \delta) + 6$

$= 4 \times 0 + 6 \times 3 + 6 = 24$

Triple sum:

$8(\alpha\beta\gamma + \alpha\beta\delta + \alpha\gamma\delta + \beta\gamma\delta) + 8(\alpha\beta + \alpha\gamma + \alpha\delta + \beta\gamma + \beta\delta + \gamma\delta)$
$+ 6(\alpha + \beta + \gamma + \delta) + 4 = -98$

Product:

$16\alpha\beta\gamma\delta + 8(\alpha\beta\gamma + \alpha\beta\delta + \alpha\gamma\delta + \beta\gamma\delta) + 4(\alpha\beta + \alpha\gamma$
$+ \alpha\delta + \beta\gamma + \beta\delta + \gamma\delta) + 2(\alpha + \beta + \gamma + \delta) + 1$

$= -97$

Hence the new equation is

$w^4 - 10w^3 + 24w^2 + 98w - 97 = 0$

Method 2

Let $w = 2x + 1$ hence $x = \dfrac{w-1}{2}$

Substituting: $\left(\dfrac{w-1}{2}\right)^4 - 3\left(\dfrac{w-1}{2}\right)^3 + 15\left(\dfrac{w-1}{2}\right) + 1 = 0$

$(w - 1)^4 - 6(w - 1)^3 + 120(w - 1) + 16 = 0$

$w^4 - 4w^3 + 6w^2 - 4w + 1 - 6(w^3 - 3w^2 + 3w - 1)$
$+ 120w - 120 + 16 = 0$

$w^4 - 10w^3 + 24w^2 + 98w - 97 = 0$

Watch out Even though the x^2 term is 0, the equation is still a quartic. Find expressions for $\Sigma\alpha$, $\Sigma\alpha\beta$, $\Sigma\alpha\beta\gamma$ and $\alpha\beta\gamma\delta$.

If you expand one pair of roots you get $(2\alpha + 1)(2\beta + 1) = 4\alpha\beta + 2\alpha + 2\beta + 1$. Each original root will appear in three of the six possible pairs, giving you this expression.

If you expand one triple of roots you get $(2\alpha + 1)(2\beta + 1)(2\gamma + 1) = 8\alpha\beta\gamma + 4(\alpha\beta + \beta\gamma + \gamma\alpha) + 2(\alpha + \beta + \gamma + \delta) + 1$. There are four possible triples, with each root appearing three times and each pair of roots appearing twice.

Expand $(2\alpha + 1)(2\beta + 1)(2\gamma + 1)(2\delta + 1)$.

Problem-solving

In many cases, it is quicker to use a substitution.

Multiply through by 16 to remove the fractions.

Expand and simplify.

Exercise 4E

1 The cubic equation $x^3 - 7x^2 + 6x + 5 = 0$ has roots α, β and γ. Find equations with roots:

a $(\alpha + 1)$, $(\beta + 1)$ and $(\gamma + 1)$

b 2α, 2β and 2γ

2 The cubic equation $3x^3 - 4x^2 - 5x + 1 = 0$ has roots α, β and γ. Find equations with roots:

a $(\alpha - 3)$, $(\beta - 3)$ and $(\gamma - 3)$

b $\dfrac{\alpha}{2}$, $\dfrac{\beta}{2}$ and $\dfrac{\gamma}{2}$

E/P **3** The cubic equation $x^3 - 3x^2 + 4x - 7 = 0$ has roots α, β and γ.
Without solving the equation, find the equation with roots $(2\alpha + 1)$, $(2\beta + 1)$ and $(2\gamma + 1)$.
Give your answer in the form $aw^3 + bw^2 + cw + d = 0$ where a, b, c and d are integers to be determined. **(5 marks)**

E/P **4** The cubic equation $x^3 + 4x^2 - 4x + 2 = 0$ has roots α, β and γ.
Without solving the equation, find the equation with roots $(2\alpha - 1)$, $(2\beta - 1)$ and $(2\gamma - 1)$.
Give your answer in the form $w^3 + pw^2 + qw + r = 0$ where p, q and r are integers to be found. **(5 marks)**

E/P **5** The cubic equation $3x^3 - x^2 + 2x - 5 = 0$ has roots α, β and γ.
Without solving the equation, find the equation with roots $(3\alpha + 1)$, $(3\beta + 1)$ and $(3\gamma + 1)$.
Give your answer in the form $aw^3 + bw^2 + cw + d = 0$ where a, b, c and d are integers to be determined. **(5 marks)**

P **6** The quartic equation $2x^4 + 4x^3 - 5x^2 + 2x - 1 = 0$ has roots α, β, γ and δ. Find equations with integer coefficients that have roots:

a 3α, 3β, 3γ and 3δ **b** $(\alpha - 1)$, $(\beta - 1)$, $(\gamma - 1)$ and $(\delta - 1)$

E/P **7** The quartic equation $x^4 + 2x^3 - 3x^2 + 4x + 5 = 0$ has roots α, β, γ and δ.
Without solving the equation, find equations with integer coefficients that have roots:

a 2α, 2β, 2γ and 2δ **(6 marks)**

b $(\alpha - 2)$, $(\beta - 2)$, $(\gamma - 2)$ and $(\delta - 2)$ **(6 marks)**

E/P **8** The quartic equation $3x^4 + 5x^3 - 4x^2 - 3x + 1 = 0$ has roots α, β, γ and δ.
Without solving the equation, find equations with integer coefficients that have roots:

a 3α, 3β, 3γ and 3δ **(6 marks)**

b $(\alpha + 1)$, $(\beta + 1)$, $(\gamma + 1)$ and $(\delta + 1)$ **(6 marks)**

Challenge

The quartic equation $2x^4 - 3x^3 + x^2 - 2x - 6 = 0$ has roots α, β, γ and δ.

a Find an equation with integer coefficients that has roots $(2\alpha + 1)$, $(2\beta + 1)$, $(2\gamma + 1)$ and $(2\delta + 1)$.

b The diagram shows the locations of the roots of the original equation on an Argand diagram:

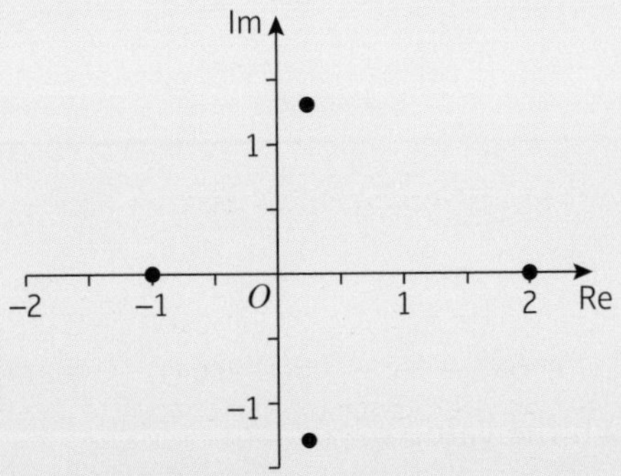

Copy this diagram and sketch the approximate locations of the roots of your new equation.

Mixed exercise 4

1 The roots of a quartic equation $ax^4 + bx^3 + cx^2 + dx + e = 0$ are $\alpha = \frac{1}{5}$, $\beta = -\frac{2}{5}$, $\gamma = -\frac{3}{5}$ and $\delta = -\frac{1}{2}$
Find integer values for a, b, c, d and e.

(E) 2 The cubic equation $x^3 + px^2 + 37x - 52 = 0$ has roots α, β and γ.

a Write down the values of $\alpha\beta + \beta\gamma + \gamma\alpha$ and $\alpha\beta\gamma$, and express p in terms of α, β and γ. **(3 marks)**

b Given that $\alpha = 3 - 2\mathrm{i}$, find the value of p. **(4 marks)**

(E) 3 The cubic equation $2x^3 + 5x^2 - 2x + q = 0$ has a root at $x = -2 + \mathrm{i}$.

a Find the other two roots of the equation. **(4 marks)**

b Hence find the value of q. **(2 marks)**

(E/P) 4 The quartic equation $x^4 - 40x^3 + 510x^2 - 2200x + 1729 = 0$ has roots α, $\alpha + 2k$, $\alpha + 4k$ and $\alpha + 6k$ for some real constant k. Solve the equation. **(7 marks)**

(E) 5 Three of the roots of the quartic equation $24x^4 - 58x^3 + 17x^2 + dx + e = 0$ are $\frac{1}{2}$, $-\frac{1}{3}$ and 2.

a Find the fourth root. **(2 marks)**

b Find the values of d and e. **(4 marks)**

(E/P) 6 The equation $x^4 + 2x^3 + mx^2 + nx + 85 = 0$, $x \in \mathbb{C}$, $m, n \in \mathbb{R}$, has roots α, β, γ and δ.
Given that $\alpha = -2 + \mathrm{i}$ and $\beta = \alpha^*$,

a show that $\gamma + \delta - 2 = 0$ and that $\gamma\delta - 17 = 0$. **(2 marks)**

b Hence find all the roots of the quartic equation and find the values of m and n. **(5 marks)**

c Show these roots on an Argand diagram. **(3 marks)**

(E) 7 A quartic equation $4x^4 - 16x^3 + 115x^2 + 4x - 29 = 0$ has roots α, β, γ and δ.

a Show that $2 - 5\mathrm{i}$ is one root of the equation. **(3 marks)**

b Without solving the equation, find the other roots. **(5 marks)**

c Show these roots on an Argand diagram. **(3 marks)**

(E) 8 The roots of the equation $2x^3 - 5x^2 + 11x - 9 = 0$ are α, β and γ.

a Write down the values of $\alpha + \beta + \gamma$, $\alpha\beta + \beta\gamma + \gamma\alpha$ and $\alpha\beta\gamma$. **(1 mark)**

b Hence find the values of:

i $\frac{1}{\alpha} + \frac{1}{\beta} + \frac{1}{\gamma}$ **(2 marks)**

ii $\alpha^2 + \beta^2 + \gamma^2$ **(2 marks)**

iii $(\alpha - 1)(\beta - 1)(\gamma - 1)$ **(3 marks)**

(E) 9 The roots of the equation $px^4 + 12x^3 + 6x^2 + 5x - 7 = 0$ are α, β, γ and δ.

a Given that $\alpha\beta\gamma\delta = -1$, write down the value of p. **(1 mark)**

b Write down the values of $\Sigma\alpha$, $\Sigma\alpha\beta$ and $\Sigma\alpha\beta\gamma$. **(1 mark)**

c Hence find the value of $\alpha^2 + \beta^2 + \gamma^2 + \delta^2$. **(3 marks)**

E **10** The roots of the equation $5x^3 + cx + 21 = 0$ are α, β and γ.

a Given that $\alpha\beta + \beta\gamma + \gamma\alpha = -6$, write down the value of c. **(1 mark)**

b Write down values for $\alpha + \beta + \gamma$ and $\alpha\beta\gamma$. **(1 mark)**

c Hence find the value of $(1 - \alpha)(1 - \beta)(1 - \gamma)$. **(3 marks)**

E/P **11** The cubic equation $2x^3 + 5x^2 + 7x - 2 = 0$ has roots α, β and γ. Without solving the equation, find the equation with roots $(3\alpha + 1)$, $(3\beta + 1)$ and $(3\gamma + 1)$. Give your answer in the form $pw^3 + qw^2 + rw + s = 0$ where p, q, r and s are integers to be found. **(5 marks)**

E/P **12** The quartic equation $6x^4 - 2x^3 - 5x^2 + 7x + 8 = 0$ has roots α, β, γ and δ. Without solving the equation, find equations with integer coefficients that have roots:

a 2α, 2β, 2γ and 2δ **(6 marks)**

b $(3\alpha - 2)$, $(3\beta - 2)$, $(3\gamma - 2)$ and $(3\delta - 2)$ **(6 marks)**

Challenge

1 The cubic equation $x^3 + 4x^2 - 5x - 7 = 0$ has roots α, β and γ. Without solving the equation, find a cubic equation that has roots $\frac{1}{\alpha + 1}$, $\frac{1}{\beta + 1}$ and $\frac{1}{\gamma + 1}$

2 The cubic equation $x^3 + 2x^2 - 3x - 5 = 0$ has roots α, β and γ. Without solving the equation, find an equation that has roots $\alpha + \beta$, $\beta + \gamma$ and $\gamma + \alpha$.

3 The quartic equation $x^4 + 2x^2 - 5x + 2 = 0$ has roots α, β, γ and δ. By using a substitution, or otherwise, find an equation that has roots $\alpha^2 + 1$, $\beta^2 + 1$, $\gamma^2 + 1$ and $\delta^2 + 1$.

Summary of key points

1 If α and β are roots of the equation $ax^2 + bx + c = 0$, then:

- $\alpha + \beta = -\frac{b}{a}$
- $\alpha\beta = \frac{c}{a}$

2 If α, β and γ are roots of the equation $ax^3 + bx^2 + cx + d = 0$, then:

- $\alpha + \beta + \gamma = \Sigma\alpha = -\frac{b}{a}$
- $\alpha\beta + \beta\gamma + \gamma\alpha = \Sigma\alpha\beta = \frac{c}{a}$
- $\alpha\beta\gamma = -\frac{d}{a}$

3 If α, β, γ and δ are roots of the equation $ax^4 + bx^3 + cx^2 + dx + e = 0$, then:

- $\alpha + \beta + \gamma + \delta = \Sigma\alpha = -\frac{b}{a}$
- $\alpha\beta + \alpha\gamma + \alpha\delta + \beta\gamma + \beta\delta + \gamma\delta = \Sigma\alpha\beta = \frac{c}{a}$
- $\alpha\beta\gamma + \alpha\beta\delta + \alpha\gamma\delta + \beta\gamma\delta = \Sigma\alpha\beta\gamma = -\frac{d}{a}$
- $\alpha\beta\gamma\delta = \frac{e}{a}$

Summary of key points

4 The rules for **reciprocals**:

- Quadratic: $\frac{1}{\alpha} + \frac{1}{\beta} = \frac{\alpha + \beta}{\alpha\beta}$
- Cubic: $\frac{1}{\alpha} + \frac{1}{\beta} + \frac{1}{\gamma} = \frac{\alpha\beta + \beta\gamma + \gamma\alpha}{\alpha\beta\gamma}$
- Quartic: $\frac{1}{\alpha} + \frac{1}{\beta} + \frac{1}{\gamma} + \frac{1}{\delta} = \frac{\alpha\beta\gamma + \beta\gamma\delta + \gamma\delta\alpha + \delta\alpha\beta}{\alpha\beta\gamma\delta}$

5 The rules for **products of powers**:

- Quadratic: $\alpha^n \times \beta^n = (\alpha\beta)^n$
- Cubic: $\alpha^n \times \beta^n \times \gamma^n = (\alpha\beta\gamma)^n$
- Quartic: $\alpha^n \times \beta^n \times \gamma^n \times \delta^n = (\alpha\beta\gamma\delta)^n$

6 The rules for **sums of squares**:

- Quadratic: $\alpha^2 + \beta^2 = (\alpha + \beta)^2 - 2\alpha\beta$
- Cubic: $\alpha^2 + \beta^2 + \gamma^2 = (\alpha + \beta + \gamma)^2 - 2(\alpha\beta + \beta\gamma + \gamma\alpha)$
- Quartic: $\alpha^2 + \beta^2 + \gamma^2 + \delta^2 = (\alpha + \beta + \gamma + \delta)^2 - 2(\alpha\beta + \alpha\gamma + \alpha\delta + \beta\gamma + \beta\delta + \gamma\delta)$

7 The rules for **sums of cubes**:

- Quadratic: $\alpha^3 + \beta^3 = (\alpha + \beta)^3 - 3\alpha\beta(\alpha + \beta)$
- Cubic: $\alpha^3 + \beta^3 + \gamma^3 = (\alpha + \beta + \gamma)^3 - 3(\alpha + \beta + \gamma)(\alpha\beta + \beta\gamma + \gamma\alpha) + 3\alpha\beta\gamma$

Volumes of revolution 5

Objectives

After completing this chapter you should be able to:

- Find the volume of revolution when a curve is rotated around the x-axis → pages 72–75
- Find the volume of revolution when a curve is rotated around the y-axis → pages 76–78
- Find more complicated volumes of revolution → pages 78–83
- Model real-life objects using volumes of revolution → pages 83–86

Prior knowledge check

1 Evaluate:

a $\int_2^4 (6x^2 - 8x)\,dx$

b $\int_4^9 \left(\frac{5}{\sqrt{x}} + x^{\frac{3}{2}}\right) dx$

c $\int_1^2 \frac{x^2 + 8x}{x^4}\,dx$ ← Pure Year 1, Chapter 13

2 Find the area of the region R bounded by the curve $y = (x + 3)(x - 1)^2$ and the x-axis.

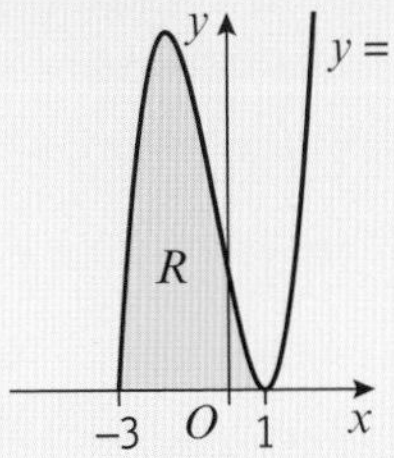

← Pure Year 1, Chapter 13

3 Find the area of the finite region bounded by the curve $y = -x^2 + 6x + 4$ and the line $x + y = 10$.

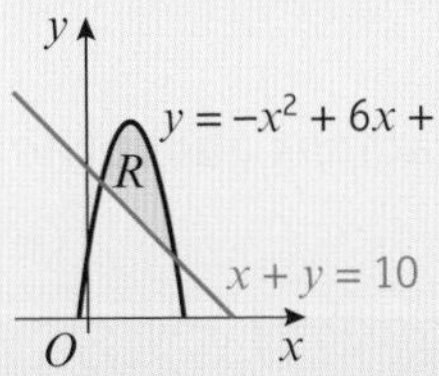

← Pure Year 1, Chapter 13

Woodworkers use lathes to create solid objects that have circular cross-sections. Solids such as this are called volumes of revolution, and you can find their volumes using calculus.

5.1 Volumes of revolution around the x-axis

You have used integration to find the area of a region R bounded by a curve, the x-axis and two vertical lines.

Notation This process is called **definite integration**. ← Pure Year 1, Sections 13.4, 13.5

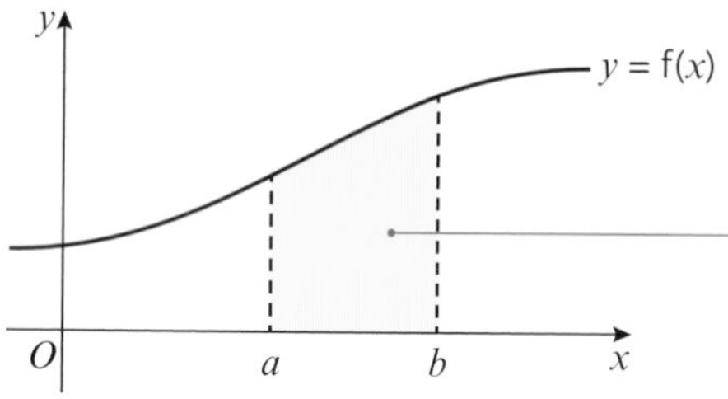

The area between a positive curve, the x-axis and the lines $x = a$ and $x = b$ is given by

$$\text{Area} = \int_a^b y\,dx$$

where $y = f(x)$ is the equation of the curve.

You can derive this formula by considering the sum of an infinite number of small strips of width δx. Each of these strips has a height of y, so the area of each strip is

$$\delta A = y\delta x$$

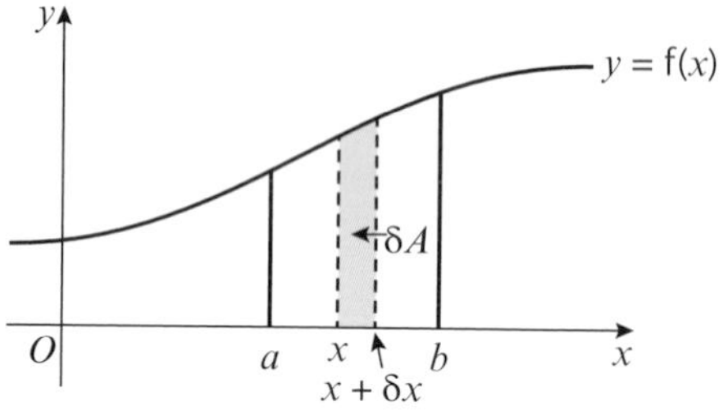

The total area is approximately the sum of these strips, or $\sum y\delta x$.

The exact area is the limit of this sum as $\delta x \to 0$, which is written as $\int y\,dx$.

You can use a similar technique to find the volume of an object created by rotating a curve around a coordinate axis. If each of these strips is rotated through 2π radians (or 360°) about the x-axis, it will form a shape that is approximately cylindrical. The volume of each cylinder will be $\pi y^2\delta x$ since it will have radius y and height δx.

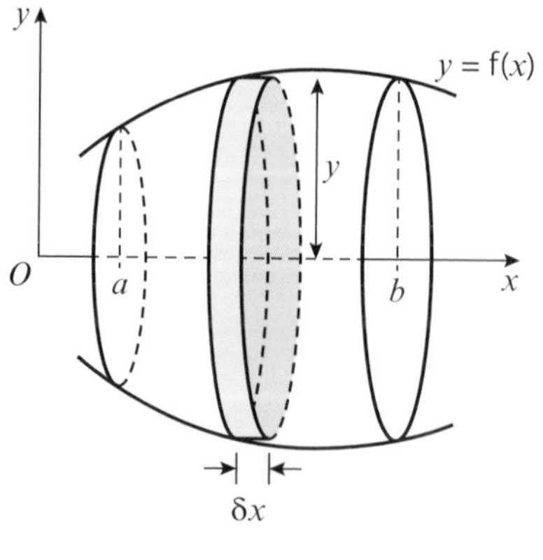

So the volume of the solid will be approximately equal to the sum of the volumes of each cylinder, or $\sum \pi y^2\delta x$. The exact volume is the limit of this sum as $\delta x \to 0$, or $\pi \int y^2\,dx$.

- **The volume of revolution formed when $y = f(x)$ is rotated through 2π radians about the x-axis between $x = a$ and $x = b$ is given by:**

  Volume $= \pi \int_a^b y^2\,dx$

Example 1

The diagram shows the region R which is bounded by the x-axis, the y-axis and the curve with equation $y = 9 - x^2$. The region is rotated through 2π radians about the x-axis. Find the exact volume of the solid generated.

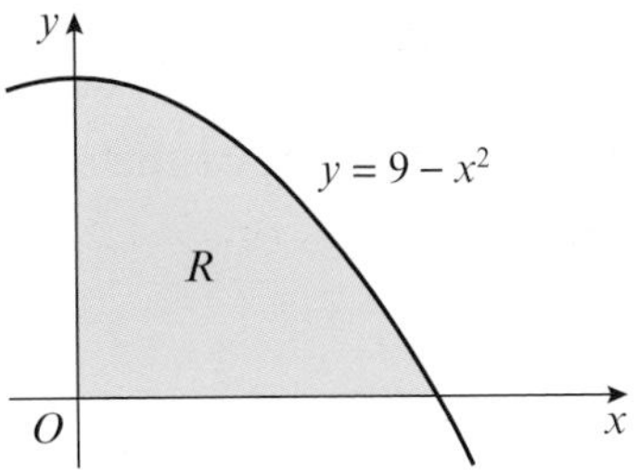

$9 - x^2 = 0$ — First find the point where the curve intersects the x-axis.

$(3 - x)(3 + x) = 0$

$x = 3$ or $x = -3$ — From the diagram, $x > 0$, therefore $x = 3$.

$V = \pi\int_0^3 (9 - x^2)^2\,dx$ — Use $V = \pi\int_a^b y^2\,dx$ with $a = 0$, $b = 3$ and $y = 9 - x^2$.

$= \pi\int_0^3 (81 - 18x^2 + x^4)\,dx$ — Simplify the integrand.

$= \pi\left[81x - 6x^3 + \frac{1}{5}x^5\right]_0^3$ — Integrate each term separately.

$= \pi\left(\left(243 - 162 + \frac{243}{5}\right) - (0 - 0 + 0)\right)$ — Substitute the limits.

$= \frac{648\pi}{5}$ — Simplify the resulting answer and write it as an exact fraction in terms of π.

Exercise 5A

1 Find the exact volume of the solid generated when each curve is rotated through 360° about the x-axis between the given limits.

a $y = 10x^2$ between $x = 0$ and $x = 2$

b $y = 5 - x$ between $x = 3$ and $x = 5$

c $y = \sqrt{x}$ between $x = 2$ and $x = 10$

d $y = 1 + \frac{1}{x^2}$ between $x = 1$ and $x = 2$

(E) 2 The curve shown in the diagram has equation $y = 5 + 4x - x^2$. The finite region R is bounded by the curve, the x-axis and the y-axis. The region is rotated through 2π radians about the x-axis to generate a solid of revolution. Find the exact volume of the solid generated. **(5 marks)**

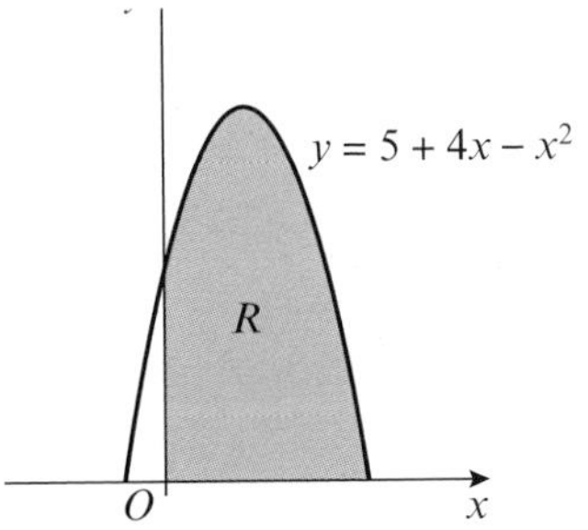

(E) **3** The diagram shows the region R which is bounded by the x-axis, the lines $x = 1$ and $x = 8$, and the curve with equation $y = 3 - \sqrt[3]{x}$. The region is rotated through 2π radians about the x-axis. Find the exact volume of the solid generated. **(5 marks)**

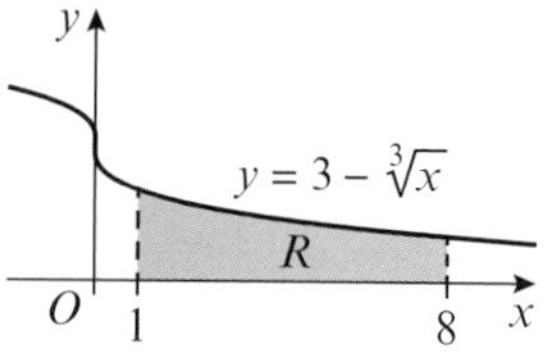

(E) **4** The diagram shows the curve C with equation $y = \sqrt{x + 2}$. The region R is bounded by the x-axis, the line $x = 2$ and C. The region is rotated through 360° about the x-axis. Find the exact volume of the solid generated. **(5 marks)**

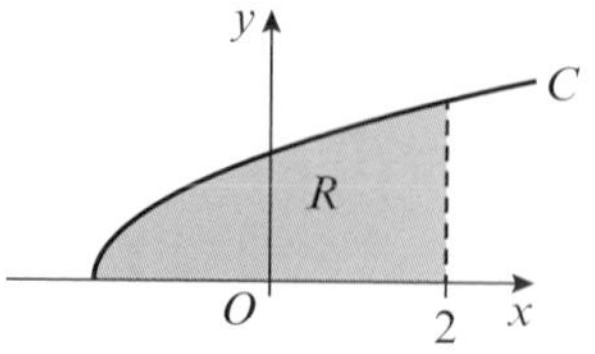

(E/P) **5** The diagram shows a sketch of the curve with equation $y = 9x^{\frac{3}{2}} - 3x^{\frac{5}{2}}$. The region R is bounded by the curve and the x-axis.

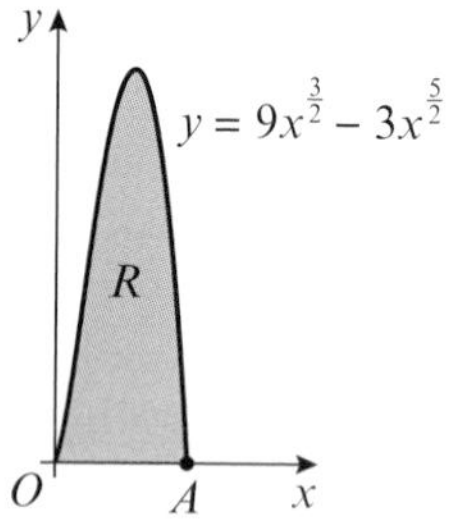

a Find the coordinates of A. **(2 marks)**

The region is rotated through 2π radians about the x-axis.

b Find the volume of the solid of revolution generated. **(5 marks)**

(E/P) **6** The curve with equation $y = \dfrac{\sqrt{3x^4 - 3}}{x^3}$ is shown in the diagram.

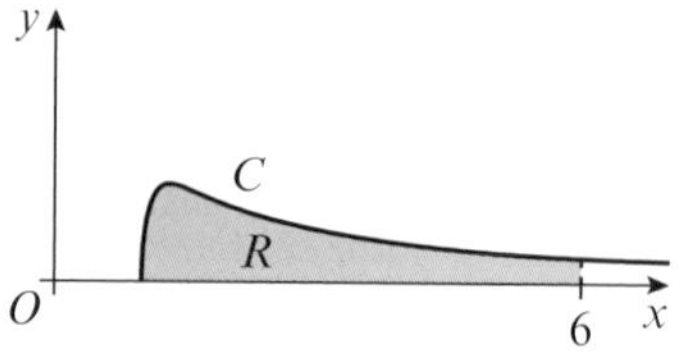

The region bounded by the curve C, the x-axis and the line $x = 6$ is shown shaded in the diagram. The region is rotated through 2π radians about the x-axis. Find the volume of the solid generated, giving your answer correct to 3 significant figures. **(6 marks)**

(P) **7** The diagram shows the curve with equation $5y^2 - x^3 = 2x - 3$. The shaded region is bounded by the curve and the line $x = 4$. The region is rotated about the x-axis to generate a solid of revolution. Find the volume of the solid generated.

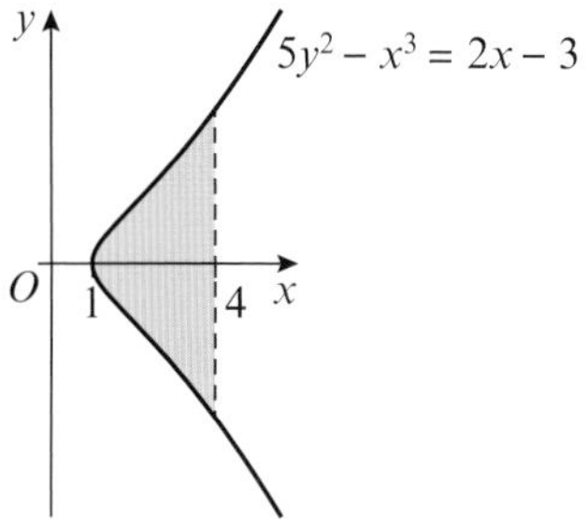

Problem-solving

Rearrange the equation to make y^2 the subject.

(E/P) **8** The curve shown in the diagram has equation $y = x\sqrt{4 - x^2}$. The finite region R is bounded by the curve, the x-axis and the line $x = a$, where $0 < a < 2$. The region is rotated through 2π radians about the x-axis to generate a solid of revolution with volume $\frac{657\pi}{160}$

Find the value of a. **(5 marks)**

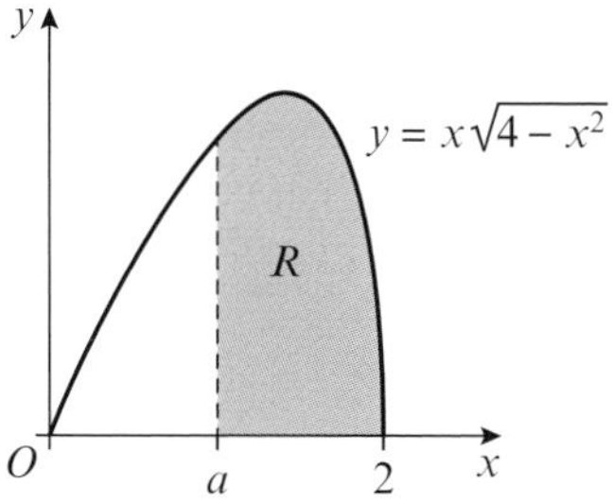

(P) **9** The diagram shows a shaded rectangular region R of length h and width r. The region R is rotated through 360° about the x-axis. Use integration to show that the volume, V, of the cylinder formed is $V = \pi r^2 h$.

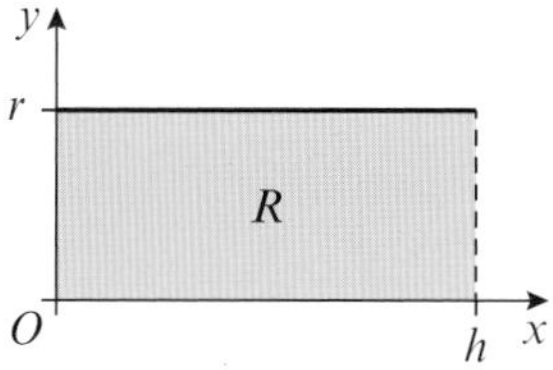

Challenge

The diagram shows the curve C with equation $y = |x^2 - 7x + 10|$. The shaded region R is bounded by the x-axis, the curve C and the lines $x = 1$ and $x = 6$. The region is rotated 2π radians about the x-axis. Find the exact volume of the solid generated.

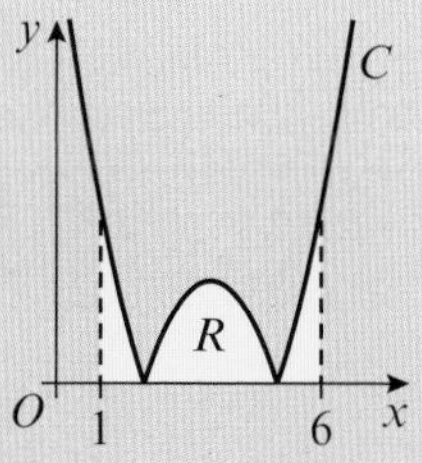

5.2 Volumes of revolution around the y-axis

You can find a volume of revolution around the y-axis by considering x as a function of y. The diagram shows a curve with equation $x = \text{f}(y)$. A small strip of height δy is rotated 2π radians about the y-axis. The volume of the cylinder created will be $\pi x^2 \delta y$ since the radius is x and the height is δy.

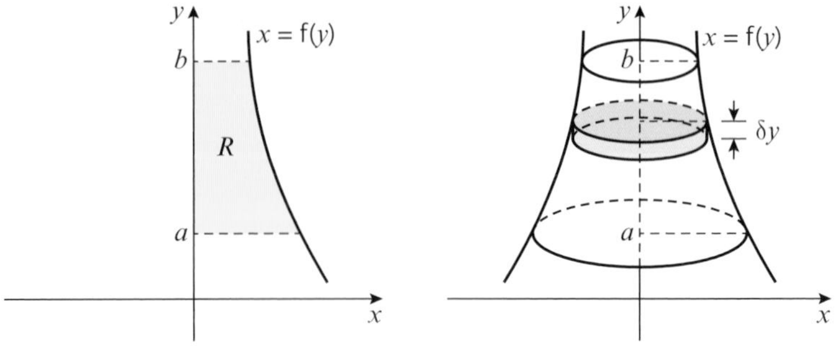

So when the whole region, R, is rotated 2π radians about the y-axis, the volume formed will be approximately equal to the sum of the volumes of each cylinder, or $\sum \pi x^2 \delta y$. The exact volume is the limit of this sum as $\delta x \to 0$, or $\pi \int x^2\,\text{d}y$.

- **The volume of revolution formed when $x = \text{f}(y)$ is rotated through 2π radians about the y-axis between $y = a$ and $y = b$ is given by**

  Volume $= \pi \int_a^b x^2\,\text{d}y$

Online Explore volumes of revolution around the x- and y-axes using GeoGebra.

Example 2

The diagram shows the curve with equation $y = \sqrt{x - 1}$. The region R is bounded by the curve, the y-axis and the lines $y = 1$ and $y = 3$.

The region is rotated through 360° about the y-axis. Find the volume of the solid generated.

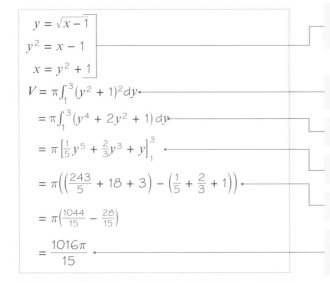

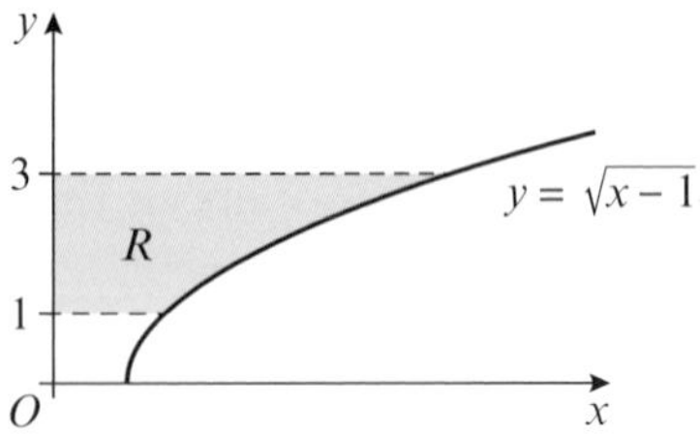

$y = \sqrt{x - 1}$

$y^2 = x - 1$

$x = y^2 + 1$ — First rearrange the equation to make x the subject.

$V = \pi \int_1^3 (y^2 + 1)^2\,\text{d}y$ — Use $V = \pi \int_a^b x^2\,\text{d}y$ with $a = 1$, $b = 3$ and $x = y^2 + 1$.

$= \pi \int_1^3 (y^4 + 2y^2 + 1)\,\text{d}y$ — Simplify the integrand.

$= \pi \left[\frac{1}{5}y^5 + \frac{2}{3}y^3 + y\right]_1^3$ — Integrate each term separately.

$= \pi\left(\left(\frac{243}{5} + 18 + 3\right) - \left(\frac{1}{5} + \frac{2}{3} + 1\right)\right)$ — Substitute the limits.

$= \pi\left(\frac{1044}{15} - \frac{28}{15}\right)$

$= \frac{1016\pi}{15}$ — Simplify the resulting answer and write it as an exact fraction in terms of π.

Exercise 5B

1 Find the exact volume of the solid generated when each curve is rotated through 360° about the y-axis between the given limits.

Hint In part **d**, rearrange the expression to make x^2 the subject.

a $x = \frac{1}{2}y + 1$ between $y = 2$ and $y = 5$

b $y = 2\sqrt{x}$ between $y = 0$ and $y = 1$

c $y = \frac{1}{x}$ between $y = 1$ and $y = 3$

d $y = 2x^2 - 4$ between $x = 5$ and $x = 11$

E 2 The curve C with equation $x = \frac{1}{2}y^2 + 1$ is shown in the diagram. The region R is bounded by the lines $y = 1$, $y = 4$, the y-axis and the curve C, as shown in the diagram. The region is rotated through 2π radians about the y-axis.

Find the volume of the solid generated. **(6 marks)**

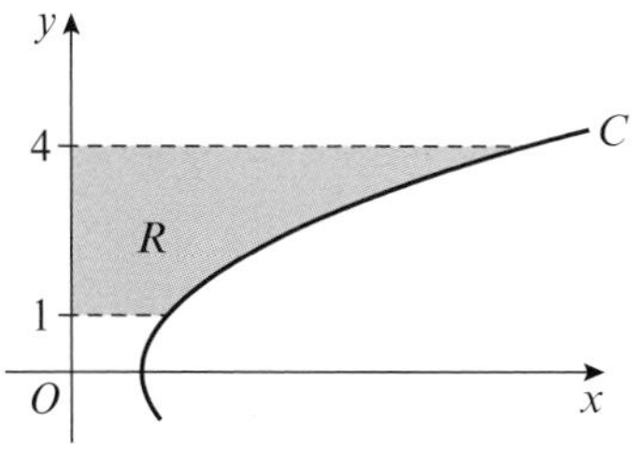

E 3 The diagram shows the finite region R, which is bounded by the curve $x = \sqrt{y} + \frac{1}{y^2}$, the lines $y = 4$, $y = 9$ and the y-axis.

a Find the exact area of the shaded region. **(3 marks)**

The region R is rotated through 2π radians about the y-axis.

b Use integration to find the volume of the solid generated. Round your answer to 2 decimal places. **(5 marks)**

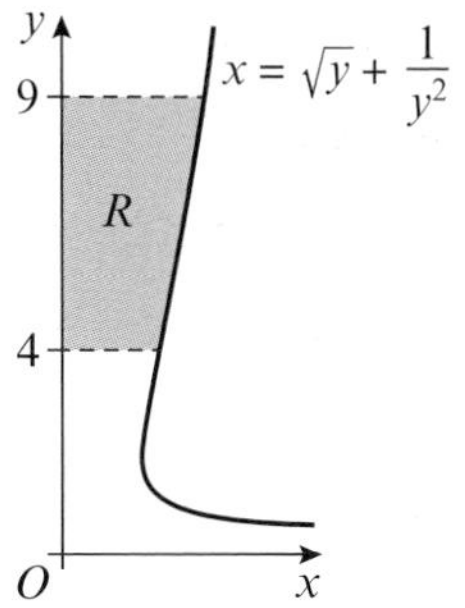

E 4 The diagram shows the finite region R, which is bounded by the curve $x = y^2 - 6y + 10$, the lines $y = 1$, $y = 4$ and the y-axis.

a Find the area of the shaded region R. **(3 marks)**

The region R is rotated through 360° about the y-axis.

b Use integration to find an exact value for the volume of the solid generated. **(5 marks)**

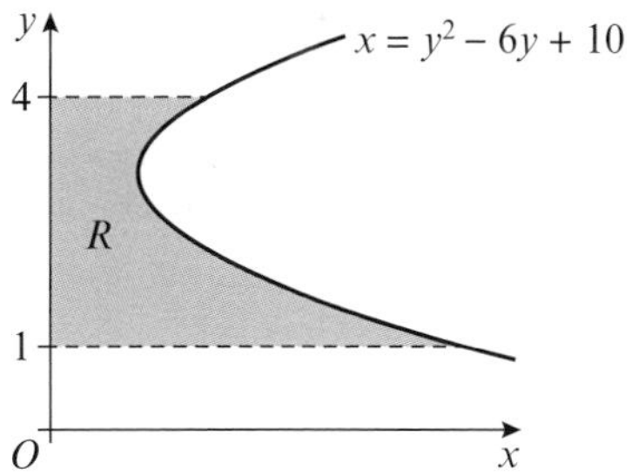

E/P 5 The curve C with equation $y = 2x^2 + 5$ is shown in the diagram. The region bounded by the y-axis, the curve C and the line $y = 10$ is shown and shaded in the diagram. The region is rotated 360° about the y-axis. Find the exact volume of the solid generated. **(6 marks)**

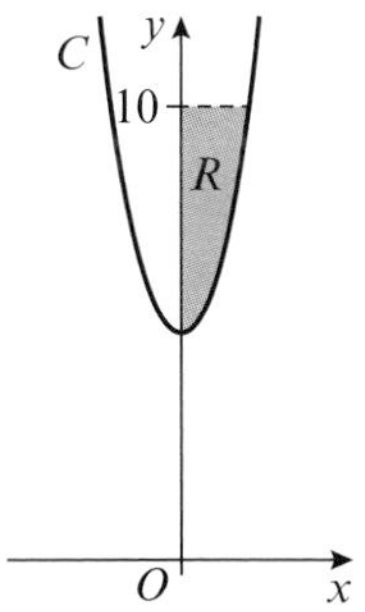

E **6** $f(x) = x^2 - 2x + 1, x \geqslant 1$

The diagram shows the finite region R bounded by the curve $y = f(x)$, the y-axis and the lines $y = 1$ and $y = 9$.

a Show that the equation $y = f(x)$ can be written as $x^2 = y + 2\sqrt{y} + 1$. **(2 marks)**

b The region R is rotated through 2π radians about the y-axis. Find the exact volume of the solid generated. **(5 marks)**

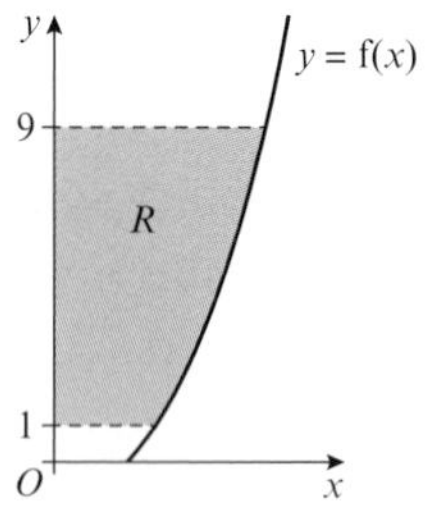

E/P **7** The diagram shows the finite region R, which is bounded by the curve $y^3 + x^2 - 2y = 4$ and the x-axis.
The region R is rotated about the y-axis to generate a solid of revolution.

Find an exact value for the volume of the solid. **(5 marks)**

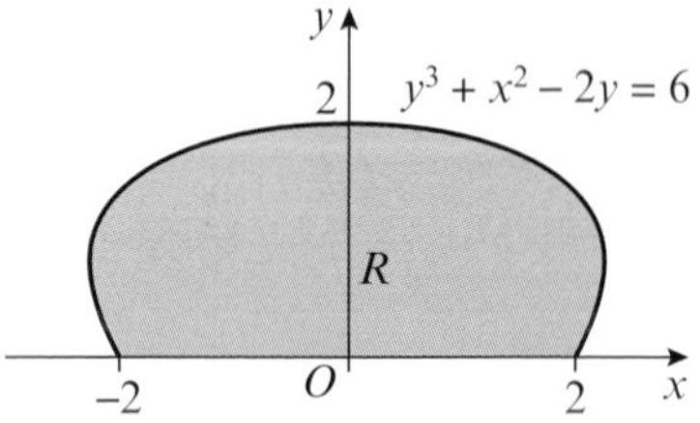

E/P **8** Part of the curve C with equation $y^2 = \dfrac{1}{2x + 1}$ is shown in the diagram.
The region R is bounded by the curve, the y-axis and the line $y = 4$.
The region R is rotated 2π radians about the y-axis.
Find the volume of the solid generated. **(5 marks)**

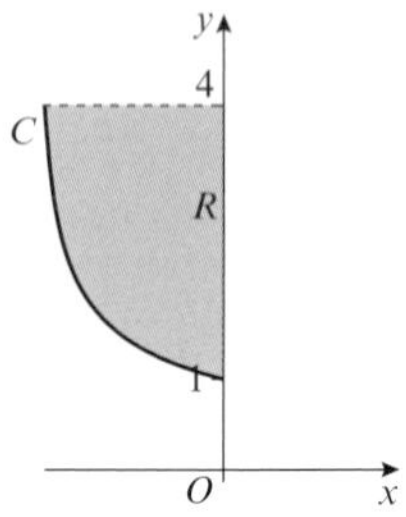

P **9** The diagram shows a shaded region R in the shape of a right-angled triangle of width r and height h. The region R is rotated through 2π radians about the y-axis. Use integration to show that the volume, V, of the cone formed is given by $V = \frac{1}{3}\pi r^2 h$.

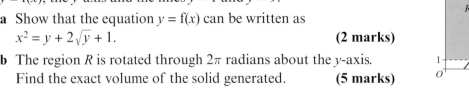

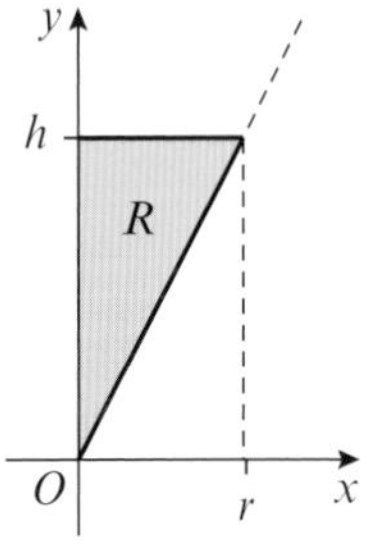

Problem-solving

Start by finding an equation for the line that forms the hypotenuse of the triangle.

5.3 Adding and subtracting volumes

You might need to solve volume of revolution problems involving the volumes of cylinders and cones. Remember these two formulae:

- **A cylinder of height h and radius r has volume $\pi r^2 h$.**
- **A cone of height h and base radius r has volume $\frac{1}{3}\pi r^2 h$.**

Example 3

The region R is bounded by the curve with equation $y = x^3 + 2$, the line $y = 5 - 2x$, and the x- and y-axes.

a Verify that the coordinates of A are (1, 3).

A solid is created by rotating the region 360° about the x-axis.

b Find the volume of this solid.

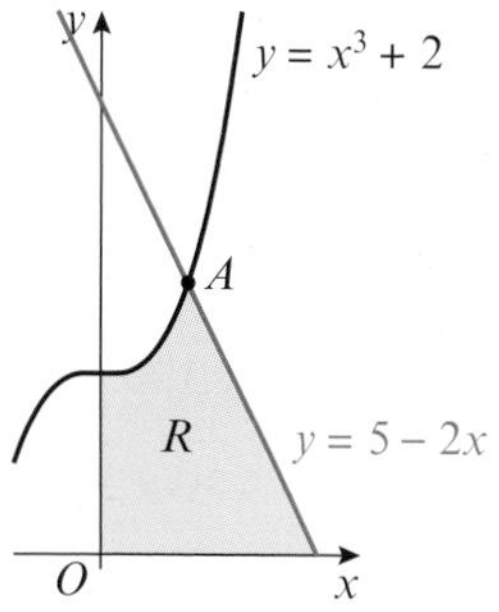

a Curve: $1^3 + 2 = 3$

Line: $5 - 2 \times 1 = 3$

So (1, 3) is the point of intersection.

Substitute $x = 1$ into each equation to show that the point (1, 3) lies on both the line and the curve.

b Find the volumes of revolution of R_1 and R_2 separately.

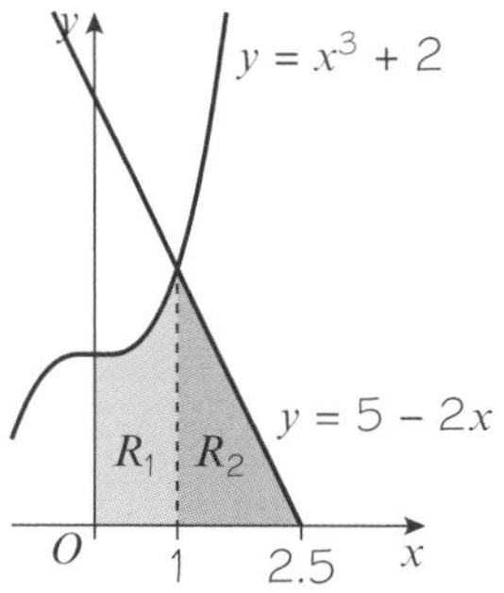

Problem-solving

Divide the original area into two separate areas. Use integration to find the volume of revolution of R_1. Then add the volume of the cone formed by rotating R_2 about the x-axis.

Volume of revolution of R_1

$V = \pi\int_0^1 (x^3 + 2)^2\,dx$ — Use $V = \pi\int_a^b y^2\,dx$ with $a = 0$, $b = 1$ and $y = x^3 + 2$.

$= \pi\int_0^1 (x^6 + 4x^3 + 4)\,dx$ — Simplify the integrand.

$= \pi\left[\frac{1}{7}x^7 + x^4 + 4x\right]_0^1$ — Integrate each term separately.

$= \pi\left(\left(\frac{1}{7} + 1 + 4\right) - (0 + 0 + 0)\right)$ — Substitute the limits.

$= \frac{36\pi}{7}$ — Simplify the resulting answer by writing it as an exact fraction in terms of π.

Volume of revolution of R_2

The line $y = 5 - 2x$ will intersect the x-axis at $x = 2.5$.

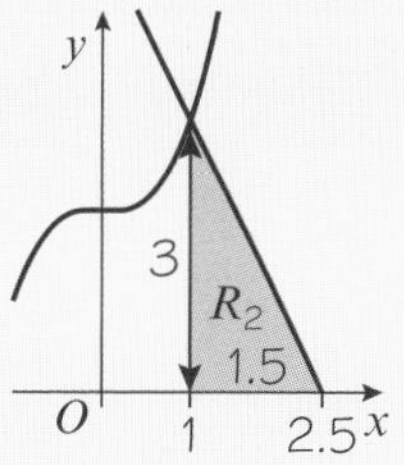

The height of the cone is $2.5 - 1 = 1.5$.

The radius of the cone is 3.

When R_2 is rotated about the x-axis, it will create a cone.

$V = \frac{1}{3}\pi \times 3^2 \times 1.5$ — The formula for the volume of a cone is $V = \frac{1}{3}\pi r^2 h$.

$V = \frac{9\pi}{2}$

The total volume is $\frac{36\pi}{7} + \frac{9\pi}{2} = \frac{135\pi}{14}$ — Add the values of R_1 and R_2 to find the total volume of revolution.

You can find more complicated volumes of revolution by subtracting one volume of revolution from another.

Example 4

The diagram shows the region R bounded by the curves with equations $y = \sqrt{x}$ and $y = \frac{1}{8x}$ and the line $x = 1$.

The region is rotated through 360° about the x-axis.
Find the exact volume of the solid generated.

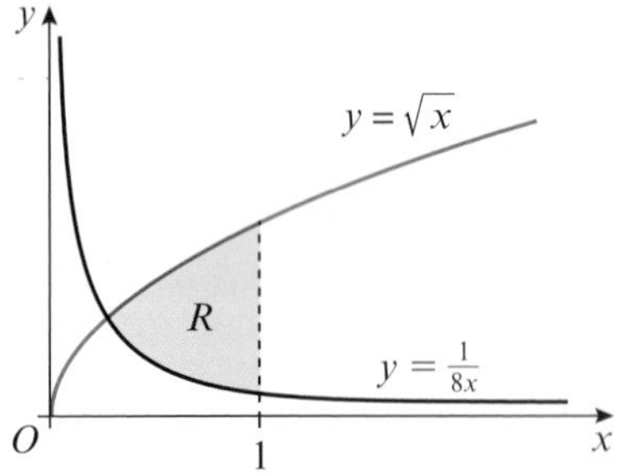

At point of intersection,

$\sqrt{x} = \frac{1}{8x}$

Solve the equations simultaneously to find the x-coordinate at the point of intersection of the two curves.

$x^{\frac{3}{2}} = \frac{1}{8}$

$x = \left(\frac{1}{8}\right)^{\frac{2}{3}} = \frac{1}{4}$

Consider volumes of revolution of R_1 and R_2 separately:

Problem-solving

Work out how you can create the necessary volume of revolution using simpler volumes of revolution. In this case, you can find the volume of revolution of the area under the $y = \sqrt{x}$ curve, then subtract the volume of revolution of the area under the $y = \frac{1}{8x}$ curve.

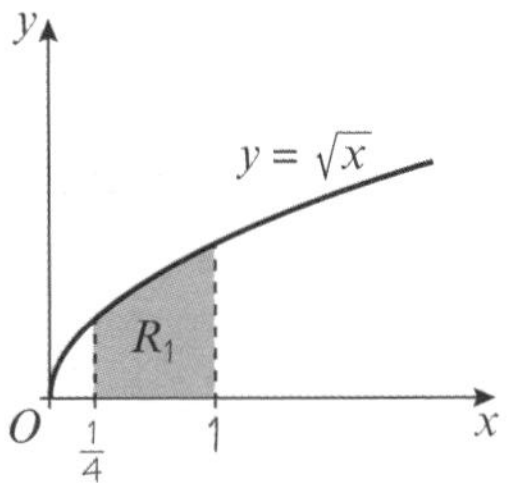

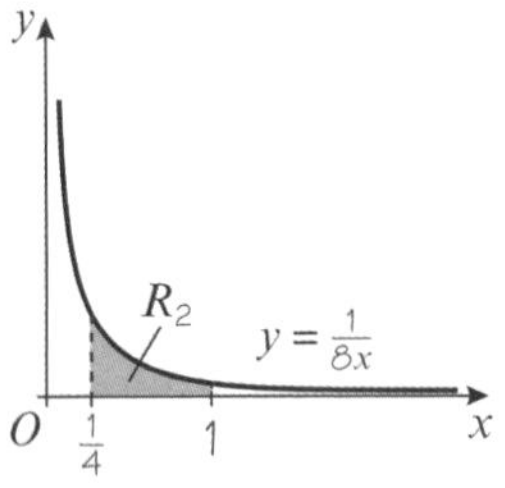

Volume of revolution of R_1:

$V_1 = \pi \int_{\frac{1}{4}}^{1} (\sqrt{x})^2 \, dx$

$= \pi \left[\frac{1}{2}x^2\right]_{\frac{1}{4}}^{1}$

$= \pi\left(\frac{1}{2} - \frac{1}{32}\right) = \frac{15\pi}{32}$

Volume of revolution of R_2:

$V_2 = \pi \int_{\frac{1}{4}}^{1} \left(\frac{1}{8x}\right)^2 dx$

$= \pi \int_{\frac{1}{4}}^{1} \frac{1}{64x^2} \, dx$

$\frac{1}{64x^2} = \frac{1}{64}x^{-2}$, so when you integrate this term becomes $\frac{\frac{1}{64}x^{-1}}{-1} = -\frac{1}{64x}$

$= \pi\left[-\frac{1}{64x}\right]_{\frac{1}{4}}^{1}$

$= \pi\left(-\frac{1}{64} - \left(-\frac{1}{16}\right)\right) = \frac{3\pi}{64}$

Volume of revolution of R:

$V = \frac{15\pi}{32} - \frac{3\pi}{64} = \frac{27\pi}{64}$

The safest way to find this volume is to work out each volume of revolution separately then subtract, as shown here. But you could also do this in one combined calculation:

$V = \pi \int_{\frac{1}{4}}^{1} (\sqrt{x})^2 \, dx - \pi \int_{\frac{1}{4}}^{1} \left(\frac{1}{8x}\right)^2 dx$

$= \pi \int_{\frac{1}{4}}^{1} \left((\sqrt{x})^2 - \left(\frac{1}{8x}\right)^2\right) dx = \pi\left[\frac{1}{2}x^2 + \frac{1}{64x}\right]_{\frac{1}{4}}^{1}$

$= \pi\left(\left(\frac{1}{2} + \frac{1}{64}\right) - \left(\frac{1}{32} + \frac{1}{16}\right)\right) = \frac{27\pi}{64}$

Exercise 5C

1 The diagram shows the line with equation $3x + 2y = 27$.

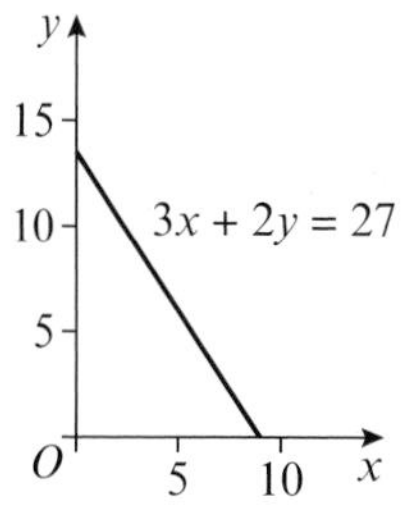

a Use integration to find the volume of revolution when the region is rotated through 360° about the x-axis.

b Use integration to find the volume of revolution when the region is rotated through 360° about the y-axis.

c Use the formula for the cone to check your answers to parts **a** and **b**. Clearly state the radius and the height in each case.

P 2 The region R is bounded by the curve with equation $y = \frac{1}{2}x^2(x + 2)$, the line $y = 16 - 4x$, and the x-axis.

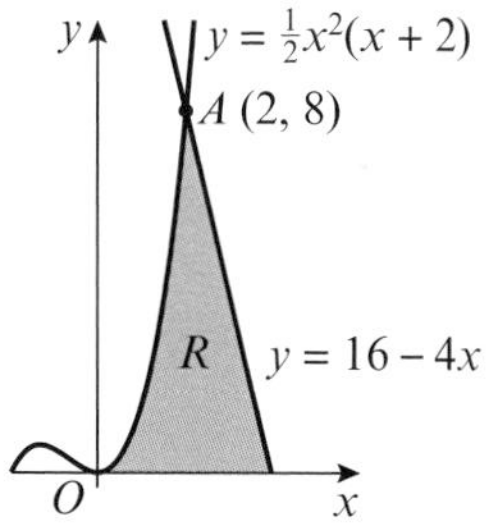

a Show that the coordinates of A are (2, 8). **(1 mark)**

A solid is created by rotating the region through 360° about the x-axis.

b Find the volume of this solid. **(6 marks)**

P 3 The region R is bounded by the curve with equation $y = -\frac{1}{2}x^2(x - 4)$ and the line $2x + y = 8$.

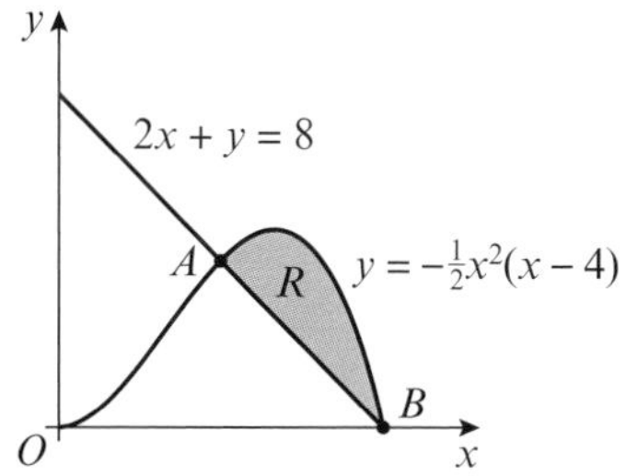

Problem-solving

You will need to find a volume of revolution then **subtract** the volume of the cone.

a Show that the coordinates of A are (2, 4) and write down the coordinates of B. **(1 mark)**

A solid is created by rotating the region through 360° about the x-axis.

b Find the volume of this solid. **(6 marks)**

P 4 The shape shown is bounded by the curve $y = \frac{1}{2}x^2$, and the lines $2x + y = 6$ and $2x - y = -6$.

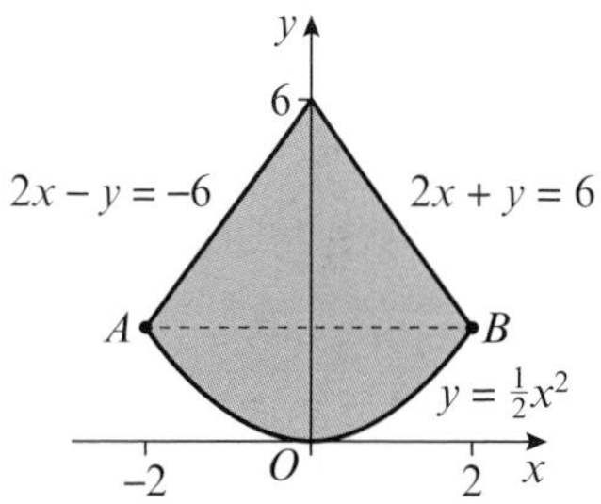

a Find the coordinates of the points A and B. **(2 marks)**

b The shape is rotated about the y-axis to generate a solid of revolution.

Find the volume of the solid generated. **(6 marks)**

E/P **5** The region R is bounded by the lines $x = -3$, $x = 3$ and $y = 3$, the curve C with equation $x^2 + y^2 = 4$ and the x-axis. The region is rotated about the y-axis to generate a solid of revolution. Find the volume of the solid generated.

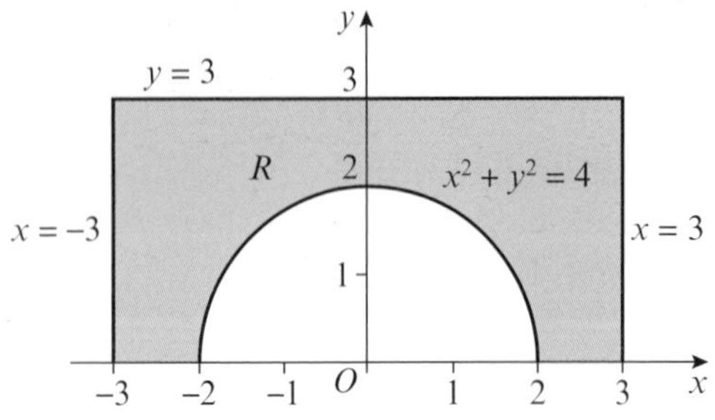

Problem-solving

Find the volume generated by rotating the curve $x^2 + y^2 = 4$ about the x-axis and subtract this from the volume of a suitable cylinder.

(6 marks)

E/P **6** The shaded region R is bounded by the curve $y = -\frac{1}{5}x^2 + 5$, the x-axis and the lines with equations $y = 4 - x$ and $y = 4 + x$.

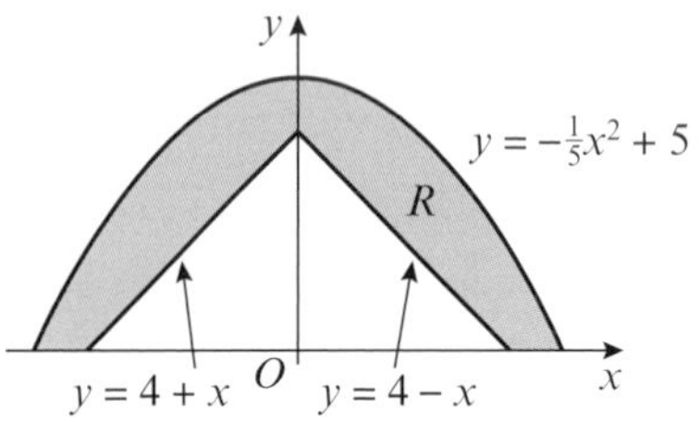

Find the volume of the solid of revolution generated when this region is rotated about the y-axis. **(8 marks)**

P **7** The shaded region is bounded by the curve C with equation $6y^2 - x^3 + 4x = 0$, $x > 0$, the straight lines $4x - 3y = 4$, $4x + 3y = 4$, and the line $x = 4$. The region is rotated about the x-axis to generate a solid of revolution. Find the exact volume of the solid generated.

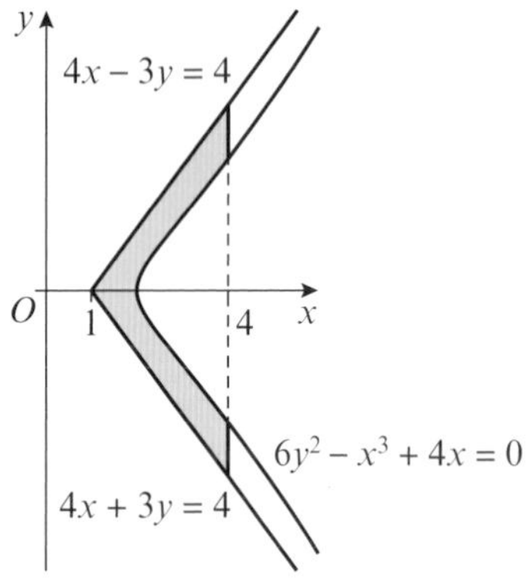

E/P **8** The shaded region is bounded by the curve with equation $y = 4 - x^2$, the curve with equation $y = \sqrt[3]{x}$, the y-axis and the line with equation $x = 1$.

The region is rotated through 360° about the x-axis. Find the exact volume of the solid generated. **(7 marks)**

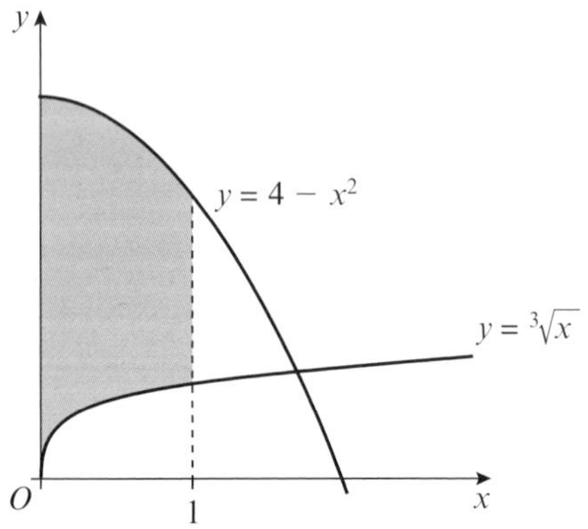

P 9 The diagram shows the region R bounded by the curve with equation $y = x^2 + 1$ and the curve with equation $x^2 + y^2 = 11$.

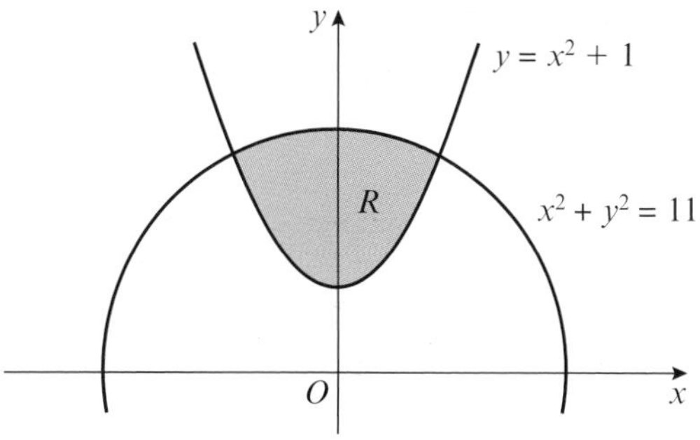

a Find the x-coordinates of the points of intersection of the two curves. **(3 marks)**

The region R is rotated through 360° about the x-axis.

b Find the volume of the solid generated, giving your answer correct to 2 decimal places. **(7 marks)**

Challenge

The shaded region shown in the diagram is bounded by the curve $y = \frac{64}{x^3}$, the line $y = \frac{40}{3} - \frac{8}{3}x$, the x-axis, and the line $y = 1$, for $0 \leqslant x \leqslant 4$. The region is rotated about the y-axis. Find an exact value for the volume of the solid generated.

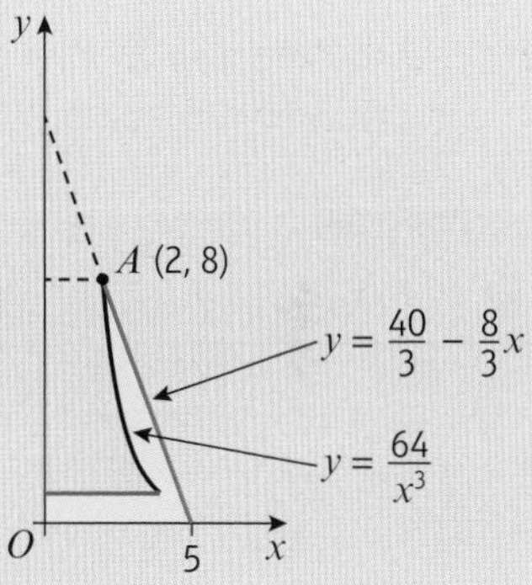

5.4 Modelling with volumes of revolution

Volumes of revolution can be used to model real-life situations.

Example 5

A manufacturer wants to cast a prototype for a new design for a pen barrel out of solid resin. The shaded region shown in the diagram is used as a model for the cross-section of the pen barrel. The region is bounded by the x-axis and the curve with equation $y = k - 100x^2$, and will be rotated around the y-axis. Each unit on the coordinate axes represents 1 cm.

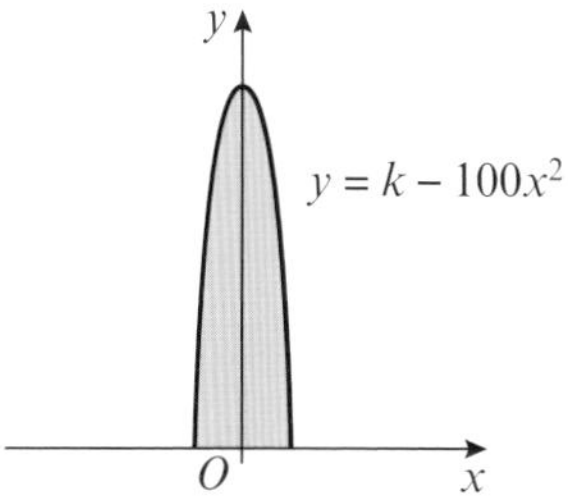

a Suggest a suitable value for k.

b Use your value of k to estimate the volume of resin needed to make the prototype.

c State one limitation of this model.

a $k = 10.$

b $y = 10 - 100x^2$

$100x^2 = 10 - y$

$x^2 = \frac{1}{10} - \frac{y}{100}$

$V = \pi\int_0^{10}\left(\frac{1}{10} - \frac{y}{100}\right)dy$

$= \pi\left[\frac{y}{10} - \frac{y^2}{200}\right]_0^{10}$

$= \pi\left(\left(\frac{10}{10} - \frac{10^2}{200}\right) - \left(\frac{0}{10} - \frac{0^2}{200}\right)\right)$

$= \pi\left(1 - \frac{1}{2}\right) = \frac{\pi}{2}$

Approximately 1.57 cm^3 of resin will be needed.

c The cross-section of the pen is unlikely to match the curve exactly.

Consider the context of the question and choose a value that makes sense. Most pens are between 10 cm and 15 cm long so any value in the range $10 \leqslant k \leqslant 15$ is sensible.

Use your value of k from part **a**.

Rearrange the expression to make x^2 the subject.

Use $V = \pi\int_a^b x^2\,dy$ with $a = 0$, $b = 10$ and $x^2 = \frac{1}{10} - \frac{y}{100}$

Give units with your answer.

Problem-solving

You can give any sensible answer that refers to the context of the question. You could also say that some resin might be wasted when the pen is made, or there might be air bubbles in the mould.

Exercise 5D

E/P **1** The diagram shows the shape of a large tent at a fair. The outside of the tent can be modelled by the equation $y^2 = -0.01x^2 + k^2$. Each unit on the coordinate axes represents 1 metre.

a Suggest a suitable value for k. **(1 mark)**

b Use your value of k to estimate the capacity of the tent. **(5 marks)**

c State one limitation of this model. **(1 mark)**

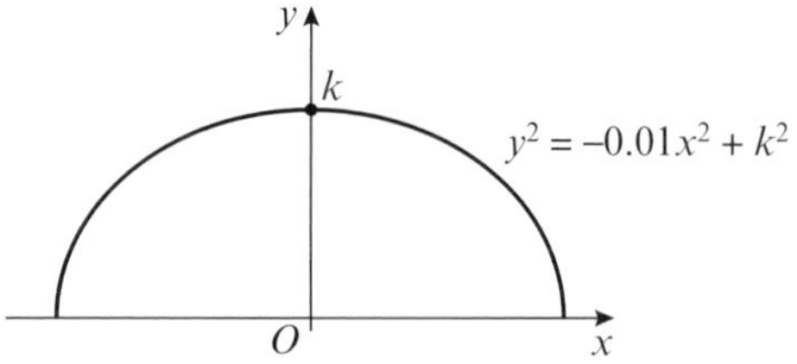

E **2** The diagram shows half of the outline of a rugby ball. The outline is modelled by the curve $y^2 = 4(16 - x)$.
The measurements shown are given in centimetres. By rotating the curve through 360° around the y-axis, find the total volume of the rugby ball. **(5 marks)**

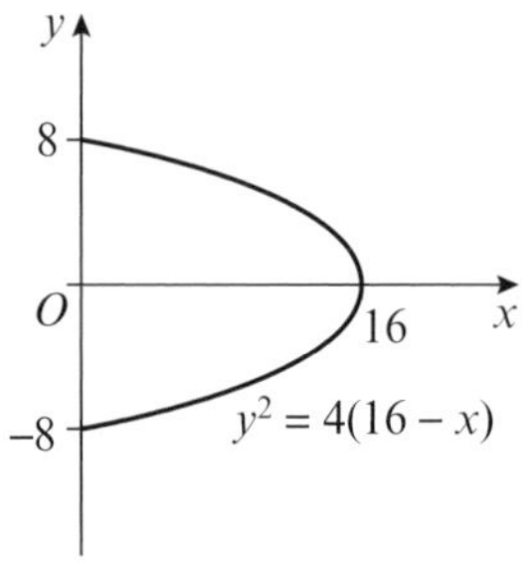

3 The cross-section of an egg can be modelled as an ellipse with equation $\frac{x^2}{9} + \frac{y^2}{4} = 1$, where the dimensions shown are in centimetres.

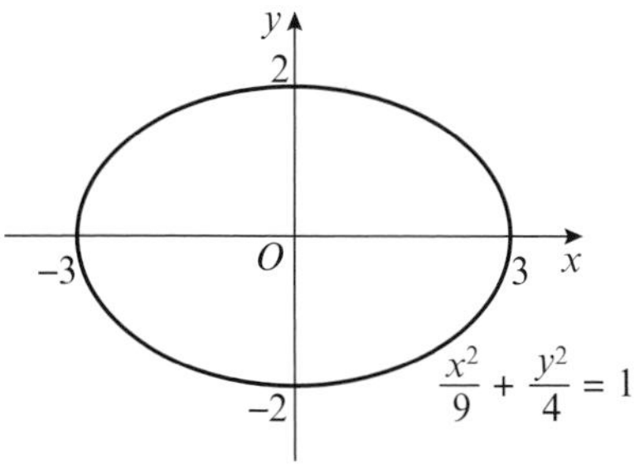

a Calculate the volume of the solid formed by rotating this curve through 360° about the x-axis.

b Show that the solid formed by rotating the curve through 360° about the y-axis has the same volume.

c Say which of these two solids most resembles an egg.

4 The diagram shows the cross-section of an egg timer, which has a height of 16 cm. The shape of the egg timer is modelled as a solid of revolution of a curve C about the y-axis. The curve C has equation $x = \sqrt[3]{y}$.

Sand flows through the egg timer at a rate of 8 cm³/min. The designer wants the egg timer to empty in 5 minutes. Calculate, to 2 decimal places, the height of sand that should be placed in the top half of the egg timer. **(5 marks)**

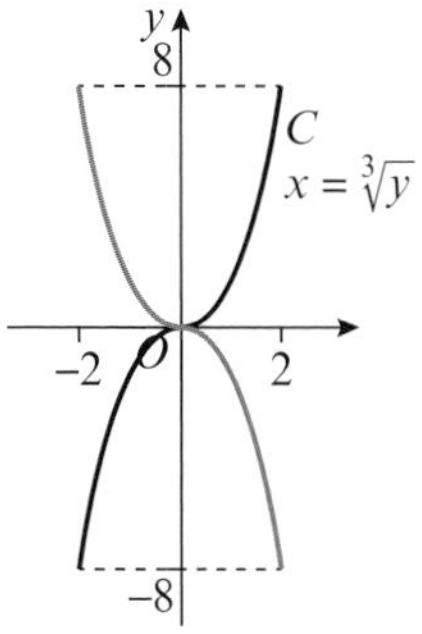

5 The diagram shows the bowl of an electric stand mixer. The height of the bowl is 18 cm. The shape of the bowl is modelled by rotating the curve with equation $y = 0.02x^3$ through 2π radians about the y-axis.

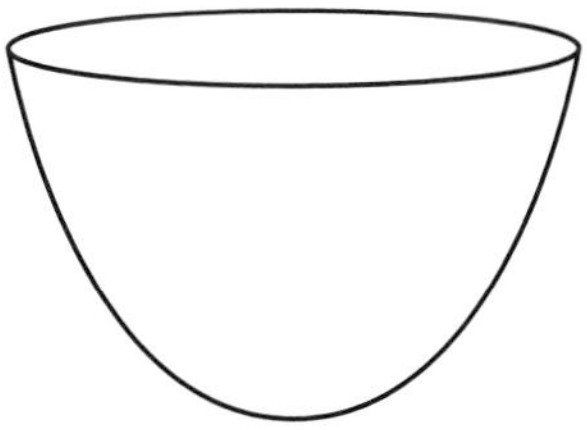

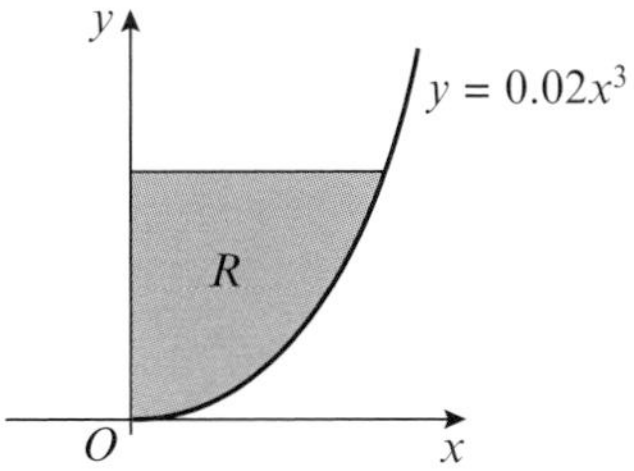

a Find the diameter of the bowl. **(2 marks)**

b Find the maximum volume of liquid that can be contained within the mixing bowl. **(4 marks)**

The mixing bowl has a paddle of height 12 cm. The paddle just touches the side of the bowl. In its starting position, the paddle forms a region R, as shown in the diagram.

c Calculate the area of the paddle. **(3 marks)**

The paddle rotates about the y-axis when the mixer is in operation.

d Find the proportion of the total volume contained within the bowl that can be mixed by the paddle. **(4 marks)**

(E/P) **6** The diagram shows a vase with a base width of 10 cm and a height of 20 cm. The edge of the vase is modelled by the equation $x = 5 - \sqrt{y}$. The vase is formed by rotating the shape through 360° about the y-axis.

a Use this model to estimate the capacity of the vase. **(5 marks)**

The vase is initially filled to a height of 10 cm. When the flowers are placed in the vase, 50 cm^3 of water is displaced.

b Determine whether the vase will overflow. **(3 marks)**

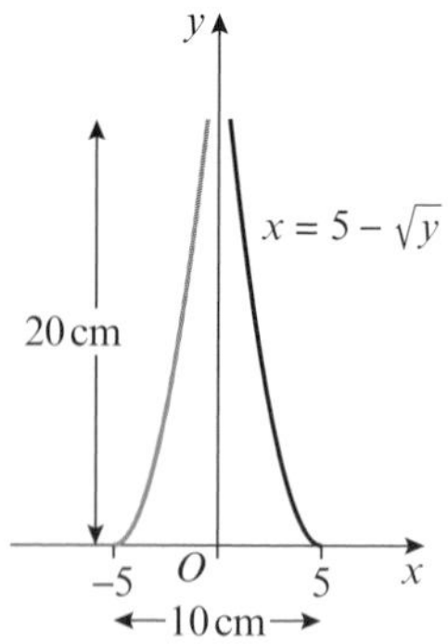

(E/P) **7** A circular spinning top is made of solid wood with a width of 18 cm and a height of 24 cm. A cross-section of the spinning top is shown in the diagram. The cross-section is formed by part of the curve with equation $y^2 = 4(x + 9)$ and the straight line with equation $y = 2x + 18$, and is symmetrical about the y-axis. The cross-section is rotated about the y-axis. Find the total volume of wood in the spinning top. **(7 marks)**

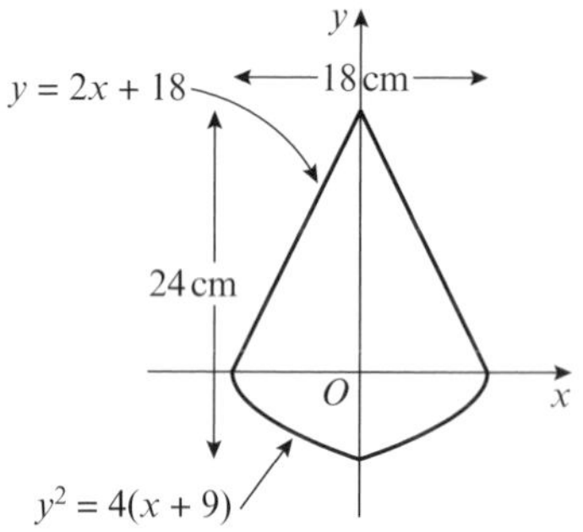

Challenge

The diagram shows the cross-section of a circular place-holder used to hold a rugby ball when penalties and conversions are kicked. The cross-section can be modelled by the straight lines with equations $3y - 4x = 24$ and $3y + 4x = 24$, and the curve with equation $y = \frac{1}{9}x^2 + 3$. The place-holder is formed by rotating this cross-section about the y-axis, and is constructed from solid plastic.

Find the volume of plastic needed to construct the place-holder.

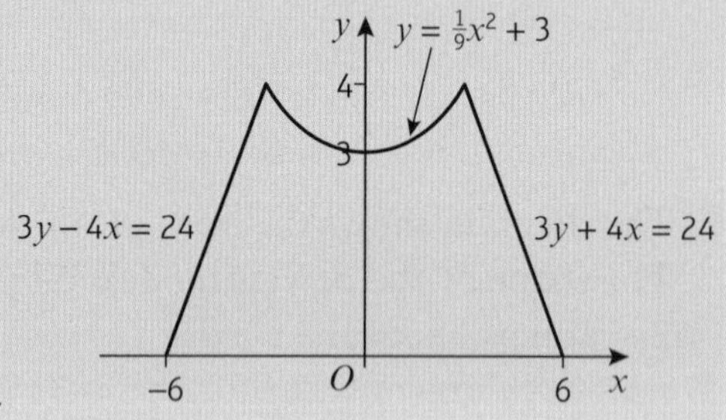

Mixed exercise 5

(E) **1** The curve shown in the diagram has equation $y = x^2\sqrt{9 - x^2}$. The finite region R is bounded by the curve and the x-axis. The region is rotated through 2π radians about the x-axis to generate a solid of revolution. Find the exact value of the volume of the solid that is generated. **(5 marks)**

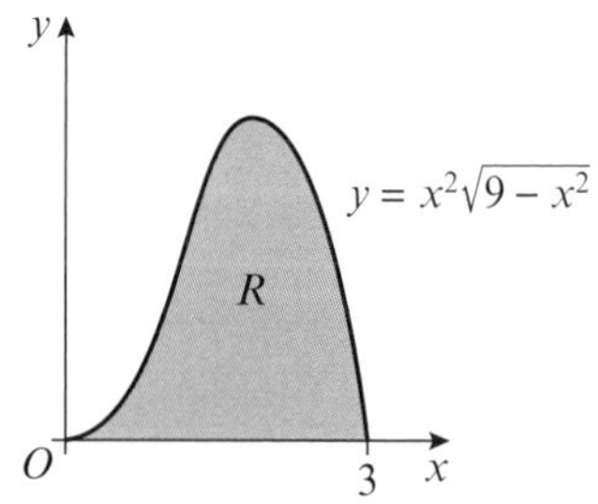

(P) **2** The diagram shows the curve with equation $2y^2 - 6\sqrt{x} + 3 = 0$. The shaded region is bounded by the curve and the line $x = 4$.

a Find the value of x at the point where the curve cuts the x-axis.

The region is rotated about the x-axis to generate a solid of revolution.

b Find the volume of the solid generated.

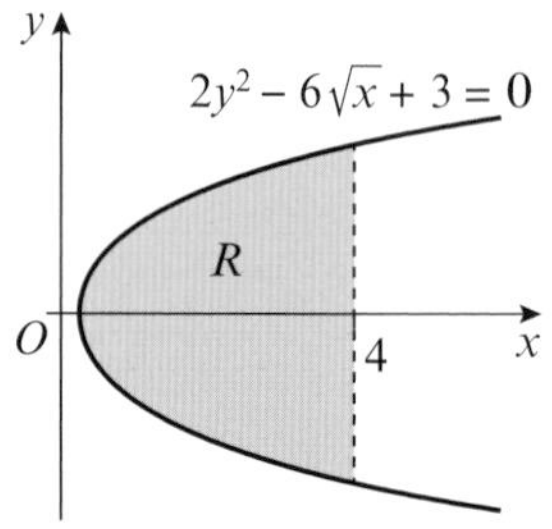

E **3** $f(x) = x^2 + 4x + 4, x \geqslant -2$

The diagram shows the finite region R bounded by the curve $y = f(x)$, the y-axis and the lines $y = 4$ and $y = 9$.

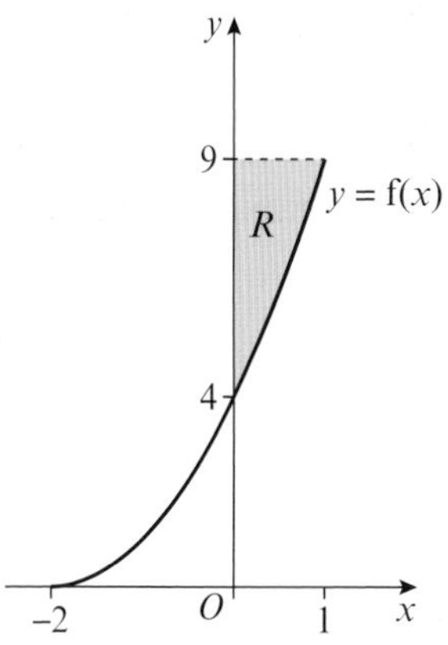

a Show that the equation $y = f(x)$ can be written as $x^2 = 4 - 4\sqrt{y} + y$. **(2 marks)**

b The region R is rotated through 2π radians about the y-axis. Find the exact volume of the solid generated. **(5 marks)**

E **4** The diagram shows the shaded region bounded by the curve with equation $y = x^2 + 3$, the line $x = 1$, the x-axis and the y-axis. Find the volume generated when the region is rotated through 2π radians:

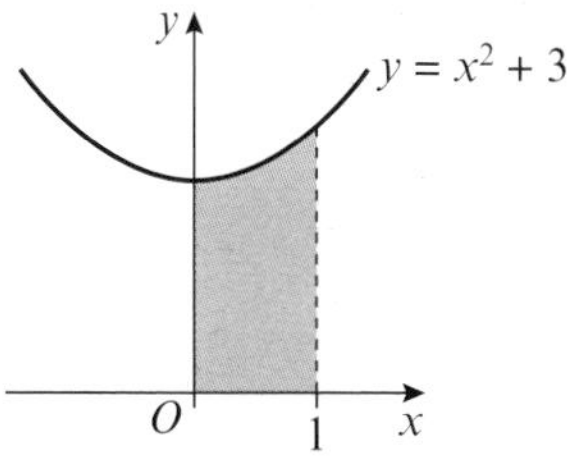

a about the x-axis **(3 marks)**

b about the y-axis. **(4 marks)**

E/P **5** The diagram shows the curve with equation $y = \frac{1}{4}x(x + 1)^2$ and the line with equation $3x + 4y = 24$. The line and the curve intersect at the point A.

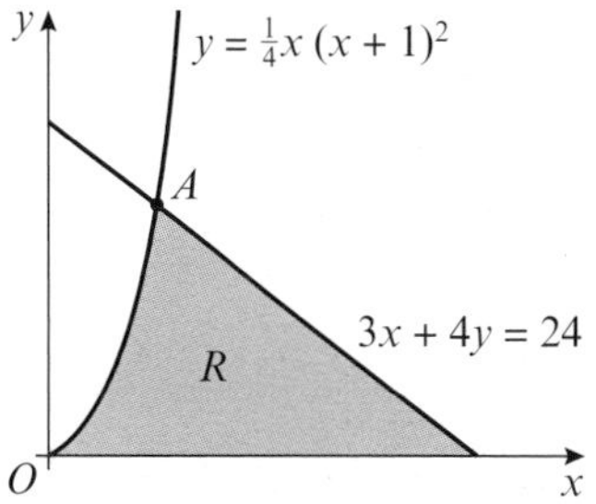

a Show that the coordinates of the point A are (2, 4.5). **(2 marks)**

The shaded region R is bounded by the curve, the line and the x-axis. The region is rotated through 2π radians about the x-axis.

b Find the exact volume of solid generated. **(6 marks)**

E/P **6** The diagram shows a cross-section of a circular golf ball trophy holder. The dimensions shown on the diagram are in centimetres. The cross-section of the trophy is formed by the lines $x = -2$, $x = 2$, the x-axis and the curve with equation $y = 0.1x^2 + 4$. The cross-section is rotated 360° about the y-axis. The trophy holder is to be cast out of solid bronze.

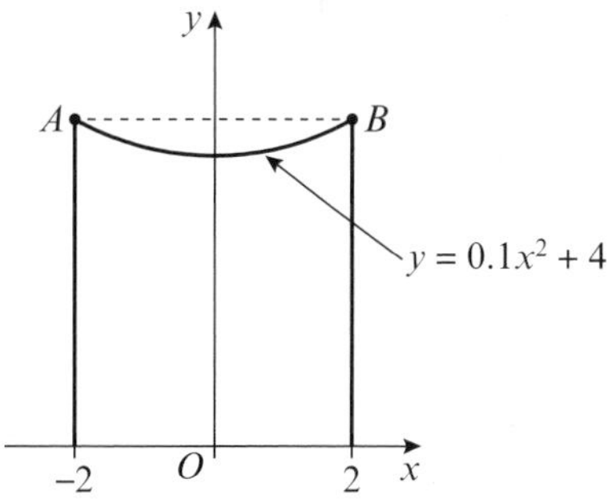

a Use this model to find the volume of bronze needed to make the trophy. **(5 marks)**

b Give one limitation of this model. **(1 mark)**

E/P **7** The diagram shows the outline of a circular mushroom. The dimensions on the diagram are in centimetres. The cap of the mushroom is modelled by the curve with equation $\frac{1}{4}x^2 - 8\sqrt{y} + 4y = 0$. The mushroom is formed by rotating the shape shown about the y-axis. Find the exact volume of the mushroom. **(6 marks)**

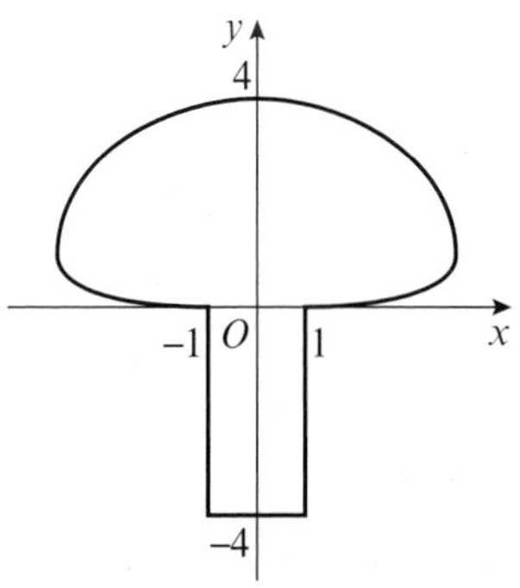

E/P **8** The shaded region is bounded by the curves with equations $y = 2x^2$ and $3y^2 + x^2 - 11y = 0$.

The shaded region is rotated 360° about the y-axis. Find the exact volume of the solid of revolution generated. **(9 marks)**

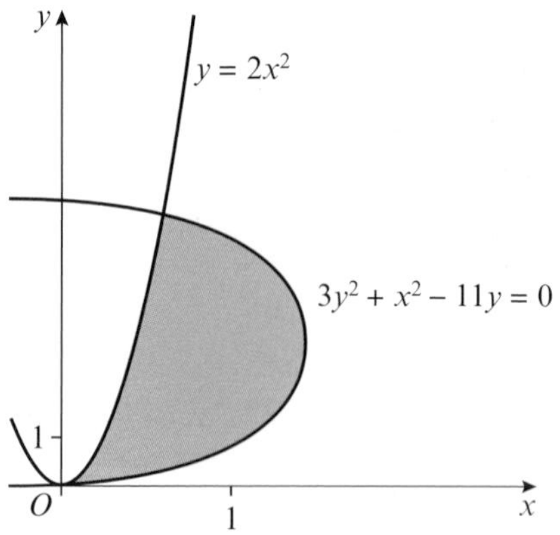

Challenge

The diagram shows a sphere of radius r and centre (0, 0).

a Show that the area of the shaded disc is $\pi(r^2 - x^2)$.

b By considering an integral over an appropriate interval, show that the volume of a sphere is $\frac{4}{3}\pi r^3$.

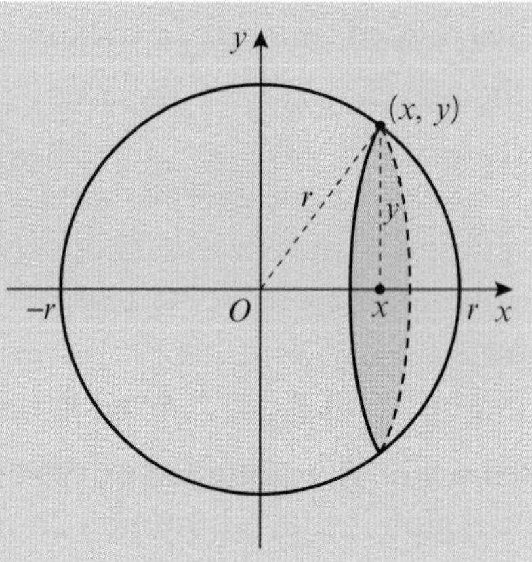

Summary of key points

1 The volume of revolution formed when $y = f(x)$ is rotated about the x-axis between $x = a$ and $x = b$ is given by

$$\textbf{Volume} = \pi \int_a^b y^2 \, dx$$

2 The volume of revolution formed when $x = f(y)$ is rotated about the y-axis between $y = a$ and $y = b$ is given by

$$\textbf{Volume} = \pi \int_a^b x^2 \, dy$$

3 A cylinder of height h and radius r has volume $\pi r^2 h$.

4 A cone of height h and base radius r has volume $\frac{1}{3}\pi r^2 h$.

Review exercise 1

(E) **1** $z_1 = 4 - 5\text{i}$ and $z_2 = p\text{i}$, where p is a real constant. Find the following, in the form $a + b\text{i}$, giving a and b in terms of p:

a $z_1 - z_2$ **(1)**

b $z_1 z_2$ **(1)**

c $\frac{z_1}{z_2}$ **(1)**

← Sections 1.1, 1.2, 1.3

(E/P) **2** $f(z) = z^3 - kz^2 + 3z$ has two imaginary roots.

a Find the range of possible values of k. **(3)**

b Given that $k = 2$, solve the equation $f(z) = 0$. **(3)**

← Section 1.1

(E) **3** The solutions to the quadratic equation $z^2 - 5z + 13 = 0$ are z_1 and z_2. Find z_1 and z_2, giving each answer in the form $a \pm ib$ where $a, b \in \mathbb{R}$. **(3)**

← Section 1.1

(E) **4** The real and imaginary parts of the complex number $z = x + \text{i}y$ satisfy the equation $(2 - \text{i})x - (1 + 3\text{i})y - 7 = 0$. Find the values of x and y. **(4)**

← Section 1.1

(E/P) **5 a** Show that the complex number $\frac{2 + 3\text{i}}{5 + \text{i}}$ can be expressed in the form $\lambda(1 + \text{i})$, stating the value of λ. **(3)**

b Hence show that $\left(\frac{2 + 3\text{i}}{5 + \text{i}}\right)^4$ is real and determine its value. **(2)**

← Sections 1.2, 1.3

(E/P) **6** $f(z) = z^3 + 5z^2 + 8z + 6$
Given that $-1 + \text{i}$ is a root of the equation $f(z) = 0$, solve $f(z) = 0$ completely. **(4)**

← Section 1.5

(E/P) **7** $f(z) = z^3 - 6z^2 + kz - 26$
Given that $f(2 - 3\text{i}) = 0$,

a find the value of k **(2)**

b find the other two roots of the equation $f(z) = 0$. **(3)**

← Section 1.5

(E) **8** $f(z) = z^4 - z^3 - 6z^2 - 20z - 16$

a Write $f(z)$ in the form
$(z^2 - 3z - 4)(z^2 + \text{b}z + \text{c})$
where b and c are real constants to be found. **(2)**

b Hence find all the solutions to the equation $f(z) = 0$. **(3)**

← Section 1.5

(E/P) **9** $g(z) = z^4 - 8z^3 + 27z^2 - 50z + 50$
Given that $g(1 - 2\text{i}) = 0$, find all the roots of the equation $g(z) = 0$. **(5)**

← Section 1.5

(E/P) **10** $f(z) = z^3 + pz^2 + qz - 12$ where p and q are real constants.
Given that α, $\frac{4}{\alpha}$ and $\alpha + \frac{4}{\alpha} + 1$ are the roots of the equation $f(z) = 0$,

a solve completely the equation $f(z) = 0$. **(5)**

b Hence find the values of p and q. **(3)**

← Sections 1.4, 4.2

(E) **11** **a** Find, in the form $p + iq$ where p and q are real, the complex number z which satisfies the equation $\frac{3z - 1}{2 - i} = \frac{4}{1 + 2i}$ **(4)**

b Show on a single Argand diagram the points which represent z and z^*. **(2)**

c Express z and z^* in modulus–argument form, giving the arguments to the nearest degree. **(3)**

← Sections 1.2, 2.1, 2.3

(E/P) **12** Given that the complex number $z = x + iy$ satisfies the equation $|z - 4i| = 1$, find the maximum and minimum possible values of $\arg z$. **(4)**

← Section 2.4

(E) **13** The complex number z is $-9 + 17i$.

a Show z on an Argand diagram. **(1)**

b Calculate $\arg z$, giving your answer in radians to two decimal places. **(2)**

c Find the complex number w for which $zw = 25 + 35i$, giving your answer in the form $p + iq$, where p and q are real. **(3)**

← Sections 1.3, 2.1, 2.2

(E) **14** $z_1 = 5 + i$, $z_2 = -2 + 3i$

a Show that $|z_1|^2 = 2|z_2|^2$. **(3)**

b Find $\arg(z_1 z_2)$. **(3)**

← Section 2.2

(E) **15** **a** Given that $z = 2 - i$, show that

$$z^2 = 3 - 4i$$ **(2)**

b Hence, or otherwise, find the roots, z_1 and z_2, of the equation

$$(z + i)^2 = 3 - 4i$$ **(3)**

c Show points representing z_1 and z_2 on a single Argand diagram. **(2)**

d Deduce that $|z_1 - z_2| = 2\sqrt{5}$. **(2)**

e Find the value of $\arg(z_1 + z_2)$. **(2)**

← Sections 1.2, 2.1, 2.2

(E) **16** The complex numbers $z_1 = 2 + 2i$ and $z_2 = 1 + 3i$ are represented on an Argand diagram by the points P and Q respectively.

a Display z_1 and z_2 on the same Argand diagram. **(2)**

b Find the exact values of $|z_1|$, $|z_2|$ and the length of PQ. **(3)**

c Hence show that $\triangle OPQ$, where O is the origin, is right angled. **(2)**

d Given that $OPQR$ is a rectangle in the Argand diagram, find the complex number z_3 represented by the point R. **(3)**

← Sections 2.1, 2.2

(E/P) **17** Show that

$$\frac{\cos 2x + i\sin 2x}{\cos 9x - i\sin 9x}$$

can be expressed in the form $\cos nx + i\sin nx$, where n is an integer to be found. **(4)**

← Section 2.3

(E/P) **18** The point P represents the complex number z in an Argand diagram.
Given that $|z - 2 + i| = 3$,

a sketch the locus of P in an Argand diagram **(2)**

b find the exact values of the maximum and minimum of $|z|$. **(2)**

← Section 2.4

(E/P) **19** Given that z satisfies $|z - 2i| = 2$,

a sketch the locus of z on an Argand diagram **(2)**

b find the maximum value of $|z|$. **(2)**

← Section 2.4

(P) **20** A complex number z is represented by the point P in an Argand diagram.

Given that $|z - 3\text{i}| = 3$,

a sketch the locus of P **(2)**

b find the complex number z which satisfies both $|z - 3\text{i}| = 3$ and $\arg(z - 3\text{i}) = \frac{3\pi}{4}$ **(3)**

← Section 2.4

(E) **21** Sketch, on an Argand diagram, the locus of the point P representing a complex number z such that

$\arg(z + 3 + \text{i}) = \frac{\pi}{2}$ **(3)**

← Section 2.4

(P) **22** The complex number z satisfies the equation $|z + 3 + \text{i}| = |z - 2 + \text{i}|$.

a Sketch the locus of z. **(2)**

b Find the minimum value of $|z|$. **(1)**

c Find a value of z that also satisfies

$\arg z = -\frac{3\pi}{4}$ **(2)**

← Section 2.4

(E) **23** Sketch, on an Argand diagram, the region which satisfies the following condition.

$\frac{\pi}{4} \leqslant \arg(z - 1) \leqslant \frac{2\pi}{3}$ **(3)**

← Section 2.5

(E/P) **24** Shade on an Argand diagram the set of points

$\left\{z \in \mathbb{C}: -\frac{\pi}{2} < \arg(z - 3 - 3\text{i}) \leqslant \frac{3\pi}{4}\right\}$

$\cap \{z \in \mathbb{C} : |z - 3\text{i}| \leqslant 3\}$ **(6)**

← Section 2.5

(E) **25** Use standard formulae to show that

$\sum_{r=1}^{n} (2r - 1)^2 = \frac{1}{3}n(4n^2 - 1)$ **(4)**

← Section 3.2

(E) **26** Use standard formulae to show that

$\sum_{r=1}^{n} r(r^2 - 3) = \frac{1}{4}n(n + 1)(n - 2)(n + 3)$ **(4)**

← Section 3.2

(E) **27** **a** Use standard formulae to show that

$\sum_{r=1}^{n} r(2r - 1) = \frac{n(n + 1)(4n - 1)}{6}$ **(4)**

b Hence, evaluate $\sum_{r=11}^{30} r(2r - 1)$ **(2)**

← Section 3.2

(E) **28** **a** Use standard formulae to show that

$\sum_{r=1}^{n} (6r^2 + 4r - 5) = n(2n^2 + 5n - 2)$ **(4)**

b Hence calculate the value of

$\sum_{r=10}^{25} (6r^2 + 4r - 5)$ **(2)**

← Section 3.2

(E) **29** **a** Use standard formulae to show that

$\sum_{r=1}^{n} r(r + 1) = \frac{1}{3}n(n + 1)(n + 2)$ **(4)**

b Hence, or otherwise, show that

$\sum_{r=n}^{3n} r(r + 1) = \frac{1}{3}n(2n + 1)(pn + q)$, stating

the values of the integers p and q. **(3)**

← Section 3.2

(E) **30** Given that

$\sum_{r=1}^{n} r^2(r - 1) = \frac{1}{12}n(n + 1)(pn^2 + qn + r)$

a find the values of p, q and r. **(4)**

b Hence evaluate $\sum_{r=50}^{100} r^2(r - 1)$ **(2)**

← Section 3.2

(E) **31** The roots of the equation $3x^3 + kx + 11 = 0$ are α, β and γ.

a Given that $\alpha\beta + \beta\gamma + \gamma\alpha = -4$, write down the value of k. **(1)**

b Write down values of $\alpha + \beta + \gamma$ and $\alpha\beta\gamma$. **(1)**

c Hence find the value of

$(1 - \alpha)(1 - \beta)(1 - \gamma)$ **(3)**

← Sections 4.2, 4.4

(E) **32** The roots of the equation
$$ax^4 + 7x^3 + 5x^2 + 3x - 4 = 0$$
are α, β, γ and δ.

a Given that $\alpha\beta\gamma\delta = -1$, write down the value of a. **(1)**

b Write down the values of $\Sigma\alpha$, $\Sigma\alpha\beta$ and $\Sigma\alpha\beta\gamma$. **(1)**

c Hence find the value of
$$\alpha^2 + \beta^2 + \gamma^2 + \delta^2$$ **(3)**

← Sections 4.3, 4.4

(E/P) **33** The cubic equation $x^3 + 3x^2 + 5x - 1 = 0$ has roots α, β and γ. Without solving the equation, find the equation with roots $(2\alpha + 1)$, $(2\beta + 1)$ and $(2\gamma + 1)$.
Give your answer in the form $pw^3 + qw^2 + rw + s = 0$ where p, q, r and s are integers to be found. **(5)**

← Section 4.5

(E/P) **34** The quartic equation $x^4 - x^3 - 2x^2 + 3x + 4 = 0$ has roots α, β, γ and δ. Without solving the equation, find equations with integer coefficients that have roots:

a 3α, 3β, 3γ and 3δ **(4)**

b $(2\alpha - 1)$, $(2\beta - 1)$, $(2\gamma - 1)$ and $(2\delta - 1)$ **(6)**

← Section 4.5

(E) **35** The curve shown in the diagram is $y = x\sqrt{1 - x^2}$

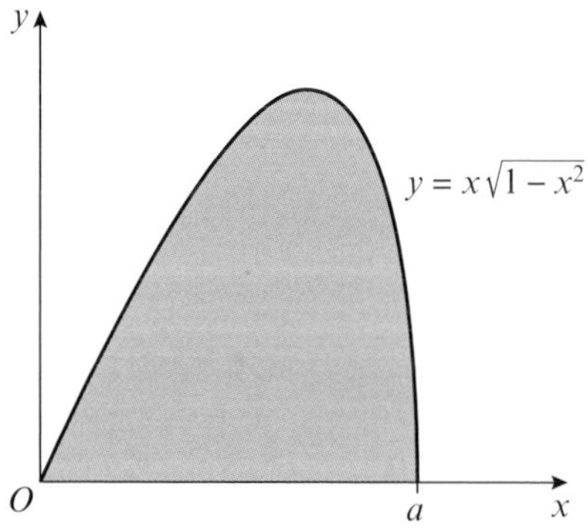

a Write down the value of a. **(1)**

The finite shaded region bounded by the curve and the x-axis is rotated through 2π radians about the x-axis.

b Find the exact volume of the solid generated. **(5)**

← Section 5.1

(E/P) **36** The curve shown in the diagram is $y = \sqrt{x^2 + 3}$

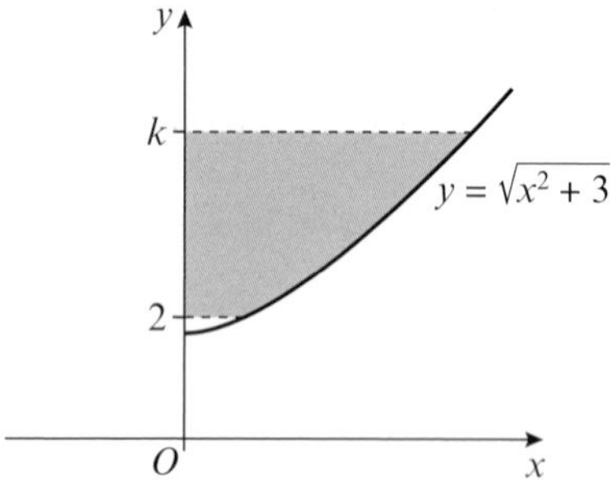

The finite region between the curve, the y-axis and the lines $y = 2$ and $y = k$ is rotated through 2π radians about the y-axis. Given that the volume of the solid generated is 48π, find the value of k. **(5)**

← Section 5.2

(E/P) **37** The graph shows the curve $y = 4 - x^2$ and the line $y = 2x + 1$.

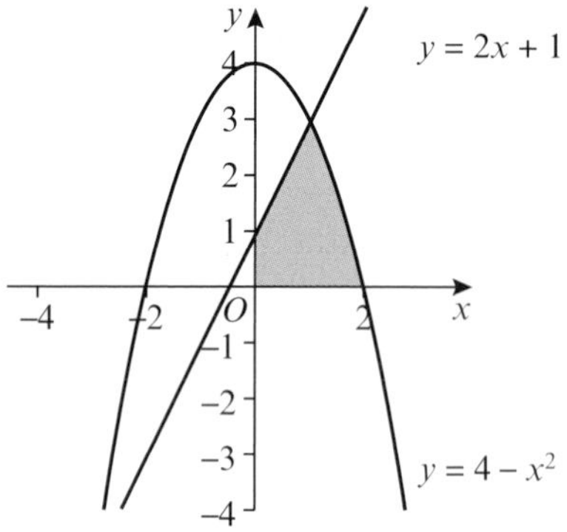

The region indicated is rotated through 2π radians about the x-axis. Find, correct to three significant figures, the volume of the solid generated. **(5)**

← Section 5.3

E/P **38** Nellie is a champion bowler and decides to make a stand for her favourite ball to rest on. Viewed from above, the stand will be circular. She models the cross-section of the stand using a curve and two lines as shown below where the dimensions are in centimetres.

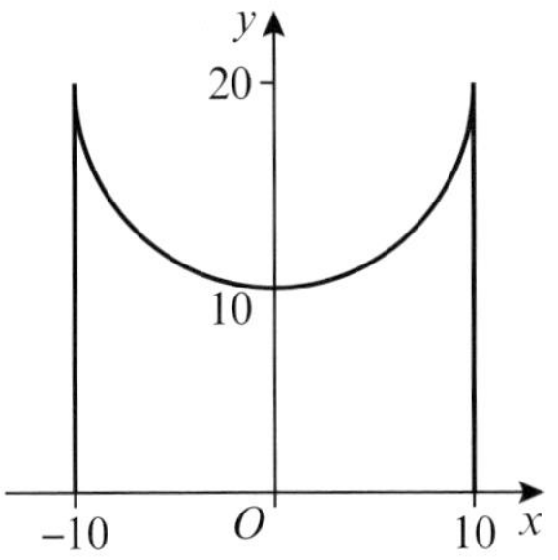

The curved section is modelled using the equation $x^2 + (y - k)^2 = 100$.

a Write down the value of k. **(1)**

b Show that the volume of revolution formed by rotating the curve about the y-axis between the lines $y = a$ and $y = b$ can be written as

$$\frac{\pi}{3}(60(b^2 - a^2) - (b^3 - a^3) - 900(b - a))$$

where $10 \leqslant a < b \leqslant 20$. **(5)**

The stand is made from a resin which costs £0.025 per cm³.

c Find, to the nearest penny, the cost of Nellie's stand. **(2)**

← Section 5.4

E/P **39** The diagram shows parts of the curves with equations $y = 12 - x^2$ and $y = 8 - 0.2x^2$.

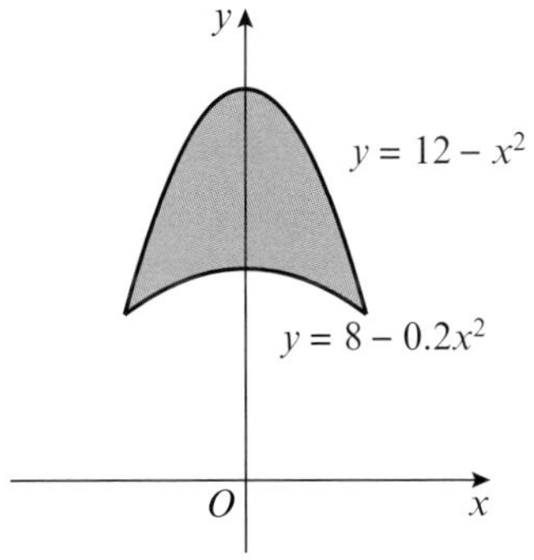

A jeweller models a gold ring as the volume of revolution formed when the area bounded by these two curves is rotated through 360° about the x-axis.

a Given that the dimensions on the diagram are in mm, state the maximum outer diameter of the ring. **(1)**

The density of gold is 19.3 g/cm³.

b Find the mass of the ring according to this model, giving your answer in grams to 1 decimal place. **(10)**

c Give one reason why the actual mass of the ring is likely to be different from your answer to part **b.** **(1)**

← Sections 5.3, 5.4

Challenge

1 In the Argand diagram the point P represents the complex number z.

Given that $\arg\left(\frac{z - 8}{z - 2}\right) = \frac{\pi}{2}$,

a sketch the locus of P

b deduce the value of $|z - 5|$. ← Chapter 2

2 The rth term of a finite series is denoted by u_r

Given that $\sum_{r=1}^{n} u_r = n^2 + 5n$,

a express u_r in terms of r

b show that $\sum_{r=n}^{2n} u_r = (n + 1)(3n + 4)$.

← Chapter 3

3 The cubic equation $x^3 - 5x^2 + 11x - 15 = 0$ has roots α, β and γ. By using a substitution, or otherwise, find an equation that has roots $\alpha^2 + 1$, $\beta^2 + 1$, and $\gamma^2 + 1$. ← Chapter 4

Matrices

6

Objectives

After completing this chapter you should be able to:

- Understand the concept of a matrix → pages 95–99
- Define the zero and identity matrices → pages 95–99
- Add and subtract matrices → pages 95–99
- Multiply a matrix by a scalar → pages 96–99
- Multiply matrices → pages 99–103
- Calculate the determinant of a matrix → pages 104–108
- Find the inverse of a matrix → pages 108–116
- Use matrices to solve systems of equations → pages 116–121
- Interpret simultaneous equations geometrically → pages 118–121

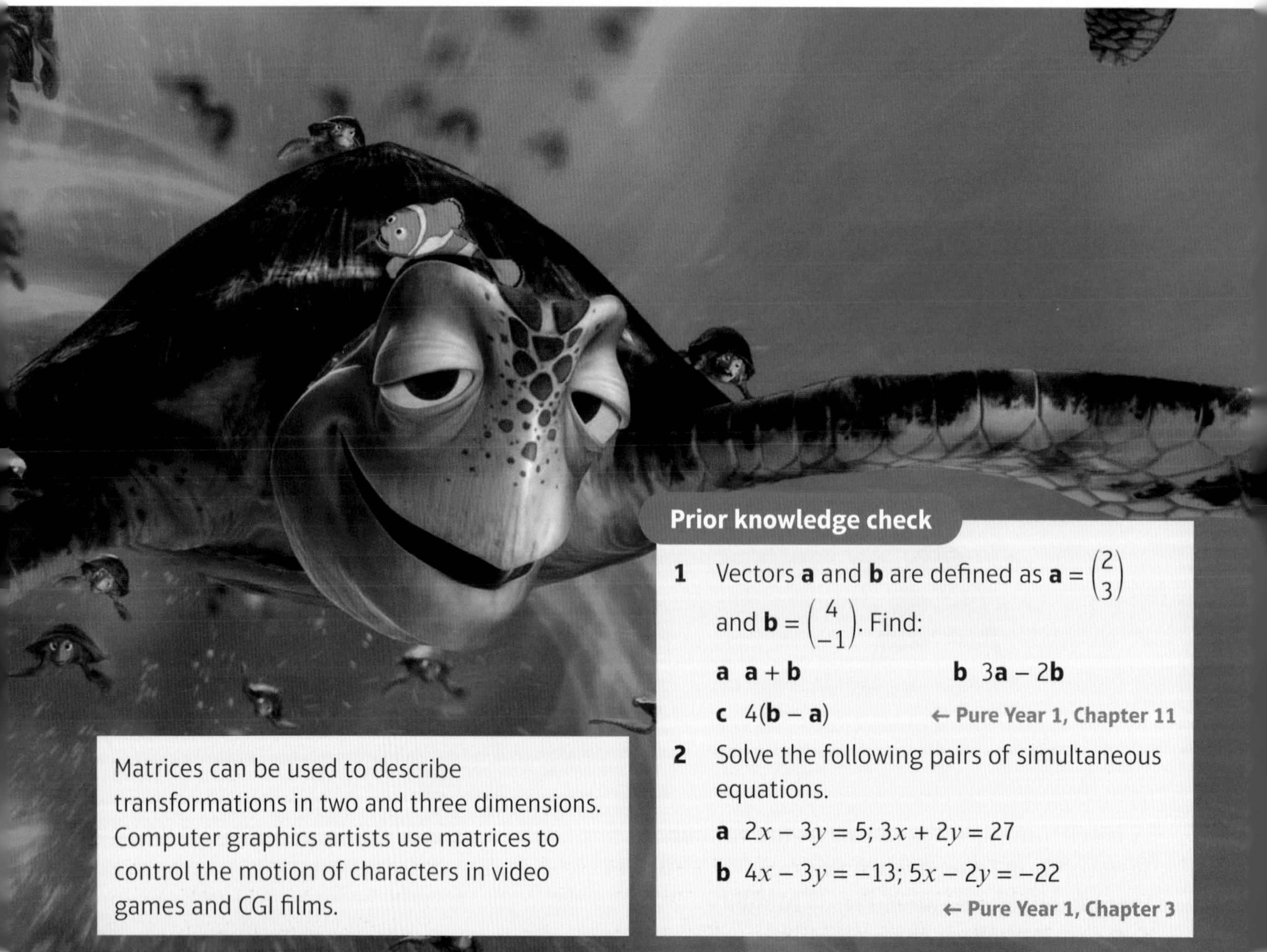

Matrices can be used to describe transformations in two and three dimensions. Computer graphics artists use matrices to control the motion of characters in video games and CGI films.

Prior knowledge check

1 Vectors **a** and **b** are defined as $\mathbf{a} = \begin{pmatrix} 2 \\ 3 \end{pmatrix}$ and $\mathbf{b} = \begin{pmatrix} 4 \\ -1 \end{pmatrix}$. Find:

a $\mathbf{a} + \mathbf{b}$ **b** $3\mathbf{a} - 2\mathbf{b}$

c $4(\mathbf{b} - \mathbf{a})$ ← Pure Year 1, Chapter 11

2 Solve the following pairs of simultaneous equations.

a $2x - 3y = 5;\ 3x + 2y = 27$

b $4x - 3y = -13;\ 5x - 2y = -22$

← Pure Year 1, Chapter 3

6.1 Introduction to matrices

A matrix is an **array** of **elements** (which are usually numbers) set out in a pair of brackets.

Links A **vector** is a simple example of a matrix with just one column.
← Pure Year 1, Chapter 11; Pure Year 2, Chapter 12

You can describe the **size** of a matrix using the number of rows and columns it contains.

For example $\begin{pmatrix} 2 & 1 \\ 4 & 0 \end{pmatrix}$ is a 2×2 matrix and $\begin{pmatrix} 1 & 4 & -1 & 1 \\ 2 & 3 & 0 & 2 \end{pmatrix}$ is a 2×4 matrix. Generally, you can refer to a matrix as $n \times m$ where n is the number of rows and m is the number of columns.

- **A square matrix is one where the numbers of rows and columns are the same.**
- **A zero matrix is one in which all of the elements are zero. The zero matrix is denoted by 0.**
- **An identity matrix is a square matrix in which the elements on the leading diagonal (starting top left) are all 1 and the remaining elements are 0. Identity matrices are denoted by $\mathbf{I}_k$ where k describes the size. The 3 × 3 identity matrix is**
  $\mathbf{I}_3 = \begin{pmatrix} 1 & 0 & 0 \\ 0 & 1 & 0 \\ 0 & 0 & 1 \end{pmatrix}$

Notation Matrices are usually represented with bold capital letters such as **M** or **A**.

Example 1

Write down the size of each matrix in the form $n \times m$.

a $\begin{pmatrix} 2 & -1 \\ 1 & 3 \end{pmatrix}$ **b** $(1 \quad 0 \quad 2)$

c $\begin{pmatrix} 4 \\ -1 \end{pmatrix}$ **d** $\begin{pmatrix} 3 & 2 \\ -1 & 1 \\ 0 & -3 \end{pmatrix}$

a $\begin{pmatrix} 2 & -1 \\ 1 & 3 \end{pmatrix}$ — There are two rows and two columns.

The size is 2×2.

b $(1 \quad 0 \quad 2)$ — There is just one row and three columns.

The size is 1×3.

c $\begin{pmatrix} 4 \\ -1 \end{pmatrix}$ — There are two rows and one column.

The size is 2×1.

d $\begin{pmatrix} 3 & 2 \\ -1 & 1 \\ 0 & -3 \end{pmatrix}$ — There are three rows and two columns.

The size is 3×2.

- **To add or subtract matrices, you add or subtract the corresponding elements in each matrix. You can only add or subtract matrices that are the same size.**

Notation Matrices which are the same size are said to be **additively conformable**.

Example 2

Find: **a** $\begin{pmatrix} 2 & -1 \\ 0 & 3 \end{pmatrix} + \begin{pmatrix} -1 & 4 \\ 5 & 3 \end{pmatrix}$

b $\begin{pmatrix} 1 & -3 & 4 \\ 2 & 1 & 1 \end{pmatrix} - \begin{pmatrix} 0 & 2 & 1 \\ 5 & 2 & 3 \end{pmatrix}$

a $\begin{pmatrix} 2 & -1 \\ 0 & 3 \end{pmatrix} + \begin{pmatrix} -1 & 4 \\ 5 & 3 \end{pmatrix}$

$= \begin{pmatrix} 1 & 3 \\ 5 & 6 \end{pmatrix}$

Top row:
$2 + -1 = 1$
$-1 + 4 = 3$

Bottom row:
$0 + 5 = 5$
$3 + 3 = 6$

b $\begin{pmatrix} 1 & -3 & 4 \\ 2 & 1 & 1 \end{pmatrix} - \begin{pmatrix} 0 & 2 & 1 \\ 5 & 2 & 3 \end{pmatrix}$

$= \begin{pmatrix} 1 & -5 & 3 \\ -3 & -1 & -2 \end{pmatrix}$

Top row:
$1 - 0 = 1$
$-3 - 2 = -5$
$4 - 1 = 3$

Bottom row:
$2 - 5 = -3$
$1 - 2 = -1$
$1 - 3 = -2$

- **To multiply a matrix by a scalar, you multiply every element in the matrix by that scalar.**

Notation A **scalar** is a number rather than a matrix. In questions on matrices, scalars will be represented by non-bold letters and numbers.

Example 3

$\mathbf{A} = \begin{pmatrix} 1 & 2 \\ -1 & 0 \end{pmatrix}$, $\mathbf{B} = (6 \;\; 0 \;\; -4)$

Find: **a** $2\mathbf{A}$ **b** $\frac{1}{2}\mathbf{B}$

c Explain why you cannot work out $\mathbf{A} + \mathbf{B}$.

a $2\mathbf{A} = \begin{pmatrix} 2 & 4 \\ -2 & 0 \end{pmatrix}$

Note that $2\mathbf{A}$ gives the same answer as $\mathbf{A} + \mathbf{A}$.

Top row:
$2 \times 1 = 2$
$2 \times 2 = 4$

Bottom row:
$2 \times -1 = -2$
$2 \times 0 = 0$

b $\frac{1}{2}\mathbf{B} = (3 \;\; 0 \;\; -2)$

$\frac{1}{2} \times 6 = 3$
$\frac{1}{2} \times 0 = 0$
$\frac{1}{2} \times (-4) = -2$

c **A** and **B** are not the same size, so you can't add them.

You could also say that **A** and **B** are not additively conformable.

Example 4

$\mathbf{A} = \begin{pmatrix} a & 0 \\ 1 & 2 \end{pmatrix}$, $\mathbf{B} = \begin{pmatrix} 1 & b \\ 0 & 3 \end{pmatrix}$, $\mathbf{C} = \begin{pmatrix} 6 & 6 \\ 1 & c \end{pmatrix}$.

Given that $\mathbf{A} + 2\mathbf{B} = \mathbf{C}$, find the values of the constants a, b and c.

$$\begin{pmatrix} a & 0 \\ 1 & 2 \end{pmatrix} + 2\begin{pmatrix} 1 & b \\ 0 & 3 \end{pmatrix} = \begin{pmatrix} 6 & 6 \\ 1 & c \end{pmatrix}$$

$$\begin{pmatrix} a+2 & 2b \\ 1 & 8 \end{pmatrix} = \begin{pmatrix} 6 & 6 \\ 1 & c \end{pmatrix}$$

If two matrices are equal, then all of their corresponding elements are equal.

$a + 2 = 6 \Rightarrow a = 4$ — Compare top left elements.

$2b = 6 \Rightarrow b = 3$ — Compare top right elements.

$c = 8$ — Compare bottom right elements.

Exercise 6A

1 Write the size of each matrix in the form $n \times m$.

a $\begin{pmatrix} 1 & 0 \\ -1 & 3 \end{pmatrix}$ **b** $\begin{pmatrix} 1 \\ 2 \end{pmatrix}$ **c** $\begin{pmatrix} 1 & 2 & 1 \\ 3 & 0 & -1 \end{pmatrix}$

d $(1 \quad 2 \quad 3)$ **e** $(3 \quad -1)$ **f** $\begin{pmatrix} 1 & 0 & 0 \\ 0 & 1 & 0 \\ 0 & 0 & 1 \end{pmatrix}$

2 Write down the 4×4 identity matrix, $\mathbf{I}_4$.

3 Two matrices **A** and **B** are given as:

$\mathbf{A} = \begin{pmatrix} 1 & 3 & a \\ 2 & -1 & 4 \end{pmatrix}$, $\mathbf{B} = \begin{pmatrix} 1 & 3 & 6 \\ b & -1 & 4 \end{pmatrix}$

If $\mathbf{A} = \mathbf{B}$, write down the values of a and b.

4 For the matrices

$\mathbf{A} = \begin{pmatrix} 2 & -1 \\ 1 & 3 \end{pmatrix}$, $\mathbf{B} = \begin{pmatrix} 4 & 1 \\ -1 & -2 \end{pmatrix}$, $\mathbf{C} = \begin{pmatrix} 6 & 0 \\ 0 & 1 \end{pmatrix}$

find:

a $\mathbf{A} + \mathbf{C}$ **b** $\mathbf{B} - \mathbf{A}$ **c** $\mathbf{A} + \mathbf{B} - \mathbf{C}$

5 For the matrices

$\mathbf{A} = \begin{pmatrix} 1 \\ 2 \end{pmatrix}$, $\mathbf{B} = (1 \quad -1)$, $\mathbf{C} = (-1 \quad 1 \quad 0)$,

$\mathbf{D} = (0 \quad 1 \quad -1)$, $\mathbf{E} = \begin{pmatrix} 3 \\ -1 \end{pmatrix}$, $\mathbf{F} = (2 \quad 1 \quad 3)$

find where possible:

a $\mathbf{A} + \mathbf{B}$ **b** $\mathbf{A} - \mathbf{E}$ **c** $\mathbf{F} - \mathbf{D} + \mathbf{C}$ **d** $\mathbf{B} + \mathbf{C}$

e $\mathbf{F} - (\mathbf{D} + \mathbf{C})$ **f** $\mathbf{A} - \mathbf{F}$ **g** $\mathbf{C} - (\mathbf{F} - \mathbf{D})$

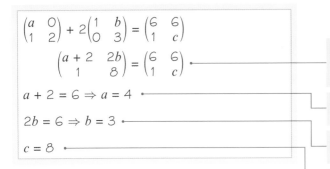

6 Given that $\begin{pmatrix} a & 2 \\ -1 & b \end{pmatrix} - \begin{pmatrix} 1 & c \\ d & -2 \end{pmatrix} = \begin{pmatrix} 5 & 0 \\ 0 & 5 \end{pmatrix}$, find the values of the constants a, b, c and d.

P **7** Given that $\begin{pmatrix} 1 & 2 & 0 \\ a & b & c \end{pmatrix} + \begin{pmatrix} a & b & c \\ 1 & 2 & 0 \end{pmatrix} = \begin{pmatrix} c & 5 & c \\ c & c & c \end{pmatrix}$, find the values of a, b and c.

8 Given that $\begin{pmatrix} 5 & 3 \\ 0 & -1 \\ 2 & 1 \end{pmatrix} + \begin{pmatrix} a & b \\ c & d \\ e & f \end{pmatrix} = \begin{pmatrix} 7 & 1 \\ 2 & 0 \\ 1 & 4 \end{pmatrix}$, find the values of a, b, c, d, e and f.

9 For the matrices $\mathbf{A} = \begin{pmatrix} 1 & -1 & 3 \\ 2 & 0 & 2 \\ 3 & 4 & 0 \end{pmatrix}$, $\mathbf{B} = \begin{pmatrix} 6 & 3 & -4 \\ 1 & 1 & 2 \\ -2 & 0 & -3 \end{pmatrix}$ and $\mathbf{C} = \begin{pmatrix} -2 & 0 & 3 \\ 2 & 8 & -6 \\ -1 & 1 & 1 \end{pmatrix}$, find:

a $\mathbf{A} + \mathbf{B}$ **b** $\mathbf{B} - \mathbf{C}$ **c** $\mathbf{C} + \mathbf{A}$

d A matrix $\mathbf{M} = \begin{pmatrix} 5 & -6 & b \\ 4 & a & 6 \\ 2 & 0 & c \end{pmatrix}$. Find the values of a, b and c if:

i $\mathbf{A} + \mathbf{M} = \begin{pmatrix} 6 & -7 & -2 \\ 6 & 3 & 8 \\ 5 & 4 & 6 \end{pmatrix}$ **ii** $\mathbf{M} - \mathbf{B} = \begin{pmatrix} -1 & -9 & -5 \\ -2 & 7 & -1 \\ -3 & 2 & 2 \end{pmatrix}$

10 For the matrices $\mathbf{A} = \begin{pmatrix} 2 & 0 \\ 4 & -6 \end{pmatrix}$, $\mathbf{B} = \begin{pmatrix} 1 \\ -1 \end{pmatrix}$, find:

a $3\mathbf{A}$ **b** $\frac{1}{2}\mathbf{A}$ **c** $2\mathbf{B}$

d Explain why it is not possible to find $\mathbf{A} - \mathbf{B}$.

11 The matrices **A** and **B** are defined as:

$\mathbf{A} = \begin{pmatrix} 3 & -2 \\ 1 & 0 \end{pmatrix}$ and $\mathbf{B} = \begin{pmatrix} 2 & 1 \\ -2 & 3 \end{pmatrix}$

Find:

a $3\mathbf{A} + 2\mathbf{B}$ **b** $2\mathbf{A} - 4\mathbf{B}$ **c** $5\mathbf{A} - 2\mathbf{B}$ **d** $\frac{1}{2}\mathbf{A} + \frac{3}{2}\mathbf{B}$

12 The matrices **M** and **N** are defined as:

$\mathbf{M} = \begin{pmatrix} 2 & 4 & -1 \\ 1 & -3 & -1 \\ 0 & 2 & 2 \end{pmatrix}$ and $\mathbf{N} = \begin{pmatrix} 6 & -2 & 5 \\ 3 & -3 & 1 \\ 1 & -1 & 0 \end{pmatrix}$

Find:

a $\mathbf{M} + 2\mathbf{N}$ **b** $3\mathbf{M} - \mathbf{N}$ **c** $4\mathbf{M} + 5\mathbf{N}$ **d** $\frac{2}{3}\mathbf{M} - \frac{1}{2}\mathbf{N}$

13 Find the value of k and the value of x such that $\begin{pmatrix} 0 & 1 \\ 2 & 0 \end{pmatrix} + k\begin{pmatrix} 0 & 2 \\ -1 & 0 \end{pmatrix} = \begin{pmatrix} 0 & 7 \\ x & 0 \end{pmatrix}$.

14 Find the values of a, b, c and d such that $2\begin{pmatrix} a & 0 \\ 1 & b \end{pmatrix} - 3\begin{pmatrix} 1 & c \\ d & -1 \end{pmatrix} = \begin{pmatrix} 3 & 3 \\ -4 & -4 \end{pmatrix}$.

15 Find the values of a, b, c and d such that $\begin{pmatrix} 5 & a \\ b & 0 \end{pmatrix} - 2\begin{pmatrix} c & 2 \\ 1 & -1 \end{pmatrix} = \begin{pmatrix} 9 & 1 \\ 3 & d \end{pmatrix}$.

(P) **16** Find the value of k such that $\begin{pmatrix} -3 \\ k \end{pmatrix} + k\begin{pmatrix} 2k \\ 2k \end{pmatrix} = \begin{pmatrix} k \\ 6 \end{pmatrix}$.

(P) **17** The matrices **A** and **B** are defined as:

$$\mathbf{A} = \begin{pmatrix} p & 0 & 0 \\ 0 & q^2 & r \\ 0 & 0 & 5 \end{pmatrix} \text{ and } \mathbf{B} = \begin{pmatrix} 2q & 0 & 0 \\ 0 & 4 & 6 \\ 0 & 0 & 2 \end{pmatrix}$$

where p, q and r are positive constants.

Given that $\mathbf{A} - k\mathbf{B} = \mathbf{I}_3$, where $\mathbf{I}_3$ is the 3×3 identity matrix, find:

a the value of k **b** the values of p, q and r.

(P) **18** The matrices **P** and **Q** are defined as:

$$\mathbf{P} = \begin{pmatrix} 0 & 2 & c \\ a & 0 & 0 \\ 0 & \text{b} & -1 \end{pmatrix} \text{ and } \mathbf{Q} = \begin{pmatrix} 0 & -1 & -1 \\ 3 & d & 0 \\ 0 & 2 & e \end{pmatrix}$$

where a, b, c, d, and e are constants.

Given that $\mathbf{P} - k\mathbf{Q} = \mathbf{0}$, where $\mathbf{0}$ is the zero matrix, find the values of a, b, c, d, e and k.

6.2 Matrix multiplication

Two matrices can be multiplied together. Unlike the operations we have seen so far, this is completely different from normal arithmetic multiplication.

- **Matrices can be multiplied together if the number of columns in the first matrix is equal to the number of rows in the second matrix.**

Notation If **AB** exists, then matrix **A** is said to be **multiplicatively conformable** with matrix **B**.

The **product matrix** will have the same number of rows as the first matrix, and the same number of columns as the second matrix.

$$\mathbf{AB} = \mathbf{C}$$

If **A** has size $n \times m$ and **B** has size $m \times k$ then the product matrix, **C**, has size $n \times k$.

The **order** in which you multiply matrices is important. This has two consequences:

- In general **AB** ≠ **BA** (even if **A** and **B** are both square matrices).
- If **AB** exists, **BA** does not necessarily exist (for example if **A** is a 3×2 matrix and **B** is a 2×4 matrix).

- **To find the product of two multiplicatively conformable matrices, you multiply the elements in each row in the left-hand matrix by the corresponding elements in each column in the right-hand matrix, then add the results together.**

$$\begin{pmatrix} 5 & -1 & 2 \\ 8 & 3 & -4 \end{pmatrix} \times \begin{pmatrix} 2 & 2 \\ 9 & -3 \\ 7 & 4 \end{pmatrix} = \begin{pmatrix} 15 & 21 \\ 15 & -9 \end{pmatrix}$$

You are multiplying a 2×3 matrix by a 3×2 matrix, so the product matrix will have size 2×2.

To find the bottom left element, work out $8 \times 2 + 3 \times 9 + (-4) \times 7 = 16 + 27 - 28 = 15$

Example 5

Given that $\mathbf{A} = \begin{pmatrix} 1 & -2 \\ 3 & 4 \end{pmatrix}$ and $\mathbf{B} = \begin{pmatrix} -3 \\ 2 \end{pmatrix}$

a find **AB** **b** explain why it is not possible to find **BA**.

a First calculate the size of AB.

$(2 \times 2) \times (2 \times 1)$ gives 2×1

$AB = \begin{pmatrix} 1 & -2 \\ 3 & 4 \end{pmatrix}\begin{pmatrix} -3 \\ 2 \end{pmatrix} = \begin{pmatrix} p \\ q \end{pmatrix}$

$p = 1 \times (-3) + (-2) \times 2 = -7$

$q = 3 \times (-3) + 4 \times 2 = -1$

So $AB = \begin{pmatrix} -7 \\ -1 \end{pmatrix}$

b BA cannot be found, since the number of columns in B is not the same as the number of rows in A.

The number of rows is two from here.

The number of columns is one from here.

The top number is the total of the first row of **A** multiplied by the first column of **B**.

The bottom number is the total of the second row of **A** multiplied by the first column of **B**.

Watch out Remember that order is important. **B** is not multiplicatively conformable with **A**, but **A** is multiplicatively conformable with **B**.

Example 6

Given that $\mathbf{A} = \begin{pmatrix} -1 & 0 \\ 2 & 3 \end{pmatrix}$ and $\mathbf{B} = \begin{pmatrix} 4 & 1 \\ 0 & -2 \end{pmatrix}$, find:

a **AB** **b** **BA**

a A is a 2×2 matrix and B is a 2×2 matrix so they can be multiplied and the product will be a 2×2 matrix.

$AB = \begin{pmatrix} -1 & 0 \\ 2 & 3 \end{pmatrix}\begin{pmatrix} 4 & 1 \\ 0 & -2 \end{pmatrix} = \begin{pmatrix} a & b \\ c & d \end{pmatrix}$

$a = (-1) \times 4 + 0 \times 0 = -4$

$b = (-1) \times 1 + 0 \times (-2) = -1$

$c = 2 \times 4 + 3 \times 0 = 8$

$d = 2 \times 1 + 3 \times (-2) = -4$

So $AB = \begin{pmatrix} -4 & -1 \\ 8 & -4 \end{pmatrix}$

b BA will also be a 2×2 matrix.

$\begin{pmatrix} 4 & 1 \\ 0 & -2 \end{pmatrix}\begin{pmatrix} -1 & 0 \\ 2 & 3 \end{pmatrix} = \begin{pmatrix} -2 & 3 \\ -4 & -6 \end{pmatrix}$

This time there are four elements to be found.

a is the total of the first row multiplied by the first column.
b is the total of the first row multiplied by the second column.
c is the total of the second row multiplied by the first column.
d is the total of the second row multiplied by the second column.

First row times first column
$4 \times (-1) + 1 \times 2 = -2$

First row times second column
$4 \times 0 + 1 \times 3 = 3$

Second row times second column
$0 \times 0 + (-2) \times 3 = -6$

Second row times first column
$0 \times (-1) + (-2) \times 2 = -4$

You can enter matrices directly into your calculator to multiply them quickly

Example 7

$\mathbf{A} = \begin{pmatrix} -1 \\ a \end{pmatrix}$ and $\mathbf{B} = (b \quad 2)$

Given that $\mathbf{BA} = (0)$, find $\mathbf{AB}$ in terms of a.

(0) is a 1 × 1 zero matrix. You could also write it as **0**.

$\mathbf{BA} = (b \quad 2)\begin{pmatrix} -1 \\ a \end{pmatrix} = (-b + 2a)$

BA is a 1 × 1 matrix.

So $\mathbf{BA} = (0)$ implies that $b = 2a$.

$\mathbf{AB} = \begin{pmatrix} -1 \\ a \end{pmatrix}(b \quad 2) = \begin{pmatrix} -b & -2 \\ ab & 2a \end{pmatrix}$

AB is a 2 × 2 matrix.

Substituting $b = 2a$ gives $\mathbf{AB} = \begin{pmatrix} -2a & -2 \\ 2a^2 & 2a \end{pmatrix}$.

Watch out Although you can multiply matrices using a calculator, you need to know how the process works so that you can deal with matrices containing unknowns.

Example 8

$\mathbf{A} = (1 \quad -1 \quad 2)$, $\mathbf{B} = (3 \quad -2)$ and $\mathbf{C} = \begin{pmatrix} 4 \\ 5 \end{pmatrix}$. Find **BCA**.

$\mathbf{BC} = (3 \quad -2)\begin{pmatrix} 4 \\ 5 \end{pmatrix} = (2)$

$(\mathbf{BC})\mathbf{A} = (2)(1 \quad -1 \quad 2) = (2 \quad -2 \quad 4)$

This product could have been calculated by first working out **CA** and then multiplying **B** by this product. In general, matrix multiplication is **associative** (meaning that the bracketing makes no difference provided the order stays the same), so (**BC**)**A** = **B**(**CA**).

Exercise 6B

1 Given the sizes of the following matrices:

Matrix	A	B	C	D	E
Size	2 × 2	1 × 2	1 × 3	3 × 2	2 × 3

find the sizes of these matrix products.

a BA **b** DE **c** CD

d ED **e** AE **f** DA

2 Use your calculator to find these products:

a $\begin{pmatrix} 1 & 2 \\ 2 & 4 \end{pmatrix}\begin{pmatrix} -1 \\ 2 \end{pmatrix}$ **b** $\begin{pmatrix} 1 & 2 \\ 3 & 4 \end{pmatrix}\begin{pmatrix} 0 & 5 \\ -1 & -2 \end{pmatrix}$

3 The matrix $\mathbf{A} = \begin{pmatrix} -1 & -2 \\ 0 & 3 \end{pmatrix}$ and the matrix $\mathbf{B} = \begin{pmatrix} 1 & 0 & 1 \\ 1 & 1 & 0 \end{pmatrix}$.

Use your calculator to find:

a AB **b** $\mathbf{A}^2$

Hint $\mathbf{A}^2$ means **A** × **A**.

4 The matrices **A**, **B** and **C** are given by:

$\mathbf{A} = \begin{pmatrix} 2 \\ 1 \end{pmatrix}$, $\mathbf{B} = \begin{pmatrix} 3 & 1 \\ -1 & 2 \end{pmatrix}$, $\mathbf{C} = (-3 \quad -2)$

Without using your calculator, determine whether or not the following products exist and find the products of those that do.

a **AB** **b** **AC** **c** **BC**

d **BA** **e** **CA** **f** **CB**

5 Find $\begin{pmatrix} 2 & a \\ 1 & -1 \end{pmatrix}\begin{pmatrix} 1 & 3 & 0 \\ 0 & -1 & 2 \end{pmatrix}$, giving your answer in terms of a.

6 Find $\begin{pmatrix} 3 & 2 \\ -1 & x \end{pmatrix}\begin{pmatrix} x & -2 \\ 1 & 3 \end{pmatrix}$, giving your answer in terms of x.

7 The matrices **A**, **B** and **C** are defined as:

$\mathbf{A} = \begin{pmatrix} 2 & -1 \\ 3 & 4 \end{pmatrix}$, $\mathbf{B} = \begin{pmatrix} 1 & 0 \\ -3 & 2 \end{pmatrix}$ and $\mathbf{C} = \begin{pmatrix} -3 & 1 \\ 1 & 2 \end{pmatrix}$.

Use your calculator to find:

a **AB** – **C** **b** **BC** + 3**A** **c** 4**B** – 3**CA**

8 The matrices **M** and **N** are defined as:

$\mathbf{M} = \begin{pmatrix} 3 & k \\ k & 1 \end{pmatrix}$ and $\mathbf{N} = \begin{pmatrix} 1 & k \\ k & -1 \end{pmatrix}$. Find, in terms of k:

a **MN** **b** **NM** **c** 3**M** – 2**N** **d** 2**MN** + 3**N**

(P) **9** The matrix $\mathbf{A} = \begin{pmatrix} 1 & 2 \\ 0 & 1 \end{pmatrix}$.

Find:

a $\mathbf{A}^2$

b $\mathbf{A}^3$

c Suggest a form for $\mathbf{A}^k$.

Links You might be asked to prove this general form for $\mathbf{A}^k$. → Section 8.3

(P) **10** The matrix $\mathbf{A} = \begin{pmatrix} a & 0 \\ b & 0 \end{pmatrix}$.

a Find, in terms of a and b, the matrix $\mathbf{A}^2$.

Given that $\mathbf{A}^2 = 3\mathbf{A}$,

b find the value of a.

11 $\mathbf{A} = (-1 \quad 3)$, $\mathbf{B} = \begin{pmatrix} 2 \\ 1 \\ 0 \end{pmatrix}$, $\mathbf{C} = \begin{pmatrix} 4 & -2 \\ 0 & -3 \end{pmatrix}$

Find: **a** **BAC** **b** $\mathbf{AC}^2$

12 $\mathbf{A} = \begin{pmatrix} 1 \\ -1 \\ 2 \end{pmatrix}$, $\mathbf{B} = (3 \quad -2 \quad -3)$

Find: **a** **ABA** **b** **BAB**

(P) **13** **a** Write down $\mathbf{I}_2$.

b Given that matrix $\mathbf{A} = \begin{pmatrix} 2 & -2 \\ 1 & 3 \end{pmatrix}$, show that $\mathbf{AI} = \mathbf{IA} = \mathbf{A}$.

(P) **14** $\mathbf{A} = \begin{pmatrix} 2 & -1 \\ 3 & 2 \end{pmatrix}$, $\mathbf{B} = \begin{pmatrix} 4 & 2 \\ -1 & 0 \end{pmatrix}$ and $\mathbf{C} = \begin{pmatrix} 1 & 2 \\ 0 & -1 \end{pmatrix}$.

Show that $\mathbf{AB} + \mathbf{AC} = \mathbf{A}(\mathbf{B} + \mathbf{C})$.

(E/P) **15** $\mathbf{A} = \begin{pmatrix} 1 & 2 \\ 3 & 1 \end{pmatrix}$ and $\mathbf{I}$ is the 2 × 2 identity matrix.

Prove that $\mathbf{A}^2 = 2\mathbf{A} + 5\mathbf{I}$. **(2 marks)**

(E) **16** A matrix **M** is given as $\mathbf{M} = \begin{pmatrix} 1 & 2 & c \\ a & -1 & 1 \\ 1 & b & 0 \end{pmatrix}$.

Find $\mathbf{M}^2$ in terms of a, b and c. **(3 marks)**

(E/P) **17** A matrix **A** is given as $\mathbf{A} = \begin{pmatrix} 1 & -1 & b \\ a & 2 & 0 \\ 1 & 0 & 3 \end{pmatrix}$.

Given that $\mathbf{A}^2 = \begin{pmatrix} -4 & -3 & -8 \\ 9 & 1 & -6 \\ 4 & -1 & 7 \end{pmatrix}$, find the values of a and b. **(3 marks)**

(E/P) **18** $\mathbf{A} = \begin{pmatrix} p & 3 \\ 6 & p \end{pmatrix}$ and $\mathbf{B} = \begin{pmatrix} q & 2 \\ 4 & q \end{pmatrix}$, where p and q are constants. Prove that $\mathbf{AB} = \mathbf{BA}$. **(3 marks)**

(E/P) **19** The matrix $\mathbf{A} = \begin{pmatrix} 3 & p \\ -4 & q \end{pmatrix}$ is such that $\mathbf{A}^2 = \mathbf{I}$. Find the values of p and q. **(3 marks)**

Challenge

A 2 × 2 matrix **A** has the property that $\mathbf{A}^2 = \mathbf{0}$. Find a possible matrix **A** such that:

a at least one of the elements in **A** is non-zero

b all of the elements in **A** are non-zero.

6.3 Determinants

You can calculate the **determinant** of a square matrix. The determinant is a scalar value associated with that matrix.

- **For a 2 × 2 matrix M =** $\begin{pmatrix} a & b \\ c & d \end{pmatrix}$**, the determinant of M is** $ad - bc$**.**
- **If det M = 0 then M is a singular matrix.**
- **If det M ≠ 0 then M is a non-singular matrix.**

Notation You can write the determinant of **M** as det **M**, |**M**| or $\begin{vmatrix} a & b \\ c & d \end{vmatrix}$. It is also sometimes written as Δ.

Links Singular matrices do not have an **inverse**. → Section 6.4

Example 9

Given that $\mathbf{A} = \begin{pmatrix} 6 & 5 \\ 1 & 2 \end{pmatrix}$, find det **A**.

det A = $ad - bc = 6 \times 2 - 5 \times 1 = 12 - 5 = 7$

Example 10

$$\mathbf{A} = \begin{pmatrix} 4 & p+2 \\ -1 & 3-p \end{pmatrix}$$

Given that **A** is singular, find the value of p.

det A = $4(3 - p) - (p + 2)(-1)$
det A = $12 - 4p + p + 2 = 14 - 3p$
A is singular so det A = 0.
$14 - 3p = 0 \Rightarrow p = \frac{14}{3}$

Watch out Although you can find the determinant using a calculator, you need to know how the process works so that you can deal with matrices containing unknowns.

Finding the determinant of a 3 × 3 matrix is more difficult.

- **You find the determinant of a 3 × 3 matrix by reducing the 3 × 3 determinant to 2 × 2 determinants using the formula:**

$$\begin{vmatrix} a & b & c \\ d & e & f \\ g & h & i \end{vmatrix} = a\begin{vmatrix} e & f \\ h & i \end{vmatrix} - b\begin{vmatrix} d & f \\ g & i \end{vmatrix} + c\begin{vmatrix} d & e \\ g & h \end{vmatrix}$$

Watch out There is a minus sign in front of the second term.

In this expression for the determinant, each of the elements a, b and c is multiplied by its **minor**.

- **The minor of an element in a 3 × 3 matrix is the determinant of the 2 × 2 matrix that remains after the row and column containing that element have been crossed out.**

Example 11

Find the minors of the elements 5 and 7 in the matrix

$$\begin{pmatrix} 5 & 0 & 2 \\ -1 & 8 & 1 \\ 6 & 7 & 3 \end{pmatrix}$$

$$\begin{pmatrix} \not{5} & \not{0} & \not{2} \\ \not{-1} & 8 & 1 \\ \not{6} & 7 & 3 \end{pmatrix}$$

To find the minor of 5, you begin by crossing out the row and the column containing 5.

$$\begin{vmatrix} 8 & 1 \\ 7 & 3 \end{vmatrix} = 8 \times 3 - 7 \times 1 = 24 - 7 = 17$$

When you have crossed out the row and the column containing 5, you are left with the elements $\begin{pmatrix} 8 & 1 \\ 7 & 3 \end{pmatrix}$ and you evaluate the determinant of this 2×2 matrix.

The minor of 5 is 17.

$$\begin{pmatrix} 5 & \not{0} & 2 \\ -1 & \not{8} & 1 \\ \not{6} & \not{7} & \not{3} \end{pmatrix}$$

To find the minor of 7, you begin by crossing out the row and the column containing 7.

$$\begin{vmatrix} 5 & 2 \\ -1 & 1 \end{vmatrix} = 5 \times 1 - 2 \times (-1) = 5 + 2 = 7$$

When you have crossed out the row and the column containing 7, you are left with the elements $\begin{pmatrix} 5 & 2 \\ -1 & 1 \end{pmatrix}$ and you evaluate the determinant of this matrix.

The minor of 7 is 7.

Example 12

Find the value of $\begin{vmatrix} 1 & 2 & 4 \\ 3 & 2 & 1 \\ -1 & 4 & 3 \end{vmatrix}$.

$$\begin{vmatrix} 1 & 2 & 4 \\ 3 & 2 & 1 \\ -1 & 4 & 3 \end{vmatrix} = 1\begin{vmatrix} 2 & 1 \\ 4 & 3 \end{vmatrix} - 2\begin{vmatrix} 3 & 1 \\ -1 & 3 \end{vmatrix} + 4\begin{vmatrix} 3 & 2 \\ -1 & 4 \end{vmatrix}$$

$$= 1(6 - 4) - 2(9 + 1) + 4(12 + 2)$$

$$= 1 \times 2 - 2 \times 10 + 4 \times 14$$

$$= 2 - 20 + 56 = 38$$

The determinant of this 2×2 matrix is the minor of the top left element.

The determinant of this 2×2 matrix is the minor of the top centre element.

The determinant of this 2×2 matrix is the minor of the top right element.

Example 13

The matrix $\mathbf{A} = \begin{pmatrix} 3 & k & 0 \\ -2 & 1 & 2 \\ 5 & 0 & k+3 \end{pmatrix}$, where k is a constant.

a Find det $\mathbf{A}$ in terms of k.

Given that $\mathbf{A}$ is singular,

b find the possible values of k.

a $\begin{vmatrix} 3 & k & 0 \\ -2 & 1 & 2 \\ 5 & 0 & k+3 \end{vmatrix} = 3\begin{vmatrix} 1 & 2 \\ 0 & k+3 \end{vmatrix} - k\begin{vmatrix} -2 & 2 \\ 5 & k+3 \end{vmatrix} + 0\begin{vmatrix} -2 & 1 \\ 5 & 0 \end{vmatrix}$

$= 3(k + 3) - k(-2(k + 3) - 10)$

$= 3k + 9 + 2k^2 + 16k$

$= 2k^2 + 19k + 9$

b As **A** is singular,

$2k^2 + 19k + 9 = 0$

$(2k + 1)(k + 9) = 0$

$k = -\frac{1}{2}, -9$

Part **b** will require you to solve det **A** = 0, so multiply this expression out, collect together terms and express the result as a quadratic.

Problem-solving

As **A** is singular, its determinant is 0. This gives a quadratic equation, which you solve, giving two possible values of k.

Exercise 6C

1 Find the determinants of the following matrices.

a $\begin{pmatrix} 3 & 4 \\ -1 & 2 \end{pmatrix}$ **b** $\begin{pmatrix} 4 & 2 \\ 1 & 2 \end{pmatrix}$ **c** $\begin{pmatrix} -2 & 1 \\ 3 & 0 \end{pmatrix}$

d $\begin{pmatrix} -4 & -4 \\ 1 & 1 \end{pmatrix}$ **e** $\begin{pmatrix} 7 & -4 \\ 0 & 3 \end{pmatrix}$ **f** $\begin{pmatrix} -1 & -1 \\ -6 & -10 \end{pmatrix}$

P **2** Find the values of a for which these matrices are singular.

a $\begin{pmatrix} a & 1+a \\ 3 & 2 \end{pmatrix}$ **b** $\begin{pmatrix} 1+a & 3-a \\ a+2 & 1-a \end{pmatrix}$ **c** $\begin{pmatrix} 2+a & 1-a \\ 1-a & a \end{pmatrix}$

E/P **3** Given that k is a real number and that $\mathbf{M} = \begin{pmatrix} -2 & 1-k \\ k-1 & k \end{pmatrix}$,

find the exact values of k for which **M** is a singular matrix. **(3 marks)**

E/P **4** $\mathbf{P} = \begin{pmatrix} 3k & 4-k \\ k-2 & -k \end{pmatrix}$, where k is a real constant.

Given that **P** is a singular matrix, find the possible values of k. **(3 marks)**

5 The matrix $\mathbf{A} = \begin{pmatrix} a & 2a \\ b & 2b \end{pmatrix}$ and the matrix $\mathbf{B} = \begin{pmatrix} 2b & -2a \\ -b & a \end{pmatrix}$.

a Find det **A** and det **B**.

b Find **AB**.

6 Use your calculator to find the values of these determinants.

a $\begin{vmatrix} 1 & 0 & 0 \\ 0 & 2 & 0 \\ 0 & 0 & 3 \end{vmatrix}$ **b** $\begin{vmatrix} 0 & 4 & 0 \\ 5 & -2 & 3 \\ 2 & 1 & 4 \end{vmatrix}$ **c** $\begin{vmatrix} 1 & 0 & 1 \\ 2 & 4 & 1 \\ 3 & 5 & 2 \end{vmatrix}$ **d** $\begin{vmatrix} 2 & -3 & 4 \\ 2 & 2 & 2 \\ 5 & 5 & 5 \end{vmatrix}$

7 Without using your calculator, find the values of these determinants.

a $\begin{vmatrix} 4 & 3 & -1 \\ 2 & -2 & 0 \\ 0 & 4 & -2 \end{vmatrix}$ **b** $\begin{vmatrix} 3 & -2 & 1 \\ 4 & 1 & -3 \\ 7 & 2 & -4 \end{vmatrix}$ **c** $\begin{vmatrix} 5 & -2 & -3 \\ 6 & 4 & 2 \\ -2 & -4 & -3 \end{vmatrix}$

P **8** The matrix $\mathbf{A} = \begin{pmatrix} 2 & 1 & -4 \\ 2k+1 & 3 & k \\ 1 & 0 & 1 \end{pmatrix}$.

Given that **A** is singular, find the value of the constant k.

P **9** The matrix $\mathbf{A} = \begin{pmatrix} 2 & -1 & 3 \\ k & 2 & 4 \\ -2 & 1 & k+3 \end{pmatrix}$, where k is a constant.

Given that the determinant of **A** is 8, find the possible values of k.

10 The matrix $\mathbf{A} = \begin{pmatrix} 2 & 5 & 3 \\ -2 & 0 & 4 \\ 3 & 10 & 8 \end{pmatrix}$ and the matrix $\mathbf{B} = \begin{pmatrix} 1 & 1 & 0 \\ 1 & 2 & 2 \\ 0 & -2 & -1 \end{pmatrix}$.

a Show that **A** is singular.

b Find **AB**.

c Show that **AB** is also singular.

E/P **11** Show that, for all values of a, b and c, the matrix $\begin{pmatrix} 0 & a & -b \\ -a & 0 & c \\ b & -c & 0 \end{pmatrix}$ is singular. **(3 marks)**

E/P **12** Show that, for all real values of x, the matrix $\begin{pmatrix} 2 & -2 & 4 \\ 3 & x & -2 \\ -1 & 3 & x \end{pmatrix}$ is non-singular. **(3 marks)**

E/P **13** Find all the values of x for which the matrix $\begin{pmatrix} x-3 & -2 & 0 \\ 1 & x & -2 \\ -2 & -1 & x+1 \end{pmatrix}$ is singular. **(4 marks)**

E/P **14** The matrix $\mathbf{M} = \begin{pmatrix} 1 & -3 \\ 2 & 1 \end{pmatrix}$ and the matrix $\mathbf{N} = \begin{pmatrix} -1 & k \\ 4 & 3 \end{pmatrix}$, where k is a constant.

a Evaluate the determinant of **M**. **(1 mark)**

b Given that the determinant of **N** is 7, find the value of k. **(2 marks)**

c Using the value of k found in part **b**, find **MN**. **(1 mark)**

d Verify that det **MN** = det **M** det **N**. **(1 mark)**

E/P **15** The matrix $\mathbf{A} = \begin{pmatrix} 2 & 1 & -1 \\ 1 & 0 & 4 \\ -4 & 2 & 1 \end{pmatrix}$ and the matrix $\mathbf{B} = \begin{pmatrix} 3 & 1 & 2 \\ k & 4 & 5 \\ 0 & 2 & 3 \end{pmatrix}$, where k is a constant.

a Evaluate the determinant of **A**. **(2 marks)**

Given that the determinant of **B** is 2,

b find the value of k. **(3 marks)**

Using the value of k found in part **b**,

c find **AB** **(2 marks)**

d verify that det **AB** = det **A** det **B**. **(2 marks)**

Challenge

a Find all the possible 2 × 2 singular matrices whose elements are the numbers 1 and –1.

b Find all the possible 2 × 2 singular matrices whose elements are the numbers 1 and 0.

Hint In part **a**, there are 8 possible matrices.

6.4 Inverting a 2 × 2 matrix

You can find the **inverse** of any non-singular matrix.

- **The inverse of a matrix M is the matrix $\mathbf{M}^{-1}$ such that $\mathbf{M}\mathbf{M}^{-1} = \mathbf{M}^{-1}\mathbf{M} = \mathbf{I}$.**

You can use the following formula to find the inverse of a 2 × 2 matrix.

- **If $\mathbf{M} = \begin{pmatrix} a & b \\ c & d \end{pmatrix}$, then $\mathbf{M}^{-1} = \frac{1}{\det \mathbf{M}}\begin{pmatrix} d & -b \\ -c & a \end{pmatrix}$.**

Note If det **M** = 0, you will not be able to find the inverse matrix, since $\frac{1}{\det \mathbf{M}}$ is undefined.

Example 14

$\mathbf{A} = \begin{pmatrix} 3 & 2 \\ -1 & 1 \end{pmatrix}$, $\mathbf{B} = \begin{pmatrix} 2 & 1 \\ 2 & 1 \end{pmatrix}$, $\mathbf{C} = \begin{pmatrix} 1 & 3 \\ 2 & 0 \end{pmatrix}$

For each of the matrices **A**, **B** and **C**, determine whether or not the matrix is singular. If the matrix is non-singular, find its inverse.

$\mathbf{A} = \begin{pmatrix} 3 & 2 \\ -1 & 1 \end{pmatrix}$ so det **A** = 3 × 1 − 2 × (−1)

Use the determinant formula with $a = 3$, $b = 2$, $c = -1$ and $d = 1$.

det **A** = 5

Since 5 ≠ 0, **A** is non-singular.

So $\mathbf{A}^{-1} = \frac{1}{5}\begin{pmatrix} 1 & -2 \\ 1 & 3 \end{pmatrix}$ or $\begin{pmatrix} 0.2 & -0.4 \\ 0.2 & 0.6 \end{pmatrix}$

Swap a and d and change the signs of b and c.

$\mathbf{A}^{-1}$ can be left in either form.

$\mathbf{B} = \begin{pmatrix} 2 & 2 \\ 1 & 1 \end{pmatrix}$ so det **B** = 2 × 1 − 1 × 2 = 0

Remember if det **B** = 0 then **B** is singular.

So **B** is singular and $\mathbf{B}^{-1}$ cannot be found.

$\mathbf{C} = \begin{pmatrix} 1 & 3 \\ 2 & 0 \end{pmatrix}$ so det $\mathbf{C} = 1 \times 0 - 3 \times 2 = -6$

This is non-zero and so **C** is a non-singular matrix.

$\mathbf{C}^{-1} = -\frac{1}{6}\begin{pmatrix} 0 & -3 \\ -2 & 1 \end{pmatrix}$ or $\begin{pmatrix} 0 & \frac{1}{2} \\ \frac{1}{3} & -\frac{1}{6} \end{pmatrix}$

Note that a determinant can be a negative number.

Swap a and d and change the signs of b and c. Then multiply by $\frac{1}{\det \mathbf{C}}$

You can find the inverse of a matrix using your calculator.

- **If A and B are non-singular matrices, then $(\mathbf{AB})^{-1} = \mathbf{B}^{-1}\mathbf{A}^{-1}$.**

Example 15

P and **Q** are non-singular matrices. Prove that $(\mathbf{PQ})^{-1} = \mathbf{Q}^{-1}\mathbf{P}^{-1}$.

Let $\mathbf{C} = (\mathbf{PQ})^{-1}$ then $(\mathbf{PQ})\mathbf{C} = \mathbf{I}$. — Use the definition of inverse $\mathbf{A}^{-1}\mathbf{A} = \mathbf{I} = \mathbf{A}\mathbf{A}^{-1}$.

$\mathbf{P}^{-1}\mathbf{PQC} = \mathbf{P}^{-1}\mathbf{I}$ — Multiply on the left by $\mathbf{P}^{-1}$.

$(\mathbf{P}^{-1}\mathbf{P})\mathbf{QC} = \mathbf{P}^{-1}\mathbf{I}$ — Remember $\mathbf{P}^{-1}\mathbf{P} = \mathbf{I}$, $\mathbf{IQ} = \mathbf{Q}$ and $\mathbf{P}^{-1}\mathbf{I} = \mathbf{P}^{-1}$.

So $\mathbf{QC} = \mathbf{P}^{-1}$

$\mathbf{Q}^{-1}\mathbf{QC} = \mathbf{Q}^{-1}\mathbf{P}^{-1}$ — Multiply on the left by $\mathbf{Q}^{-1}$.

$\mathbf{IC} = \mathbf{Q}^{-1}\mathbf{P}^{-1}$ — Use $\mathbf{Q}^{-1}\mathbf{Q} = \mathbf{I}$.

$\mathbf{C} = \mathbf{Q}^{-1}\mathbf{P}^{-1}$

So $(\mathbf{PQ})^{-1} = \mathbf{Q}^{-1}\mathbf{P}^{-1}$ as required.

Example 16

A and **B** are non-singular 2×2 matrices such that $\mathbf{BAB} = \mathbf{I}$.

a Prove that $\mathbf{A} = \mathbf{B}^{-1}\mathbf{B}^{-1}$.

Given that $\mathbf{B} = \begin{pmatrix} 2 & 5 \\ 1 & 3 \end{pmatrix}$:

b find the matrix **A** such that $\mathbf{BAB} = \mathbf{I}$.

a $\mathbf{BAB} = \mathbf{I}$

$\mathbf{B}^{-1}\mathbf{BAB} = \mathbf{B}^{-1}\mathbf{I}$ — Multiply on the left by $\mathbf{B}^{-1}$.

$(\mathbf{B}^{-1}\mathbf{B})\mathbf{AB} = \mathbf{B}^{-1}\mathbf{I}$ — Remember $\mathbf{B}^{-1}\mathbf{B} = \mathbf{I}$ and $\mathbf{B}^{-1}\mathbf{I} = \mathbf{B}^{-1}$.

$\mathbf{AB} = \mathbf{B}^{-1}$

$\mathbf{AB}\mathbf{B}^{-1} = \mathbf{B}^{-1}\mathbf{B}^{-1}$ — Multiply on the right by $\mathbf{B}^{-1}$ and remember $\mathbf{B}\mathbf{B}^{-1} = \mathbf{I}$.

$\mathbf{AI} = \mathbf{B}^{-1}\mathbf{B}^{-1}$

And hence $\mathbf{A} = \mathbf{B}^{-1}\mathbf{B}^{-1}$ as required.

b $\mathbf{B} = \begin{pmatrix} 2 & 5 \\ 1 & 3 \end{pmatrix}$ so $\det \mathbf{B} = 2 \times 3 - 5 \times 1 = 1$

So $\mathbf{B}^{-1} = \frac{1}{1}\begin{pmatrix} 3 & -5 \\ -1 & 2 \end{pmatrix} = \begin{pmatrix} 3 & -5 \\ -1 & 2 \end{pmatrix}$

First find $\mathbf{B}^{-1}$.

From part **a**,

$$\mathbf{A} = \mathbf{B}^{-1}\mathbf{B}^{-1} = \begin{pmatrix} 3 & -5 \\ -1 & 2 \end{pmatrix}\begin{pmatrix} 3 & -5 \\ -1 & 2 \end{pmatrix}$$

Use the result from part **a** and matrix multiplication to find **A**.

$$\mathbf{A} = \begin{pmatrix} 14 & -25 \\ -5 & 9 \end{pmatrix}$$

Exercise 6D

1 Determine which of these matrices are singular and which are non-singular. For those that are non-singular find the inverse matrix.

a $\begin{pmatrix} 3 & -1 \\ -4 & 2 \end{pmatrix}$ **b** $\begin{pmatrix} 3 & 3 \\ -1 & -1 \end{pmatrix}$ **c** $\begin{pmatrix} 2 & 5 \\ 0 & 0 \end{pmatrix}$

d $\begin{pmatrix} 1 & 2 \\ 3 & 5 \end{pmatrix}$ **e** $\begin{pmatrix} 6 & 3 \\ 4 & 2 \end{pmatrix}$ **f** $\begin{pmatrix} 4 & 3 \\ 6 & 2 \end{pmatrix}$

2 Find inverses of these matrices, giving your answers in terms of a and b.

a $\begin{pmatrix} a & 1+a \\ 1+a & 2+a \end{pmatrix}$ **b** $\begin{pmatrix} 2a & 3b \\ -a & -b \end{pmatrix}$

(P) **3 a** Given that $\mathbf{ABC} = \mathbf{I}$, prove that $\mathbf{B}^{-1} = \mathbf{CA}$.

b Given that $\mathbf{A} = \begin{pmatrix} 0 & 1 \\ -1 & -6 \end{pmatrix}$ and $\mathbf{C} = \begin{pmatrix} 2 & 1 \\ -3 & -1 \end{pmatrix}$, find **B**.

4 a Given that $\mathbf{AB} = \mathbf{C}$, find an expression for **B**, in terms of **A** and **C**.

b Given further that $\mathbf{A} = \begin{pmatrix} 2 & -1 \\ 4 & 3 \end{pmatrix}$ and $\mathbf{C} = \begin{pmatrix} 3 & 6 \\ 1 & 22 \end{pmatrix}$, find **B**.

5 a Given that $\mathbf{BAC} = \mathbf{B}$, where **B** is a non-singular matrix, find an expression for **A** in terms of **C**.

b When $\mathbf{C} = \begin{pmatrix} 5 & 3 \\ 3 & 2 \end{pmatrix}$, find **A**.

6 The matrix $\mathbf{A} = \begin{pmatrix} 2 & -1 \\ -4 & 3 \end{pmatrix}$ and $\mathbf{AB} = \begin{pmatrix} 4 & 7 & -8 \\ -8 & -13 & 18 \end{pmatrix}$. Find the matrix **B**.

(P) **7** The matrix $\mathbf{B} = \begin{pmatrix} 5 & -4 \\ 2 & 1 \end{pmatrix}$ and $\mathbf{AB} = \begin{pmatrix} 11 & -1 \\ -8 & 9 \\ -2 & -1 \end{pmatrix}$. Find the matrix **A**.

8 The matrix $\mathbf{A} = \begin{pmatrix} 3a & b \\ 4a & 2b \end{pmatrix}$, where a and b are non-zero constants.

a Find $\mathbf{A}^{-1}$, giving your answer in terms of a and b.

The matrix $\mathbf{B} = \begin{pmatrix} -a & b \\ 3a & 2b \end{pmatrix}$ and the matrix $\mathbf{X}$ is given by $\mathbf{B} = \mathbf{XA}$.

b Find $\mathbf{X}$, giving your answer in terms of a and b.

9 The non-singular matrices $\mathbf{A}$ and $\mathbf{B}$ are such that $\mathbf{AB} = \mathbf{BA}$, and $\mathbf{ABA} = \mathbf{B}$.

a Prove that $\mathbf{A}^2 = \mathbf{I}$. **(3 marks)**

Given that $\mathbf{A} = \begin{pmatrix} 0 & 1 \\ 1 & 0 \end{pmatrix}$, by considering a matrix $\mathbf{B}$ of the form $\begin{pmatrix} a & b \\ c & d \end{pmatrix}$,

b show that $a = d$ and $b = c$. **(3 marks)**

10 $\mathbf{M} = \begin{pmatrix} 2 & 3 \\ k & -1 \end{pmatrix}$ where k is a constant.

a For which values of k does $\mathbf{M}$ have an inverse? **(2 marks)**

b Given that $\mathbf{M}$ is non-singular, find $\mathbf{M}^{-1}$ in terms of k. **(3 marks)**

11 Given that $\mathbf{A} = \begin{pmatrix} 4 & p \\ -2 & -2 \end{pmatrix}$ where p is a constant and $p \neq 4$,

a find $\mathbf{A}^{-1}$ in terms of p. **(2 marks)**

b Given that $\mathbf{A} + \mathbf{A}^{-1} = \begin{pmatrix} 5 & \frac{9}{2} \\ -3 & -4 \end{pmatrix}$, find the value of p. **(3 marks)**

12 $\mathbf{M} = \begin{pmatrix} k & -3 \\ 4 & k+3 \end{pmatrix}$ where k is a real constant.

a Find det $\mathbf{M}$ in terms of k. **(2 marks)**

b Show that $\mathbf{M}$ is non-singular for all values of k. **(3 marks)**

c Given that $10\mathbf{M}^{-1} + \mathbf{M} = \mathbf{I}$ where $\mathbf{I}$ is the 2×2 identity matrix, find the value of k. **(3 marks)**

13 Given that $\mathbf{A} = \begin{pmatrix} a & 2 \\ 3 & 2a \end{pmatrix}$ where a is a real constant,

a find $\mathbf{A}^{-1}$ in terms of a **(3 marks)**

b write down two values of a for which $\mathbf{A}^{-1}$ does not exist. **(1 mark)**

6.5 Inverting a 3 × 3 matrix

Finding the inverse of a 3 × 3 matrix is more complicated. You need to know the following definition.

- **The transpose of a matrix is found by interchanging the rows and the columns.**

Notation The transpose of the matrix **M** is written as $\mathbf{M}^{\mathrm{T}}$.

For example, if $\mathbf{A} = \begin{pmatrix} 1 & 4 \\ 2 & 3 \end{pmatrix}$, $\mathbf{A}^{\mathrm{T}} = \begin{pmatrix} 1 & 2 \\ 4 & 3 \end{pmatrix}$.

- **Finding the inverse of a 3 × 3 matrix A usually consists of the following 5 steps.**

Step 1 Find the determinant of A, det A.

Step 2 Form the matrix of the minors of A. In this chapter, the symbol M is used for the matrix of the minors unless this causes confusion with another matrix in the question.

In forming the matrix of minors, M, each of the nine elements of the matrix A is replaced by its minor.

Step 3 From the matrix of minors, form the matrix of cofactors by changing the signs of some elements of the matrix of minors according to the rule of alternating signs illustrated by the pattern

$$\begin{pmatrix} + & - & + \\ - & + & - \\ + & - & + \end{pmatrix}$$

You leave the elements of the matrix of minors corresponding to the + signs in this pattern unchanged. You change the signs of the elements corresponding to the – signs.

Notation A cofactor is a minor with its appropriate sign.

In this chapter, the symbol **C** is used for the matrix of the cofactors unless this causes confusion with another matrix in the question.

Step 4 Write down the transpose, $\mathbf{C}^{\mathrm{T}}$, of the matrix of cofactors.

Step 5 The inverse of the matrix A is given by the formula

$$\mathbf{A}^{-1} = \frac{1}{\det \mathbf{A}} \mathbf{C}^{\mathrm{T}}.$$

Each element of the matrix $\mathbf{C}^{\mathrm{T}}$ is divided by the determinant of **A**.

Example 17

The matrix $\mathbf{A} = \begin{pmatrix} 1 & 3 & 1 \\ 0 & 4 & 1 \\ 2 & -1 & 0 \end{pmatrix}$. Find $\mathbf{A}^{-1}$.

Step 1

$\det \mathbf{A} = 1\begin{vmatrix} 4 & 1 \\ -1 & 0 \end{vmatrix} - 3\begin{vmatrix} 0 & 1 \\ 2 & 0 \end{vmatrix} + 1\begin{vmatrix} 0 & 4 \\ 2 & -1 \end{vmatrix}$

$= 1(0 + 1) - 3(0 - 2) + 1(0 - 8)$

$= 1 + 6 - 8 = -1$

The first step of finding the inverse of a matrix is to evaluate its determinant.

Step 2

$$\mathbf{M} = \begin{pmatrix} \begin{vmatrix} 4 & 1 \\ -1 & 0 \end{vmatrix} & \begin{vmatrix} 0 & 1 \\ 2 & 0 \end{vmatrix} & \begin{vmatrix} 0 & 4 \\ 2 & -1 \end{vmatrix} \\ \begin{vmatrix} 3 & 1 \\ -1 & 0 \end{vmatrix} & \begin{vmatrix} 1 & 1 \\ 2 & 0 \end{vmatrix} & \begin{vmatrix} 1 & 3 \\ 2 & -1 \end{vmatrix} \\ \begin{vmatrix} 3 & 1 \\ 4 & 1 \end{vmatrix} & \begin{vmatrix} 1 & 1 \\ 0 & 1 \end{vmatrix} & \begin{vmatrix} 1 & 3 \\ 0 & 4 \end{vmatrix} \end{pmatrix}$$

$$= \begin{pmatrix} 1 & -2 & -8 \\ 1 & -2 & -7 \\ -1 & 1 & 4 \end{pmatrix}$$

The second step is to form the matrix of minors. The minor of an element is found by deleting the row and the column in which the element lies, then finding the determinant of the resulting 2×2 matrix.

For example, to find the minor of 4 in $\begin{pmatrix} 1 & 3 & 1 \\ 0 & 4 & 1 \\ 2 & -1 & 0 \end{pmatrix}$, delete the row and column containing 4, $\begin{pmatrix} 1 & \not{3} & 1 \\ \not{0} & \not{4} & \not{1} \\ 2 & \not{-1} & 0 \end{pmatrix}$. The minor is the determinant of the elements left, $\begin{vmatrix} 1 & 1 \\ 2 & 0 \end{vmatrix} = -2$.

Step 3

$$\mathbf{C} = \begin{pmatrix} 1 & 2 & -8 \\ -1 & -2 & 7 \\ -1 & -1 & 4 \end{pmatrix}$$

You find the matrix of cofactors by adjusting the signs of the minors using the pattern $\begin{pmatrix} + & - & + \\ - & + & - \\ + & - & + \end{pmatrix}$. Here you leave the elements $\begin{pmatrix} 1 & & -8 \\ & -2 & \\ -1 & & 4 \end{pmatrix}$ unchanged but change the signs of $\begin{pmatrix} & -2 & \\ 1 & & -7 \\ & 1 & \end{pmatrix}$.

Step 4

$$\mathbf{C}^{\mathrm{T}} = \begin{pmatrix} 1 & -1 & -1 \\ 2 & -2 & -1 \\ -8 & 7 & 4 \end{pmatrix}$$

Step 5

$$\mathbf{A}^{-1} = \frac{1}{\det \mathbf{A}}\mathbf{C}^{\mathrm{T}} = \frac{1}{-1}\begin{pmatrix} 1 & -1 & -1 \\ 2 & -2 & -1 \\ -8 & 7 & 4 \end{pmatrix}$$

$$= \begin{pmatrix} -1 & 1 & 1 \\ -2 & 2 & 1 \\ 8 & -7 & -4 \end{pmatrix}$$

You divide each term of the transpose of the matrix of cofactors, $\mathbf{C}^{\mathrm{T}}$, by the determinant of $\mathbf{A}$, -1.

Example 18

The matrix $\mathbf{A} = \begin{pmatrix} 3 & 2 & -2 \\ -2 & k & 0 \\ -1 & -3 & 3 \end{pmatrix}$, $k \neq 0$. Find $\mathbf{A}^{-1}$.

Step 1

$\det \mathbf{A} = 3\begin{vmatrix} k & 0 \\ -3 & 3 \end{vmatrix} - 2\begin{vmatrix} -2 & 0 \\ -1 & 3 \end{vmatrix} + (-2)\begin{vmatrix} -2 & k \\ -1 & -3 \end{vmatrix}$

$= 3(3k - 0) - 2(-6 - 0) - 2(6 + k)$

$= 9k + 12 - 12 - 2k = 7k$

Watch out Make sure you understand the steps needed to find the inverse of a 3×3 matrix. You won't be able to use your calculator if the matrix contains unknowns.

As you are given that $k \neq 0$, the matrix is non-singular and the inverse can be found.

Step 2

$$M = \begin{pmatrix} \begin{vmatrix} k & 0 \\ -3 & 3 \end{vmatrix} & \begin{vmatrix} -2 & 0 \\ -1 & 3 \end{vmatrix} & \begin{vmatrix} -2 & k \\ -1 & -3 \end{vmatrix} \\ \begin{vmatrix} 2 & -2 \\ -3 & 3 \end{vmatrix} & \begin{vmatrix} 3 & -2 \\ -1 & 3 \end{vmatrix} & \begin{vmatrix} 3 & 2 \\ -1 & -3 \end{vmatrix} \\ \begin{vmatrix} 2 & -2 \\ k & 0 \end{vmatrix} & \begin{vmatrix} 3 & -2 \\ -2 & 0 \end{vmatrix} & \begin{vmatrix} 3 & 2 \\ -2 & k \end{vmatrix} \end{pmatrix}$$

$$= \begin{pmatrix} 3k & -6 & k+6 \\ 0 & 7 & -7 \\ 2k & -4 & 3k+4 \end{pmatrix}$$

The second step is to find the matrix of minors in terms of k.

Step 3

$$C = \begin{pmatrix} 3k & 6 & k+6 \\ 0 & 7 & 7 \\ 2k & 4 & 3k+4 \end{pmatrix}$$

You obtain the matrix of the cofactors from the matrix of the minors by changing the signs of the elements corresponding to the – signs in the pattern $\begin{pmatrix} + & - & + \\ - & + & - \\ + & - & + \end{pmatrix}$.

Step 4

$$C^T = \begin{pmatrix} 3k & 0 & 2k \\ 6 & 7 & 4 \\ k+6 & 7 & 3k+4 \end{pmatrix}$$

Step 5

$$A^{-1} = \frac{1}{\det A} C^T = \frac{1}{7k}\begin{pmatrix} 3k & 0 & 2k \\ 6 & 7 & 4 \\ k+6 & 7 & 3k+4 \end{pmatrix}$$

You can leave the answer in this form or write the inverse matrix as $\begin{pmatrix} \frac{3}{7} & 0 & \frac{2}{7} \\ \frac{6}{7k} & \frac{1}{k} & \frac{4}{7k} \\ \frac{k+6}{7k} & \frac{1}{k} & \frac{3k+4}{7k} \end{pmatrix}$.

Example 19

The matrix $\mathbf{A} = \begin{pmatrix} -2 & 3 & -3 \\ 0 & 1 & 0 \\ 1 & -1 & 2 \end{pmatrix}$ and the matrix $\mathbf{B}$ is such that $(\mathbf{AB})^{-1} = \begin{pmatrix} 8 & -17 & 9 \\ -5 & 10 & -6 \\ -3 & 5 & -4 \end{pmatrix}$.

a Show that $\mathbf{A}^{-1} = \mathbf{A}$.

b Find $\mathbf{B}^{-1}$.

a $$A^2 = \begin{pmatrix} -2 & 3 & -3 \\ 0 & 1 & 0 \\ 1 & -1 & 2 \end{pmatrix}\begin{pmatrix} -2 & 3 & -3 \\ 0 & 1 & 0 \\ 1 & -1 & 2 \end{pmatrix}$$

$$= \begin{pmatrix} 4+0-3 & -6+3+3 & 6+0-6 \\ 0+0+0 & 0+1+0 & 0+0+0 \\ -2+0+2 & 3-1-2 & -3+0+4 \end{pmatrix}$$

$$= \begin{pmatrix} 1 & 0 & 0 \\ 0 & 1 & 0 \\ 0 & 0 & 1 \end{pmatrix} = I$$

$AA = I$

$A^{-1}AA = A^{-1}$

$A = A^{-1}$ as required

Problem-solving

Proving $\mathbf{A} = \mathbf{A}^{-1}$ is equivalent to proving $\mathbf{A}^2 = \mathbf{I}$. You still need to add working to show that $\mathbf{A}^2 = \mathbf{I}$ implies that $\mathbf{A} = \mathbf{A}^{-1}$.

Multiply both sides by $\mathbf{A}^{-1}$.

Since $\mathbf{A}^{-1}\mathbf{AA} = (\mathbf{A}^{-1}\mathbf{A})\mathbf{A} = \mathbf{IA} = \mathbf{A}$.

b $(AB)^{-1} = B^{-1}A^{-1}$

$(AB)^{-1}A = B^{-1}A^{-1}A = B^{-1}I = B^{-1}$

$B^{-1} = (AB)^{-1}A$

$$= \begin{pmatrix} 8 & -17 & 9 \\ -5 & 10 & -6 \\ -3 & 5 & -4 \end{pmatrix}\begin{pmatrix} -2 & 3 & -3 \\ 0 & 1 & 0 \\ 1 & -1 & 2 \end{pmatrix}$$

$$= \begin{pmatrix} -16+0+9 & 24-17-9 & -24+0+18 \\ 10+0-6 & -15+10+6 & 15+0-12 \\ 6+0-4 & -9+5+4 & 9+0-8 \end{pmatrix}$$

$$= \begin{pmatrix} -7 & -2 & -6 \\ 4 & 1 & 3 \\ 2 & 0 & 1 \end{pmatrix}$$

Multiply both sides of this formula on the right by **A** and use $\mathbf{A}^{-1}\mathbf{A} = \mathbf{I}$ to obtain an expression for $\mathbf{B}^{-1}$ in terms of $(\mathbf{AB})^{-1}$ and **A**, both of which you already know.

You could check your answer by multiplying these matrices quickly using your calculator.

Notation If $\mathbf{A}^{-1} = \mathbf{A}$, then the matrix **A** is said to be **self-inverse**.

Exercise 6E

1 Use your calculator to find the inverses of these matrices.

a $\begin{pmatrix} 1 & 0 & 0 \\ 0 & 2 & 1 \\ 0 & 1 & 2 \end{pmatrix}$ **b** $\begin{pmatrix} 1 & 0 & 0 \\ 0 & 2 & 0 \\ 0 & 0 & 3 \end{pmatrix}$ **c** $\begin{pmatrix} 1 & 0 & 0 \\ 0 & \frac{3}{5} & -\frac{4}{5} \\ 0 & \frac{4}{5} & \frac{3}{5} \end{pmatrix}$

2 Without using a calculator, find the inverses of these matrices.

a $\begin{pmatrix} 1 & -3 & 2 \\ 0 & -2 & 1 \\ 3 & 0 & 2 \end{pmatrix}$ **b** $\begin{pmatrix} 2 & 3 & 2 \\ 3 & -2 & 1 \\ 2 & 1 & 1 \end{pmatrix}$ **c** $\begin{pmatrix} 3 & 2 & -7 \\ 1 & -3 & 1 \\ 0 & 2 & -2 \end{pmatrix}$

3 The matrix $\mathbf{A} = \begin{pmatrix} 1 & 0 & 1 \\ 0 & 1 & 0 \\ 2 & 0 & 1 \end{pmatrix}$ and the matrix $\mathbf{B} = \begin{pmatrix} 2 & 1 & -1 \\ 1 & 0 & 1 \\ 1 & 2 & 1 \end{pmatrix}$.

a Find $\mathbf{A}^{-1}$. **b** Find $\mathbf{B}^{-1}$.

Given that $(\mathbf{AB})^{-1} = \begin{pmatrix} -\frac{2}{3} & \frac{1}{2} & \frac{1}{2} \\ 1 & -\frac{1}{2} & -\frac{1}{2} \\ \frac{2}{3} & \frac{1}{2} & -\frac{1}{2} \end{pmatrix}$

c verify that $\mathbf{B}^{-1}\mathbf{A}^{-1} = (\mathbf{AB})^{-1}$.

(P) **4** The matrix $\mathbf{A} = \begin{pmatrix} 2 & 0 & 3 \\ k & 1 & 1 \\ 1 & 1 & 4 \end{pmatrix}$.

a Show that $\det \mathbf{A} = 3(k + 1)$. **(3 marks)**

b Given that $k \neq -1$, find $\mathbf{A}^{-1}$. **(4 marks)**

E/P **5** The matrix $\mathbf{A} = \begin{pmatrix} 5 & a & 4 \\ b & -7 & 8 \\ 2 & -2 & c \end{pmatrix}$.

Given that $\mathbf{A} = \mathbf{A}^{-1}$, find the values of the constants a, b and c. **(6 marks)**

6 The matrix $\mathbf{A} = \begin{pmatrix} 2 & -1 & 1 \\ 4 & -3 & 0 \\ -3 & 3 & 1 \end{pmatrix}$.

a Show that $\mathbf{A}^3 = \mathbf{I}$.

b Hence find $\mathbf{A}^{-1}$.

7 The matrix $\mathbf{A} = \begin{pmatrix} 1 & 1 & 0 \\ 3 & -3 & 1 \\ 0 & 3 & 2 \end{pmatrix}$.

a Show that $\mathbf{A}^3 = 13\mathbf{A} - 15\mathbf{I}$.

b Deduce that $15\mathbf{A}^{-1} = 13\mathbf{I} - \mathbf{A}^2$.

c Hence find $\mathbf{A}^{-1}$.

8 The matrix $\mathbf{A} = \begin{pmatrix} 2 & 0 & 1 \\ 4 & 3 & -2 \\ 0 & 3 & -4 \end{pmatrix}$.

a Show that $\mathbf{A}$ is singular.

The matrix $\mathbf{C}$ is the matrix of the cofactors of $\mathbf{A}$.

b Find $\mathbf{C}$.

c Show that $\mathbf{AC}^{\mathrm{T}} = \mathbf{0}$.

E/P **9** $\mathbf{M} = \begin{pmatrix} 2 & k & 3 \\ -1 & 2 & 1 \\ 1 & -1 & -1 \end{pmatrix}$ where k is a real constant.

a For which values of k does $\mathbf{M}$ have an inverse? **(2 marks)**

b Given that $\mathbf{M}$ is non-singular, find $\mathbf{M}^{-1}$ in terms of k. **(4 marks)**

E/P **10** $\mathbf{A} = \begin{pmatrix} p & 2p & 3 \\ 4 & -1 & 1 \\ 1 & -2 & 0 \end{pmatrix}$ where p is a real constant.

Given that $\mathbf{A}$ is non-singular, find $\mathbf{A}^{-1}$ in terms of p. **(4 marks)**

6.6 Solving systems of equations using matrices

You can use the inverse of a 3 × 3 matrix to solve a system of three simultaneous linear equations in three unknowns.

- **If** $\mathbf{A}\begin{pmatrix} x \\ y \\ z \end{pmatrix} = \mathbf{v}$ **then** $\begin{pmatrix} x \\ y \\ z \end{pmatrix} = \mathbf{A}^{-1}\mathbf{v}$**.**

If **A** is non-singular, a unique solution for $\begin{pmatrix} x \\ y \\ z \end{pmatrix}$ can be found for any vector **v**.

Example 20

Use an inverse matrix to solve the simultaneous equations:

$$-x + 6y - 2z = 21$$
$$6x - 2y - z = -16$$
$$-2x + 3y + 5z = 24$$

Write the system of equations using matrices:

$$\begin{pmatrix} -1 & 6 & -2 \\ 6 & -2 & -1 \\ -2 & 3 & 5 \end{pmatrix}\begin{pmatrix} x \\ y \\ z \end{pmatrix} = \begin{pmatrix} 21 \\ -16 \\ 24 \end{pmatrix}$$

Find the inverse of the left-hand matrix:

$$\frac{1}{189}\begin{pmatrix} 7 & 36 & 10 \\ 28 & 9 & 13 \\ -14 & 9 & 34 \end{pmatrix}$$

Left-multiply the right-hand matrix by this inverse:

$$\frac{1}{189}\begin{pmatrix} 7 & 36 & 10 \\ 28 & 9 & 13 \\ -14 & 9 & 34 \end{pmatrix}\begin{pmatrix} 21 \\ -16 \\ 24 \end{pmatrix} = \frac{1}{189}\begin{pmatrix} -189 \\ 756 \\ 378 \end{pmatrix} = \begin{pmatrix} -1 \\ 4 \\ 2 \end{pmatrix}$$

Hence $x = -1$, $y = 4$ and $z = 2$.

You can confirm that this is equivalent to the original equations by multiplying out the left-hand side:

$$\begin{pmatrix} -1 & 6 & -2 \\ 6 & -2 & -1 \\ -2 & 3 & 5 \end{pmatrix}\begin{pmatrix} x \\ y \\ z \end{pmatrix} = \begin{pmatrix} -x + 6y - 2z \\ 6x - 2y - z \\ -2x + 3y + 5z \end{pmatrix}$$

Use your calculator to find the inverse matrix and then to multiply the matrices.

Problem-solving

Once you have found the inverse matrix, you could use it to solve a similar system of equations with a different answer vector.

Example 21

A colony of 1000 mole-rats is made up of adult males, adult females and youngsters. Originally there were 100 more adult females than adult males.

After one year:

- the number of adult males had increased by 2%
- the number of adult females had increased by 3%
- the number of youngsters had decreased by 4%
- the total number of mole-rats had decreased by 20

Form and solve a matrix equation to find out how many of each type of mole-rat were in the original colony.

Problem-solving

Assign a letter to each unknown value, then work your way through the question text using the information to formulate equations.

x = number of adult males
y = number of adult females
z = number of youngsters

$$x + y + z = 1000$$
$$x - y = -100$$
$$1.02x + 1.03y + 0.96z = 980$$

This represents the total number of mole-rats in the original colony.

There are 100 more adult females than adult males.

This equation represents the number of mole-rats of each type after 1 year. You could also consider the percentage changes in each population and the total overall change of −20:

$0.02x + 0.03y - 0.04z = -20$

So $\begin{pmatrix} 1 & 1 & 1 \\ 1 & -1 & 0 \\ 1.02 & 1.03 & 0.96 \end{pmatrix}\begin{pmatrix} x \\ y \\ z \end{pmatrix} = \begin{pmatrix} 1000 \\ -100 \\ 980 \end{pmatrix}$

Convert the three equations into a matrix equation.

$\mathbf{A}^{-1} = \frac{1}{13}\begin{pmatrix} -96 & 7 & 100 \\ -96 & -6 & 100 \\ 205 & -1 & -200 \end{pmatrix}$

Use your calculator to find $\mathbf{A}^{-1}$.

$\begin{pmatrix} x \\ y \\ z \end{pmatrix} = \mathbf{A}^{-1}\begin{pmatrix} 1000 \\ -100 \\ 980 \end{pmatrix}$

$= \frac{1}{13}\begin{pmatrix} -96 & 7 & 100 \\ -96 & -6 & 100 \\ 205 & -1 & -200 \end{pmatrix}\begin{pmatrix} 1000 \\ -100 \\ 980 \end{pmatrix}$

$\begin{pmatrix} x \\ y \\ z \end{pmatrix} = \begin{pmatrix} 100 \\ 200 \\ 700 \end{pmatrix}$

Use your calculator to multiply $\mathbf{A}^{-1}$ by the answer vector.

There were 100 adult males, 200 adult females and 700 youngsters in the original colony.

Write your answer in the context of the question, and check that it makes sense. You could also check that your answer matches the information given in the question. For example, the initial number of mole-rats is 100 + 200 + 700 = 1000, as given in the question.

You need to be able to determine whether a system of three linear equations in three unknowns is **consistent** or **inconsistent**.

- **A system of linear equations is consistent if there is at least one set of values that satisfies all the equations simultaneously. Otherwise, it is inconsistent.**

If the matrix corresponding to a set of linear equations is non-singular, then the system has one unique solution and is consistent. However, if the matrix is singular, there are two possibilities: either the system is consistent and has infinitely many solutions, or it is inconsistent and has no solutions.

You can visualise the different situations by considering the points of intersection of the planes corresponding to each equation. Here are some of the different possible configurations:

Links An equation in the form $ax + by + cz = d$ is the equation of a plane in three dimensions.
→ Section 9.2

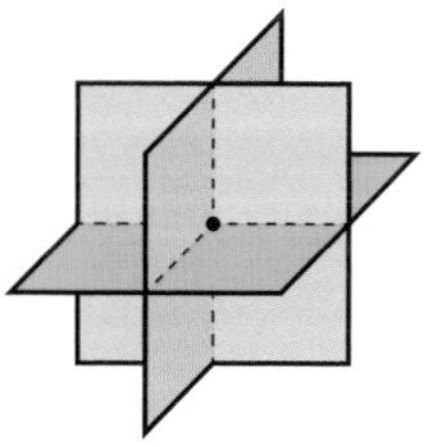

The planes meet at a **point**. The system of equations is **consistent** and has **one solution** represented by this point. This is the only case when the corresponding matrix is **non-singular**.

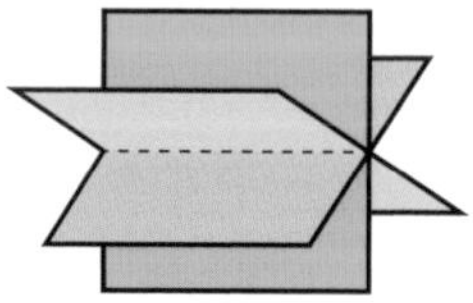

The planes form a **sheaf**. The system of equations is **consistent** and has **infinitely many solutions** represented by the line of intersection of the three planes.

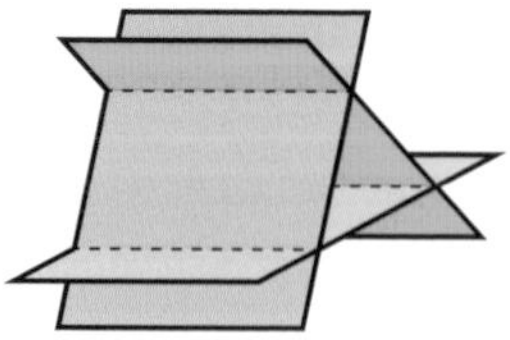

The planes form a **prism**. The system of equations is **inconsistent** and has **no solutions**.

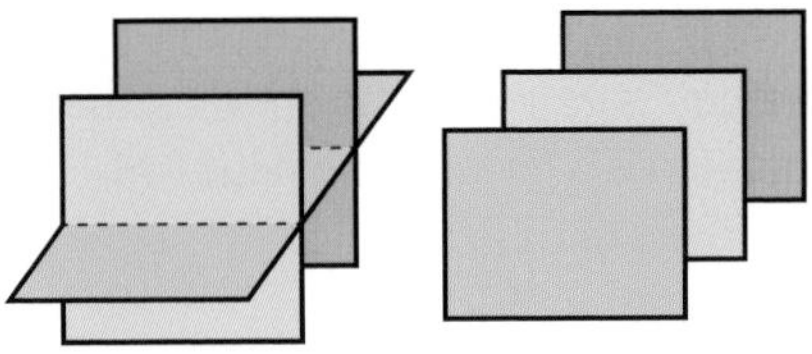

Two or more of the planes are parallel and non-identical. The system of equations is **inconsistent** and has **no solutions**.

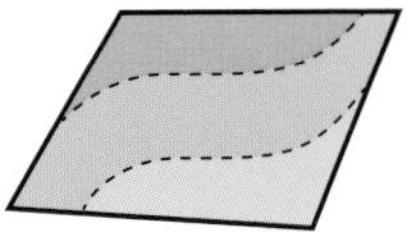

All three equations represent the same plane. In this case the system of equations is **consistent** and has **infinitely many solutions**.

Hint If one row of the corresponding matrix is a **linear multiple** of another row, then these two rows will represent parallel planes.

Example 22

A system of equations is shown below:

$$3x - ky - 6z = k$$
$$kx + 3y + 3z = 2$$
$$-3x - y + 3z = -2$$

For each of the following values of k, determine whether the system of equations is consistent or inconsistent. If the system is consistent, determine whether there is a unique solution or an infinity of solutions. In each case, identify the geometric configuration of the planes corresponding to each value of k.

a $k = 0$ **b** $k = 1$ **c** $k = -6$

a $k = 0$: $\begin{vmatrix} 3 & 0 & -6 \\ 0 & 3 & 3 \\ -3 & -1 & 3 \end{vmatrix} = -18$

Use your calculator to find the determinant of the corresponding matrix.

The corresponding matrix is non-singular so the system is consistent and has a unique solution. The planes meet at a single point.

Problem-solving

If the matrix is singular, you need to consider the equations to determine whether the system is consistent. Eliminate one of the variables from two different pairs of equations. If the resulting two equations are consistent then the system will be consistent.

b $k = 1$: $\begin{vmatrix} 3 & -1 & -6 \\ 1 & 3 & 3 \\ -3 & -1 & 3 \end{vmatrix} = 0$

$3x - y - 6z = 1$ (1)

$x + 3y + 3z = 2$ (2)

$-3x - y + 3z = -2$ (3)

(1) + 2 × (2): $5x + 5y = 5$ (4)

(2) − (3): $4x + 4y = 4$ (5)

Eliminate z from equations (1) and (2) to form equation (4), and eliminate z from equations (2) and (3) to form equation (5).

Equations (4) and (5) are consistent so the system is consistent and has an infinity of solutions. The planes meet in a sheaf.

Equations (4) and (5) are consistent because one is a linear multiple of the other. Any values of x and y that satisfy one equation will also satisfy the other.

c $k = -6$: $\begin{vmatrix} 3 & 6 & -6 \\ -6 & 3 & 3 \\ -3 & -1 & 3 \end{vmatrix} = 0$

$$\begin{aligned} 3x + 6y - 6z &= -6 && (1) \\ -6x + 3y + 3z &= 2 && (2) \\ -3x - y + 3z &= -2 && (3) \end{aligned}$$

(1) + (3): $5y - 3z = -8$ (4)

2 × (1) + (2): $15y - 9z = -10$ (5)

Equations (4) and (5) are inconsistent so the system is inconsistent and has no solutions. The planes form a prism.

Problem-solving

This method is equivalent to showing that one equation is a linear combination of the other two. In this case $4 \times (1) + 3 \times (2) = -5 \times (3)$.

Eliminate x in two different ways to get two equations in y and z.

$3 \times (4)$ gives $15y - 9z = -24$. There are no values of y and z that can satisfy this equation and equation (5) simultaneously.

Exercise 6F

1 Solve the following systems of equations using inverse matrices.

a $\begin{aligned} 2x - 6y + 4z &= 32 \\ 3x + 2y - 9z &= -49 \\ -2x + 4y + z &= -3 \end{aligned}$

b $\begin{aligned} -4x + 6y - 2z &= -22 \\ 3x + 3y - 2z &= 1 \\ -6x - 7y + 3z &= 3 \end{aligned}$

c $\begin{aligned} 4x + 7y - 2z &= 21 \\ -10x - y - 7z &= 0 \\ -2x + y - 4z &= 9 \end{aligned}$

d $\begin{aligned} -3x - 6y + 4z &= -23 \\ -3x + 6y - 10z &= -1 \\ 3x + 7y - 3z &= 27 \end{aligned}$

E/P **2** Three planes A, B and C are defined by the following equations.

A: $x - 3y - 4z = 3$

B: $6x + 5y - 7z = 30$

C: $x + 4y + 6z = -3$

By constructing and solving a suitable matrix equation, show that these three planes intersect at a single point and find the coordinates of that point. **(5 marks)**

E/P **3** Phyllis invested £3000 across three savings accounts, A, B and C. She invested £190 more in account A than in account B.
After two years, account A had increased in value by 1%, account B had increased in value by 2.5% and account C had decreased in value by 1.5%. The total value of Phyllis's savings had increased by £41.
Form and solve a matrix equation to find out how much money was invested by Phyllis in each account. **(7 marks)**

E/P **4** A colony of bats is made up of brown bats, grey bats and black bats. Initially there are 2000 bats and there are 250 more brown bats than grey bats.
After one year:

- the number of brown bats had fallen by 1%
- the number of grey bats had fallen by 2%

- the number of black bats had increased by 4%
- overall there were 40 more bats

Form and solve a matrix equation to find out how many of each colour bat there were in the initial colony. **(7 marks)**

E/P **5** Three planes A, B and C are defined by the following equations:

$$A: \quad x + ay + 2z = a$$
$$B: \quad x - y - z = a$$
$$C: \quad x + 4y + 4z = 0$$

Given that the planes do not meet at a single point,

a find the value of a **(4 marks)**

b determine whether the three equations form a consistent system, and give a geometric interpretation of your answer. **(4 marks)**

Hint If the three planes do not meet at a single point, the corresponding 3×3 matrix must be singular.

E/P **6** The matrix $\mathbf{M}$ is given by $\mathbf{M} = \begin{pmatrix} 1 & 4 & q \\ 2 & 3 & -3 \\ q & q & -2 \end{pmatrix}$

a Given that det $\mathbf{M} = 0$, show that $q^2 + 9q - 10 = 0$. **(4 marks)**

A system of simultaneous equations is shown below:

$$x + 4y + qz = -16$$
$$2x + 3y - 3z = \tfrac{1}{2}q$$
$$qx + qy - 2z = -2$$

b For each of the following values of q, determine whether the system of equations is consistent or inconsistent. If the system is consistent, determine whether there is a unique solution or an infinity of solutions. In each case, identify the geometric configuration of the planes corresponding to each equation.

i $q = -10$ **ii** $q = 2$ **iii** $q = 1$ **(7 marks)**

Mixed exercise 6

P **1** The matrix $\mathbf{A} = \begin{pmatrix} 1 & -3 \\ 2 & 1 \end{pmatrix}$ and $\mathbf{AB} = \begin{pmatrix} 4 & 1 & 9 \\ 1 & 9 & 4 \end{pmatrix}$. Find the matrix $\mathbf{B}$.

E/P **2** The matrix $\mathbf{A} = \begin{pmatrix} a & b \\ 2a & 3b \end{pmatrix}$, where a and b are non-zero constants.

a Find $\mathbf{A}^{-1}$, giving your answer in terms of a and b. **(2 marks)**

The matrix $\mathbf{Y} = \begin{pmatrix} a & 2b \\ 2a & b \end{pmatrix}$ and the matrix $\mathbf{X}$ is given by $\mathbf{XA} = \mathbf{Y}$.

b Find $\mathbf{X}$, giving your answer in terms of a and b. **(3 marks)**

(E) **3** The 2 × 2, non-singular matrices **A**, **B** and **X** satisfy **XB** = **BA**.

a Find an expression for **X** in terms of **A** and **B**. **(1 mark)**

b Given that $\mathbf{A} = \begin{pmatrix} 5 & 3 \\ 0 & -2 \end{pmatrix}$ and $\mathbf{B} = \begin{pmatrix} 2 & 1 \\ -1 & -1 \end{pmatrix}$, find **X**. **(2 marks)**

(E/P) **4** A matrix **A** is given as $\mathbf{A} = \begin{pmatrix} a & 2 & -1 \\ -1 & 1 & -1 \\ b & 2 & 1 \end{pmatrix}$.

Given that $\mathbf{A}^2 = \begin{pmatrix} -4 & 2 & -4 \\ -5 & -3 & -1 \\ 4 & 10 & -4 \end{pmatrix}$, find the values of a and b. **(3 marks)**

(E/P) **5** $\mathbf{A} = \begin{pmatrix} 1 & 0 & 2 \\ t & 3 & 1 \\ -2 & -1 & 1 \end{pmatrix}$

Given that **A** is singular, find the value of t. **(3 marks)**

(E/P) **6** $\mathbf{M} = \begin{pmatrix} 1 & 0 & 0 \\ x & 2 & 0 \\ 3 & 1 & 1 \end{pmatrix}$

Find $\mathbf{M}^{-1}$ in terms of x. **(4 marks)**

(E/P) **7** $\mathbf{A} = \begin{pmatrix} k & -2 \\ -4 & k \end{pmatrix}$ where k is a real constant.

a For which values of k does **A** have an inverse? **(2 marks)**

b Given that **A** is non-singular, find $\mathbf{A}^{-1}$ in terms of k. **(3 marks)**

(E/P) **8** $\mathbf{B} = \begin{pmatrix} k & 6 \\ -1 & k-2 \end{pmatrix}$ where k is a real constant.

a Find det **B** in terms of k. **(2 marks)**

b Show that **B** is non-singular for all values of k. **(3 marks)**

c Given that $21\mathbf{B}^{-1} + \mathbf{B} = -8\mathbf{I}$ where **I** is the 2 × 2 identity matrix, find the value of k. **(3 marks)**

(E/P) **9** Given that $\mathbf{M} = \begin{pmatrix} 2 & -m \\ m & -1 \end{pmatrix}$ where m is a real constant,

a write down two values of m such that **M** is singular **(2 marks)**

b find $\mathbf{M}^{-1}$ in terms of m, given that **M** is non-singular. **(3 marks)**

(E/P) **10** $\mathbf{A} = \begin{pmatrix} 3 & 4 & 5 \\ 1 & a & -1 \\ -2 & 1 & 1 \end{pmatrix}$ where a is a real constant.

a For which values of a does **A** have an inverse? **(2 marks)**

b Given that **A** is non-singular, find $\mathbf{A}^{-1}$ in terms of a. **(4 marks)**

P **11** Three planes A, B and C are defined by the following equations.

$$A: \quad x + y + z = 6$$
$$B: \quad x - 4y + 2z = -2$$
$$C: \quad 2x + y - 3z = 0$$

By constructing and solving a suitable matrix equation, show that these three planes intersect at a single point and find the coordinates of that point. **(5 marks)**

P **12** A sheep farmer has three types of sheep: Hampshire, Dorset horn and Wiltshire horn. Initially his flock had 2500 sheep in it. There were 300 more Hampshire sheep than Wiltshire horn.

After one year:

- the number of Hampshire sheep had increased by 6%
- the number of Dorset horn had increased by 4%
- the number of Wiltshire horn had increased by 3%
- overall the flock size had increased by 110

Form and solve a matrix equation to find out how many of each type of sheep there were in the initial flock. **(7 marks)**

P **13** **a** Determine the values of the real constants a and b for which there are infinitely many solutions to the simultaneous equations

$$2x + 3y + z = 6$$
$$-x + y + 2z = 7$$
$$ax + y + 4z = b$$

(6 marks)

b Give a geometric interpretation of the three planes formed by these equations. **(1 mark)**

Challenge

Given that **A** and **B** are 2×2 matrices, prove that $\det(\mathbf{AB}) = \det \mathbf{A} \det \mathbf{B}$.

Summary of key points

1 A **square matrix** is one where the numbers of rows and columns are the same.

2 A zero matrix is one in which all of the numbers are zero. The zero matrix is denoted by 0.

3 An identity matrix is a square matrix in which the numbers in the leading diagonal (starting top left) are 1 and all the rest are 0. Identity matrices are denoted by $\mathbf{I}_k$ where k describes the size. The 3 × 3 identity matrix is $\mathbf{I}_3 = \begin{pmatrix} 1 & 0 & 0 \\ 0 & 1 & 0 \\ 0 & 0 & 1 \end{pmatrix}$

4 To add or subtract matrices, you add or subtract the corresponding elements in each matrix. You can only add or subtract matrices that are the same size.

5 To multiply a matrix by a scalar, you multiply every element in the matrix by that scalar.

6
- Matrices can be multiplied together if the number of columns in the first matrix is equal to the number of rows in the second matrix. In this case the first is said to be multiplicatively conformable with the second.
- To find the product of two multiplicatively conformable matrices, you multiply the elements in each row in the left-hand matrix by the corresponding elements in each column in the right-hand matrix, then add the results together.

7 For a 2 × 2 matrix $\mathbf{M} = \begin{pmatrix} a & b \\ c & d \end{pmatrix}$, the determinant of **M** is $ad - bc$.

8
- If det **M** = 0 then **M** is a **singular** matrix.
- If det **M** ≠ 0 then **M** is a **non-singular** matrix.

9 You find the determinant of a 3 × 3 matrix by reducing the 3 × 3 determinant to 2 × 2 determinants using the formula:

$$\begin{vmatrix} a & b & c \\ d & e & f \\ g & h & i \end{vmatrix} = a\begin{vmatrix} e & f \\ h & i \end{vmatrix} - b\begin{vmatrix} d & f \\ g & i \end{vmatrix} + c\begin{vmatrix} d & e \\ g & h \end{vmatrix}$$

10 The **minor** of an element in a 3 × 3 matrix is the determinant of the 2 × 2 matrix that remains after the row and column containing that element have been crossed out.

11 The **inverse** of a matrix **M** is the matrix $\mathbf{M}^{-1}$ such that $\mathbf{M}\mathbf{M}^{-1} = \mathbf{M}^{-1}\mathbf{M} = \mathbf{I}$.

12 If $\mathbf{M} = \begin{pmatrix} a & b \\ c & d \end{pmatrix}$, then $\mathbf{M}^{-1} = \frac{1}{\det\mathbf{M}}\begin{pmatrix} d & -b \\ -c & a \end{pmatrix}$.

13 If **A** and **B** are non-singular matrices, then $(\mathbf{AB})^{-1} = \mathbf{B}^{-1}\mathbf{A}^{-1}$.

14 The **transpose** of a matrix is found by interchanging the rows and the columns.

15 Finding the inverse of a 3 × 3 matrix **A** usually consists of the following 5 steps.

Step 1 Find the determinant of **A**, det **A**.

Step 2 Form the matrix of the minors of **A**, **M**.

In forming the matrix **M**, each of the nine elements of the matrix **A** is replaced by its minor.

Step 3 From the matrix of minors, form the matrix of **cofactors**, **C**, by changing the signs of some elements of the matrix of minors according to the **rule of alternating signs** illustrated by the pattern

$$\begin{pmatrix} + & - & + \\ - & + & - \\ + & - & + \end{pmatrix}$$

You leave the elements of the matrix of minors corresponding to the + signs in this pattern unchanged. You change the signs of the elements corresponding to the – signs.

Step 4 Write down the transpose, $\mathbf{C}^{\mathrm{T}}$, of the matrix of cofactors.

Step 5 The inverse of the matrix **A** is given by the formula

$$\mathbf{A}^{-1} = \frac{1}{\det \mathbf{A}}\mathbf{C}^{\mathrm{T}}$$

16 If $\mathbf{A}\begin{pmatrix} x \\ y \\ z \end{pmatrix} = \mathbf{v}$ then $\begin{pmatrix} x \\ y \\ z \end{pmatrix} = \mathbf{A}^{-1}\mathbf{v}$.

17 A system of linear equations is **consistent** if there is at least one set of values that satisfies all the equations simultaneously. Otherwise, it is **inconsistent**.

Linear transformations

7

Objectives

After completing this chapter you should be able to:

- Understand the properties of linear transformations and represent them using matrices → pages 127–130
- Perform reflections and rotations using matrices → pages 131–136
- Carry out enlargements and stretches using matrices → pages 136–140
- Find the coordinates of invariant points and the equations of invariant lines → pages 131–140
- Carry out successive transformations using matrix products → pages 140–144
- Understand linear transformations in three dimensions → pages 144–147
- Use inverse matrices to reverse linear transformations → pages 148–150

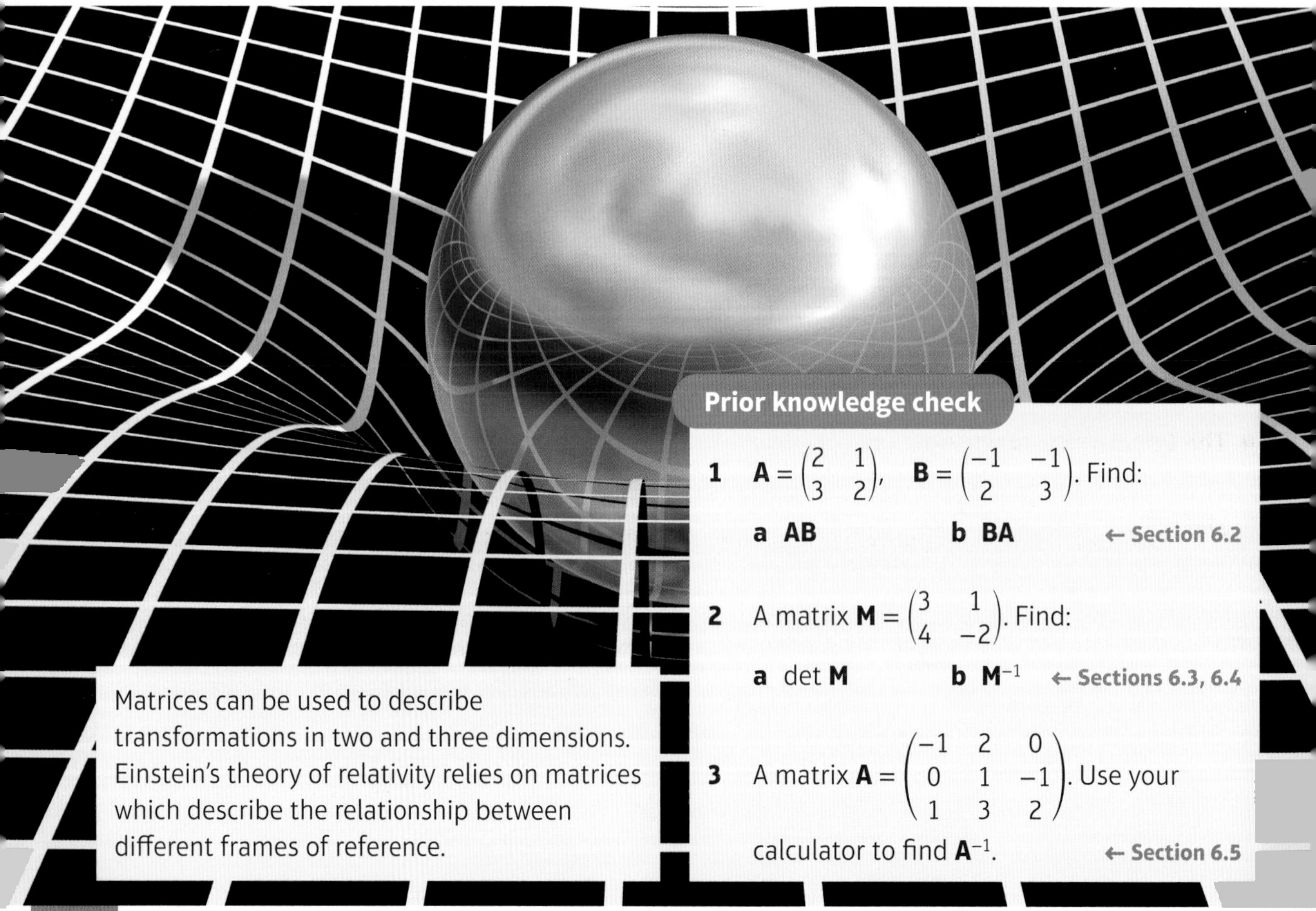

Matrices can be used to describe transformations in two and three dimensions. Einstein's theory of relativity relies on matrices which describe the relationship between different frames of reference.

Prior knowledge check

1 $\mathbf{A} = \begin{pmatrix} 2 & 1 \\ 3 & 2 \end{pmatrix}$, $\mathbf{B} = \begin{pmatrix} -1 & -1 \\ 2 & 3 \end{pmatrix}$. Find:

a $\mathbf{AB}$ **b** $\mathbf{BA}$ ← Section 6.2

2 A matrix $\mathbf{M} = \begin{pmatrix} 3 & 1 \\ 4 & -2 \end{pmatrix}$. Find:

a $\det \mathbf{M}$ **b** $\mathbf{M}^{-1}$ ← Sections 6.3, 6.4

3 A matrix $\mathbf{A} = \begin{pmatrix} -1 & 2 & 0 \\ 0 & 1 & -1 \\ 1 & 3 & 2 \end{pmatrix}$. Use your calculator to find $\mathbf{A}^{-1}$. ← Section 6.5

7.1 Linear transformations in two dimensions

You can define a transformation in two dimensions by describing how a general point with position vector $\begin{pmatrix} x \\ y \end{pmatrix}$ is transformed. The new point is called the **image**.

Example 1

The three transformations S, T and U are defined as follows. Find the image of the point (2, 3) under each of these transformations.

$$S: \begin{pmatrix} x \\ y \end{pmatrix} \mapsto \begin{pmatrix} x + 4 \\ y - 1 \end{pmatrix} \qquad T: \begin{pmatrix} x \\ y \end{pmatrix} \mapsto \begin{pmatrix} 2x - y \\ x + y \end{pmatrix} \qquad U: \begin{pmatrix} x \\ y \end{pmatrix} \mapsto \begin{pmatrix} 2y \\ -x^2 \end{pmatrix}$$

$S: \begin{pmatrix} 2 \\ 3 \end{pmatrix} \mapsto \begin{pmatrix} 2 + 4 \\ 3 - 1 \end{pmatrix} = \begin{pmatrix} 6 \\ 2 \end{pmatrix}$. The image is (6, 2)

Substitute $x = 2$ and $y = 3$ into the expressions for the image of the general point.

$T: \begin{pmatrix} 2 \\ 3 \end{pmatrix} \mapsto \begin{pmatrix} 2 \times 2 - 3 \\ 2 + 3 \end{pmatrix} = \begin{pmatrix} 1 \\ 5 \end{pmatrix}$. The image is (1, 5)

Substituting $x = 2$ and $y = 3$ into $2x - y$ gives 1 and into $x + y$ gives 5.

$U: \begin{pmatrix} 2 \\ 3 \end{pmatrix} \mapsto \begin{pmatrix} 2 \times 3 \\ -2^2 \end{pmatrix} = \begin{pmatrix} 6 \\ -4 \end{pmatrix}$. The image is (6, −4)

When $x = 2$, $-x^2 = -2^2 = -4$

A **linear transformation** has the special properties that the transformation only involves linear terms in x and y. In the example above, T is a linear transformation while S and U are not.

Watch out S represents a translation, but this is not a linear transformation since you can't write $x + 4$ in the form $ax + by$.

- **Linear transformations always map the origin onto itself.**
- **Any linear transformation can be represented by a matrix.**
- **The linear transformation $T: \begin{pmatrix} x \\ y \end{pmatrix} \mapsto \begin{pmatrix} ax + by \\ cx + dy \end{pmatrix}$ can be represented by the matrix $\mathbf{M} = \begin{pmatrix} a & b \\ c & d \end{pmatrix}$**

since $\begin{pmatrix} a & b \\ c & d \end{pmatrix}\begin{pmatrix} x \\ y \end{pmatrix} = \begin{pmatrix} ax + by \\ cx + dy \end{pmatrix}$.

Note You can transform any point P by multiplying the transformation matrix by the position vector of P.

Example 2

Find matrices to represent these linear transformations.

a $T: \begin{pmatrix} x \\ y \end{pmatrix} \mapsto \begin{pmatrix} 2y + x \\ 3x \end{pmatrix}$ **b** $V: \begin{pmatrix} x \\ y \end{pmatrix} \mapsto \begin{pmatrix} -2y \\ 3x + y \end{pmatrix}$

a Transformation T is equivalent to

$T: \begin{pmatrix} x \\ y \end{pmatrix} \mapsto \begin{pmatrix} 1x + 2y \\ 3x + 0y \end{pmatrix}$

Write the transformation in the form $\begin{pmatrix} ax + by \\ cx + dy \end{pmatrix}$.

so the matrix is $\begin{pmatrix} 1 & 2 \\ 3 & 0 \end{pmatrix}$.

Use the coefficients of x and y to form the matrix.

b Transformation V is equivalent to

$V: \begin{pmatrix} x \\ y \end{pmatrix} \mapsto \begin{pmatrix} 0x - 2y \\ 3x + y \end{pmatrix}$

Write the transformation in the form $\begin{pmatrix} ax + by \\ cx + dy \end{pmatrix}$.

so the matrix is $\begin{pmatrix} 0 & -2 \\ 3 & 1 \end{pmatrix}$.

Use the coefficients of x and y to form the matrix.

Example 3

a The square S has coordinates (1, 1), (3, 1), (3, 3) and (1, 3). Find the vertices of the image of S under the transformation given by the matrix $\mathbf{M} = \begin{pmatrix} -1 & 2 \\ 2 & 1 \end{pmatrix}$.

b Sketch S and the image of S on a coordinate grid.

a $\begin{pmatrix} -1 & 2 \\ 2 & 1 \end{pmatrix}\begin{pmatrix} 1 \\ 1 \end{pmatrix} = \begin{pmatrix} 1 \\ 3 \end{pmatrix}$

Write each point as a column vector and then use the usual rule for multiplying matrices.

$\begin{pmatrix} -1 & 2 \\ 2 & 1 \end{pmatrix}\begin{pmatrix} 3 \\ 1 \end{pmatrix} = \begin{pmatrix} -1 \\ 7 \end{pmatrix}$

$\begin{pmatrix} -1 & 2 \\ 2 & 1 \end{pmatrix}\begin{pmatrix} 3 \\ 3 \end{pmatrix} = \begin{pmatrix} 3 \\ 9 \end{pmatrix}$

$\begin{pmatrix} -1 & 2 \\ 2 & 1 \end{pmatrix}\begin{pmatrix} 1 \\ 3 \end{pmatrix} = \begin{pmatrix} 5 \\ 5 \end{pmatrix}$

The vertices of the image of S, S', lie at (1, 3), (−1, 7), (3, 9) and (5, 5).

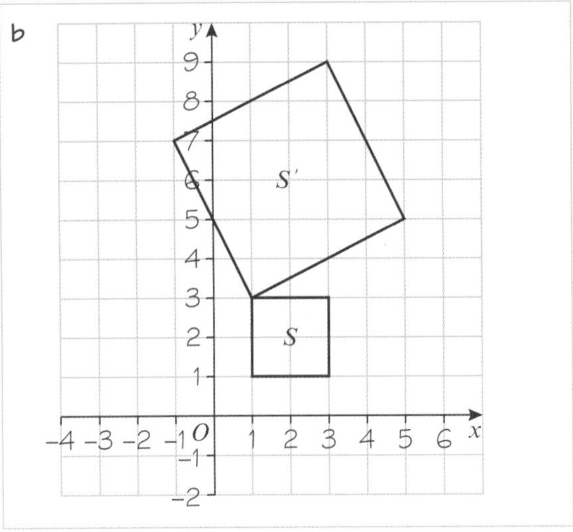

Notation S' is used to denote the image of S after a transformation.

Exercise 7A

1 Which of the following transformations are linear transformations?

a $P: \begin{pmatrix} x \\ y \end{pmatrix} \mapsto \begin{pmatrix} 2x \\ y+1 \end{pmatrix}$ **b** $Q: \begin{pmatrix} x \\ y \end{pmatrix} \mapsto \begin{pmatrix} x^2 \\ y \end{pmatrix}$ **c** $R: \begin{pmatrix} x \\ y \end{pmatrix} \mapsto \begin{pmatrix} 2x+y \\ x+xy \end{pmatrix}$

d $S: \begin{pmatrix} x \\ y \end{pmatrix} \mapsto \begin{pmatrix} y \\ -x \end{pmatrix}$ **e** $T: \begin{pmatrix} x \\ y \end{pmatrix} \mapsto \begin{pmatrix} y+3 \\ x+3 \end{pmatrix}$ **f** $U: \begin{pmatrix} x \\ y \end{pmatrix} \mapsto \begin{pmatrix} 2x \\ 3y-2x \end{pmatrix}$

2 Identify which of these are linear transformations and give their matrix representations. Give reasons to explain why the other transformations are not linear.

a $S: \begin{pmatrix} x \\ y \end{pmatrix} \mapsto \begin{pmatrix} 2x-y \\ 3x \end{pmatrix}$ **b** $T: \begin{pmatrix} x \\ y \end{pmatrix} \mapsto \begin{pmatrix} 2y+1 \\ x-1 \end{pmatrix}$ **c** $U: \begin{pmatrix} x \\ y \end{pmatrix} \mapsto \begin{pmatrix} xy \\ 0 \end{pmatrix}$

d $V: \begin{pmatrix} x \\ y \end{pmatrix} \mapsto \begin{pmatrix} 2y \\ -x \end{pmatrix}$ **e** $W: \begin{pmatrix} x \\ y \end{pmatrix} \mapsto \begin{pmatrix} y \\ x \end{pmatrix}$

3 Identify which of these are linear transformations and give their matrix representations. Give reasons to explain why the other transformations are not linear.

a $S: \begin{pmatrix} x \\ y \end{pmatrix} \mapsto \begin{pmatrix} x^2 \\ y^2 \end{pmatrix}$ **b** $T: \begin{pmatrix} x \\ y \end{pmatrix} \mapsto \begin{pmatrix} -y \\ x \end{pmatrix}$ **c** $U: \begin{pmatrix} x \\ y \end{pmatrix} \mapsto \begin{pmatrix} x-y \\ x-y \end{pmatrix}$

d $V: \begin{pmatrix} x \\ y \end{pmatrix} \mapsto \begin{pmatrix} 0 \\ 0 \end{pmatrix}$ **e** $W: \begin{pmatrix} x \\ y \end{pmatrix} \mapsto \begin{pmatrix} x \\ y \end{pmatrix}$

4 Find matrix representations for these linear transformations:

a $P: \begin{pmatrix} x \\ y \end{pmatrix} \mapsto \begin{pmatrix} y+2x \\ -y \end{pmatrix}$ **b** $Q: \begin{pmatrix} x \\ y \end{pmatrix} \mapsto \begin{pmatrix} -y \\ x+2y \end{pmatrix}$

5 The triangle T has vertices at (−1, 1), (2, 3) and (5, 1).
Find the vertices of the image of T under the transformations represented by these matrices:

a $\begin{pmatrix} -1 & 0 \\ 0 & 1 \end{pmatrix}$ **b** $\begin{pmatrix} 1 & 4 \\ 0 & -2 \end{pmatrix}$ **c** $\begin{pmatrix} 0 & -2 \\ 2 & 0 \end{pmatrix}$

6 The square S has vertices at (−1, 0), (0, 1), (1, 0) and (0, −1).
Find the vertices of the image of S under the transformations represented by these matrices:

a $\begin{pmatrix} 2 & 0 \\ 0 & 3 \end{pmatrix}$ **b** $\begin{pmatrix} 1 & -1 \\ 1 & 1 \end{pmatrix}$ **c** $\begin{pmatrix} 1 & 1 \\ 1 & -1 \end{pmatrix}$

7 The rectangle R has vertices at (2, 1), (4, 1), (4, 2) and (2, 2).

a Find the vertices of the image of R under the transformation represented by the matrix $\begin{pmatrix} -1 & 0 \\ 0 & -1 \end{pmatrix}$.

b Sketch R and its image, R', on a coordinate grid.

c Describe fully the transformation that maps R onto R'.

8 A quadrilateral Q has coordinates (1, 0), (4, 2), (3, 4) and (0, 2).

a Find the vertices of the image of Q under the transformation represented by the matrix $\begin{pmatrix} 2 & 0 \\ 0 & 2 \end{pmatrix}$.

b Sketch Q and its image, Q', on a coordinate grid.

c Describe fully the transformation that maps Q onto Q'.

(P) **9** A square S has coordinates (−1, 0), (−3, 0), (−3, 2) and (−1, 2).

a Find the vertices of the image of S under the transformation represented by the matrix $\begin{pmatrix} 0 & 2 \\ 2 & 0 \end{pmatrix}$.

b Sketch S and the image of S on a coordinate grid.

c Describe fully the two transformations that map S onto S'.

10 A triangle T has vertices (4, 1), (4, 3) and (1, 3).

a Find the vertices of the image of T under the transformation represented by the matrix $\begin{pmatrix} 1 & 0 \\ 0 & 1 \end{pmatrix}$.

b Describe the effect of the transformation represented by the matrix $\begin{pmatrix} 1 & 0 \\ 0 & 1 \end{pmatrix}$.

Challenge

A transformation T is defined as $T: \begin{pmatrix} x \\ y \end{pmatrix} \mapsto \begin{pmatrix} 2x - 3y \\ x + y \end{pmatrix}$.

a Show that $T\begin{pmatrix} kx \\ ky \end{pmatrix} = kT\begin{pmatrix} x \\ y \end{pmatrix}$.

b Show that $T\left(\begin{pmatrix} x_1 \\ y_1 \end{pmatrix} + \begin{pmatrix} x_2 \\ y_2 \end{pmatrix}\right) = T\begin{pmatrix} x_1 \\ y_1 \end{pmatrix} + T\begin{pmatrix} x_2 \\ y_2 \end{pmatrix}$.

Notation $T\begin{pmatrix} x \\ y \end{pmatrix}$ is used to denote the image of the point $\begin{pmatrix} x \\ y \end{pmatrix}$ after the transformation T.

Note All linear transformations have these properties.

7.2 Reflections and rotations

Any linear transformation can be defined by the effect it has on unit vectors $\begin{pmatrix}1\\0\end{pmatrix}$ and $\begin{pmatrix}0\\1\end{pmatrix}$.

The transformation represented by the matrix $\mathbf{M} = \begin{pmatrix}a & b\\c & d\end{pmatrix}$ will map $\begin{pmatrix}1\\0\end{pmatrix}$ to $\begin{pmatrix}a\\c\end{pmatrix}$ and $\begin{pmatrix}0\\1\end{pmatrix}$ to $\begin{pmatrix}b\\d\end{pmatrix}$.

You can visualise this transformation by considering the unit square:

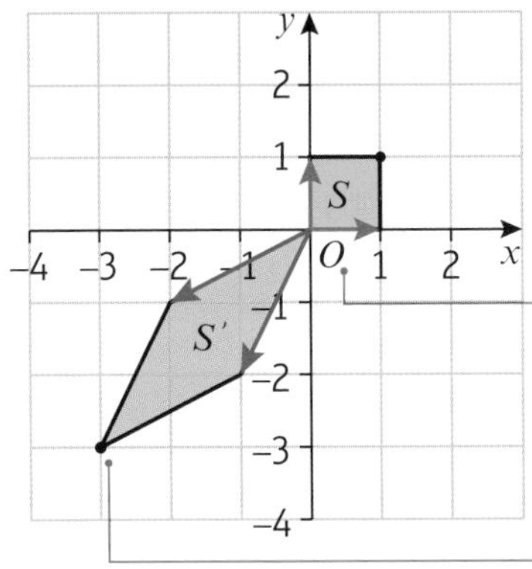

The linear transformation $\begin{pmatrix}-2 & -1\\-1 & -2\end{pmatrix}$ stretches and rotates the unit square S to produce the image S'.

The origin does not change under a linear transformation.

$T(\mathbf{a} + \mathbf{b}) = T\mathbf{a} + T\mathbf{b}$ so you can add the two image vectors to find the fourth vertex of the new shape. This means that the entire transformation can be defined by the images of the two unit vectors $\begin{pmatrix}1\\0\end{pmatrix}$ and $\begin{pmatrix}0\\1\end{pmatrix}$.

Points which are mapped onto themselves under the given transformation are called **invariant points**. Lines which map onto themselves are called **invariant lines**.

The only invariant point in the above transformation is the origin, which is always an invariant point of any linear transformation. The above transformation has invariant lines $y = x$ and $y = -x$.

Example 4

The transformation U, represented by the 2×2 matrix **Q**, is a reflection in the y-axis.

a Write down the matrix **Q**.

b Write down the equations of three different invariant lines of this transformation.

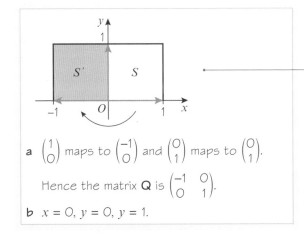

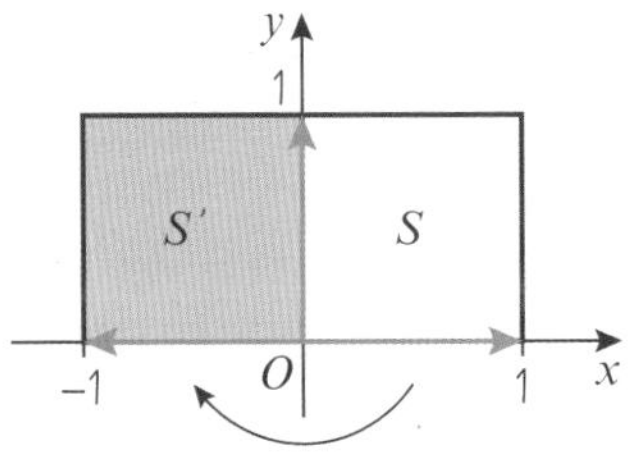

Consider the unit square, and the effect that the transformation has on the unit vectors $\begin{pmatrix}1\\0\end{pmatrix}$ and $\begin{pmatrix}0\\1\end{pmatrix}$. This will completely define the transformation.

a $\begin{pmatrix}1\\0\end{pmatrix}$ maps to $\begin{pmatrix}-1\\0\end{pmatrix}$ and $\begin{pmatrix}0\\1\end{pmatrix}$ maps to $\begin{pmatrix}0\\1\end{pmatrix}$.

Hence the matrix **Q** is $\begin{pmatrix}-1 & 0\\0 & 1\end{pmatrix}$.

b $x = 0$, $y = 0$, $y = 1$.

Problem-solving

Each point on the mirror line is invariant, so this line is an invariant line. Any line perpendicular to the mirror line will also be an invariant line, although the points on the line will not, in general, be invariant points.

- **A reflection in the y-axis is represented by the matrix $\begin{pmatrix} -1 & 0 \\ 0 & 1 \end{pmatrix}$. Points on the y-axis are invariant points, and the lines $x = 0$ and $y = k$ for any value of k are invariant lines.**
- **A reflection in the x-axis is represented by the matrix $\begin{pmatrix} 1 & 0 \\ 0 & -1 \end{pmatrix}$. Points on the x-axis are invariant points, and the lines $y = 0$ and $x = k$ for any value of k are invariant lines.**

Example 5

$\mathbf{P} = \begin{pmatrix} 0 & -1 \\ -1 & 0 \end{pmatrix}$

a Describe fully the single geometrical transformation U represented by the matrix **P**.

b Given that U maps the point with coordinates (a, b) onto the point with coordinates $(3 + 2a, b + 1)$, find the values of a and b.

a $\begin{pmatrix} 1 \\ 0 \end{pmatrix}$ maps to $\begin{pmatrix} 0 \\ -1 \end{pmatrix}$ and $\begin{pmatrix} 0 \\ 1 \end{pmatrix}$ maps to $\begin{pmatrix} -1 \\ 0 \end{pmatrix}$,
hence the transformation U represented by matrix **P** is a reflection in the line $y = -x$.

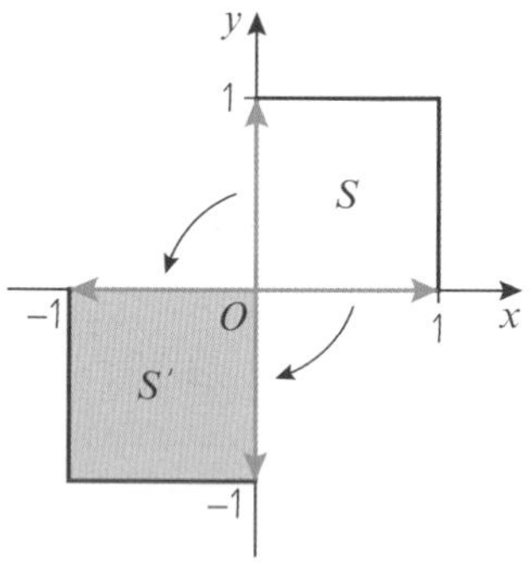

You can visualise the transformation by sketching the effect it has on the unit square.

b $\begin{pmatrix} 0 & -1 \\ -1 & 0 \end{pmatrix}\begin{pmatrix} a \\ b \end{pmatrix} = \begin{pmatrix} 3 + 2a \\ b + 1 \end{pmatrix}$

$\begin{pmatrix} -b \\ -a \end{pmatrix} = \begin{pmatrix} 3 + 2a \\ b + 1 \end{pmatrix}$

So $-b = 3 + 2a$ (1)

and $-a = b + 1$ (2)

Solving simultaneously gives $a = -2$ and $b = 1$.

Problem-solving

Write a matrix equation to show the transformation, then solve the resulting equations simultaneously to find the values of a and b.

You can solve two equations in two unknowns quickly using your calculator.

- **A reflection in the line $y = x$ is represented by the matrix $\begin{pmatrix} 0 & 1 \\ 1 & 0 \end{pmatrix}$. Points on the line $y = x$ are invariant points, and the lines $y = x$ and $y = -x + k$ for any value of k are invariant lines.**
- **A reflection in the line $y = -x$ is represented by the matrix $\begin{pmatrix} 0 & -1 \\ -1 & 0 \end{pmatrix}$. Points on the line $y = -x$ are invariant points, and the lines $y = -x$ and $y = x + k$ for any value of k are invariant lines.**

Example 6

The transformation U, represented by the 2×2 matrix $\mathbf{P}$, is a rotation of 180° about the point (0, 0).

a Write down the matrix $\mathbf{P}$.

b Show that the line $y = 3x$ is invariant under this transformation.

a The given rotation is shown in the diagram:

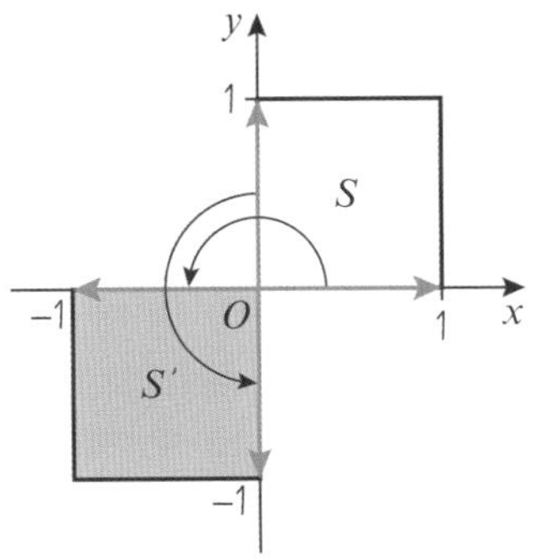

If you need to find the matrix that represents a given transformation, it can help to draw a sketch transforming the unit square. Remember the transformation is defined by its effect on the unit vectors.

$\begin{pmatrix}1\\0\end{pmatrix}$ maps to $\begin{pmatrix}-1\\0\end{pmatrix}$ and $\begin{pmatrix}0\\1\end{pmatrix}$ maps to $\begin{pmatrix}0\\-1\end{pmatrix}$.

Hence the matrix is $\begin{pmatrix}-1 & 0\\0 & -1\end{pmatrix}$.

b $\begin{pmatrix}-1 & 0\\0 & -1\end{pmatrix}\begin{pmatrix}x\\3x\end{pmatrix} = \begin{pmatrix}-x\\-3x\end{pmatrix}$

Since $-3x = 3(-x)$, this point lies on the line $y = 3x$.

So points on the line $y = 3x$ are mapped to points on the line $y = 3x$.

Hence $y = 3x$ is an invariant line.

Problem-solving

Write a general point on the line $y = 3x$ as $\begin{pmatrix}x\\y\end{pmatrix} = \begin{pmatrix}x\\3x\end{pmatrix}$. Apply the transformation to this point, then show that the image also lies on the line. Watch out: although the line is invariant, the only invariant **point** on the line is the origin.

You need to be able to write down the matrix representing a rotation about any angle.

- **The matrix representing a rotation through angle θ anticlockwise about the origin is $\begin{pmatrix}\cos\theta & -\sin\theta\\ \sin\theta & \cos\theta\end{pmatrix}$**

  The only invariant point is the origin (0, 0). For $\theta \neq 180°$, there are no invariant lines. For $\theta = 180°$, any line passing through the origin is an invariant line.

Note This general rotation matrix is given in the formulae booklet.

Example 7

$$\mathbf{M} = \begin{pmatrix} -\frac{\sqrt{2}}{2} & -\frac{\sqrt{2}}{2} \\ \frac{\sqrt{2}}{2} & -\frac{\sqrt{2}}{2} \end{pmatrix}$$

a Describe geometrically the rotation represented by $\mathbf{M}$.

b A square S has vertices at (1, 0), (2, 0), (2, 1) and (1, 1). Find the coordinates of the vertices of the image of S under the transformation described by $\mathbf{M}$.

a $\cos 135° = -\frac{\sqrt{2}}{2}$ and $\sin 135° = \frac{\sqrt{2}}{2}$ so **M** is a rotation, anticlockwise, through 135° about (0, 0).

You are told that the **M** represents a rotation, so compare the matrix with $\begin{pmatrix} \cos\theta & -\sin\theta \\ \sin\theta & \cos\theta \end{pmatrix}$.

b Apply the matrix to each vertex of S in turn:

$$\begin{pmatrix} -\frac{\sqrt{2}}{2} & -\frac{\sqrt{2}}{2} \\ \frac{\sqrt{2}}{2} & -\frac{\sqrt{2}}{2} \end{pmatrix}\begin{pmatrix} 1 \\ 0 \end{pmatrix} = \begin{pmatrix} -\frac{\sqrt{2}}{2} \\ \frac{\sqrt{2}}{2} \end{pmatrix}$$

$$\begin{pmatrix} -\frac{\sqrt{2}}{2} & -\frac{\sqrt{2}}{2} \\ \frac{\sqrt{2}}{2} & -\frac{\sqrt{2}}{2} \end{pmatrix}\begin{pmatrix} 2 \\ 0 \end{pmatrix} = \begin{pmatrix} -\sqrt{2} \\ \sqrt{2} \end{pmatrix}$$

$$\begin{pmatrix} -\frac{\sqrt{2}}{2} & -\frac{\sqrt{2}}{2} \\ \frac{\sqrt{2}}{2} & -\frac{\sqrt{2}}{2} \end{pmatrix}\begin{pmatrix} 2 \\ 1 \end{pmatrix} = \begin{pmatrix} -\frac{3\sqrt{2}}{2} \\ \frac{\sqrt{2}}{2} \end{pmatrix}$$

$$\begin{pmatrix} -\frac{\sqrt{2}}{2} & -\frac{\sqrt{2}}{2} \\ \frac{\sqrt{2}}{2} & -\frac{\sqrt{2}}{2} \end{pmatrix}\begin{pmatrix} 1 \\ 1 \end{pmatrix} = \begin{pmatrix} -\sqrt{2} \\ 0 \end{pmatrix}$$

Online Explore rotations of the unit square using GeoGebra.

Work out the position vector of each vertex in the image of S.

The vertices of S' are $\left(-\frac{\sqrt{2}}{2}, \frac{\sqrt{2}}{2}\right)$, $(-\sqrt{2}, \sqrt{2})$, $\left(-\frac{3\sqrt{2}}{2}, \frac{\sqrt{2}}{2}\right)$ and $(-\sqrt{2}, 0)$.

Watch out Read the question carefully. Give your final answers as coordinates, not position vectors.

Exercise 7B

1 **a** Write down the matrix representing a reflection in the x-axis.
A triangle has vertices at $A = (1, 3)$, $B = (3, 3)$ and $C = (3, 2)$.

b Use matrices to show that the images of these vertices after a reflection in the x-axis are $A' = (1, -3)$, $B' = (3, -3)$ and $C' = (3, -2)$.

2 **a** Write down the matrix representing a reflection in the line $y = -x$.
A rectangle has vertices at $P = (1, 1)$, $Q = (1, 3)$, $R = (2, 3)$ and $S = (2, 1)$.

b Use matrices to show that the images of these vertices after a reflection in the line $y = -x$ are $P' = (-1, -1)$, $Q' = (-3, -1)$, $R' = (-3, -2)$ and $S' = (-1, -2)$.

3 Find the matrices that represent the following rotations.

a 90° anticlockwise about (0, 0)

b 270° anticlockwise about (0, 0)

c 45° anticlockwise about (0, 0)

d 210° anticlockwise about (0, 0)

e 135° clockwise about (0, 0)

Watch out The rotation matrix is for angles measured anticlockwise, so make sure you convert the clockwise angle to its equivalent anticlockwise angle.

4 A triangle has vertices at $A = (1, 1)$, $B = (4, 1)$ and $C = (4, 2)$. Find the exact coordinates of the vertices of the triangle after a rotation through:

a 90° anticlockwise about (0, 0) **b** 150° anticlockwise about (0, 0)

5 A rectangle has vertices at $P = (2, 2)$, $Q = (2, 3)$, $R = (4, 3)$ and $S = (4, 2)$. Find the exact coordinates of the vertices of the rectangle after a rotation through:

a 270° anticlockwise about (0, 0) **b** 135° clockwise about (0, 0)

P **6** $\mathbf{A} = \begin{pmatrix} 1 & 0 \\ 0 & -1 \end{pmatrix}$ and $\mathbf{B} = \begin{pmatrix} 0 & 1 \\ -1 & 0 \end{pmatrix}$

a Write down fully the transformations represented by the matrices **A** and **B**. **(4 marks)**

b The point (3, 2) is transformed by matrix **A**. Write down the coordinates of the image of this point. **(1 mark)**

c The point (a, b) is transformed onto the point $(a - 3b, 2a - 2b)$ by matrix **B**. Find the values of a and b. **(3 marks)**

E **7** $\mathbf{M} = \begin{pmatrix} -\frac{1}{\sqrt{2}} & \frac{1}{\sqrt{2}} \\ -\frac{1}{\sqrt{2}} & -\frac{1}{\sqrt{2}} \end{pmatrix}$

a Write down fully the transformation represented by matrix **M**. **(2 marks)**

b The transformation represented by **M** maps the point (p, q) onto the point C with coordinates $(-\sqrt{2}, -2\sqrt{2})$. Find the values of p and q. **(4 marks)**

c Use your calculator to find $\mathbf{M}^3$. **(1 mark)**

d Point C is mapped onto the point D by the transformation represented by $\mathbf{M}^3$. Find the coordinates of point D and describe fully the transformation represented by $\mathbf{M}^3$. **(2 marks)**

E/P **8** **a** Describe fully the transformation represented by the matrix $\mathbf{A} = \begin{pmatrix} 0 & 1 \\ 1 & 0 \end{pmatrix}$. **(2 marks)**

b Write down the equations of three different invariant lines under this transformation. **(2 marks)**

c Write down $\mathbf{A}^{50}$. **(1 mark)**

E/P **9** The matrix $\begin{pmatrix} a & b \\ c & -0.5 \end{pmatrix}$ represents an anticlockwise rotation about the origin, through an angle θ.

a Write down the value of a. **(1 mark)**

b Find two possible values of θ, and write down the matrix corresponding to each rotation. **(3 marks)**

E/P **10** **a** Write down the matrix representing a rotation through 270° clockwise about (0, 0). **(1 mark)**

b A point (a, b) transformed using this matrix is such that its image is the point $(a - 5b, 4b)$. Find the values of a and b. **(3 marks)**

Challenge

Prove that the general matrix representing a rotation through angle θ anticlockwise about the origin is $\begin{pmatrix} \cos\theta & -\sin\theta \\ \sin\theta & \cos\theta \end{pmatrix}$.

7.3 Enlargements and stretches

You can describe enlargements and stretches using linear transformations.

Example 8

$\mathbf{M} = \begin{pmatrix} 3 & 0 \\ 0 & 2 \end{pmatrix}$

a Find the image T' of a triangle T with vertices (1, 1), (1, 2) and (2, 2) under the transformation represented by **M**.

b Sketch T and T' on the same set of coordinate axes.

c Describe geometrically the transformation represented by **M**.

a $\begin{pmatrix} 3 & 0 \\ 0 & 2 \end{pmatrix}\begin{pmatrix} 1 \\ 1 \end{pmatrix} = \begin{pmatrix} 3 \\ 2 \end{pmatrix}$

$\begin{pmatrix} 3 & 0 \\ 0 & 2 \end{pmatrix}\begin{pmatrix} 1 \\ 2 \end{pmatrix} = \begin{pmatrix} 3 \\ 4 \end{pmatrix}$

$\begin{pmatrix} 3 & 0 \\ 0 & 2 \end{pmatrix}\begin{pmatrix} 2 \\ 2 \end{pmatrix} = \begin{pmatrix} 6 \\ 4 \end{pmatrix}$

Use matrix multiplication to find the image of each vertex under the transformation.

The coordinates of the image are (3, 2), (3, 4) and (6, 4).

b

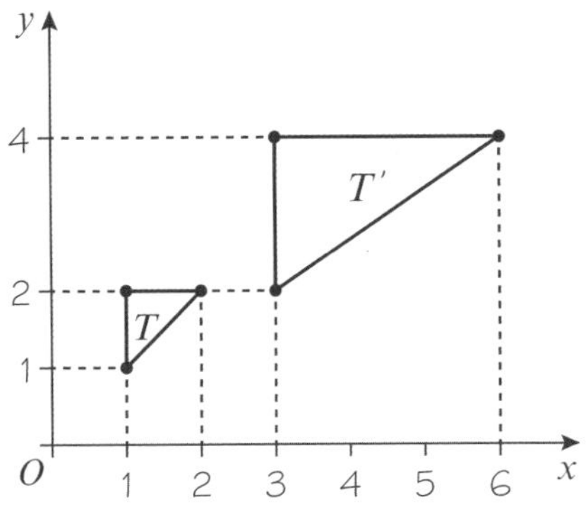

c The triangle has been stretched by a scale factor of 3 parallel to the x-axis and by a scale factor of 2 parallel to the y-axis.

Online Explore enlargements and stretches of triangle T using GeoGebra.

Every x-coordinate has been multiplied by 3, and every y-coordinate has been doubled. Note that this is not the same as an enlargement, because the triangle has been stretched by different factors in the x- and y-directions.

- **A transformation represented by the matrix $\begin{pmatrix} a & 0 \\ 0 & b \end{pmatrix}$ is a stretch of scale factor a parallel to the x-axis and a stretch of scale factor b parallel to the y-axis.**

  In the case where $a = b$, the transformation is an enlargement with scale factor a.

Note A stretch parallel to the x-axis only will have matrix $\begin{pmatrix} a & 0 \\ 0 & 1 \end{pmatrix}$.

A stretch parallel to the y-axis only will have matrix $\begin{pmatrix} 1 & 0 \\ 0 & b \end{pmatrix}$.

- **For any stretch in this form, the x- and y-axes are invariant lines and the origin is an invariant point.**
- **For a stretch parallel to the x-axis only, points on the y-axis are invariant points, and any line parallel to the x-axis is an invariant line.**
- **For a stretch parallel to the y-axis only, points on the x-axis are invariant points, and any line parallel to the y-axis is an invariant line.**

Watch out For a stretch in both directions, although the axes are invariant lines, the points on the axes are not themselves invariant. Any point on an axis (apart from the origin) will map to a **different** point on the same axis.

Reflections and rotations of 2D shapes both preserve the area of a shape. When a shape is stretched, its area can increase or decrease. You can use the determinant of the matrix representing this transformation to work out the scale factor for the change in area.

- **For a linear transformation represented by matrix M, det M represents the scale factor for the change in area. This is sometimes called the area scale factor.**

Watch out If the determinant of the matrix **M** is negative, the shape has been reflected.

Example 9

$\mathbf{M} = \begin{pmatrix} 2 & 0 \\ 0 & 4 \end{pmatrix}$

a Describe fully the transformation represented by **M**.

b A triangle T has vertices at (1, 0), (4, 0) and (4, 2). Find the area of the triangle.

c The triangle T is transformed using the matrix **M**. Use the determinant of **M** to find the area of the image of T.

a The matrix is of the form $\begin{pmatrix} a & 0 \\ 0 & b \end{pmatrix}$ so it is a stretch parallel to the x-axis, with scale factor 2 and a stretch parallel to the y-axis, with scale factor 4.

b Using $A = \frac{1}{2} \times \text{base} \times \text{height}$:

$$A = \frac{1}{2} \times 3 \times 2 = 3$$

c det M = 8

det **M** = 2 × 4 − 0 × 0 = 8 ← Section 6.3

The area of the image of T is 3 × 8 = 24.

The area increases by a factor of det **M**.

Exercise 7C

1 Write down the matrices representing the following linear transformations.

a A stretch with scale factor 4 parallel to the x-axis

b A stretch with scale factor 3 parallel to the y-axis

c An enlargement with scale factor 2

d A stretch with scale factor 5 parallel to the x-axis and a stretch scale factor $\frac{1}{2}$ parallel to the y-axis

2 For each of the transformations in question **1**, write down the area scale factor.

Hint In an enlargement, 'scale factor' refers to the linear scale factor of the enlargement so the area scale factor will be the square of this value.

3 The unit square is transformed using the matrix $\begin{pmatrix} 3 & 0 \\ 0 & 4 \end{pmatrix}$.

a Write down the coordinates of any invariant points.

b Work out the area of the resulting rectangle.

4 Write down the matrices representing the following transformations.

a A stretch with scale factor −2 parallel to the x-axis

b A stretch with scale factor −3 parallel to the x-axis and scale factor 4 parallel to the y-axis

c An enlargement with scale factor $-\frac{1}{2}$

(P) **5** Prove that the line $y = kx$ is invariant under the transformation $\begin{pmatrix} a & 0 \\ 0 & a \end{pmatrix}$, where a and k are real constants.

(E/P) **6** $\mathbf{M} = \begin{pmatrix} 2 & 0 \\ 0 & -3 \end{pmatrix}$

a Describe fully the transformation represented by **M**. **(2 marks)**

A 2D shape with area k is transformed using the transformation represented by **M**.

b Given that the image of the shape has area 24, find the value of k. **(2 marks)**

7 A triangle with vertices at (1, 3), (5, 3) and (5, 2) is transformed using the matrix $\begin{pmatrix} 3 & 0 \\ 0 & 3 \end{pmatrix}$.

a Find the coordinates of the vertices of the resulting image.

b Find the area of the new triangle.

8 A rectangle has vertices at (2, 0), (4, 0), (4, 5) and (2, 5). The rectangle is transformed by a stretch with scale factor 2 parallel to the x-axis and a stretch with scale factor −3 parallel to the y-axis.

a Find the coordinates of vertices of the resulting image.

b Find the area of the new rectangle.

(E/P) **9** $\mathbf{A} = \begin{pmatrix} 2\sqrt{5} & 0 \\ 0 & 2\sqrt{5} \end{pmatrix}$

a Describe fully the transformation represented by the matrix **A**. **(2 marks)**

b A triangle T has coordinates $(a, 1)$, (4, 1) and (4, 3). Given that T is transformed using matrix **A**, and the area of the resulting triangle is 60, find the value of a. **(3 marks)**

(E/P) **10** $\mathbf{M} = \begin{pmatrix} p & 1 \\ p & q \end{pmatrix}$ where p and q are constants and $q > 0$.

a Find $\mathbf{M}^2$ in terms of p and q. **(3 marks)**

Given that $\mathbf{M}^2$ represents an enlargement with centre (0, 0) and scale factor 6,

b find the values of p and q. **(3 marks)**

(E/P) **11** $\mathbf{A} = \begin{pmatrix} 5 & -1 \\ 2 & -2 \end{pmatrix}$ and $\mathbf{B} = \begin{pmatrix} 2 & 1 \\ 2 & 5 \end{pmatrix}$

a Find the matrix **M** where **M** = **AB**. **(1 mark)**

b Describe fully the transformation represented by the matrix **M**. **(3 marks)**

c A triangle T has vertices at (2, 1), (6, 1) and (6, k). Given that T is transformed using matrix **M**, and the resulting triangle has area 320, find the value of k. **(4 marks)**

(E) **12** $\mathbf{M} = \begin{pmatrix} -1 & \sqrt{2} \\ -\sqrt{2} & -1 \end{pmatrix}$

A pentagon P of area 12 is transformed using matrix **M**. Find the area of the image of the pentagon P'.

Hint You will learn how to describe transformations such as this in **Section 7.4**.

(2 marks)

E/P **13** A triangle T has vertices at the points $A = (k, 1)$, $B = (4, 1)$ and $C = (4, k)$ where k is an integer constant.

Triangle T is transformed by the matrix $\begin{pmatrix} 4 & -1 \\ k & 2 \end{pmatrix}$.

Given that triangle T has a right angle at B, and the area of the image triangle T' is 10, find the value of k. **(5 marks)**

E **14** A triangle T has vertices at (0, 0), (7, 7) and (3, −2).

a Write down the matrix, **M**, which represents a rotation through 45° anticlockwise about (0, 0). **(1 mark)**

b Find the exact coordinates of the image of T when T is transformed using **M**. **(3 marks)**

c Show that $\det \mathbf{M} = 1$. **(2 marks)**

d Hence find the area of the original triangle T. **(1 mark)**

Challenge

A transformation U is represented by the matrix $\mathbf{P} = \begin{pmatrix} 7 & 0 \\ 0 & 0 \end{pmatrix}$.

a Find det **P**.

b Show that any point in the xy-plane is mapped onto the x-axis by U.

Problem-solving

A non-zero singular 2 × 2 matrix maps any point in the plane onto a straight line through the origin.

7.4 Successive transformations

You can use matrix products to represent combinations of transformations.

- **The matrix PQ represents the transformation Q, with matrix Q followed by the transformation P, with matrix P.**

Example 10

a Find the 2 × 2 matrix **T** that represents a rotation through 90° anticlockwise about the origin followed by a reflection in the line $y = x$.

b Describe the single transformation represented by **T**.

a Rotation 90° anticlockwise about the origin: $\begin{pmatrix} 0 & -1 \\ 1 & 0 \end{pmatrix}$

Reflection in the line $y = x$: $\begin{pmatrix} 0 & 1 \\ 1 & 0 \end{pmatrix}$

Write down the two matrices for the single transformations.

$\mathbf{T} = \begin{pmatrix} 0 & 1 \\ 1 & 0 \end{pmatrix}\begin{pmatrix} 0 & -1 \\ 1 & 0 \end{pmatrix} = \begin{pmatrix} 1 & 0 \\ 0 & -1 \end{pmatrix}$

Find the matrix product to determine the single matrix that has the same effect as these combined transformations. Make sure you apply the matrix product in the right order.

b T represents reflection in the line $y = 0$.

Check where **T** maps the position vectors $\begin{pmatrix} 1 \\ 0 \end{pmatrix}$ and $\begin{pmatrix} 0 \\ 1 \end{pmatrix}$.

Example 11

$$\mathbf{M} = \begin{pmatrix} -2\sqrt{2} & -2\sqrt{2} \\ 2\sqrt{2} & -2\sqrt{2} \end{pmatrix}$$

The matrix **M** represents an enlargement with scale factor k followed by a rotation through angle θ anticlockwise about the origin.

a Find the value of k.

b Find the value of θ.

a $\det \mathbf{M} = 16$ — Use your calculator to find det **M**.

Area scale factor = 16

$k = \sqrt{16} = 4$ — k is the linear scale factor. The rotation does not affect area, so $k = \sqrt{\det \mathbf{M}}$.

b $\begin{pmatrix} -2\sqrt{2} & -2\sqrt{2} \\ 2\sqrt{2} & -2\sqrt{2} \end{pmatrix} = \begin{pmatrix} \cos\theta & -\sin\theta \\ \sin\theta & \cos\theta \end{pmatrix}\begin{pmatrix} 4 & 0 \\ 0 & 4 \end{pmatrix}$ — If the enlargement matrix is **Q** and the rotation matrix is **P** then **M** = **PQ**.

$= \begin{pmatrix} 4\cos\theta & -4\sin\theta \\ 4\sin\theta & 4\cos\theta \end{pmatrix}$

$4\cos\theta = -2\sqrt{2}$

$\cos\theta = \frac{-2\sqrt{2}}{4} = -\frac{\sqrt{2}}{2}$

$\theta = 135°$ or $\theta = 225°$ — Use one element to find possible values of θ.

Check using the lower-left element:

$\sin\theta = \frac{\sqrt{2}}{2}$ so $\theta = 135°$

Watch out You will need to use one of the sin elements to check which angle is correct. $\sin\theta$ is positive so choose the angle in the second quadrant.

Exercise 7D

1 $\mathbf{A} = \begin{pmatrix} -1 & 0 \\ 0 & -1 \end{pmatrix}$, $\mathbf{B} = \begin{pmatrix} 0 & -1 \\ -1 & 0 \end{pmatrix}$, $\mathbf{C} = \begin{pmatrix} 2 & 0 \\ 0 & 2 \end{pmatrix}$

Find these matrix products and describe the single transformation represented by each product.

a **AB** **b** **BA** **c** **AC** **d** $\mathbf{A}^2$ **e** $\mathbf{C}^2$

2 A = rotation of 90° anticlockwise about (0, 0) B = rotation of 180° about (0, 0)

C = reflection in the x-axis D = reflection in the y-axis

a Find matrix representations of each of the four transformations A, B, C and D.

b Use matrix products to identify the single geometric transformation represented by each of these combinations.

i Reflection in the x-axis followed by a rotation of 180° about (0, 0)

ii Rotation of 180° about (0, 0) followed by a reflection in the x-axis

iii Reflection in the y-axis followed by reflection in the x-axis

iv Reflection in the y-axis followed by rotation of 90° anticlockwise about (0, 0)

v Rotation of 180° about (0, 0) followed by a second rotation of 180° about (0, 0)

vi Reflection in the x-axis followed by rotation of 90° anticlockwise about (0, 0) followed by a reflection in the y-axis

vii Reflection in the y-axis followed by rotation of 180° about (0, 0) followed by a reflection in the x-axis

3 $\mathbf{R} = \begin{pmatrix} 3 & 0 \\ 0 & -2 \end{pmatrix}$, $\mathbf{S} = \begin{pmatrix} 0 & 1 \\ -1 & 0 \end{pmatrix}$ and $\mathbf{T} = \begin{pmatrix} 5 & 0 \\ 0 & 5 \end{pmatrix}$

Find these matrix products and, where possible, use your knowledge of the standard forms of transformation matrices to find the single transformation represented by the products:

a **RS** **b** **RT** **c** **TS** **d** **TR** **e** **ST** **f** **RST**

4 A is a stretch with scale factor 2 parallel to the x-axis, and scale factor 3 parallel to the y-axis.
B is an enlargement with scale factor −2.
C is an enlargement with scale factor 4.

a Write down the matrices representing each of the transformations A, B and C.

b Find the single 2 × 2 matrix representing each of the following combined transformations:

i B followed by A **ii** C followed by A

iii B followed by C **iv** C followed by C

v C followed by B followed by A

Hint If C is represented by matrix **M**, then C followed by C will be represented by $\mathbf{M}^2$.

(P) **5** Use matrices to show that a refection in the y-axis followed by a reflection in the line $y = -x$ is equivalent to a rotation of 90° anticlockwise about (0, 0).

(E/P) **6** A student makes the following claim:

> If T is a reflection in the x-axis and U is a rotation 90° anticlockwise about the origin then T followed by U is the same as U followed by T.

Show, using matrix multiplication, that the student is incorrect. **(4 marks)**

(E/P) **7** $\mathbf{P} = \begin{pmatrix} -4 & 0 \\ 0 & 2 \end{pmatrix}$ and $\mathbf{Q} = \begin{pmatrix} k & 0 \\ 0 & k \end{pmatrix}$, where k is a constant.

a Find the matrix product **PQ**. **(2 marks)**

b Describe, in terms of k, the single transformation represented by **PQ**. **(2 marks)**

c Show that for any value of k, **PQ** = **QP**. **(2 marks)**

(E/P) **8** $\mathbf{A} = \begin{pmatrix} 3 & 0 \\ 0 & 4 \end{pmatrix}$

a Find the matrix $\mathbf{A}^2$. **(1 mark)**

b Describe fully the transformation represented by $\mathbf{A}^2$. **(2 marks)**

$\mathbf{B} = \begin{pmatrix} a & 0 \\ 0 & b \end{pmatrix}$ where a and b are constants.

c Find the general matrix $\mathbf{B}^2$ and state, in terms of a and b, the transformation represented by this matrix. **(3 marks)**

E/P **9** The matrix **R** is given by $\begin{pmatrix} \frac{1}{\sqrt{2}} & \frac{-1}{\sqrt{2}} \\ \frac{1}{\sqrt{2}} & \frac{1}{\sqrt{2}} \end{pmatrix}$

a Find $\mathbf{R}^2$. **(1 mark)**

b Describe the geometric transformation represented by $\mathbf{R}^2$. **(2 marks)**

c Hence describe the geometric transformation represented by **R**. **(1 marks)**

d Write down $\mathbf{R}^8$. **(1 mark)**

E/P **10** $\mathbf{M} = \begin{pmatrix} -\frac{3}{\sqrt{2}} & \frac{3}{\sqrt{2}} \\ -\frac{3}{\sqrt{2}} & -\frac{3}{\sqrt{2}} \end{pmatrix}$

The matrix **M** represents an enlargement with scale factor k ($k < 0$) followed by a rotation of angle θ anticlockwise about the origin.

a Find the value of k. **(2 marks)**

b Find the value of θ. **(3 marks)**

E/P **11** $\mathbf{A} = \begin{pmatrix} 0 & 1 \\ -1 & 0 \end{pmatrix}$ and $\mathbf{B} = \begin{pmatrix} 5 & 0 \\ 0 & 5 \end{pmatrix}$

A triangle T is transformed using matrix **B**. The image is then transformed using matrix **A**. Given that the area of the image, T' is 75, find the area of T. **(3 marks)**

E/P **12** The transformation T is a rotation through 225° anticlockwise about the origin.

a Write down the matrix representing this transformation. **(1 mark)**

The transformation U is a reflection in the line $y = x$.

b Write down the matrix representing this transformation. **(1 mark)**

c Find the matrix representing the combined transformation of U followed by T. **(2 marks)**

E/P **13** $\mathbf{A} = \begin{pmatrix} k & \sqrt{3} \\ \sqrt{3} & -k \end{pmatrix}$ where k is a constant.

a Find, in terms of k, the matrix $\mathbf{A}^2$. **(2 marks)**

b Describe fully the transformation represented by $\mathbf{A}^2$. **(2 marks)**

E/P **14** $\mathbf{P} = \begin{pmatrix} a & b \\ b & -a \end{pmatrix}$ where a and b are constants.

Show that the general matrix $\mathbf{P}^2$ represents an enlargement, and write down, in terms of a and b, the scale factor of the enlargement. **(3 marks)**

Challenge

a Given that $\mathbf{P} = \begin{pmatrix} \cos\theta & -\sin\theta \\ \sin\theta & \cos\theta \end{pmatrix}$, show algebraically that $\mathbf{P}^2 = \begin{pmatrix} \cos 2\theta & -\sin 2\theta \\ \sin 2\theta & \cos 2\theta \end{pmatrix}$.

b Interpret this result geometrically.

7.5 Linear transformations in three dimensions

Any linear transformation in three dimensions can be defined by the effect it has on the unit vectors $\begin{pmatrix} 1 \\ 0 \\ 0 \end{pmatrix}$, $\begin{pmatrix} 0 \\ 1 \\ 0 \end{pmatrix}$ and $\begin{pmatrix} 0 \\ 0 \\ 1 \end{pmatrix}$. The transformation represented by the matrix $\mathbf{M} = \begin{pmatrix} a & b & c \\ d & e & f \\ g & h & i \end{pmatrix}$ will map $\begin{pmatrix} 1 \\ 0 \\ 0 \end{pmatrix}$ to $\begin{pmatrix} a \\ d \\ g \end{pmatrix}$, $\begin{pmatrix} 0 \\ 1 \\ 0 \end{pmatrix}$ to $\begin{pmatrix} b \\ e \\ h \end{pmatrix}$ and $\begin{pmatrix} 0 \\ 0 \\ 1 \end{pmatrix}$ to $\begin{pmatrix} c \\ f \\ i \end{pmatrix}$.

You need to be able to carry out transformations in three dimensions that are reflections in the planes $x = 0$, $y = 0$ or $z = 0$.

Links In three dimensions the coordinate axes are labelled x, y and z. ← **Pure Year 2, Chapter 12**

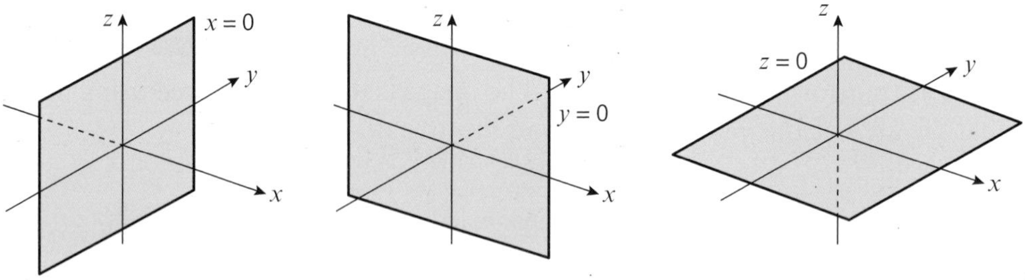

Example 12

A transformation U, in three dimensions, represents a reflection in the plane $z = 0$.

a Write down the 3 × 3 matrix that represents this transformation.

b Find the image of the point (−1, 2, 3) under this transformation.

a The point with position vector $\begin{pmatrix} 1 \\ 0 \\ 0 \end{pmatrix}$ lies in the xy-plane so stays where it is.

The point with position vector $\begin{pmatrix} 0 \\ 1 \\ 0 \end{pmatrix}$ lies in the xy-plane so stays where it is.

The plane $z = 0$ contains the x- and y-axes. Hence points in the xy-plane are invariant.

The point with position vector $\begin{pmatrix}0\\0\\1\end{pmatrix}$ reflects in the xy-plane to the point with position vector $\begin{pmatrix}0\\0\\-1\end{pmatrix}$.

Hence the matrix representing the transformation is $\begin{pmatrix}1&0&0\\0&1&0\\0&0&-1\end{pmatrix}$.

b $\begin{pmatrix}1&0&0\\0&1&0\\0&0&-1\end{pmatrix}\begin{pmatrix}-1\\2\\3\end{pmatrix}=\begin{pmatrix}-1\\2\\-3\end{pmatrix}$

The coordinates of the image are (−1, 2, −3).

The x- and y-coordinates stay the same, and the sign of the z-coordinate changes.

- **A reflection in the plane $x = 0$ is represented by the matrix** $\begin{pmatrix}-1&0&0\\0&1&0\\0&0&1\end{pmatrix}$.

- **A reflection in the plane $y = 0$ is represented by the matrix** $\begin{pmatrix}1&0&0\\0&-1&0\\0&0&1\end{pmatrix}$.

- **A reflection in the plane $z = 0$ is represented by the matrix** $\begin{pmatrix}1&0&0\\0&1&0\\0&0&-1\end{pmatrix}$.

You also need to be able to carry out rotations about one of the coordinate axes in three dimensions.

- **A rotation, angle θ, about the x-axis is represented by the matrix** $\begin{pmatrix}1&0&0\\0&\cos\theta&-\sin\theta\\0&\sin\theta&\cos\theta\end{pmatrix}$.

- **A rotation, angle θ, about the y-axis is represented by the matrix** $\begin{pmatrix}\cos\theta&0&\sin\theta\\0&1&0\\-\sin\theta&0&\cos\theta\end{pmatrix}$.

- **A rotation, angle θ, about the z-axis is represented by the matrix** $\begin{pmatrix}\cos\theta&-\sin\theta&0\\\sin\theta&\cos\theta&0\\0&0&1\end{pmatrix}$.

Online Explore rotations about the coordinate axes using GeoGebra.

Note In all cases, θ is the angle measured anticlockwise when facing in the positive direction.

Example 13

$$\mathbf{M} = \begin{pmatrix} \frac{\sqrt{3}}{2} & 0 & \frac{1}{2} \\ 0 & 1 & 0 \\ -\frac{1}{2} & 0 & \frac{\sqrt{3}}{2} \end{pmatrix}$$

a Describe the transformation represented by **M**.

b Find the image of the point with coordinates (−1, −2, 1) under the transformation represented by **M**.

a Rotation about the y-axis.

State the axis of rotation. You need to be familiar with the general forms of matrices for rotations about each of the coordinate axes.

$\cos\theta = \frac{\sqrt{3}}{2}$ so $\theta = 30°$ or $330°$

Use the top-left element to determine two possible angles.

$\sin\theta = \frac{1}{2}$ so $\theta = 30°$

$\sin\theta$ is positive so choose the angle in the first quadrant.

The transformation is a rotation anticlockwise about the y-axis through an angle of 30°.

b $$\begin{pmatrix} \frac{\sqrt{3}}{2} & 0 & \frac{1}{2} \\ 0 & 1 & 0 \\ -\frac{1}{2} & 0 & \frac{\sqrt{3}}{2} \end{pmatrix}\begin{pmatrix} -1 \\ -2 \\ 1 \end{pmatrix} = \begin{pmatrix} \frac{1-\sqrt{3}}{2} \\ -2 \\ \frac{1+\sqrt{3}}{2} \end{pmatrix}$$

Use your calculator to find this matrix product.

The coordinates of the image are

$\left(\frac{1-\sqrt{3}}{2}, -2, \frac{1+\sqrt{3}}{2}\right)$

Exercise 7E

1 Write down the matrices representing the following transformations.

a Reflection in the plane $x = 0$

b Reflection in the plane $y = 0$

c Rotation of 180° about the y-axis

d Rotation of 90° anticlockwise about the z-axis

e Rotation of 270° anticlockwise about the y-axis

f Rotation of 300° anticlockwise about the x-axis

2 Describe the transformations represented by the following matrices.

a $\begin{pmatrix} 1 & 0 & 0 \\ 0 & 1 & 0 \\ 0 & 0 & -1 \end{pmatrix}$ **b** $\begin{pmatrix} 0 & 0 & 1 \\ 0 & 1 & 0 \\ -1 & 0 & 0 \end{pmatrix}$ **c** $\begin{pmatrix} -\frac{\sqrt{2}}{2} & -\frac{\sqrt{2}}{2} & 0 \\ \frac{\sqrt{2}}{2} & -\frac{\sqrt{2}}{2} & 0 \\ 0 & 0 & 1 \end{pmatrix}$

(P) **3** $\mathbf{M} = \begin{pmatrix} 1 & 0 & 0 \\ 0 & 0 & -1 \\ 0 & 1 & 0 \end{pmatrix}$

a Determine the single transformation represented by the matrix **M**. **(3 marks)**

b The point $A = (3, -1, 4)$ is transformed using this matrix. Find the coordinates of the image of A. **(1 mark)**

c The point $B = (a, -a, 2a - 1)$ is transformed to the point with coordinates $(a, a - 5, -a)$ using matrix **M**. Find the value of a. **(3 marks)**

(P) **4** **P** is the matrix representing a rotation of 120° anticlockwise about the z-axis.

a Write down the matrix **P**. **(1 mark)**

b A point $Q = (3, -1, 0)$ is transformed using the matrix **P**. Find the coordinates of the image of Q. **(1 mark)**

c A point $R = (k, 0, k)$ is transformed using matrix **P**. Find, in terms of k, the exact coordinates of the image of R. **(3 marks)**

(E) **5** **A** is the matrix representing a reflection in the plane $x = 0$ and **B** is the matrix representing a reflection in plane $y = 0$.

a Write down the matrices **A** and **B**. **(2 marks)**

b The point $P\ (a, b, c)$ is transformed using matrix **A**. Find the coordinates of P' in terms of a, b and c. **(2 marks)**

c P' is transformed using matrix **B**. Find the coordinates of the image of P' in terms of a, b and c. **(2 marks)**

(E) **6** $\mathbf{M} = \begin{pmatrix} -\frac{\sqrt{3}}{2} & 0 & -\frac{1}{2} \\ 0 & 1 & 0 \\ \frac{1}{2} & 0 & -\frac{\sqrt{3}}{2} \end{pmatrix}$

a Find the transformation represented by matrix **M**. **(3 marks)**

b The point with coordinates $(k, -k, 0)$ is transformed using matrix **M**. Find, in terms of k, the exact coordinates of the image of this point. **(3 marks)**

(P) **7** **a** Write down the matrix representing a rotation of 315° anticlockwise about the y-axis.

A tetrahedron T has vertices at $(1, 0, -1)$, $(1, 1, -1)$, $(3, 2, 3)$ and $(0, 0, 0)$.

b Find the images of the vertices of the tetrahedron under the transformation described in part **a**.

c Hence find the volume of T.

Hint The formula for the volume of a tetrahedron is $\frac{1}{3} \times$ base area $\times$ height.

Challenge

a Find the 3×3 matrix representing a reflection in the plane $x = 0$ followed by a reflection in the plane $y = 0$.

b Find the 3×3 matrix representing a rotation of 45° anticlockwise about the x-axis followed by a reflection in the plane $x = 0$.

7.6 The inverse of a linear transformation

Since $\mathbf{A}^{-1}\mathbf{A} = \mathbf{I}$, you can use inverse matrices to reverse the effect of a linear transformation.

- **The transformation described by the matrix $\mathbf{A}^{-1}$ has the effect of reversing the transformation described by the matrix A.**

Links Reflections are **self-inverse**.
← Section 6.4, 6.5

Example 14

The matrix $\mathbf{A} = \begin{pmatrix} 2 & 4 \\ -2 & -5 \end{pmatrix}$ represents a transformation T.

Given that T maps point P with coordinates (x, y) onto the point P' with coordinates (6, 10),

a find the coordinates of P.

The matrix **B** represents a transformation U. Given that the transformation T followed by the transformation U is equivalent to a reflection in the line $y = x$,

b find **B**.

a $\mathbf{A}^{-1} = \begin{pmatrix} 2.5 & 2 \\ -1 & -1 \end{pmatrix}$

This represents the inverse transformation to T. You can find it quickly using your calculator.

$\begin{pmatrix} 2 & 4 \\ -2 & -5 \end{pmatrix}\begin{pmatrix} x \\ y \end{pmatrix} = \begin{pmatrix} 6 \\ 10 \end{pmatrix}$

$\mathbf{A}\begin{pmatrix} x \\ y \end{pmatrix} = \begin{pmatrix} 6 \\ 10 \end{pmatrix}$

$\begin{pmatrix} x \\ y \end{pmatrix} = \begin{pmatrix} 2.5 & 2 \\ -1 & -1 \end{pmatrix}\begin{pmatrix} 6 \\ 10 \end{pmatrix}$

Left multiply both sides by $\mathbf{A}^{-1}$:
$\mathbf{A}^{-1}\mathbf{A}\begin{pmatrix} x \\ y \end{pmatrix} = \mathbf{A}^{-1}\begin{pmatrix} 6 \\ 10 \end{pmatrix}$ so $\begin{pmatrix} x \\ y \end{pmatrix} = \mathbf{A}^{-1}\begin{pmatrix} 6 \\ 10 \end{pmatrix}$

$= \begin{pmatrix} 35 \\ -16 \end{pmatrix}$

P has coordinates (35, −16).

b $\mathbf{BA} = \begin{pmatrix} 0 & 1 \\ 1 & 0 \end{pmatrix}$

The matrix representing T followed by U is **BA**. This is equal to $\begin{pmatrix} 0 & 1 \\ 1 & 0 \end{pmatrix}$, which is the matrix for a reflection in the line $y = x$.

$\mathbf{B} = \begin{pmatrix} 0 & 1 \\ 1 & 0 \end{pmatrix}\begin{pmatrix} 2.5 & 2 \\ -1 & -1 \end{pmatrix} = \begin{pmatrix} -1 & -1 \\ 2.5 & 2 \end{pmatrix}$

Right multiply both sides by $\mathbf{A}^{-1}$. Remember that the order is important.
$(\mathbf{BA})\mathbf{A}^{-1} = \begin{pmatrix} 0 & 1 \\ 1 & 0 \end{pmatrix}\mathbf{A}^{-1}$, so $\mathbf{B} = \begin{pmatrix} 0 & 1 \\ 1 & 0 \end{pmatrix}\mathbf{A}^{-1}$

Exercise 7F

1 The matrix $\mathbf{R} = \begin{pmatrix} 0 & -1 \\ 1 & 0 \end{pmatrix}$

 a Give a geometrical interpretation of the transformation represented by **R**.

 b Find $\mathbf{R}^{-1}$.

 c Give a geometrical interpretation of the transformation represented by $\mathbf{R}^{-1}$.

2 a The matrix $\mathbf{S} = \begin{pmatrix} -1 & 0 \\ 0 & -1 \end{pmatrix}$

i Give a geometrical interpretation of the transformation represented by **S**.

ii Show that $\mathbf{S}^2 = \mathbf{I}$.

iii Give a geometrical interpretation of the transformation represented by $\mathbf{S}^{-1}$.

b The matrix $\mathbf{T} = \begin{pmatrix} 0 & -1 \\ -1 & 0 \end{pmatrix}$

i Give a geometrical interpretation of the transformation represented by **T**.

ii Show that $\mathbf{T}^2 = \mathbf{I}$.

iii Give a geometrical interpretation of the transformation represented by $\mathbf{T}^{-1}$.

c Calculate det **S** and det **T** and comment on their values in the light of the transformations they represent.

3 The matrix **A** represents a reflection in the line $y = x$ and the matrix **B** represents an anticlockwise rotation of 270° about (0, 0).

a Find the matrix $\mathbf{C} = \mathbf{BA}$ and interpret it geometrically.

b Find $\mathbf{C}^{-1}$ and give a geometrical interpretation of the transformation represented by $\mathbf{C}^{-1}$.

c Find the matrix $\mathbf{D} = \mathbf{AB}$ and interpret it geometrically.

d Find $\mathbf{D}^{-1}$ and give a geometrical interpretation of the transformation represented by $\mathbf{D}^{-1}$.

E **4** The matrix $\mathbf{A} = \begin{pmatrix} 1 & 2 \\ -3 & -2 \end{pmatrix}$ represents a transformation T.

Given that T maps point P with coordinates (x, y) onto the point P' with coordinates (5, 8),

a find the coordinates of P. **(2 marks)**

The matrix **B** represents a transformation U. Given that the transformation T followed by the transformation U is equivalent to a reflection in the line $y = -x$,

b find **B**. **(2 marks)**

E **5** $\mathbf{E} = \begin{pmatrix} 4 & 0 \\ 0 & 4 \end{pmatrix}$

a State the transformation represented by matrix **E**. **(1 mark)**

b Use your calculator to find $\mathbf{E}^{-1}$. **(1 mark)**

A triangle T transformed using matrix **E** is such that the coordinates of the image are (4, 6), (9, 7) and (3, 1).

c Using your answer to part **b**, find the coordinates of the vertices of T. **(2 marks)**

E/P **6** $\mathbf{M} = \begin{pmatrix} a & 0 \\ 0 & b \end{pmatrix}$ where a and b are non-zero constants.

a Find, in terms of a and b, the matrix $\mathbf{M}^{-1}$. **(2 marks)**

b The point $D = (p, q)$ maps to the point (−6, 8) under the transformation represented by **M**. Find, in terms of a and b, the coordinates of D. **(3 marks)**

(E) **7** $\mathbf{R} = \begin{pmatrix} \frac{\sqrt{3}}{2} & \frac{1}{2} \\ -\frac{1}{2} & \frac{\sqrt{3}}{2} \end{pmatrix}$

a Describe fully the transformation represented by **R**. **(2 marks)**

b The point (p, q) is mapped onto $\left(\frac{1 - 2\sqrt{3}}{2}, \frac{2 + \sqrt{3}}{2}\right)$ under the transformation represented by **R**. Find the values of p and q. **(3 marks)**

(E/P) **8** $\mathbf{A} = \begin{pmatrix} 2 & 4 \\ 1 & 3 \end{pmatrix}$ and $\mathbf{B} = \begin{pmatrix} 0 & -1 \\ -1 & 0 \end{pmatrix}$

Triangle T is transformed to triangle T' using the matrix **AB**.
Find the matrix **P** such that triangle T' is mapped to T. **(3 marks)**

(E/P) **9** $\mathbf{A} = \begin{pmatrix} 6 & -2 \\ -4 & 1 \end{pmatrix}$

The transformation represented by **A** maps the point P onto the point Q.
Given that Q has coordinates (a, b), find the coordinates of P in terms of a and b. **(4 marks)**

(E) **10** $\mathbf{M} = \begin{pmatrix} 1 & 0 & 1 \\ 0 & 2 & -2 \\ 1 & 3 & -1 \end{pmatrix}$

The point (a, b, c) maps to the point $(3, 2, -1)$ under **M**. Find $\mathbf{M}^{-1}$ and hence find the exact values of a, b and c. **(4 marks)**

(E/P) **11** The 3×3 matrix **P** represents a reflection in the plane $x = 0$. The matrix **Q** represents an anticlockwise rotation through 90° about the z-axis.

a Find the matrix $\mathbf{M} = \mathbf{PQ}$ and interpret it geometrically. **(3 marks)**

b Use your calculator to find $\mathbf{M}^{-1}$ and give a geometrical interpretation. **(2 marks)**

c Find the matrix $\mathbf{N} = \mathbf{QP}$ and interpret it geometrically. **(3 marks)**

d Show that $\mathbf{N}^{-1} = \mathbf{P}^{-1}\mathbf{Q}^{-1}$ and interpret geometrically the matrix $\mathbf{N}^{-1}$. **(3 marks)**

(E/P) **12** The matrix $\mathbf{A} = \begin{pmatrix} 1 & 3 & -1 \\ -1 & 1 & 2 \\ 2 & 0 & -1 \end{pmatrix}$ represents a transformation T, in three dimensions.

The matrix **B** represents a transformation U. Given that transformation T followed by transformation U represents a reflection in the xy-plane, find **B**. **(4 marks)**

Mixed exercise 7

/P **1** The matrix **Y** represents an anticlockwise rotation of 90° about (0, 0).

a Find **Y**. **(1 mark)**

The matrices **A** and **B** are such that **AB** = **Y**. Given that $\mathbf{B} = \begin{pmatrix} 3 & 2 \\ 2 & 1 \end{pmatrix}$,

b find **A**. **(3 marks)**

c Describe the transformation represented by the matrix **ABABABAB** and write down its 2 × 2 matrix. **(2 marks)**

E **2** The matrix **R** represents a reflection in the x-axis and the matrix **E** represents an enlargement with scale factor 2 and centre (0, 0).

a Find the matrix **C** = **ER** and give a geometrical interpretation of the transformation **C** represents. **(4 marks)**

b Find $\mathbf{C}^{-1}$ and give a geometrical interpretation of the transformation represented by $\mathbf{C}^{-1}$. **(2 marks)**

/P **3** $\mathbf{P} = \begin{pmatrix} 0 & k \\ k & 0 \end{pmatrix}$

Given that **P** represents a transformation T, followed by a reflection in the line $y = x$,

a find, in terms of k, a matrix representing T. **(4 marks)**

b Given that the point (−3, −2) maps to point (9, 6) under T, find the value of k. **(2 marks)**

c Given that the line $y = mx$, where m is a real constant, is invariant under T, find the two possible values of m. **(3 marks)**

/P **4** The matrix $\mathbf{M} = \begin{pmatrix} 2\sqrt{3} & -2 \\ 2 & 2\sqrt{3} \end{pmatrix}$ represents a rotation followed by an enlargement.

a Find the scale factor of the enlargement. **(2 marks)**

b Find the angle of rotation. **(3 marks)**

A point P is mapped onto a point P' under **M**. Given that the coordinates of P' are (a, b),

c find, in terms of a and b, the coordinates of P. **(4 marks)**

E **5** $\mathbf{A} = \begin{pmatrix} 0 & 1 \\ 1 & 0 \end{pmatrix}$ and $\mathbf{B} = \begin{pmatrix} 0 & 1 \\ -1 & 0 \end{pmatrix}$

a Describe fully the transformations represented by the matrices **A** and **B**. **(4 marks)**

b The point (p, q) is transformed by the matrix product **AB**. Give the coordinates of the image of this point in terms of p and q. **(2 marks)**

/P **6** $\mathbf{M} = \begin{pmatrix} -4 & 3 \\ 1 & -2 \end{pmatrix}$

A triangle T has vertices at $(k, 2)$, (6, 2) and (6, 7).

a Given that T is transformed using matrix **M**, and the area of the resulting triangle is 110, find the two possible values of k. **(3 marks)**

b Show that the line $x + 3y = 0$ is invariant under this transformation. **(4 marks)**

(E) 7 $\mathbf{A} = \begin{pmatrix} -1 & 0 \\ 0 & 1 \end{pmatrix}$ and $\mathbf{B} = \begin{pmatrix} 4 & 0 \\ 0 & 3 \end{pmatrix}$

a Find the matrix $\mathbf{P} = \mathbf{AB}$. **(1 mark)**

A triangle T is transformed using matrix $\mathbf{P}$.

b Given that the area of T' is 60, find the area of T. **(2 marks)**

(E/P) 8 $\mathbf{A} = \begin{pmatrix} -\frac{\sqrt{3}}{2} & -\frac{1}{2} & 0 \\ \frac{1}{2} & -\frac{\sqrt{3}}{2} & 0 \\ 0 & 0 & 1 \end{pmatrix}$

a Find the transformation represented by matrix $\mathbf{A}$. **(3 marks)**

b The point with coordinates $(a, b, -a)$ is transformed using matrix $\mathbf{M}$. Find, in terms of a and b, the exact coordinates of the image of this point. **(3 marks)**

(E/P) 9 $\mathbf{P} = \begin{pmatrix} a & 0 \\ 0 & a \end{pmatrix}$ where a is a non-zero constant.

a Find, in terms of a, the matrix $\mathbf{P}^{-1}$. **(2 marks)**

b The point A maps to the point (4, 7) under the transformation represented by $\mathbf{P}$. Find, in terms of a, the coordinates of A. **(3 marks)**

(E/P) 10 The matrix $\mathbf{P} = \begin{pmatrix} -1 & 2 \\ -5 & 8 \end{pmatrix}$ represents a transformation U.

A triangle T is transformed by transformation U followed by an anticlockwise rotation through 90° about the origin. The resulting image is labelled T'.

Find a matrix $\mathbf{M}$ representing a linear transformation that maps T' back onto T. **(3 marks)**

(E) 11 $\mathbf{A} = \begin{pmatrix} -1 & 0 \\ 0 & -1 \end{pmatrix}$ and $\mathbf{B} = \begin{pmatrix} 4 & -1 \\ 3 & -2 \end{pmatrix}$

The transformation represented by $\mathbf{B}$ followed by the transformation represented by $\mathbf{A}$ is equivalent to the transformation represented by matrix $\mathbf{P}$.

a Find $\mathbf{P}$. **(1 mark)**

Triangle T is transformed to the triangle T' by the transformation represented by $\mathbf{P}$.

Given that the area of the triangle T' is 35,

b find the area of triangle T. **(3 marks)**

Triangle T' is transformed to the original triangle T by the transformation represented by matrix $\mathbf{Q}$.

c Find $\mathbf{Q}$. **(2 marks)**

(E) 12 $\mathbf{M} = \begin{pmatrix} 1 & 0 & 0 \\ 0 & -\frac{\sqrt{2}}{2} & \frac{\sqrt{2}}{2} \\ 0 & -\frac{\sqrt{2}}{2} & -\frac{\sqrt{2}}{2} \end{pmatrix}$

The point (a, b, c) maps to the point (0, 1, 1) under $\mathbf{M}$. Find $\mathbf{M}^{-1}$ and hence find the exact values of a, b and c. **(4 marks)**

P **13** The matrix $\mathbf{A} = \begin{pmatrix} 1 & -1 & 0 \\ 2 & 0 & 4 \\ 3 & -2 & -1 \end{pmatrix}$ represents a transformation T.

Given that T maps point P with coordinates (x, y, z) onto the point P' with coordinates $(3, 7, 4)$,

a find the coordinates of P. **(3 marks)**

The matrix **B** represents a transformation U. Given that the transformation T followed by the transformation U is equivalent to a rotation through 90° about the x-axis,

b find **B**. **(3 marks)**

Challenge

1 Find the 3 × 3 matrix representing the single transformation that is equivalent to a reflection in the plane $y = 0$, followed by a rotation of 90° about the x-axis, followed by a reflection in the plane $z = 0$.

2 a Show that the transformation described by $\begin{pmatrix} 0 & 1 \\ 0 & 1 \end{pmatrix}$ maps any point in the plane onto the line $y = x$.

b Find the matrix representing the linear transformation that maps any point in the plane onto the straight line $y = mx$.

c Explain why, in general, the transformation that maps any point in the plane onto the straight line $ax + by = c$ is **not** a linear transformation.

Summary of key points

1 • Linear transformations always map the origin onto itself.

• Any linear transformation can be represented by a matrix.

2 The linear transformation $T: \begin{pmatrix} x \\ y \end{pmatrix} \mapsto \begin{pmatrix} ax + by \\ cx + dy \end{pmatrix}$ can be represented by the matrix $\mathbf{M} = \begin{pmatrix} a & b \\ c & d \end{pmatrix}$

since $\begin{pmatrix} a & b \\ c & d \end{pmatrix} \begin{pmatrix} x \\ y \end{pmatrix} = \begin{pmatrix} ax + by \\ cx + dy \end{pmatrix}$.

3 A reflection in the y-axis is represented by the matrix $\begin{pmatrix} -1 & 0 \\ 0 & 1 \end{pmatrix}$. Points on the y-axis are invariant points, and the lines $x = 0$ and $y = k$ for any value of k are invariant lines.

4 A reflection in the x-axis is represented by the matrix $\begin{pmatrix} 1 & 0 \\ 0 & -1 \end{pmatrix}$. Points on the x-axis are invariant points, and the lines $y = 0$ and $x = k$ for any value of k are invariant lines.

5 A reflection in the line $y = x$ is represented by the matrix $\begin{pmatrix} 0 & 1 \\ 1 & 0 \end{pmatrix}$. Points on the line $y = x$ are invariant points, and the lines $y = x$ and $y = -x + k$ for any value of k are invariant lines.

6 A reflection in the line $y = -x$ is represented by the matrix $\begin{pmatrix} 0 & -1 \\ -1 & 0 \end{pmatrix}$. Points on the line $y = -x$ are invariant points, and the lines $y = -x$ and $y = x + k$ for any value of k are invariant lines.

7 The matrix representing a rotation through angle θ anticlockwise about the origin is $\begin{pmatrix} \cos\theta & -\sin\theta \\ \sin\theta & \cos\theta \end{pmatrix}$.
The only invariant point is the origin (0, 0). For $\theta \neq 180°$, there are no invariant lines. For $\theta = 180°$, any line passing through the origin is an invariant line.

8 A transformation represented by the matrix $\begin{pmatrix} a & 0 \\ 0 & b \end{pmatrix}$ is a stretch of scale factor a parallel to the x-axis and a stretch of scale factor b parallel to the y-axis.
In the case where $a = b$, the transformation is an enlargement with scale factor a.

9 For any stretch of the above form, the x- and y-axes are invariant lines and the origin is an invariant point.

10 For a stretch parallel to the x-axis only, points on the y-axis are invariant points, and any line parallel to the x-axis is an invariant line.

11 For a stretch parallel to the y-axis only, points on the x-axis are invariant points, and any line parallel to the y-axis is an invariant line.

12 For a linear transformation represented by matrix **M**, det **M** represents the scale factor for the change in area. This is sometimes called the **area scale factor**.

13 The matrix **PQ** represents the transformation Q, with matrix **Q**, followed by the transformation P, with matrix **P**.

14 A reflection in the plane $x = 0$ is represented by the matrix $\begin{pmatrix} -1 & 0 & 0 \\ 0 & 1 & 0 \\ 0 & 0 & 1 \end{pmatrix}$.

15 A reflection in the plane $y = 0$ is represented by the matrix $\begin{pmatrix} 1 & 0 & 0 \\ 0 & -1 & 0 \\ 0 & 0 & 1 \end{pmatrix}$.

16 A reflection in the plane $z = 0$ is represented by the matrix $\begin{pmatrix} 1 & 0 & 0 \\ 0 & 1 & 0 \\ 0 & 0 & -1 \end{pmatrix}$.

17 A rotation, angle θ, anticlockwise about the x-axis is represented by the matrix

$$\begin{pmatrix} 1 & 0 & 0 \\ 0 & \cos\theta & -\sin\theta \\ 0 & \sin\theta & \cos\theta \end{pmatrix}$$

18 A rotation, angle θ, anticlockwise about the y-axis is represented by the matrix

$$\begin{pmatrix} \cos\theta & 0 & \sin\theta \\ 0 & 1 & 0 \\ -\sin\theta & 0 & \cos\theta \end{pmatrix}$$

19 A rotation, angle θ, anticlockwise about the z-axis is represented by the matrix

$$\begin{pmatrix} \cos\theta & -\sin\theta & 0 \\ \sin\theta & \cos\theta & 0 \\ 0 & 0 & 1 \end{pmatrix}$$

20 The transformation described by the matrix $\mathbf{A}^{-1}$ has the effect of reversing the transformation described by the matrix **A**.

Proof by induction

8

Objectives

After completing this chapter you should be able to:

- Understand the principle of proof by mathematical induction and prove results about sums of series → pages 156–159
- Prove results about divisibility using induction → pages 160–162
- Prove results about matrices using induction → pages 162–164

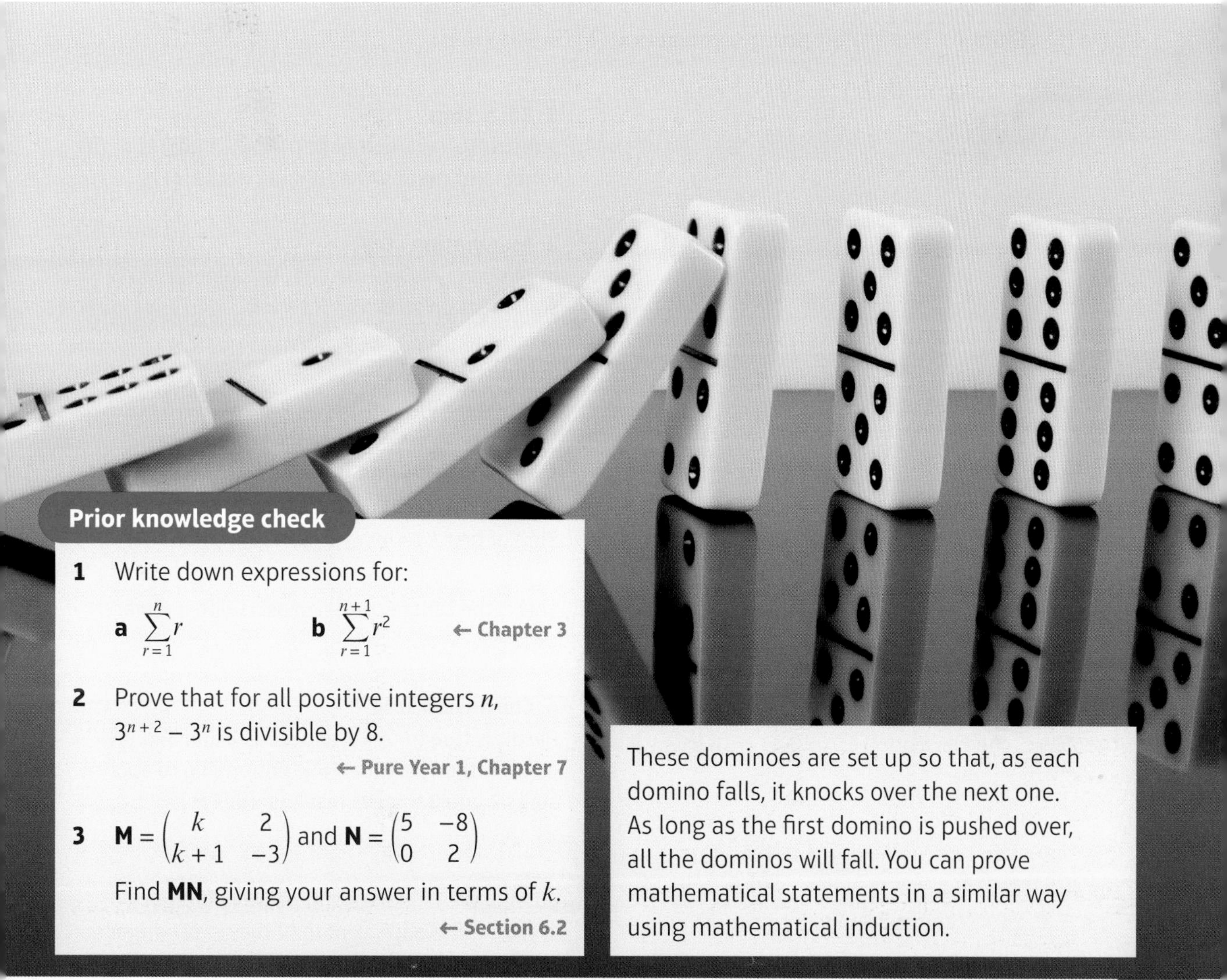

Prior knowledge check

1 Write down expressions for:

a $\sum_{r=1}^{n} r$ **b** $\sum_{r=1}^{n+1} r^2$ ← **Chapter 3**

2 Prove that for all positive integers n, $3^{n+2} - 3^n$ is divisible by 8.

← **Pure Year 1, Chapter 7**

3 $\mathbf{M} = \begin{pmatrix} k & 2 \\ k+1 & -3 \end{pmatrix}$ and $\mathbf{N} = \begin{pmatrix} 5 & -8 \\ 0 & 2 \end{pmatrix}$

Find **MN**, giving your answer in terms of k.

← **Section 6.2**

These dominoes are set up so that, as each domino falls, it knocks over the next one. As long as the first domino is pushed over, all the dominos will fall. You can prove mathematical statements in a similar way using mathematical induction.

8.1 Proof by mathematical induction

- **You can use proof by induction to prove that a general statement is true for all positive integers.**

Watch out You need to carry out **both** the basis step **and** the inductive step in order to complete the proof: carrying out just one of these is not sufficient to prove the general statement.

- **Proof by mathematical induction usually consists of the following four steps:**

Step 1: Basis: **Prove the general statement is true for $n = 1$.**

Step 2: Assumption: **Assume the general statement is true for $n = k$.**

Step 3: Inductive: **Show that the general statement is then true for $n = k + 1$.**

Step 4: Conclusion: **The general statement is then true for all positive integers, n.**

This method of proof is often useful for proving results about sums of series.

Links You can prove the general results for $\sum_{r=1}^{n} r$, $\sum_{r=1}^{n} r^2$ and $\sum_{r=1}^{n} r^3$ by induction. ← Chapter 3

Example 1

Prove by induction that for all positive integers n, $\sum_{r=1}^{n}(2r - 1) = n^2$.

$n = 1$: LHS $= \sum_{r=1}^{1}(2r - 1) = 2(1) - 1 = 1$

RHS $= 1^2 = 1$

As LHS = RHS, the summation formula is true for $n = 1$.

1. Basis step
Substitute $n = 1$ into both the LHS and RHS of the formula to check if the formula works for $n = 1$.

Assume that the summation formula is true for $n = k$:

$\sum_{r=1}^{k}(2r - 1) = k^2$

2. Assumption step
In this step you assume that the general statement given is true for $n = k$.

With $n = k + 1$ terms the summation formula becomes:

$$\sum_{r=1}^{k+1}(2r - 1) = \sum_{r=1}^{k}(2r - 1) + 2(k + 1) - 1$$
$$= k^2 + (2(k + 1) - 1)$$
$$= k^2 + (2k + 2 - 1)$$
$$= k^2 + 2k + 1$$
$$= (k + 1)^2$$

3. Inductive step
Sum to k terms plus the $(k + 1)$th term.

This is the $(k + 1)$th term.

Sum of first k terms is k^2 by assumption.

This is the same expression as n^2 with n replaced by $k + 1$.

Therefore, the summation formula is true when $n = k + 1$.

If the summation formula is true for $n = k$ then it is shown to be true for $n = k + 1$. As the result is true for $n = 1$, it is now also true for all $n \in \mathbb{Z}^+$ by mathematical induction.

4. Conclusion step
Result is true for $n = 1$ and steps **2** and **3** imply result is then true for $n = 2$. Continuing to apply steps **2** and **3** implies result is true for $n = 3, 4, 5, \ldots$ etc.

Notation $\mathbb{Z}^+$ is the set of **positive integers**, 1, 2, 3, …. It is equivalent to $\mathbb{N}$, the set of natural numbers.

Example 2

Prove by induction that for all positive integers n, $\sum_{r=1}^{n} r^2 = \frac{1}{6}n(n+1)(2n+1)$.

$n = 1$: LHS $= \sum_{r=1}^{1} r^2 = 1^2 = 1$

RHS $= \frac{1}{6}(1)(2)(3) = \frac{6}{6} = 1$

As LHS = RHS, the summation formula is true for $n = 1$.

1. Basis step
Substitute $n = 1$ into both the LHS and RHS of the formula to check if the formula works for $n = 1$.

Assume that the summation formula is true for $n = k$:

$$\sum_{r=1}^{k} r^2 = \frac{1}{6}k(k+1)(2k+1)$$

2. Assumption step
In this step you assume that the result given in the question is true for $n = k$.

With $n = k + 1$ terms the summation formula becomes:

$$\sum_{r=1}^{k+1} r^2 = \sum_{r=1}^{k} r^2 + (k+1)^2$$
$$= \frac{1}{6}k(k+1)(2k+1) + (k+1)^2$$
$$= \frac{1}{6}(k+1)(k(2k+1) + 6(k+1))$$
$$= \frac{1}{6}(k+1)(2k^2 + k + 6k + 6)$$
$$= \frac{1}{6}(k+1)(2k^2 + 7k + 6)$$
$$= \frac{1}{6}(k+1)(k+2)(2k+3)$$
$$= \frac{1}{6}(k+1)((k+1)+1)(2(k+1)+1)$$

3. Inductive step

Sum to k terms plus the $(k + 1)$th term.

Rearrange to get same expression as $\frac{1}{6}n(n+1)(2n+1)$ with n replaced by $k + 1$.

Therefore, the summation formula is true when $n = k + 1$.

If the summation formula is true for $n = k$, then it is shown to be true for $n = k + 1$. As the result is true for $n = 1$, it is therefore true for all $n \in \mathbb{Z}^+$ by mathematical induction.

4. Conclusion step
Result is true for $n = 1$ and steps **2** and **3** imply result is then true for $n = 2$. Continuing to apply steps **2** and **3** implies result is true for $n = 3, 4, 5,$ … etc.

Example 3

Prove by induction that for all positive integers n, $\sum_{r=1}^{n} r2^r = 2(1 + (n-1)2^n)$.

$n = 1$: LHS $= \sum_{r=1}^{1} r\,2^r = 1(2)^1 = 2$

RHS $= 2(1 + (1-1)2^1) = 2(1) = 2$

1. Basis step

As LHS = RHS, the summation formula is true for $n = 1$.

Assume that the summation formula is true for $n = k$:

$$\sum_{r=1}^{k} r\,2^r = 2(1 + (k-1)2^k).$$

2. Assumption step

With $n = k + 1$ terms the summation formula becomes:

$$\sum_{r=1}^{k+1} r2^r = \sum_{r=1}^{k} r2^r + (k+1)2^{k+1}$$
$$= 2(1 + (k-1)2^k) + (k+1)2^{k+1}$$
$$= 2 + 2(k-1)2^k + (k+1)2^{k+1}$$
$$= 2 + (k-1)2^{k+1} + (k+1)2^{k+1}$$
$$= 2 + (k - 1 + k + 1)2^{k+1}$$
$$= 2 + 2k2^{k+1}$$
$$= 2(1 + k2^{k+1})$$
$$= 2(1 + ((k+1) - 1)2^{k+1})$$

Therefore, the summation formula is true when $n = k + 1$.

If the summation formula is true for $n = k$, then it is shown to be true for $n = k + 1$. As the result is true for $n = 1$, it is now also true for all $n \in \mathbb{Z}^+$ by mathematical induction.

3. Inductive step

$2^1 \times 2^k = 2^{k+1}$

This is the same expression as $2(1 + (n - 1)2^n)$ with n replaced by $k + 1$.

4. Conclusion step

Watch out Induction can prove that a given statement is true for all $n \in \mathbb{Z}^+$, but it does not help you derive statements.

Exercise 8A

(E/P) **1** Prove by induction that for any positive integer n, $\sum_{r=1}^{n} r = \frac{1}{2}n(n+1)$. **(5 marks)**

(E/P) **2** Prove by induction that for any positive integer n, $\sum_{r=1}^{n} r^3 = \frac{1}{4}n^2(n+1)^2$. **(5 marks)**

(E/P) **3** **a** Prove by induction that for any positive integer n:

$$\sum_{r=1}^{n} r(r-1) = \frac{1}{3}n(n+1)(n-1)$$ **(6 marks)**

b Hence deduce an expression, in terms of n, for $\sum_{r=1}^{2n+1} r(r-1)$. **(3 marks)**

(E/P) **4** **a** Prove by induction that, for any positive integer n:

$$\sum_{r=1}^{n} r(3r-1) = n^2(n+1)$$ **(6 marks)**

b Hence use the standard result for $\sum_{r=1}^{n} r^3$ to find a value of n such that $\sum_{r=1}^{n} r^3 = 4\sum_{r=1}^{n} r(3r-1)$. **(5 marks)**

(P) **5** Prove by induction that for any positive integer n,

a $\sum_{r=1}^{n} \left(\frac{1}{2}\right)^r = 1 - \frac{1}{2^n}$ **b** $\sum_{r=1}^{n} r(r!) = (n+1)! - 1$ **c** $\sum_{r=1}^{n} \frac{4}{r(r+2)} = \frac{n(3n+5)}{(n+1)(n+2)}$

E/P **6** The box below shows a student's attempts to prove $\left(\sum_{r=1}^{n} r\right)^2 = \sum_{r=1}^{n} r^2$ using induction.

> Let $n = 1$. Then LHS $= \left(\sum_{r=1}^{1} r^2\right) = (1)^2 = 1$, and RHS $= \sum_{r=1}^{1} r^2 = 1^2 = 1$, so that LHS = RHS (Basis step).
>
> Now we assume the statement is true for $n = k$:
>
> $$\left(\sum_{r=1}^{k} r\right)^2 = \sum_{r=1}^{k} r^2$$
>
> and so for $n = k + 1$ the statement is
>
> $$\left(\sum_{r=1}^{k+1} r\right)^2 = \sum_{r=1}^{k+1} r^2$$
>
> Hence, by the principle of mathematical induction, the statement is true for all $n \in \mathbb{Z}^+$.

a Identify the error made in the proof. **(2 marks)**

b Give a counter-example to show that the original statement is not true. **(1 mark)**

E/P **7** A student claims that $\sum_{r=1}^{n} r = \frac{1}{2}(n^2 + n + 1)$, and produces the following proof.

> Assume that the statement is true for $n = k$:
>
> $$\sum_{r=1}^{k} r = \frac{1}{2}(k^2 + k + 1)$$
>
> When $n = k + 1$:
>
> $$\sum_{r=1}^{k+1} r = \sum_{r=1}^{k} r + (k + 1)$$
> $$= \frac{1}{2}(k^2 + k + 1) + (k + 1)$$
> $$= \frac{1}{2}(k^2 + k + 1 + 2(k + 1))$$
> $$= \frac{1}{2}((k^2 + 2k + 1) + (k + 1) + 1)$$
> $$= \frac{1}{2}((k + 1)^2 + (k + 1) + 1)$$
>
> This is the original formula but with $n = k + 1$. Hence, by the principle of mathematical induction, the statement is true for all $n \in \mathbb{Z}^+$.

a Identify the error made in the proof. **(2 marks)**

b Give a counter-example to show that the original statement is not true. **(1 mark)**

Challenge

Prove by induction that for all positive integers n,

$\sum_{r=1}^{n} (-1)^r r^2 = \frac{1}{2}(-1)^n n(n + 1)$

Hint $\sum_{r=1}^{n} (-1)^r r^2 = -1^2 + 2^2 - 3^2 + 4^2 - 5^2 + \ldots$

8.2 Proving divisibility results

You can use proof by induction to prove that a given expression is divisible by a certain integer.

Example 4

Prove by induction that for all positive integers n, $3^{2n} + 11$ is divisible by 4.

Let $f(n) = 3^{2n} + 11$, where $n \in \mathbb{Z}^+$.
$f(1) = 3^{2(1)} + 11 = 9 + 11 = 20 = 4(5)$, which is divisible by 4.
$f(n)$ is divisible by 4 when $n = 1$.

1. Basis step

Assume true for $n = k$, so that
$f(k) = 3^{2k} + 11$ is divisible by 4.

2. Assumption step

$f(k + 1) = 3^{2(k+1)} + 11$
$= 3^{2k} \times 3^2 + 11$
$= 9(3^{2k}) + 11$
$f(k + 1) - f(k) = (9(3^{2k}) + 11) - (3^{2k} + 11)$
$= 8(3^{2k})$
$= 4(2(3^{2k}))$
$f(k + 1) = f(k) + 4(2(3^{2k}))$
Therefore $f(n)$ is divisible by 4 when $n = k + 1$.

3. Inductive step

As both $f(k)$ and $4(2(3^{2k}))$ are divisible by 4 then the sum of these two must also be divisible by 4.

If $f(n)$ is divisible by 4 when $n = k$, then it has been shown that $f(n)$ is also divisible by 4 when $n = k + 1$. As $f(n)$ is divisible by 4 when $n = 1$, $f(n)$ is also divisible by 4 for all $n \in \mathbb{Z}^+$ by mathematical induction.

4. Conclusion step

Problem-solving

When proving that an expression $f(n)$ is divisible by r, you can complete the induction step by showing that $f(k + 1) - f(k)$ is divisible by r.

Example 5

Prove by induction that for all positive integers n, $n^3 - 7n + 9$ is divisible by 3.

Let $f(n) = n^3 - 7n + 9$, where $n \in \mathbb{Z}^+$.
$f(1) = 1 - 7 + 9 = 3$, which is divisible by 3.
$f(n)$ is divisible by 3 when $n = 1$.

1. Basis step

Assume true for $n = k$, so that
$f(k) = k^3 - 7k + 9$ is divisible by 3.

2. Assumption step

$f(k + 1) = (k + 1)^3 - 7(k + 1) + 9$
$= k^3 + 3k^2 + 3k + 1 - 7(k + 1) + 9$
$= k^3 + 3k^2 + 3k + 1 - 7k - 7 + 9$
$= k^3 + 3k^2 - 4k + 3$

3. Inductive step

Use the binomial theorem or multiply out three brackets.

$f(k+1) - f(k) = (k^3 + 3k^2 - 4k + 3)$
$- (k^3 - 7k + 9)$
$= 3k^2 + 3k - 6$
$= 3(k^2 + k - 2)$
$f(k+1) = f(k) + 3(k^2 + k - 2)$
Therefore $f(n)$ is divisible by 3 when $n = k + 1$.
If $f(n)$ is divisible by 3 when $n = k$, then it has been shown that $f(n)$ is also divisible by 3 when $n = k + 1$. As $f(n)$ is divisible by 3 when $n = 1$, $f(n)$ is also divisible by 3 for all $n \in \mathbb{Z}^+$ by mathematical induction.

As both $f(k)$ and $3(k^2 + k - 2)$ are divisible by 3 then their sum must also be divisible by 3.

4. Conclusion step

Example 6

Prove by induction that for all positive integers n, $11^{n+1} + 12^{2n-1}$ is divisible by 133.

1. Basis step

Let $f(n) = 11^{n+1} + 12^{2n-1}$, where $n \in \mathbb{Z}^+$.
$f(1) = 11^2 + 12 = 133$, which is divisible by 133.
$f(n)$ is divisible by 133 when $n = 1$.

2. Assumption step

Assume true for $n = k$, so that
$f(k) = 11^{k+1} + 12^{2k-1}$ is divisible by 133.

3. Inductive step

$f(k+1) = 11^{k+1+1} + 12^{2(k+1)-1}$
$= 11^{k+1}(11)^1 + 12^{2k-1}(12)^2$
$= 11(11^{k+1}) + 144(12^{2k-1})$

$12^{2(k+1)-1} = 12^{2k+2-1}$
$= 12^{2k-1+2}$
$= 12^{2k-1}(12)^2$

$f(k+1) - f(k) = (11(11^{k+1}) + 144(12^{2k-1}))$
$- (11^{k+1} + 12^{2k-1})$
$= 10(11^{k+1}) + 143(12^{2k-1})$
$= 10(11^{k+1}) + 10(12^{2k-1})$
$+ 133(12^{2k-1})$
$= 10(11^{k+1} + 12^{2k-1})$
$+ 133(12^{2k-1})$
$f(k+1) = f(k) + 10(11^{k+1} + 12^{2k-1})$
$+ 133(12^{2k-1})$
$= f(k) + 10f(k) + 133(12^{2k-1})$
$= 11f(k) + 133(12^{2k-1})$

Problem-solving

Always keep an eye on what you are trying to prove. You need to show that this expression is divisible by 133, so write $143(12^{2k-1})$ as $10(12^{2k-1}) + 133(12^{2k-1})$.

As both $11f(k)$ and $133(12^{2k-1})$ are divisible by 133 then their sum must also be divisible by 133.

4. Conclusion step

Therefore $f(n)$ is divisible by 133 when $n = k + 1$.
If $f(n)$ is divisible by 133 when $n = k$, then it has been shown that $f(n)$ is also divisible by 133 when $n = k + 1$. As $f(n)$ is divisible by 133 when $n = 1$, $f(n)$ is also divisible by 133 for all $n \in \mathbb{Z}^+$ by mathematical induction.

Exercise 8B

(P) **1** Prove by induction that for all positive integers n:

a $8^n - 1$ is divisible by 7
b $3^{2n} - 1$ is divisible by 8
c $5^n + 9^n + 2$ is divisible by 4
d $2^{4n} - 1$ is divisible by 15
e $3^{2n-1} + 1$ is divisible by 4
f $n^3 + 6n^2 + 8n$ is divisible by 3
g $n^3 + 5n$ is divisible by 6
h $2^n(3^{2n}) - 1$ is divisible by 17

(E/P) **2** $f(n) = 13^n - 6^n$

a Show that $f(k + 1) = 6f(k) + 7(13^k)$. **(3 marks)**

b Hence, or otherwise, prove by induction that for all positive integers n, $f(n)$ is divisible by 7. **(4 marks)**

(E/P) **3** $g(n) = 5^{2n} - 6n + 8$

a Show that $g(k + 1) = 25g(k) + 9(16k - 22)$. **(3 marks)**

b Hence, or otherwise, prove by induction that for all positive integers n, $g(n)$ is divisible by 9. **(4 marks)**

(E/P) **4** Prove by induction that for all positive integers n, $8^n - 3^n$ is divisible by 5. **(6 marks)**

(E/P) **5** Prove by induction that for all positive integers n, $3^{2n+2} + 8n - 9$ is divisible by 8. **(6 marks)**

(E/P) **6** Prove by induction that for all positive integers n, $2^{6n} + 3^{2n-2}$ is divisible by 5. **(6 marks)**

8.3 Proving statements involving matrices

You can use matrix multiplication to prove results involving powers of matrices.

Example 7

Prove by induction that for all positive integers n, $\begin{pmatrix} 1 & -1 \\ 0 & 2 \end{pmatrix}^n = \begin{pmatrix} 1 & 1 - 2^n \\ 0 & 2^n \end{pmatrix}$.

$n = 1$: LHS $= \begin{pmatrix} 1 & -1 \\ 0 & 2 \end{pmatrix}^1 = \begin{pmatrix} 1 & -1 \\ 0 & 2 \end{pmatrix}$

RHS $= \begin{pmatrix} 1 & 1 - 2^1 \\ 0 & 2^1 \end{pmatrix} = \begin{pmatrix} 1 & -1 \\ 0 & 2 \end{pmatrix}$

As LHS = RHS, the matrix equation is true for $n = 1$.

1. Basis step

Substitute $n = 1$ into both the LHS and RHS of the formula to check to see if the formula works for $n = 1$.

Assume that the matrix equation is true for $n = k$:

$\begin{pmatrix} 1 & -1 \\ 0 & 2 \end{pmatrix}^k = \begin{pmatrix} 1 & 1 - 2^k \\ 0 & 2^k \end{pmatrix}$

2. Assumption step

In this step you assume that the general statement given is true for $n = k$.

With $n = k + 1$ the matrix equation becomes

$$\begin{pmatrix} 1 & -1 \\ 0 & 2 \end{pmatrix}^{k+1} = \begin{pmatrix} 1 & -1 \\ 0 & 2 \end{pmatrix}^{k} \begin{pmatrix} 1 & -1 \\ 0 & 2 \end{pmatrix}$$

$$= \begin{pmatrix} 1 & 1 - 2^k \\ 0 & 2^k \end{pmatrix} \begin{pmatrix} 1 & -1 \\ 0 & 2 \end{pmatrix}$$

$$= \begin{pmatrix} 1 + 0 & -1 + 2 - 2(2^k) \\ 0 + 0 & 0 + 2(2^k) \end{pmatrix}$$

$$= \begin{pmatrix} 1 & 1 - 2^{k+1} \\ 0 & 2^{k+1} \end{pmatrix}$$

Therefore the matrix equation is true when $n = k + 1$.

If the matrix equation is true for $n = k$, then it is shown to be true for $n = k + 1$. As the matrix equation is true for $n = 1$, it is also true for all $n \in \mathbb{Z}^+$ by mathematical induction.

3. Inductive step

Use the assumption step.

As this is a proof, you should show your working for each element in the matrix multiplication.

← Section 6.2

This is the right-hand side of the original equation with n replaced by $k + 1$.

4. Conclusion step

Example 8

Prove by induction that for all positive integers n, $\begin{pmatrix} -2 & 9 \\ -1 & 4 \end{pmatrix}^n = \begin{pmatrix} -3n + 1 & 9n \\ -n & 3n + 1 \end{pmatrix}$.

$n = 1$: LHS $= \begin{pmatrix} -2 & 9 \\ -1 & 4 \end{pmatrix}^1 = \begin{pmatrix} -2 & 9 \\ -1 & 4 \end{pmatrix}$

RHS $= \begin{pmatrix} -3(1) + 1 & 9(1) \\ -(1) & 3(1) + 1 \end{pmatrix} = \begin{pmatrix} -2 & 9 \\ -1 & 4 \end{pmatrix}$

As LHS = RHS, the matrix equation is true for $n = 1$.

Assume that the matrix equation is true for $n = k$:

$$\begin{pmatrix} -2 & 9 \\ -1 & 4 \end{pmatrix}^k = \begin{pmatrix} -3k + 1 & 9k \\ -k & 3k + 1 \end{pmatrix}$$

With $n = k + 1$ the matrix equation becomes

$$\begin{pmatrix} -2 & 9 \\ -1 & 4 \end{pmatrix}^{k+1} = \begin{pmatrix} -2 & 9 \\ -1 & 4 \end{pmatrix}^{k} \begin{pmatrix} -2 & 9 \\ -1 & 4 \end{pmatrix}$$

$$= \begin{pmatrix} -3k + 1 & 9k \\ -k & 3k + 1 \end{pmatrix} \begin{pmatrix} -2 & 9 \\ -1 & 4 \end{pmatrix}$$

$$= \begin{pmatrix} 6k - 2 - 9k & -27k + 9 + 36k \\ 2k - 3k - 1 & -9k + 12k + 4 \end{pmatrix}$$

$$= \begin{pmatrix} -3k - 2 & 9k + 9 \\ -k - 1 & 3k + 4 \end{pmatrix}$$

$$= \begin{pmatrix} -3(k + 1) + 1 & 9(k + 1) \\ -(k + 1) & 3(k + 1) + 1 \end{pmatrix}$$

Therefore the matrix equation is true when $n = k + 1$.

1. Basis step

Substitute $n = 1$ into both the LHS and RHS of the formula to check to see if the formula works for $n = 1$.

2. Assumption step

In this step you assume that the general statement given is true for $n = k$.

3. Inductive step

Use the assumption step.

This is the right-hand side of the original equation with n replaced by $k + 1$.

If the matrix equation is true for $n = k$, then it is shown to be true for $n = k + 1$. As the matrix equation is true for $n = 1$, it is also true for all $n \in \mathbb{Z}^+$ by mathematical induction.

4. Conclusion step

Exercise 8C

E/P **1** Prove by induction that for all positive integers n,

$$\begin{pmatrix} 1 & 2 \\ 0 & 1 \end{pmatrix}^n = \begin{pmatrix} 1 & 2n \\ 0 & 1 \end{pmatrix}$$ **(6 marks)**

E/P **2** Prove by induction that for all positive integers n,

$$\begin{pmatrix} 3 & -4 \\ 1 & -1 \end{pmatrix}^n = \begin{pmatrix} 2n + 1 & -4n \\ n & -2n + 1 \end{pmatrix}$$ **(6 marks)**

E/P **3** Prove by induction that for all positive integers n,

$$\begin{pmatrix} 2 & 0 \\ 1 & 1 \end{pmatrix}^n = \begin{pmatrix} 2^n & 0 \\ 2^n - 1 & 1 \end{pmatrix}$$ **(6 marks)**

E/P **4** **a** Prove by induction that for all positive integers n,

$$\begin{pmatrix} 5 & -8 \\ 2 & -3 \end{pmatrix}^n = \begin{pmatrix} 4n + 1 & -8n \\ 2n & 1 - 4n \end{pmatrix}$$ **(6 marks)**

b Hence find the value of n such that:

$$\begin{pmatrix} -1 & 3 \\ -2 & 5 \end{pmatrix}\begin{pmatrix} 5 & -8 \\ 2 & -3 \end{pmatrix}^n = \begin{pmatrix} 11 & -21 \\ 10 & -19 \end{pmatrix}$$ **(4 marks)**

E/P **5** The matrix $\mathbf{M} = \begin{pmatrix} 2 & 5 \\ 0 & 1 \end{pmatrix}$

a Prove by induction that for all positive integers n,

$$\mathbf{M}^n = \begin{pmatrix} 2^n & 5(2^n - 1) \\ 0 & 1 \end{pmatrix}$$ **(6 marks)**

b Hence find an expression for $(\mathbf{M}^n)^{-1}$ in terms of n. **(4 marks)**

Challenge

Prove by induction that for all $n \in \mathbb{Z}^+$,

$$\begin{pmatrix} 3 & 1 & 0 \\ 0 & 1 & 0 \\ 0 & -1 & 4 \end{pmatrix}^n = \begin{pmatrix} 3^n & \frac{3^n - 1}{2} & 0 \\ 0 & 1 & 0 \\ 0 & \frac{1 - 4^n}{3} & 4^n \end{pmatrix}$$

Mixed exercise 8

E/P **1** Prove by induction that $9^n - 1$ is divisible by 8 for all positive integers n. **(6 marks)**

P **2** The matrix **B** is given by $\mathbf{B} = \begin{pmatrix} 1 & 0 \\ 0 & 3 \end{pmatrix}$.

a Find $\mathbf{B}^2$ and $\mathbf{B}^3$.

b Use your answer to part **a** to suggest a general statement for $\mathbf{B}^n$, for all positive integers n.

c Prove by induction that your answer to part **b** is correct.

E/P **3** Prove by induction that for all positive integers n, $\sum_{r=1}^{n}(3r + 4) = \frac{1}{2}n(3n + 11)$. **(6 marks)**

E/P **4** The matrix **A** is given by $\mathbf{A} = \begin{pmatrix} 9 & 16 \\ -4 & -7 \end{pmatrix}$.

a Prove by induction that $\mathbf{A}^n = \begin{pmatrix} 8n + 1 & 16n \\ -4n & 1 - 8n \end{pmatrix}$ for all positive integers n. **(6 marks)**

The matrix **B** is given by $\mathbf{B} = (\mathbf{A}^n)^{-1}$.

b Hence find **B** in terms of n. **(4 marks)**

E/P **5** The function f is defined by $f(n) = 5^{2n-1} + 1$, where n is a positive integer.

a Show that $f(n + 1) - f(n) = \mu(5^{2n-1})$, where μ is an integer to be determined. **(3 marks)**

b Hence prove by induction that $f(n)$ is divisible by 6. **(4 marks)**

E/P **6** Prove by induction that $7^n + 4^n + 1$ is divisible by 6 for all positive integers n. **(6 marks)**

E/P **7** Prove by induction that for all positive integers n, $\sum_{r=1}^{n} r(r + 4) = \frac{1}{6}n(n + 1)(2n + 13)$. **(6 marks)**

E/P **8** **a** Prove by induction that for all positive integers n:

$$\sum_{r=1}^{2n} r^2 = \frac{1}{3}n(2n + 1)(4n + 1)$$ **(6 marks)**

b Given that $\sum_{r=1}^{2n} r^2 = k\sum_{r=1}^{n} r^2$, show that k must satisfy $n = \frac{2 - k}{k - 8}$ **(5 marks)**

E/P **9** The matrix $\mathbf{M} = \begin{pmatrix} 2c & 1 \\ 0 & c \end{pmatrix}$ for some positive constant c

a Prove by induction that for all positive integers n:

$$\mathbf{M}^n = c^n\begin{pmatrix} 2^n & \frac{2^n - 1}{c} \\ 0 & 1 \end{pmatrix}$$ **(7 marks)**

b Given that $\det(\mathbf{M}^n) = 50^n$, find the value of c. **(5 marks)**

Challenge

$\mathbf{M} = \begin{pmatrix} \cos\theta & -\sin\theta \\ \sin\theta & \cos\theta \end{pmatrix}$

a Prove by induction that for all positive integers n, $\mathbf{M}^n = \begin{pmatrix} \cos n\theta & -\sin n\theta \\ \sin n\theta & \cos n\theta \end{pmatrix}$.

b Interpret this result geometrically by describing the linear transformations represented by **M** and $\mathbf{M}^n$.

Summary of key points

1 You can use **proof by induction** to prove that a general statement is true for all positive integers.

2 Proof by mathematical induction usually consists of the following four steps:

- **Basis:** Show the general statement is true for $n = 1$.
- **Assumption:** Assume that the general statement is true for $n = k$.
- **Inductive:** Show the general statement is true for $n = k + 1$.
- **Conclusion:** State that the general statement is then true for all positive integers, n.

Vectors

9

Objectives

After completing this chapter you should be able to:

- Understand and use the vector and Cartesian forms of the equation of a straight line in three dimensions → pages 168–175
- Understand and use the vector and Cartesian forms of the equation of a plane → pages 175–178
- Calculate the scalar product for two 3D vectors → pages 178–184
- Calculate the angle between two vectors, two lines, a line and a plane, or two planes → pages 184–189
- Understand and use the scalar product form of the equation of a plane → pages 185–189
- Determine whether two lines meet and determine the point of intersection → pages 189–192
- Calculate the perpendicular distance between: two lines, a point and a line, or a point and a plane → pages 193–201

Prior knowledge check

1 Find $\overrightarrow{BC}$ given that:

a $\overrightarrow{AC} = \begin{pmatrix} 2 \\ 3 \end{pmatrix}$ and $\overrightarrow{AB} = \begin{pmatrix} 7 \\ 1 \end{pmatrix}$

b $\overrightarrow{AB} = \begin{pmatrix} -1 \\ 5 \\ 4 \end{pmatrix}$ and $\overrightarrow{CA} = \begin{pmatrix} -6 \\ 0 \\ 3 \end{pmatrix}$

← Pure Year 2, Chapter 12

2 Find the exact distance between the points with coordinates:

a (3, −2) and (−1, 4)

b (1, 3, −2) and (−3, 2, −5)

← Pure Year 2, Chapter 12

3 Given $\mathbf{a} = 4\mathbf{i} - 3\mathbf{j} + 2\mathbf{k}$, find:

a $|\mathbf{a}|$

b the unit vector in the direction of **a**.

4 The lines l_1 and l_2 have equations l_1: $3x - 4y = 7$ and l_2: $2x + 5y = -3$. Find the coordinates of the point of intersection of l_1 and l_2.

← Pure Year 1, Chapter 5

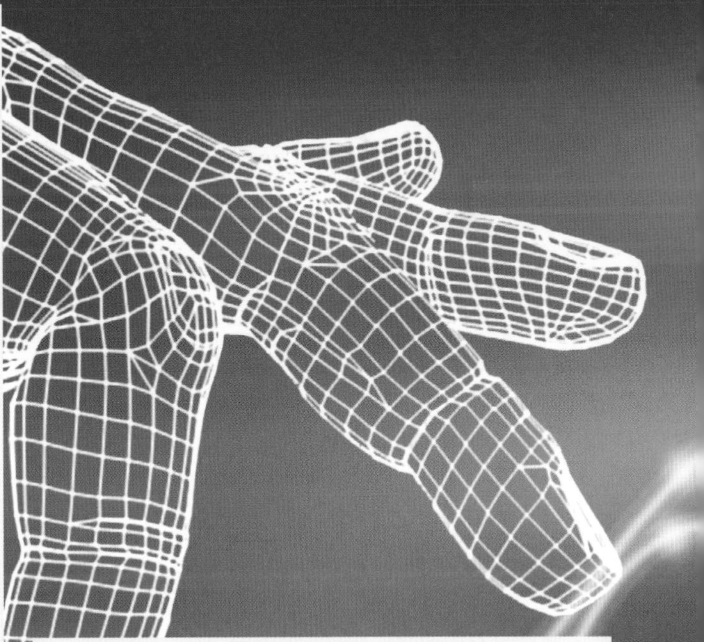

Vectors can be used to describe points, lines and planes in 3D. Computer graphics artists use **3D vectors** to define shapes based on polygons. By creating a shape from thousands of polygons you can create the illusion of a smoothly curved surface.

9.1 Equation of a line in three dimensions

You need to know how to write the equation of a straight line in vector form.

Suppose a straight line passes through a given point A, with position vector **a**, and is parallel to the given vector **b**. Only one such line is possible. Let R be an arbitrary point on the line, with position vector **r**.

Since $\overrightarrow{AR}$ is parallel to **b**, $\overrightarrow{AR} = \lambda\mathbf{b}$, where λ is a scalar.
The vector **b** is called the **direction vector** of the line.
So the position vector **r** can be written as $\mathbf{a} + \lambda\mathbf{b}$.

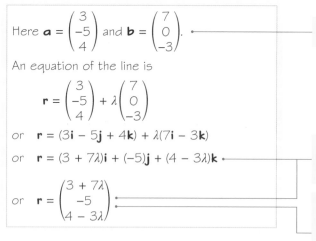

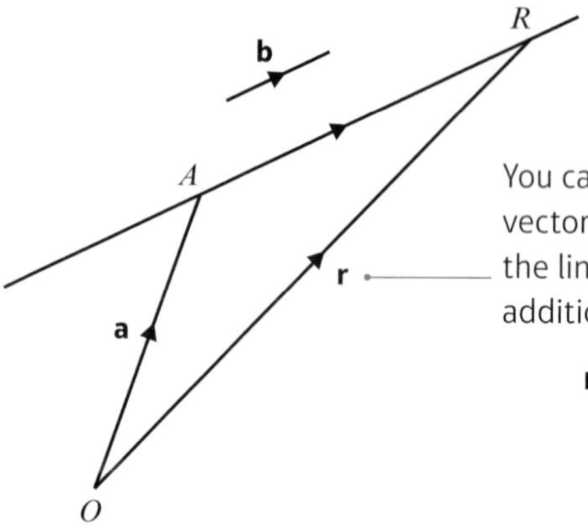

You can find the position vector of any point R on the line by using vector addition ($\triangle OAR$):

$\mathbf{r} = \mathbf{a} + \overrightarrow{AR}$

- **A vector equation of a straight line passing through the point A with position vector a, and parallel to the vector b is**

  $$\mathbf{r} = \mathbf{a} + \lambda\mathbf{b}$$

  where λ is a scalar parameter.

Notation **r** is the position vector of a general point on the line. Scalar parameters in vector equations are often given Greek letters such as λ (lambda) and μ (mu).

By taking different values of the parameter λ, you can find the position vectors of different points that lie on the straight line.

Online Explore the vector equation of a line using GeoGebra.

Example 1

Find a vector equation of the straight line which passes through the point A, with position vector $3\mathbf{i} - 5\mathbf{j} + 4\mathbf{k}$, and is parallel to the vector $7\mathbf{i} - 3\mathbf{k}$.

Here $\mathbf{a} = \begin{pmatrix} 3 \\ -5 \\ 4 \end{pmatrix}$ and $\mathbf{b} = \begin{pmatrix} 7 \\ 0 \\ -3 \end{pmatrix}$.

b is the direction vector.

An equation of the line is

$$\mathbf{r} = \begin{pmatrix} 3 \\ -5 \\ 4 \end{pmatrix} + \lambda\begin{pmatrix} 7 \\ 0 \\ -3 \end{pmatrix}$$

or $\mathbf{r} = (3\mathbf{i} - 5\mathbf{j} + 4\mathbf{k}) + \lambda(7\mathbf{i} - 3\mathbf{k})$

or $\mathbf{r} = (3 + 7\lambda)\mathbf{i} + (-5)\mathbf{j} + (4 - 3\lambda)\mathbf{k}$

You sometimes need to show the separate x, y, z components in terms of λ.

or $\mathbf{r} = \begin{pmatrix} 3 + 7\lambda \\ -5 \\ 4 - 3\lambda \end{pmatrix}$

*You can represent a 3D vector using column notation, $\begin{pmatrix} x \\ y \\ z \end{pmatrix}$, or using **ijk**-notation, $x\mathbf{i} + y\mathbf{j} + z\mathbf{k}$.*

← Pure Year 2, Chapter 12

Now suppose a straight line passes through two given points C and D, with position vectors **c** and **d** respectively. Again, only one such line is possible.

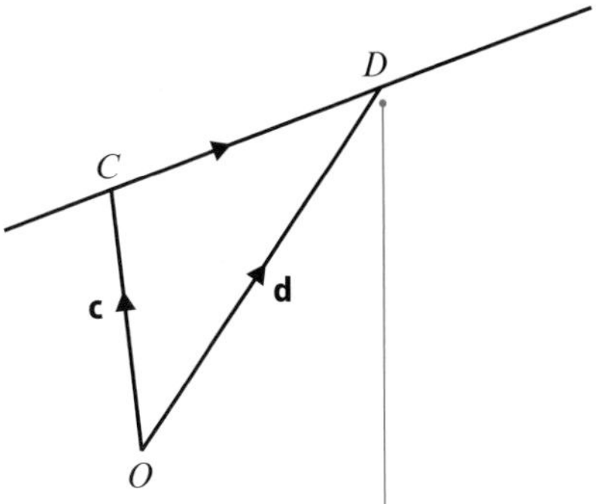

You can use $\overrightarrow{CD}$ as a direction vector for the line: $\overrightarrow{CD} = \mathbf{d} - \mathbf{c}$.

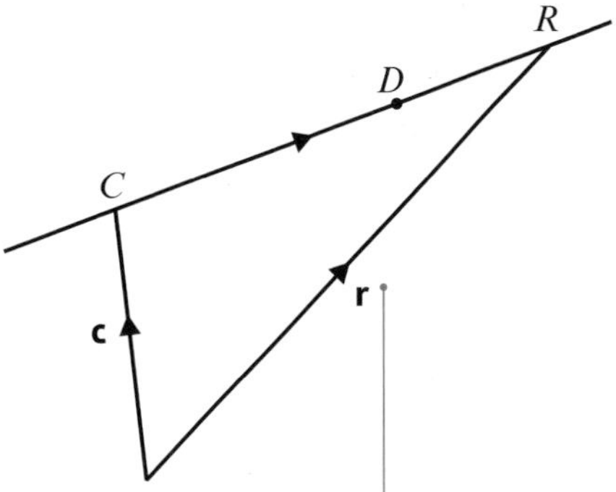

You can now use one of the two given points and the direction vector to form an equation for the straight line.

- **A vector equation of a straight line passing through the points C and D, with position vectors c and d respectively, is**

  $$\mathbf{r} = \mathbf{c} + \lambda(\mathbf{d} - \mathbf{c})$$

  where λ is a scalar parameter.

Note You can use any point on the straight line as the initial point in the vector equation. An alternative vector equation for this line would be $\mathbf{r} = \mathbf{d} + \lambda(\mathbf{d} - \mathbf{c})$.

Example 2

Find a vector equation of the straight line which passes through the points A and B, with coordinates (4, 5, −1) and (6, 3, 2) respectively.

$\mathbf{a} = \begin{pmatrix} 4 \\ 5 \\ -1 \end{pmatrix}$ $\mathbf{b} = \begin{pmatrix} 6 \\ 3 \\ 2 \end{pmatrix}$ — Write down the position vectors of A and B.

$\mathbf{b} - \mathbf{a} = \begin{pmatrix} 6 \\ 3 \\ 2 \end{pmatrix} - \begin{pmatrix} 4 \\ 5 \\ -1 \end{pmatrix} = \begin{pmatrix} 2 \\ -2 \\ 3 \end{pmatrix}$ — Find a direction vector for the line.

Use one of the given points to form the equation.

$\mathbf{r} = \begin{pmatrix} 4 \\ 5 \\ -1 \end{pmatrix} + t\begin{pmatrix} 2 \\ -2 \\ 3 \end{pmatrix}$

You don't have to use λ for the parameter. In this example, the parameter is represented by the letter t.

or $\mathbf{r} = (4\mathbf{i} + 5\mathbf{j} - \mathbf{k}) + t(2\mathbf{i} - 2\mathbf{j} + 3\mathbf{k})$

or $\mathbf{r} = (4 + 2t)\mathbf{i} + (5 - 2t)\mathbf{j} + (-1 + 3t)\mathbf{k}$

or $\mathbf{r} = \begin{pmatrix} 4 + 2t \\ 5 - 2t \\ -1 + 3t \end{pmatrix}$

You can give your answer in any of these forms.

Example 3

The straight line l has vector equation $\mathbf{r} = (3\mathbf{i} + 2\mathbf{j} - 5\mathbf{k}) + t(\mathbf{i} - 6\mathbf{j} - 2\mathbf{k})$.
Given that the point $(a, b, 0)$ lies on l, find the value of a and the value of b.

$\mathbf{r} = \begin{pmatrix} 3 + t \\ 2 - 6t \\ -5 - 2t \end{pmatrix}$

You can write the equation in this form.

$-5 - 2t = 0$

$t = -\frac{5}{2}$

Use the z-coordinate (which is equal to zero) to find the value of t.

$a = 3 + t = \frac{1}{2}$

$b = 2 - 6t = 17$

$a = \frac{1}{2}$ and $b = 17$

Find a and b using the value of t.

Example 4

The straight line l has vector equation $\mathbf{r} = (2\mathbf{i} + 5\mathbf{j} - 3\mathbf{k}) + \lambda(6\mathbf{i} - 2\mathbf{j} + 4\mathbf{k})$.
Show that another vector equation of l is $\mathbf{r} = (8\mathbf{i} + 3\mathbf{j} + \mathbf{k}) + \mu(3\mathbf{i} - \mathbf{j} + 2\mathbf{k})$.

Use the equation $\mathbf{r} = \begin{pmatrix} 2 \\ 5 \\ -3 \end{pmatrix} + \lambda \begin{pmatrix} 6 \\ -2 \\ 4 \end{pmatrix}$.

When $\lambda = 1$, $\mathbf{r} = \begin{pmatrix} 8 \\ 3 \\ 1 \end{pmatrix}$, so the point $(8, 3, 1)$ lies on l.

To show that $(8\mathbf{i} + 3\mathbf{j} + \mathbf{k})$ lies on l, find a value of λ that gives this point. It is often easier to work in column vectors.

$\begin{pmatrix} 6 \\ -2 \\ 4 \end{pmatrix} = 2\begin{pmatrix} 3 \\ -1 \\ 2 \end{pmatrix}$ so these two vectors are parallel.

If one vector is a scalar multiple of another then the vectors are parallel.

So an alternative form of the equation is

$\mathbf{r} = \begin{pmatrix} 8 \\ 3 \\ 1 \end{pmatrix} + \mu\begin{pmatrix} 3 \\ -1 \\ 2 \end{pmatrix}$

Watch out Using the same value of the parameter in each equation will give **different** points on the line. You should use a different letter for the parameter of the second equation.

You also need to be able to write the equation of a line in three dimensions in **Cartesian form**. This means that the equation is given in terms of coordinates relative to the x-, y- and z-axes.

- **If $\mathbf{a} = \begin{pmatrix} a_1 \\ a_2 \\ a_3 \end{pmatrix}$ and $\mathbf{b} = \begin{pmatrix} b_1 \\ b_2 \\ b_3 \end{pmatrix}$ the equation of the line with vector equation $\mathbf{r} = \mathbf{a} + \lambda\mathbf{b}$ can be given in Cartesian form as**

$$\frac{x - a_1}{b_1} = \frac{y - a_2}{b_2} = \frac{z - a_3}{b_3}$$

Each of the three expressions is equal to λ.

Example 5

With respect to the fixed origin O, the line l is given by the equation

$$\mathbf{r} = \begin{pmatrix} a_1 \\ a_2 \\ a_3 \end{pmatrix} + \lambda \begin{pmatrix} b_1 \\ b_2 \\ b_3 \end{pmatrix}$$

a Prove that a Cartesian form of the equation of l is

$$\frac{x - a_1}{b_1} = \frac{y - a_2}{b_2} = \frac{z - a_3}{b_3}$$

b Hence find a Cartesian equation of the line with equation $\mathbf{r} = \begin{pmatrix} 4 \\ 3 \\ -2 \end{pmatrix} + \lambda \begin{pmatrix} -1 \\ 2 \\ 5 \end{pmatrix}$.

a $\begin{pmatrix} x \\ y \\ z \end{pmatrix} = \begin{pmatrix} a_1 + \lambda b_1 \\ a_2 + \lambda b_2 \\ a_3 + \lambda b_3 \end{pmatrix}$

Write the position vector of the general point on the line as $\mathbf{r} = \begin{pmatrix} x \\ y \\ z \end{pmatrix}$.

$x = a_1 + \lambda b_1$, $y = a_2 + \lambda b_2$, $z = a_3 + \lambda b_3$

Use the vector equation of the line to write expressions for x, y and z in terms of λ.

Rearranging,

$\lambda = \frac{x - a_1}{b_1}$, $\lambda = \frac{y - a_2}{b_2}$, $\lambda = \frac{z - a_3}{b_3}$

Make λ the subject of each equation.

So $\frac{x - a_1}{b_1} = \frac{y - a_2}{b_2} = \frac{z - a_3}{b_3}$

For any point on the line, the value of λ is a constant, so equate the three different expressions for λ.

b $\frac{x - 4}{-1} = \frac{y - 3}{2} = \frac{z + 2}{5}$

If you need to convert between vector and Cartesian forms you can quote this result without proof in your exam. Be careful with the signs on the top of each fraction.

Example 6

The line l has equation $\mathbf{r} = \begin{pmatrix} -2 \\ 1 \\ 4 \end{pmatrix} + \lambda \begin{pmatrix} 1 \\ -2 \\ 1 \end{pmatrix}$, and the point P has position vector $\begin{pmatrix} 2 \\ 1 \\ 3 \end{pmatrix}$.

a Show that P does not lie on l.

Given that a circle, centre P, intersects l at points A and B, and that A has position vector $\begin{pmatrix} 0 \\ -3 \\ 6 \end{pmatrix}$,

b find the position vector of B.

a $\mathbf{r} = \begin{pmatrix} -2 + \lambda \\ 1 - 2\lambda \\ 4 + \lambda \end{pmatrix}$

If $P(2, 1, 3)$ lies on the line then

$2 = -2 + \lambda \Rightarrow \lambda = 4$

$1 = 1 - 2\lambda \Rightarrow \lambda = 0$

$3 = 4 + \lambda \Rightarrow \lambda = -1$

so P does not lie on l.

b $\overrightarrow{AP} = \begin{pmatrix} 2 \\ 1 \\ 3 \end{pmatrix} - \begin{pmatrix} 0 \\ -3 \\ 6 \end{pmatrix} = \begin{pmatrix} 2 \\ 4 \\ -3 \end{pmatrix}$

$|\overrightarrow{AP}| = \sqrt{2^2 + 4^2 + (-3)^2} = \sqrt{29}$

The position vector of B is $\begin{pmatrix} -2 + \lambda \\ 1 - 2\lambda \\ 4 + \lambda \end{pmatrix}$.

$\overrightarrow{BP} = \begin{pmatrix} 2 \\ 1 \\ 3 \end{pmatrix} - \begin{pmatrix} -2 + \lambda \\ 1 - 2\lambda \\ 4 + \lambda \end{pmatrix} = \begin{pmatrix} 4 - \lambda \\ 2\lambda \\ -1 - \lambda \end{pmatrix}$

$(4 - \lambda)^2 + 4\lambda^2 + (-1 - \lambda)^2 = 29$

$16 - 8\lambda + \lambda^2 + 4\lambda^2 + 1 + 2\lambda + \lambda^2 = 29$

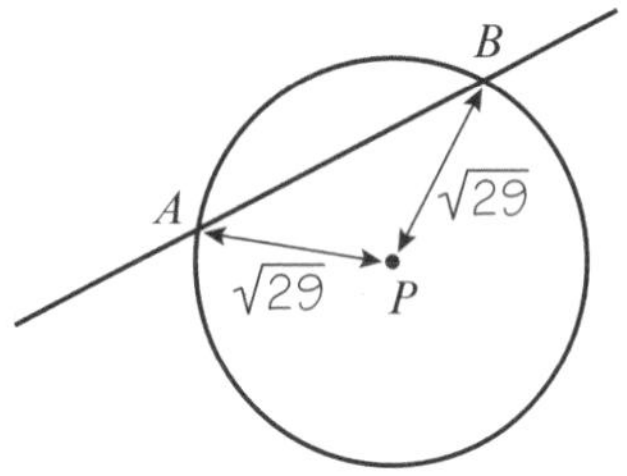

$6\lambda^2 - 6\lambda + 17 = 29$

$6\lambda^2 - 6\lambda - 12 = 0$

$\lambda^2 - \lambda - 2 = 0$

$(\lambda - 2)(\lambda + 1) = 0$

So $\lambda = 2$ or $\lambda = -1$

$\lambda = 2$ gives $\begin{pmatrix} 0 \\ -3 \\ 6 \end{pmatrix}$. This is the position vector of point A.

$\lambda = -1$ gives $\begin{pmatrix} -3 \\ 3 \\ 3 \end{pmatrix}$. This is the position vector of point B.

Problem-solving

It is often useful to write the general point on a line as a single vector. You can write each component in the form $a + \lambda b$.

If P lies on l, there is one value of λ that satisfies all 3 equations. You only need to show that two of these equations are not consistent to show that P does not lie on l.

The distance between the points with position vectors $\begin{pmatrix} a_1 \\ a_2 \\ a_3 \end{pmatrix}$ and $\begin{pmatrix} b_1 \\ b_2 \\ b_3 \end{pmatrix}$ is $\sqrt{(b_1 - a_1)^2 + (b_2 - a_2)^2 + (b_3 - a_3)^2}$. As P is the centre of the circle and A lies on the circle, the radius of the circle is $\sqrt{29}$.

← Pure Year 2, Chapter 12

Use the general point on the line to represent the position vector of B.

B lies on the circle so the length $|\overrightarrow{BP}| = \sqrt{29}$.

Solve the resulting quadratic equation to find two possible values of λ. One will correspond to point A, and the other will correspond to point B.

Substitute values of λ into $\begin{pmatrix} -2 + \lambda \\ 1 - 2\lambda \\ 4 + \lambda \end{pmatrix}$. Check that one of the values gives the position vector of A. The other value must give the position vector for B.

Exercise 9A

1 For the following pairs of vectors, find a vector equation of the straight line which passes through the point, with position vector **a**, and is parallel to the vector **b**:

a $\mathbf{a} = 6\mathbf{i} + 5\mathbf{j} - \mathbf{k}$, $\mathbf{b} = 2\mathbf{i} - 3\mathbf{j} - \mathbf{k}$

b $\mathbf{a} = 2\mathbf{i} + 5\mathbf{j}$, $\mathbf{b} = \mathbf{i} + \mathbf{j} + \mathbf{k}$

c $\mathbf{a} = -7\mathbf{i} + 6\mathbf{j} + 2\mathbf{k}$, $\mathbf{b} = 3\mathbf{i} + \mathbf{j} + 2\mathbf{k}$

d $\mathbf{a} = \begin{pmatrix} 2 \\ 0 \\ 4 \end{pmatrix}$, $\mathbf{b} = \begin{pmatrix} -3 \\ 2 \\ 1 \end{pmatrix}$

e $\mathbf{a} = \begin{pmatrix} 6 \\ -11 \\ 2 \end{pmatrix}$, $\mathbf{b} = \begin{pmatrix} 0 \\ 5 \\ -2 \end{pmatrix}$

2 For the points P and Q with position vectors **p** and **q** respectively, find:

i the vector $\overrightarrow{PQ}$

ii a vector equation of the straight line that passes through P and Q

a $\mathbf{p} = 3\mathbf{i} - 4\mathbf{j} + 2\mathbf{k}$, $\mathbf{q} = 5\mathbf{i} + 3\mathbf{j} - \mathbf{k}$

b $\mathbf{p} = 2\mathbf{i} + \mathbf{j} - 3\mathbf{k}$, $\mathbf{q} = 4\mathbf{i} - 2\mathbf{j} + \mathbf{k}$

c $\mathbf{p} = \mathbf{i} - 2\mathbf{j} + 4\mathbf{k}$, $\mathbf{q} = -2\mathbf{i} - 3\mathbf{j} + 2\mathbf{k}$

d $\mathbf{p} = \begin{pmatrix} 3 \\ -1 \\ 4 \end{pmatrix}$, $\mathbf{q} = \begin{pmatrix} -2 \\ 3 \\ 1 \end{pmatrix}$

e $\mathbf{p} = \begin{pmatrix} 4 \\ -2 \\ 3 \end{pmatrix}$, $\mathbf{q} = \begin{pmatrix} -2 \\ 2 \\ 4 \end{pmatrix}$

3 Find a vector equation of the line which is parallel to the z-axis and passes through the point $(4, -3, 8)$.

4 a Find a vector equation of the line which passes through the points:

i $(2, 1, 9)$ and $(4, -1, 8)$

ii $(-3, 5, 0)$ and $(7, 2, 2)$

iii $(1, 11, -4)$ and $(5, 9, 2)$

iv $(-2, -3, -7)$ and $(12, 4, -3)$

b Write down a Cartesian equation in the form $\frac{x - a_1}{b_1} = \frac{y - a_2}{b_2} = \frac{z - a_3}{b_3}$ for each line in part **a**.

P 5 The point $(1, p, q)$ lies on the line l. Find the values of p and q, given that the equation of l is:

a $\mathbf{r} = \begin{pmatrix} 2 \\ -3 \\ 1 \end{pmatrix} + \lambda\begin{pmatrix} 1 \\ -4 \\ -9 \end{pmatrix}$

b $\mathbf{r} = \begin{pmatrix} -4 \\ 6 \\ -1 \end{pmatrix} + \lambda\begin{pmatrix} 2 \\ -5 \\ -8 \end{pmatrix}$

c $\mathbf{r} = \begin{pmatrix} 16 \\ -9 \\ -10 \end{pmatrix} + \lambda\begin{pmatrix} 3 \\ 2 \\ 1 \end{pmatrix}$

P 6 The line l_1 has equation $\mathbf{r} = \begin{pmatrix} 2 \\ 1 \\ -3 \end{pmatrix} + \lambda\begin{pmatrix} -1 \\ 2 \\ 4 \end{pmatrix}$. The line l_2 has equation $\frac{x - 4}{2} = \frac{y + 1}{-4} = \frac{z - 3}{-8}$

Show that l_1 and l_2 are parallel.

P 7 Show that the line l_1 with equation $\mathbf{r} = (3 + 2\lambda)\mathbf{i} + (2 - 3\lambda)\mathbf{j} + (-1 + 4\lambda)\mathbf{k}$ is parallel to the line l_2 which passes through the points $A(5, 4, -1)$ and $B(3, 7, -5)$.

8 Show that the points $A(-3, -4, 5)$, $B(3, -1, 2)$ and $C(9, 2, -1)$ are collinear.

Hint Points are said to be **collinear** if they all lie on the same straight line.

9 Show that the points with position vectors $\begin{pmatrix} 1 \\ 7 \\ -2 \end{pmatrix}$, $\begin{pmatrix} 3 \\ -1 \\ 8 \end{pmatrix}$ and $\begin{pmatrix} 10 \\ 4 \\ 0 \end{pmatrix}$ do not lie on the same straight line.

(E/P) **10** The points $P(2, 0, 4)$, $Q(a, 5, 1)$ and $R(3, 10, b)$, where a and b are constants, are collinear. Find the values of a and b. **(5 marks)**

(E) **11** The line l_1 has equation

$\mathbf{r} = (8\mathbf{i} - 5\mathbf{j} + 4\mathbf{k}) + \lambda(3\mathbf{i} + \mathbf{j} - 6\mathbf{k})$

A is the point on l_1 such that $\lambda = -2$.

The line l_2 passes through A and is parallel to the line with equation

$\mathbf{r} = (10\mathbf{i} + 3\mathbf{j} - 9\mathbf{k}) + \lambda(2\mathbf{i} - 4\mathbf{j} + \mathbf{k})$

Find an equation for l_2. **(6 marks)**

(E/P) **12** The point A with coordinates $(4, a, 0)$ lies on the line L with vector equation

$\mathbf{r} = (10\mathbf{i} + 8\mathbf{j} - 12\mathbf{k}) + \lambda(\mathbf{i} - \mathbf{j} + b\mathbf{k})$

where a and b are constants.

a Find the values of a and b. **(3 marks)**

The point X lies on L where $\lambda = -1$.

b Find the coordinates of X. **(1 mark)**

(E) **13** The line l has equation $\mathbf{r} = \begin{pmatrix} 3 \\ -5 \\ 9 \end{pmatrix} + \lambda\begin{pmatrix} 1 \\ 2 \\ -2 \end{pmatrix}$.

A and B are the points on l with $\lambda = 5$ and $\lambda = 2$ respectively.
Find the distance AB. **(4 marks)**

(E) **14** The line l has equation $\mathbf{r} = \begin{pmatrix} 1 \\ -2 \\ 3 \end{pmatrix} + \lambda\begin{pmatrix} 2 \\ 1 \\ -1 \end{pmatrix}$.

C and A are the points on l with $\lambda = 4$ and $\lambda = 3$ respectively.
A circle has centre C and intersects l at the points A and B.
Find the position vector of B. **(3 marks)**

(E/P) **15** The line l has equation $x - 5 = \frac{y + 1}{3} = \frac{z - 6}{-2}$

A circle C has centre $(4, -1, 2)$ and radius $3\sqrt{5}$.
Given that C intersects l at two distinct points, A and B, find the coordinates of A and B. **(7 marks)**

> **Problem-solving**
> Write $x - 5$ as $\frac{x - 5}{1}$ and convert the equation of the line into vector form.

(E/P) **16** The line l_1 has equation $\mathbf{r} = \begin{pmatrix} -4 \\ 6 \\ 5 \end{pmatrix} + \lambda\begin{pmatrix} 1 \\ -1 \\ 1 \end{pmatrix}$. A and B are the points on l_1 with $\lambda = 2$ and $\lambda = 5$ respectively.

a Find the position vectors of A and B. **(2 marks)**

The point P has position vector $\begin{pmatrix} 0 \\ 2 \\ 3 \end{pmatrix}$.

The line l_2 passes through the point P and is parallel to the line l_1.

b Find a vector equation of the line l_2. **(2 marks)**

The points C and D both lie on line l_2 such that $AB = AC = AD$.

c Show that P is the midpoint of CD. **(7 marks)**

P 17 A tightrope is modelled as a line segment between points with coordinates (2, 3, 8) and (22, 18, 8), relative to a fixed origin O, where the units of distance are metres. Two support cables are anchored to a fixed point A on the wire. The other ends of the cables are anchored
to points with coordinates (14, 1, 0) and (6, 17, 0) respectively.

a Given that the support cables are both 12 m long, find the coordinates of A. **(8 marks)**

b Give one criticism of this model. **(1 mark)**

9.2 Equation of a plane in three dimensions

The equation of a plane can be written in vector form.

Suppose a plane passes through a given point A, with position vector **a**. Let R be an arbitrary point on the plane, with position vector **r**.

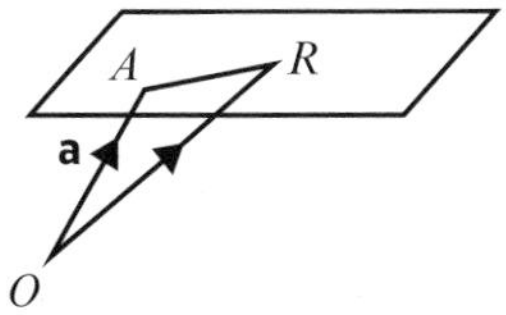

Then, using the triangle law, $\mathbf{r} = \mathbf{a} + \overrightarrow{AR}$.

Since $\overrightarrow{AR}$ lies in the plane, it can be written as $\lambda\mathbf{b} + \mu\mathbf{c}$, where **b** and **c** are non-parallel vectors in the plane and where λ and μ are scalars.

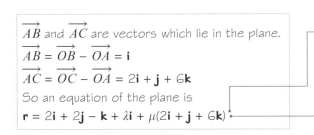

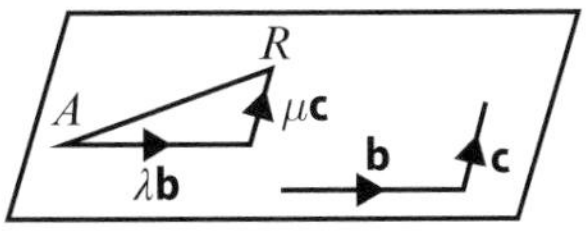

So the position vector **r** can be written as $\mathbf{r} = \mathbf{a} + \lambda\mathbf{b} + \mu\mathbf{c}$.

- **The vector equation of a plane is $\mathbf{r} = \mathbf{a} + \lambda\mathbf{b} + \mu\mathbf{c}$, where:**
  - **r is the position vector of a general point in the plane**
  - **a is the position vector of a point in the plane**
  - **b and c are non-parallel, non-zero vectors in the plane**
  - **λ and μ are scalars**

Example 7

Find, in the form $\mathbf{r} = \mathbf{a} + \lambda\mathbf{b} + \mu\mathbf{c}$, an equation of the plane that passes through the points $A(2, 2, -1)$, $B(3, 2, -1)$ and $C(4, 3, 5)$.

$\overrightarrow{AB}$ and $\overrightarrow{AC}$ are vectors which lie in the plane.

$\overrightarrow{AB} = \overrightarrow{OB} - \overrightarrow{OA} = \mathbf{i}$

$\overrightarrow{AC} = \overrightarrow{OC} - \overrightarrow{OA} = 2\mathbf{i} + \mathbf{j} + 6\mathbf{k}$

So an equation of the plane is

$\mathbf{r} = 2\mathbf{i} + 2\mathbf{j} - \mathbf{k} + \lambda\mathbf{i} + \mu(2\mathbf{i} + \mathbf{j} + 6\mathbf{k})$

There are many other forms of this answer which are also correct. You could use $3\mathbf{i} + 2\mathbf{j} - \mathbf{k}$ or $4\mathbf{i} + 3\mathbf{j} + 5\mathbf{k}$ instead of $2\mathbf{i} + 2\mathbf{j} - \mathbf{k}$ in the equation.

You could write this equation as

$$\mathbf{r} = \begin{pmatrix} 2 \\ 2 \\ -1 \end{pmatrix} + \lambda\begin{pmatrix} 1 \\ 0 \\ 0 \end{pmatrix} + \mu\begin{pmatrix} 2 \\ 1 \\ 6 \end{pmatrix}$$

Example 8

Verify that the point P with position vector $\begin{pmatrix} 2 \\ 2 \\ -1 \end{pmatrix}$ lies in the plane with vector equation

$$\mathbf{r} = \begin{pmatrix} 3 \\ 4 \\ -2 \end{pmatrix} + \lambda\begin{pmatrix} 2 \\ 1 \\ 1 \end{pmatrix} + \mu\begin{pmatrix} 1 \\ -1 \\ 2 \end{pmatrix}$$

$$\mathbf{r} = \begin{pmatrix} 3 + 2\lambda + \mu \\ 4 + \lambda - \mu \\ -2 + \lambda + 2\mu \end{pmatrix}$$

The position vector of any point on the plane can be written in this form.

If P lies on the plane,

$$\begin{pmatrix} 2 \\ 2 \\ -1 \end{pmatrix} = \begin{pmatrix} 3 + 2\lambda + \mu \\ 4 + \lambda - \mu \\ -2 + \lambda + 2\mu \end{pmatrix}$$

$2 = 3 + 2\lambda + \mu$ so $2\lambda + \mu = -1$ (1)

$2 = 4 + \lambda - \mu$ so $\lambda - \mu = -2$ (2)

$-1 = -2 + \lambda + 2\mu$ so $\lambda + 2\mu = 1$ (3)

Solving equations (2) and (3) simultaneously,

(3) − (2): $3\mu = 3$ so $\mu = 1$

Sub in (2): $\lambda - 1 = -2$ so $\lambda = -1$

Check in equation (1):

$2\lambda + \mu = -2 + 1 = -1$ so P lies in the plane.

Problem-solving

If the point P lies on the plane then there will be values of λ and μ that satisfy **all three** of these equations simultaneously. Solve one pair of equations simultaneously, then check that the solutions satisfy the third equation.

The direction of a plane can be described by giving a **normal vector, n**. This is a vector that is perpendicular to the plane.

One normal vector can describe an infinite number of parallel planes, so the normal vector on its own is not enough information to define a plane uniquely.

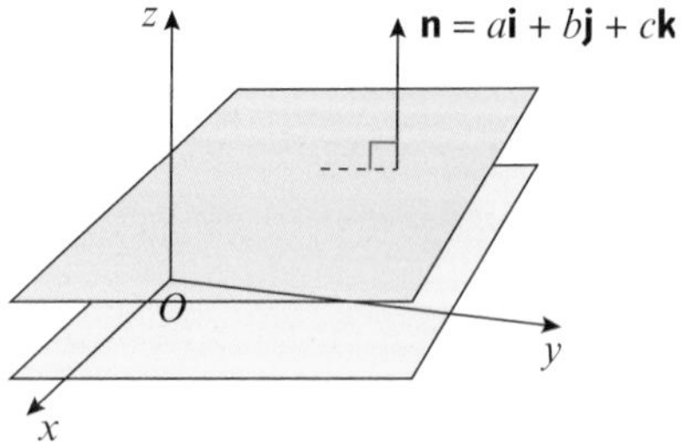

- **A Cartesian equation of a plane in three dimensions can be written in the form $ax + by + cz = d$ where a, b, c and d are constants, and $\begin{pmatrix} a \\ b \\ c \end{pmatrix}$ is the normal vector to the plane.**

Note Compare this equation to the Cartesian equation of a line in two dimensions: $ax + by = c$.

Online Explore the vector and Cartesian equations of a plane using GeoGebra.

You can derive this result using the **scalar product**, which you will learn about later in this chapter.

Example 9

The plane Π is perpendicular to the normal vector $\mathbf{n} = 3\mathbf{i} - 2\mathbf{j} + \mathbf{k}$ and passes through the point P with position vector $8\mathbf{i} + 4\mathbf{j} - 7\mathbf{k}$. Find a Cartesian equation of Π.

Notation Planes are often represented by the capital Greek letter pi, Π.

$3x - 2y + z = d$

The general equation is $ax + by + cz = d$ where the normal vector is $\begin{pmatrix} a \\ b \\ c \end{pmatrix} = \begin{pmatrix} 3 \\ -2 \\ 1 \end{pmatrix}$.

$3 \times 8 - 2 \times 4 + 1 \times (-7) = 9$

Substitute the values of x, y and z for point P into this question to find the value of d.

So $d = 9$ and the Cartesian equation of Π is $3x - 2y + z = 9$.

Exercise 9B

1 Find, in the form $\mathbf{r} = \mathbf{a} + \lambda\mathbf{b} + \mu\mathbf{c}$, an equation of the plane that passes through the points:

a $(1, 2, 0)$, $(3, 1, -1)$ and $(4, 3, 2)$
b $(3, 4, 1)$, $(-1, -2, 0)$ and $(2, 1, 4)$
c $(2, -1, -1)$, $(3, 1, 2)$ and $(4, 0, 1)$
d $(-1, 1, 3)$, $(-1, 2, 5)$ and $(0, 4, 4)$.

2 The plane Π is perpendicular to the normal vector $\begin{pmatrix} -1 \\ 3 \\ 2 \end{pmatrix}$ and passes through the point with position vector $\begin{pmatrix} 4 \\ -2 \\ 6 \end{pmatrix}$. Find a Cartesian equation of Π.

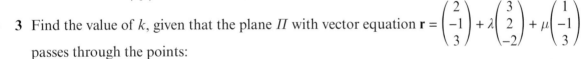

(P) **3** Find the value of k, given that the plane Π with vector equation $\mathbf{r} = \begin{pmatrix} 2 \\ -1 \\ 3 \end{pmatrix} + \lambda\begin{pmatrix} 3 \\ 2 \\ -2 \end{pmatrix} + \mu\begin{pmatrix} 1 \\ -1 \\ 3 \end{pmatrix}$ passes through the points:

a $(7, -1, k)$
b $(1, k, 11)$
c $(k, -4, 10)$
d $(10, k, -k)$

4 A Cartesian equation of the plane Π is $2x - 3y + 5z = 1$.

a Verify that the plane passes through the point: **i** $(1, 2, 1)$ **ii** $(2, -4, -3)$
b Write down an equation of a normal vector to the plane.

5 The line l is normal to the plane Π with Cartesian equation $5x - 3y - 4z = 9$ and passes through the point $(2, 3, -2)$. Find:

a a vector equation of l
b a Cartesian equation of l

6 The diagram shows a cube with a vertex at the origin and sides of length 3.
Find Cartesian equations for the planes containing each face of the cube.

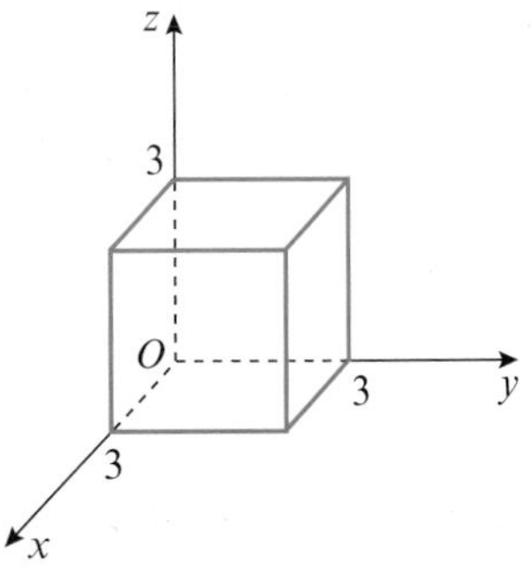

E/P **7** Show that the points (2, 2, 3), (1, 5, 3), (4, 3, −1) and (3, 6, −1) are coplanar. **(6 marks)**

> **Notation** Points are said to be **coplanar** if they all lie on the same plane.

E/P **8** Show that the points (2, 3, 4), (2, −1, 3), (5, 3, −2) and (−1, −9, 8) are **not** coplanar. **(6 marks)**

E/P **9** The plane Π has vector equation $\mathbf{r} = 3\mathbf{i} - 2\mathbf{j} + \mathbf{k} + \lambda(-2\mathbf{i} + 3\mathbf{j} + 5\mathbf{k}) + \mu(4\mathbf{i} + 2\mathbf{j} - 3\mathbf{k})$.
The point A lies on Π such that $\lambda = 1$ and $\mu = 2$.

a Find the position vector of A. **(2 marks)**

Point B has coordinates (1, −7, −1).

b Show that B lies on Π. **(2 marks)**

The line l passes through points A and B.

c Find a vector equation of l. **(3 marks)**

The point C lies on l such that $|\overrightarrow{OA}| = |\overrightarrow{OC}|$.

d Find the position vector of C. **(3 marks)**

> **Challenge**
>
> A plane has vector equation $\mathbf{r} = 2\mathbf{i} + 3\mathbf{j} + \lambda(\mathbf{i} - 2\mathbf{j} + \mathbf{k}) + \mu(2\mathbf{i} - \mathbf{j} + 3\mathbf{k})$.
> A line has vector equation $\mathbf{r} = (2\mathbf{i} + 6\mathbf{j} + \mathbf{k}) + t(5\mathbf{i} - 7\mathbf{j} + 6\mathbf{k})$.
> Show that the line lies entirely within the plane.

9.3 Scalar product

You need to know the definition of the scalar product of two vectors in either two or three dimensions, and how it can be used to find the angle between two vectors. To define the scalar product you need to know how to find the **angle between two vectors**.

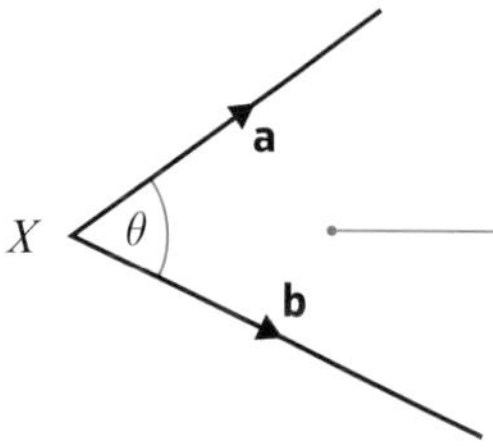

On the diagram, the angle between the vectors **a** and **b** is θ.
Notice that **a** and **b** are both directed away from the point X.

Example 10

Find the angle between the vectors **a** and **b** on the diagram.

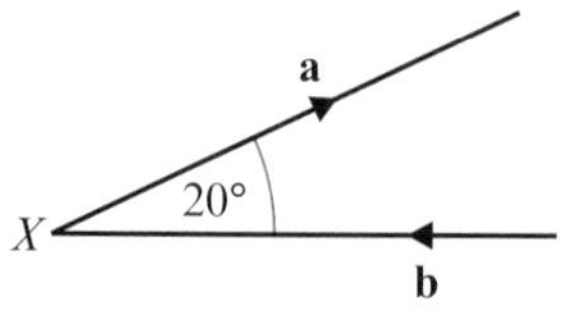

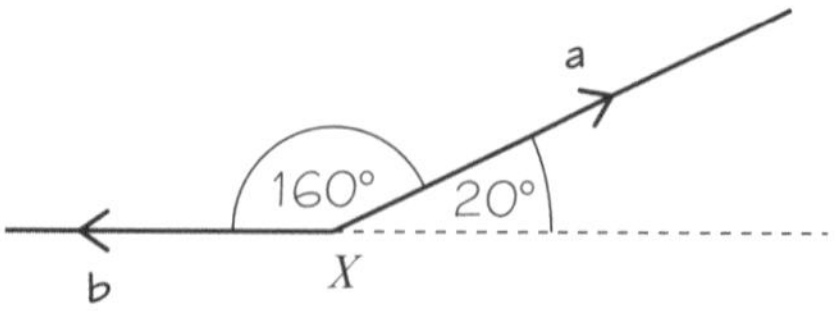

The angle between **a** and **b** is 180° − 20° = 160°.

For the correct angle, **a** and **b** must both be pointing away from X, so re-draw to show this.

- **The scalar product of two vectors a and b is written as a.b, and defined as**

  $\mathbf{a}.\mathbf{b} = |\mathbf{a}||\mathbf{b}| \cos \theta$

  where θ is the angle between a and b.

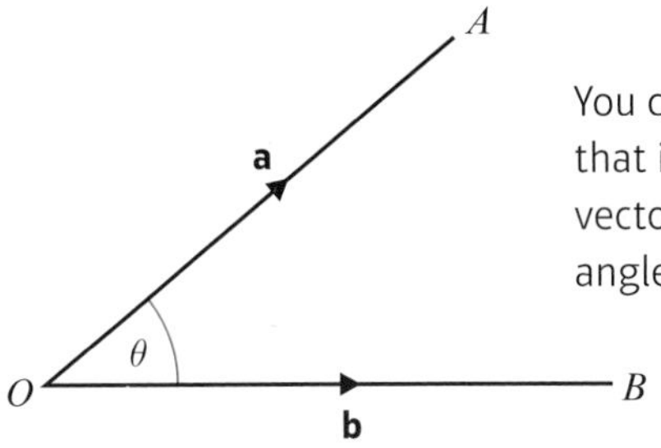

You can see from this diagram that if **a** and **b** are the position vectors of A and B, then the angle between **a** and **b** is $\angle AOB$.

Notation The scalar product is often called the **dot product**. You say 'a dot b'.

Online Use GeoGebra to consider the scalar product as the component of one vector in the direction of another.

- **If a and b are the position vectors of the points A and B, then** $\cos(\angle AOB) = \dfrac{\mathbf{a}.\mathbf{b}}{|\mathbf{a}||\mathbf{b}|}$

If two vectors **a** and **b** are perpendicular, the angle between them is 90°.

Since $\cos 90° = 0$, $\mathbf{a}.\mathbf{b} = |\mathbf{a}||\mathbf{b}| \cos 90° = 0$.

- **The non-zero vectors a and b are perpendicular if and only if a.b = 0.**

If **a** and **b** are parallel, the angle between them is 0°.

- **If a and b are parallel, $\mathbf{a}.\mathbf{b} = |\mathbf{a}||\mathbf{b}|$. In particular, $\mathbf{a}.\mathbf{a} = |\mathbf{a}|^2$.**

Example 11

Find the values of

a $\mathbf{i}.\mathbf{j}$ **b** $\mathbf{k}.\mathbf{k}$ **c** $(4\mathbf{j}).\mathbf{k} + (3\mathbf{i}).(3\mathbf{i})$

a $\mathbf{i}.\mathbf{j} = 1 \times 1 \times \cos 90° = 0$

i and j are unit vectors (magnitude 1), and are perpendicular.

b $\mathbf{k}.\mathbf{k} = 1 \times 1 \times \cos 0° = 1$

k is a unit vector (magnitude 1) and the angle between **k** and itself is 0°.

c $(4\mathbf{j}).\mathbf{k} + (3\mathbf{i}).(3\mathbf{i})$
$= (4 \times 1 \times \cos 90°) + (3 \times 3 \times \cos 0°)$
$= 0 + 9 = 9$

Example 12

Given that $\mathbf{a} = \begin{pmatrix} a_1 \\ a_2 \\ a_3 \end{pmatrix}$ and $\mathbf{b} = \begin{pmatrix} b_1 \\ b_2 \\ b_3 \end{pmatrix}$, prove that $\mathbf{a}.\mathbf{b} = a_1b_1 + a_2b_2 + a_3b_3$.

$$
\begin{aligned}
\mathbf{a}.\mathbf{b} &= (a_1\mathbf{i} + a_2\mathbf{j} + a_3\mathbf{k}).(b_1\mathbf{i} + b_2\mathbf{j} + b_3\mathbf{k})\\
&= a_1\mathbf{i}.(b_1\mathbf{i} + b_2\mathbf{j} + b_3\mathbf{k})\\
&\quad + a_2\mathbf{j}.(b_1\mathbf{i} + b_2\mathbf{j} + b_3\mathbf{k})\\
&\quad + a_3\mathbf{k}.(b_1\mathbf{i} + b_2\mathbf{j} + b_3\mathbf{k})\\
&= (a_1\mathbf{i}).(b_1\mathbf{i}) + (a_1\mathbf{i}).(b_2\mathbf{j}) + (a_1\mathbf{i}).(b_3\mathbf{k})\\
&\quad + (a_2\mathbf{j}).(b_1\mathbf{i}) + (a_2\mathbf{j}).(b_2\mathbf{j}) + (a_2\mathbf{j}).(b_3\mathbf{k})\\
&\quad + (a_3\mathbf{k}).(b_1\mathbf{i}) + (a_3\mathbf{k}).(b_2\mathbf{j}) + (a_3\mathbf{k}).(b_3\mathbf{k})\\
&= (a_1b_1)\mathbf{i}.\mathbf{i} + (a_1b_2)\mathbf{i}.\mathbf{j} + (a_1b_3)\mathbf{i}.\mathbf{k}\\
&\quad + (a_2b_1)\mathbf{j}.\mathbf{i} + (a_2b_2)\mathbf{j}.\mathbf{j} + (a_2b_3)\mathbf{j}.\mathbf{k}\\
&\quad + (a_3b_1)\mathbf{k}.\mathbf{i} + (a_3b_2)\mathbf{k}.\mathbf{j} + (a_3b_3)\mathbf{k}.\mathbf{k}\\
&= a_1b_1 + a_2b_2 + a_3b_3
\end{aligned}
$$

Use the results for parallel and perpendicular unit vectors:

$\mathbf{i}.\mathbf{i} = \mathbf{j}.\mathbf{j} = \mathbf{k}.\mathbf{k} = 1$

$\mathbf{i}.\mathbf{j} = \mathbf{i}.\mathbf{k} = \mathbf{j}.\mathbf{i} = \mathbf{j}.\mathbf{k} = \mathbf{k}.\mathbf{i} = \mathbf{k}.\mathbf{j} = 0$

The above example leads to a simple formula for finding the scalar product of two vectors given in Cartesian component form:

- **If $\mathbf{a} = a_1\mathbf{i} + a_2\mathbf{j} + a_3\mathbf{k}$ and $\mathbf{b} = b_1\mathbf{i} + b_2\mathbf{j} + b_3\mathbf{k}$,**

$$\mathbf{a}.\mathbf{b} = \begin{pmatrix} a_1 \\ a_2 \\ a_3 \end{pmatrix} . \begin{pmatrix} b_1 \\ b_2 \\ b_3 \end{pmatrix} = a_1b_1 + a_2b_2 + a_3b_3$$

You can use this result without proof in your exam.

Example 13

Given that $\mathbf{a} = 8\mathbf{i} - 5\mathbf{j} - 4\mathbf{k}$ and $\mathbf{b} = 5\mathbf{i} + 4\mathbf{j} - \mathbf{k}$,

a Find **a.b**.

b Find the angle between **a** and **b**, giving your answer in degrees to 1 decimal place.

a $\mathbf{a}.\mathbf{b} = \begin{pmatrix} 8 \\ -5 \\ -4 \end{pmatrix} . \begin{pmatrix} 5 \\ 4 \\ -1 \end{pmatrix}$

$= (8 \times 5) + (-5 \times 4) + (-4 \times -1)$

$= 40 - 20 + 4$

$= 24$

Write in column vector form.

Use $\mathbf{a}.\mathbf{b} = a_1b_1 + a_2b_2 + a_3b_3$.

b $\mathbf{a}.\mathbf{b} = |\mathbf{a}||\mathbf{b}| \cos\theta$

$|\mathbf{a}| = \sqrt{8^2 + (-5)^2 + (-4)^2} = \sqrt{105}$

$|\mathbf{b}| = \sqrt{5^2 + 4^2 + (-1)^2} = \sqrt{42}$

$\sqrt{105}\sqrt{42}\cos\theta = 24$

$\cos\theta = \dfrac{24}{\sqrt{105}\sqrt{42}}$

$\theta = 68.8°$ (1 d.p.)

Use the scalar product definition.

Find the modulus of **a** and of **b**.

Use $\mathbf{a}.\mathbf{b} = |\mathbf{a}||\mathbf{b}| \cos\theta$.

Example 14

Given that $\mathbf{a} = -\mathbf{i} + \mathbf{j} + 3\mathbf{k}$ and $\mathbf{b} = 7\mathbf{i} - 2\mathbf{j} + 2\mathbf{k}$, find the angle between **a** and **b**, giving your answer in degrees to 1 decimal place.

$$\mathbf{a.b} = \begin{pmatrix}-1\\1\\3\end{pmatrix}.\begin{pmatrix}7\\-2\\2\end{pmatrix} = -7 - 2 + 6 = -3$$

$$|\mathbf{a}| = \sqrt{(-1)^2 + 1^2 + 3^2} = \sqrt{11}$$

$$|\mathbf{b}| = \sqrt{7^2 + (-2)^2 + 2^2} = \sqrt{57}$$

For the scalar product formula, you need to find **a.b**, |**a**| and |**b**|.

$$\sqrt{11}\sqrt{57}\cos\theta = -3$$

Use $\mathbf{a.b} = |\mathbf{a}||\mathbf{b}|\cos\theta$.

$$\cos\theta = \frac{-3}{\sqrt{11}\sqrt{57}}$$

The cosine is negative, so the angle is obtuse.

$$\theta = 96.9° \text{ (1 d.p.)}$$

Example 15

Given that the vectors $\mathbf{a} = 2\mathbf{i} - 6\mathbf{j} + \mathbf{k}$ and $\mathbf{b} = 5\mathbf{i} + 2\mathbf{j} + \lambda\mathbf{k}$ are perpendicular, find the value of λ.

$$\mathbf{a.b} = \begin{pmatrix}2\\-6\\1\end{pmatrix}.\begin{pmatrix}5\\2\\\lambda\end{pmatrix}$$

$$= 10 - 12 + \lambda$$

$$= -2 + \lambda$$

Find the scalar product.

$$-2 + \lambda = 0$$

For perpendicular vectors, the scalar product is zero.

$$\lambda = 2$$

Example 16

Given that $\mathbf{a} = -2\mathbf{i} + 5\mathbf{j} - 4\mathbf{k}$ and $\mathbf{b} = 4\mathbf{i} - 8\mathbf{j} + 5\mathbf{k}$, find a vector which is perpendicular to both **a** and **b**.

$$\mathbf{a}.\begin{pmatrix}x\\y\\z\end{pmatrix} = 0 \text{ and } \mathbf{b}.\begin{pmatrix}x\\y\\z\end{pmatrix} = 0$$

Both scalar products are zero.

$$\begin{pmatrix}-2\\5\\-4\end{pmatrix}.\begin{pmatrix}x\\y\\z\end{pmatrix} = 0 \text{ and } \begin{pmatrix}4\\-8\\5\end{pmatrix}.\begin{pmatrix}x\\y\\z\end{pmatrix} = 0$$

$$-2x + 5y - 4z = 0 \qquad (1)$$

$$4x - 8y + 5z = 0 \qquad (2)$$

Let $z = 1$

$-2x + 5y = 4$ (from 1)

$4x - 8y = -5$ (from 2)

So $x = \frac{7}{4}$, $y = \frac{3}{2}$ and $z = 1$

A possible vector is $\frac{7}{4}\mathbf{i} + \frac{3}{2}\mathbf{j} + \mathbf{k}$.

Another possible vector is

$4(\frac{7}{4}\mathbf{i} + \frac{3}{2}\mathbf{j} + \mathbf{k}) = 7\mathbf{i} + 6\mathbf{j} + 4\mathbf{k}$

Choose a (non-zero) value for z (or for x, or for y).

Solving simutaneously gives $x = \frac{7}{4}$ and $y = \frac{3}{2}$

You can multiply by a scalar constant to find another vector which is also perpendicular to both **a** and **b**.

Example 17

The points A, B and C have coordinates (2, −1, 1), (5, 1, 7) and (6, −3, 1) respectively.

a Find $\overrightarrow{AB}.\overrightarrow{AC}$

b Hence, or otherwise, find the area of triangle ABC.

a $\overrightarrow{AB} = \begin{pmatrix} 3 \\ 2 \\ 6 \end{pmatrix}$ and $\overrightarrow{AC} = \begin{pmatrix} 4 \\ -2 \\ 0 \end{pmatrix}$

$\overrightarrow{AB}.\overrightarrow{AC} = 3 \times 4 + 2 \times (-2) + 6 \times 0 = 8$

b $|\overrightarrow{AB}| = \sqrt{3^2 + 2^2 + 6^2} = 7$

$|\overrightarrow{AC}| = \sqrt{4^2 + (-2)^2 + 0^2} = 2\sqrt{5}$

$\cos(\angle BAC) = \frac{\overrightarrow{AB}.\overrightarrow{AC}}{|\overrightarrow{AB}||\overrightarrow{AC}|}$

$= \frac{8}{7 \times 2\sqrt{5}}$

$= 0.2555...$

$\angle BAC = 75.1937...°$

Area $= \frac{1}{2}|\overrightarrow{AB}||\overrightarrow{AC}|\sin(\angle BAC)$

$= \frac{1}{2} \times 7 \times 2\sqrt{5}\sin(75.1937...°)$

$= 15.13$ (2 d.p.)

Use the scalar product to find the angle between $\overrightarrow{AB}$ and $\overrightarrow{AC}$. Then use area $= \frac{1}{2}ab\sin\theta$ to find the area of the triangle.

Problem-solving

You could find $\angle BAC$ by finding the lengths AB, BC and AC and using the cosine rule, but it is quicker to use a vector method.

Exercise 9C

1 The vectors **a** and **b** each have magnitude 3, and the angle between **a** and **b** is 60°. Find **a.b**.

2 For each pair of vectors, find **a.b**:

a $\mathbf{a} = 5\mathbf{i} + 2\mathbf{j} + 3\mathbf{k}$, $\mathbf{b} = 2\mathbf{i} - \mathbf{j} - 2\mathbf{k}$
b $\mathbf{a} = 10\mathbf{i} - 7\mathbf{j} + 4\mathbf{k}$, $\mathbf{b} = 3\mathbf{i} - 5\mathbf{j} - 12\mathbf{k}$
c $\mathbf{a} = \mathbf{i} + \mathbf{j} - \mathbf{k}$, $\mathbf{b} = -\mathbf{i} - \mathbf{j} + 4\mathbf{k}$
d $\mathbf{a} = 2\mathbf{i} - \mathbf{k}$, $\mathbf{b} = 6\mathbf{i} - 5\mathbf{j} - 8\mathbf{k}$
e $\mathbf{a} = 3\mathbf{j} + 9\mathbf{k}$, $\mathbf{b} = \mathbf{i} - 12\mathbf{j} + 4\mathbf{k}$

3 In each part, find the angle between **a** and **b**, giving your answer in degrees to 1 decimal place:

a $\mathbf{a} = 3\mathbf{i} + 7\mathbf{j}$, $\mathbf{b} = 5\mathbf{i} + \mathbf{j}$
b $\mathbf{a} = 2\mathbf{i} - 5\mathbf{j}$, $\mathbf{b} = 6\mathbf{i} + 3\mathbf{j}$
c $\mathbf{a} = \mathbf{i} - 7\mathbf{j} + 8\mathbf{k}$, $\mathbf{b} = 12\mathbf{i} + 2\mathbf{j} + \mathbf{k}$
d $\mathbf{a} = -\mathbf{i} - \mathbf{j} + 5\mathbf{k}$, $\mathbf{b} = 11\mathbf{i} - 3\mathbf{j} + 4\mathbf{k}$
e $\mathbf{a} = 6\mathbf{i} - 7\mathbf{j} + 12\mathbf{k}$, $\mathbf{b} = -2\mathbf{i} + \mathbf{j} + \mathbf{k}$
f $\mathbf{a} = 4\mathbf{i} + 5\mathbf{k}$, $\mathbf{b} = 6\mathbf{i} - 2\mathbf{j}$
g $\mathbf{a} = -5\mathbf{i} + 2\mathbf{j} - 3\mathbf{k}$, $\mathbf{b} = 2\mathbf{i} - 2\mathbf{j} - 11\mathbf{k}$
h $\mathbf{a} = \mathbf{i} + \mathbf{j} + \mathbf{k}$, $\mathbf{b} = \mathbf{i} - \mathbf{j} + \mathbf{k}$

4 Find the value, or values, of λ for which the given vectors are perpendicular:

a $3\mathbf{i} + 5\mathbf{j}$ and $\lambda\mathbf{i} + 6\mathbf{j}$
b $2\mathbf{i} + 6\mathbf{j} - \mathbf{k}$ and $\lambda\mathbf{i} - 4\mathbf{j} - 14\mathbf{k}$
c $3\mathbf{i} + \lambda\mathbf{j} - 8\mathbf{k}$ and $7\mathbf{i} - 5\mathbf{j} + \mathbf{k}$
d $9\mathbf{i} - 3\mathbf{j} + 5\mathbf{k}$ and $\lambda\mathbf{i} + \lambda\mathbf{j} + 3\mathbf{k}$
e $\lambda\mathbf{j} + 3\mathbf{j} - 2\mathbf{k}$ and $\lambda\mathbf{i} + \lambda\mathbf{j} + 5\mathbf{k}$

5 Find, to the nearest tenth of a degree, the angle that the vector $9\mathbf{i} - 5\mathbf{j} + 3\mathbf{k}$ makes with:

a the positive x-axis
b the positive y-axis

6 Find, to the nearest tenth of a degree, the angle that the vector $\mathbf{i} + 11\mathbf{j} - 4\mathbf{k}$ makes with:

a the positive y-axis
b the positive z-axis

7 The angle between the vectors $\mathbf{i} + \mathbf{j} + \mathbf{k}$ and $2\mathbf{i} + \mathbf{j} + \mathbf{k}$ is θ. Calculate the exact value of $\cos\theta$.

8 The angle between the vectors $\mathbf{i} + 3\mathbf{j}$ and $\mathbf{j} + \lambda\mathbf{k}$ is 60°. Show that $\lambda = \pm\sqrt{\frac{13}{5}}$

9 Find a vector which is perpendicular to both **a** and **b**, where:

a $\mathbf{a} = \mathbf{i} + \mathbf{j} - 3\mathbf{k}$, $\mathbf{b} = 5\mathbf{i} - 2\mathbf{j} - \mathbf{k}$
b $\mathbf{a} = 2\mathbf{i} + 3\mathbf{j} - 4\mathbf{k}$, $\mathbf{b} = \mathbf{i} - 6\mathbf{j} + 3\mathbf{k}$
c $\mathbf{a} = 4\mathbf{i} - 4\mathbf{j} - \mathbf{k}$, $\mathbf{b} = -2\mathbf{i} - 9\mathbf{j} + 6\mathbf{k}$

10 The points A and B have position vectors $2\mathbf{i} + 5\mathbf{j} + \mathbf{k}$ and $6\mathbf{i} + \mathbf{j} - 2\mathbf{k}$ respectively, and O is the origin. Calculate each of the angles in $\triangle OAB$, giving your answers in degrees to 1 decimal place.

11 The points A, B and C have coordinates (1, 3, 1), (2, 7, −3) and (4, −5, 2) respectively.

a Find the exact lengths of AB and BC.
b Calculate, to one decimal place, the size of $\angle ABC$.

(P) 12 Given that the points A and B have coordinates (7, 4, 4) and (2, 2, 1) respectively,

a find the value of $\cos \angle AOB$, where O is the origin **(4 marks)**
b show that the area of $\triangle AOB$ is $\frac{\sqrt{53}}{2}$ **(3 marks)**

(P) **13** AB is a diameter of a circle centred at the origin O, and P is a point on the circumference of the circle. By considering the position vectors of A, B and P, prove that AP is perpendicular to BP.

Problem-solving

This is a vector proof of the fact that the angle in a semi-circle is 90°.

(E/P) **14** Points A, B and C have coordinates $(5, -1, 0)$, $(2, 4, 10)$ and $(6, -1, 4)$ respectively.

a Find the vectors $\overrightarrow{CA}$ and $\overrightarrow{CB}$. **(2 marks)**

b Find the area of the triangle ABC. **(4 marks)**

c Point D is such that A, B, C and D are the vertices of a parallelogram. Find the coordinates of three possible positions of D. **(3 marks)**

d Write down the area of the parallelogram. **(1 mark)**

(E/P) **15** The points P, Q and R have coordinates $(1, -1, 6)$, $(-2, 5, 4)$ and $(0, 3, -5)$ respectively.

a Show that PQ is perpendicular to QR. **(3 marks)**

b Hence find the centre and radius of the circle that passes through points P, Q and R. **(3 marks)**

Challenge

1 Using the definition $\mathbf{a.b} = |\mathbf{a}||\mathbf{b}| \cos\theta$, prove that $\mathbf{a.b} = \mathbf{b.a}$.

2 The diagram shows arbitrary vectors **a**, **b** and **c**, and the vector **b** + **c**.

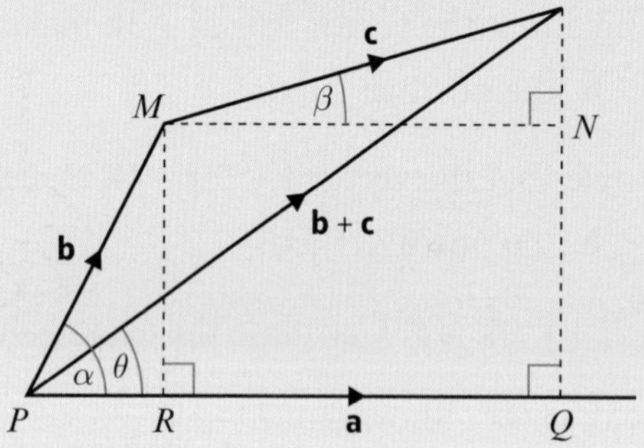

a Show that:

i $\mathbf{a.(b + c)} = |\mathbf{a}| \times PQ$

ii $\mathbf{a.b} = |\mathbf{a}| \times PR$

iii $\mathbf{a.c} = |\mathbf{a}| \times RQ$

b Hence prove that $\mathbf{a.(b + c)} = \mathbf{a.b} + \mathbf{a.c}$.

9.4 Calculating angles between lines and planes

If two straight lines in three dimensions intersect, then you can calculate the size of the angle between them using the scalar product.

- **The acute angle θ between two intersecting straight lines is given by**

$$\cos\theta = \left|\frac{\mathbf{a.b}}{|\mathbf{a}||\mathbf{b}|}\right|$$

where a and b are direction vectors of the lines.

Watch out The modulus signs around the whole expression ensure you get an acute angle. If you need to work out the size of an **obtuse** angle between two lines, use the formula then subtract the resulting acute angle from 180°.

Example 18

The lines l_1 and l_2 have vector equations $\mathbf{r} = (2\mathbf{i} + \mathbf{j} + \mathbf{k}) + t(3\mathbf{i} - 8\mathbf{j} - \mathbf{k})$ and $\mathbf{r} = (7\mathbf{i} + 4\mathbf{j} + \mathbf{k}) + s(2\mathbf{i} + 2\mathbf{j} + 3\mathbf{k})$ respectively.

Given that l_1 and l_2 intersect, find the size of the acute angle between the lines to one decimal place.

$\mathbf{a} = \begin{pmatrix} 3 \\ -8 \\ -1 \end{pmatrix}$ and $\mathbf{b} = \begin{pmatrix} 2 \\ 2 \\ 3 \end{pmatrix}$

Use the direction vectors.

$\cos\theta = \left|\dfrac{\mathbf{a}.\mathbf{b}}{|\mathbf{a}||\mathbf{b}|}\right|$

$\mathbf{a}.\mathbf{b} = \begin{pmatrix} 3 \\ -8 \\ -1 \end{pmatrix}.\begin{pmatrix} 2 \\ 2 \\ 3 \end{pmatrix}$

$= 6 - 16 - 3 = -13$

$|\mathbf{a}| = \sqrt{3^2 + (-8)^2 + (-1)^2} = \sqrt{74}$

$|\mathbf{b}| = \sqrt{2^2 + 2^2 + 3^2} = \sqrt{17}$

$\cos\theta = \left|\dfrac{-13}{\sqrt{74}\,\sqrt{17}}\right|$

$\theta = 68.5°$ (1 d.p.)

Use the formula. Be careful with the modulus signs. If $\cos\theta$ is positive then θ will be an acute angle, as required.

You can use the scalar product to write a vector equation of a plane efficiently.

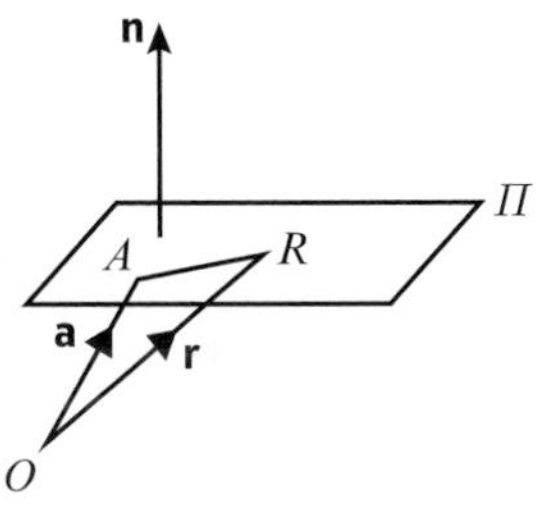

Suppose a plane Π passes through a given point A, with position vector **a**, and that the normal vector **n** is perpendicular to the plane. Let R be an arbitrary point on the plane, with position vector **r**.

Then, $\overrightarrow{AR} = \mathbf{r} - \mathbf{a}$

As $\overrightarrow{AR}$ is a vector which lies in the plane, $\overrightarrow{AR}$ is perpendicular to **n** so $\overrightarrow{AR}.\mathbf{n} = 0$.

This means $(\mathbf{r} - \mathbf{a}).\mathbf{n} = 0$

You can rewrite this as $\mathbf{r}.\mathbf{n} = \mathbf{a}.\mathbf{n}$

Since **a** is a fixed point, **a.n** is a scalar constant, k, and the equation of the plane Π is $\mathbf{r}.\mathbf{n} = k$.

- **The scalar product form of the equation of a plane is $\mathbf{r}.\mathbf{n} = k$ where $k = \mathbf{a}.\mathbf{n}$ for any point in the plane with position vector a.**

Example 19

The plane Π passes through the point A and is perpendicular to the vector **n**.

Given that $\overrightarrow{OA} = \begin{pmatrix} 2 \\ 3 \\ -5 \end{pmatrix}$ and $\mathbf{n} = \begin{pmatrix} 3 \\ 1 \\ -1 \end{pmatrix}$ where O is the origin, find an equation of the plane:

a in scalar product form **b** in Cartesian form

a $\mathbf{r.n} = k$, where $k = \mathbf{a.n}$

$$\mathbf{a.n} = \begin{pmatrix} 2 \\ 3 \\ -5 \end{pmatrix} \cdot \begin{pmatrix} 3 \\ 1 \\ -1 \end{pmatrix}$$

$= 2 \times 3 + 3 \times 1 + (-5) \times (-1)$

$= 6 + 3 + 5 = 14$

So a scalar product form of the equation of Π is $\mathbf{r}.\begin{pmatrix} 3 \\ 1 \\ -1 \end{pmatrix} = 14$.

b $\begin{pmatrix} x \\ y \\ z \end{pmatrix} \cdot \begin{pmatrix} 3 \\ 1 \\ -1 \end{pmatrix} = 14$

So a Cartesian form of equation of Π is $3x + y - z = 14$

Use $\mathbf{r.n} = k$ where $k = \mathbf{a.n}$ for any point in the plane with position vector $\mathbf{a}$.

Problem-solving

You can convert between scalar product form and Cartesian form quickly by writing the general position vector of a point in the plane as $\mathbf{r} = \begin{pmatrix} x \\ y \\ z \end{pmatrix}$.

You need to be able to calculate the angle between a line and a plane.

Example 20

Find the acute angle between the line l with equation $\mathbf{r} = 2\mathbf{i} + \mathbf{j} - 5\mathbf{k} + \lambda(3\mathbf{i} + 4\mathbf{j} - 12\mathbf{k})$ and the plane with equation $\mathbf{r}.(2\mathbf{i} - 2\mathbf{j} - \mathbf{k}) = 2$.

Online Explore the angle between a line and a plane using GeoGebra.

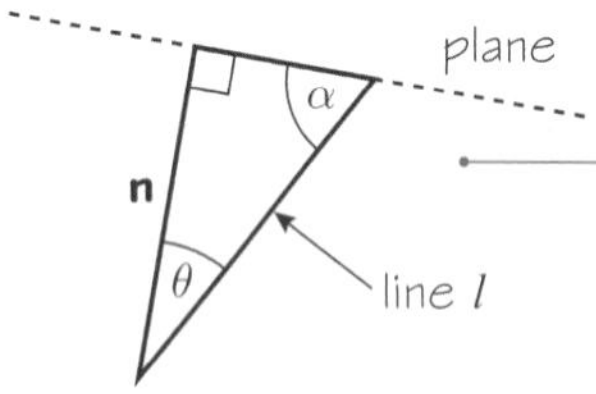

Draw a diagram showing the line, the plane and the normal to the plane. Let the required angle be α and show α and θ in your diagram.

The normal to the plane is in the direction $\mathbf{n} = 2\mathbf{i} - 2\mathbf{j} - \mathbf{k}$.

The angle between this normal and the line l is θ,

where $\cos\theta = \dfrac{(3\mathbf{i} + 4\mathbf{j} - 12\mathbf{k}).(2\mathbf{i} - 2\mathbf{j} - \mathbf{k})}{\sqrt{3^2 + 4^2 + (-12)^2}\sqrt{2^2 + (-2)^2 + (-1)^2}}$

$= \dfrac{10}{13 \times 3} = \dfrac{10}{39}$

First find the angle between the given line and the normal to the plane.

So the angle between the plane and the line l is α where $\alpha + \theta = 90°$.

So $\sin\alpha = \dfrac{10}{39}$ and $\alpha = 14.9°$.

Subtract the angle θ from 90°, to give angle α, or use the trigonometric connection that $\cos\theta = \sin\alpha$.

- **The acute angle θ between the line with equation $\mathbf{r} = \mathbf{a} + \lambda\mathbf{b}$ and the plane with equation $\mathbf{r.n} = k$ is given by the formula**

$$\sin\theta = \left|\frac{\mathbf{b.n}}{|\mathbf{b}||\mathbf{n}|}\right|$$

You need to be able to calculate the angle between two planes.

Example 21

Find the acute angle between the planes with equations $\mathbf{r}.(4\mathbf{i} + 4\mathbf{j} - 7\mathbf{k}) = 13$ and $\mathbf{r}.(7\mathbf{i} - 4\mathbf{j} + 4\mathbf{k}) = 6$ respectively.

Online Visualise the angle between two planes using GeoGebra.

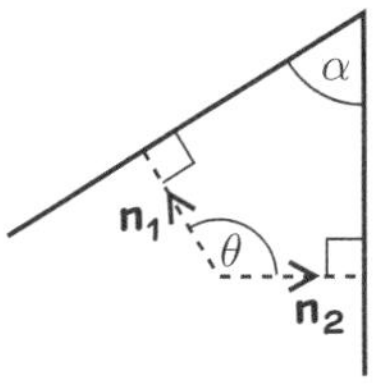

Draw a diagram showing the planes and the normals to the planes. Let the required angle be α and show α and θ in your diagram.

The normals to the planes are in the directions $\mathbf{n}_1 = 4\mathbf{i} + 4\mathbf{j} - 7\mathbf{k}$, and $\mathbf{n}_2 = 7\mathbf{i} - 4\mathbf{j} + 4\mathbf{k}$.

The angle between these normals is θ, where

$$\cos\theta = \frac{(4\mathbf{i} + 4\mathbf{j} - 7\mathbf{k}).(7\mathbf{i} - 4\mathbf{j} + 4\mathbf{k})}{\sqrt{4^2 + 4^2 + (-7)^2}\sqrt{7^2 + (-4)^2 + 4^2}}$$

$$= \frac{28 - 16 - 28}{\sqrt{16 + 16 + 49}\sqrt{49 + 16 + 16}}$$

$$= -\frac{16}{81}$$

First find the angle between the normals to the planes.

So $\theta = 101.4°$

So the angle between the planes is $180 - 101.4 = 78.6°$

Subtract the angle θ from 180°, to give angle α.

- **The acute angle θ between the plane with equation $\mathbf{r}.\mathbf{n}_1 = k_1$ and the plane with equation $\mathbf{r}.\mathbf{n}_2 = k_2$ is given by the formula**

$$\cos\theta = \left|\frac{\mathbf{n}_1.\mathbf{n}_2}{|\mathbf{n}_1||\mathbf{n}_2|}\right|$$

Exercise 9D

1 Given that each pair of lines intersect, find, to 1 decimal place, the acute angle between the lines with vector equations:

a $\mathbf{r} = (2\mathbf{i} + \mathbf{j} + \mathbf{k}) + \lambda(3\mathbf{i} - 5\mathbf{j} - \mathbf{k})$ and $\mathbf{r} = (7\mathbf{i} + 4\mathbf{j} + \mathbf{k}) + \mu(2\mathbf{i} + \mathbf{j} - 9\mathbf{k})$

b $\mathbf{r} = (\mathbf{i} - \mathbf{j} + 7\mathbf{k}) + \lambda(-2\mathbf{i} - \mathbf{j} + 3\mathbf{k})$ and $\mathbf{r} = (8\mathbf{i} + 5\mathbf{j} - \mathbf{k}) + \mu(-4\mathbf{i} - 2\mathbf{j} + \mathbf{k})$

c $\mathbf{r} = (3\mathbf{i} + 5\mathbf{j} - \mathbf{k}) + \lambda(\mathbf{i} + \mathbf{j} + \mathbf{k})$ and $\mathbf{r} = (-\mathbf{i} + 11\mathbf{j} + 5\mathbf{k}) + \mu(2\mathbf{i} - 7\mathbf{j} + 3\mathbf{k})$

d $\mathbf{r} = (\mathbf{i} + 6\mathbf{j} - \mathbf{k}) + \lambda(8\mathbf{i} - \mathbf{j} - 2\mathbf{k})$ and $\mathbf{r} = (6\mathbf{i} + 9\mathbf{j}) + \mu(\mathbf{i} + 3\mathbf{j} - 7\mathbf{k})$

e $\mathbf{r} = (2\mathbf{i} + \mathbf{k}) + \lambda(11\mathbf{i} + 5\mathbf{j} - 3\mathbf{k})$ and $\mathbf{r} = (\mathbf{i} + \mathbf{j}) + \mu(-3\mathbf{i} + 5\mathbf{j} + 4\mathbf{k})$

2 Find, in the form $\mathbf{r}.\mathbf{n} = k$, an equation of the plane that passes through the point with position vector $\mathbf{a}$ and is perpendicular to the vector $\mathbf{n}$ where:

a $\mathbf{a} = \mathbf{i} - \mathbf{j} - \mathbf{k}$ and $\mathbf{n} = 2\mathbf{i} + \mathbf{j} + \mathbf{k}$

b $\mathbf{a} = \mathbf{i} + 2\mathbf{j} + \mathbf{k}$ and $\mathbf{n} = 5\mathbf{i} - \mathbf{j} - 3\mathbf{k}$

c $\mathbf{a} = 2\mathbf{i} - 3\mathbf{k}$ and $\mathbf{n} = \mathbf{i} + 3\mathbf{j} + 4\mathbf{k}$

d $\mathbf{a} = 4\mathbf{i} - 2\mathbf{j} + \mathbf{k}$ and $\mathbf{n} = 4\mathbf{i} + \mathbf{j} - 5\mathbf{k}$

3 Find a Cartesian equation for each of the planes in question 2.

4 A plane has equation $\mathbf{r}.\mathbf{n} = k$, where $\mathbf{n} = \begin{pmatrix} n_1 \\ n_2 \\ n_3 \end{pmatrix}$. Find a Cartesian equation of the plane.

5 Find the acute angle between the line with equation $\mathbf{r} = 2\mathbf{i} + \mathbf{j} - 5\mathbf{k} + \lambda(4\mathbf{i} + 4\mathbf{j} + 7\mathbf{k})$ and the plane with equation $\mathbf{r}.(2\mathbf{i} + \mathbf{j} - 2\mathbf{k}) = 13$.

6 Find the acute angle between the line with equation $\mathbf{r} = -\mathbf{i} - 7\mathbf{j} + 13\mathbf{k} + \lambda(3\mathbf{i} + 4\mathbf{j} - 12\mathbf{k})$ and the plane with equation $\mathbf{r}.(4\mathbf{i} - 4\mathbf{j} - 7\mathbf{k}) = 9$.

7 Find the acute angle between the planes with equations $\mathbf{r}.(\mathbf{i} + 2\mathbf{j} - 2\mathbf{k}) = 1$ and $\mathbf{r}.(-4\mathbf{i} + 4\mathbf{j} + 7\mathbf{k}) = 7$ respectively.

8 Find the acute angle between the planes with equations $\mathbf{r}.(3\mathbf{i} - 4\mathbf{j} + 12\mathbf{k}) = 9$ and $\mathbf{r}.(5\mathbf{i} - 12\mathbf{k}) = 7$ respectively.

(P) **9** The straight lines l_1 and l_2 have vectors equations $\mathbf{r} = (\mathbf{i} + 4\mathbf{j} + 2\mathbf{k}) + \lambda(8\mathbf{i} + 5\mathbf{j} + \mathbf{k})$ and $\mathbf{r} = (\mathbf{i} + 4\mathbf{j} + 2\mathbf{k}) + \mu(3\mathbf{i} + \mathbf{j})$ respectively, and P is the point with coordinates (1, 4, 2).

a Show that the point $Q(9, 9, 3)$ lies on l_1.

Given that l_1 and l_2 intersect, find:

b the cosine of the acute angle between l_1 and l_2

c the possible coordinates of the point R, such that R lies on l_2 and $PQ = PR$.

(E/P) **10** The lines l_1 and l_2 have Cartesian equations $\frac{x-6}{-1} = \frac{y+3}{2} = \frac{z+2}{3}$ and $\frac{x+5}{2} = \frac{y-15}{-3} = \frac{z-3}{1}$ respectively.

a Show that the point $A(3, 3, 7)$ lies on both l_1 and l_2. **(3 marks)**

b Find the size of the acute angle between the lines at A. **(4 marks)**

(E) **11** The lines l_1 and l_2 have vector equations $\mathbf{r} = \begin{pmatrix} 1 \\ 3 \\ 3 \end{pmatrix} + \lambda\begin{pmatrix} 3 \\ 2 \\ -1 \end{pmatrix}$ and $\mathbf{r} = \begin{pmatrix} 3 \\ 5 \\ -2 \end{pmatrix} + \mu\begin{pmatrix} -4 \\ -3 \\ 1 \end{pmatrix}$.

The point A is on l_1 where $\lambda = 3$ and the point B is on l_2 where $\mu = -2$. Find the size of the acute angle between AB and l_1. **(6 marks)**

(E/P) **12 a** Show that the points $A(3, 5, -1)$, $B(2, -2, 4)$, $C(4, 3, 0)$ and $D(1, 4, -3,)$ are not coplanar. **(6 marks)**

b Find the angle between the plane containing A, B and C and the line segment AD. **(4 marks)**

(E/P) **13** A regular tetrahedron has vertices A, B, C and D, with coordinates (0, 0, 0), (0, 1, 1), (1, 1, 0) and (1, 0, 1) respectively. Show that the angle between any two adjacent faces of the tetrahedron is $\arccos\left(\frac{1}{3}\right)$.

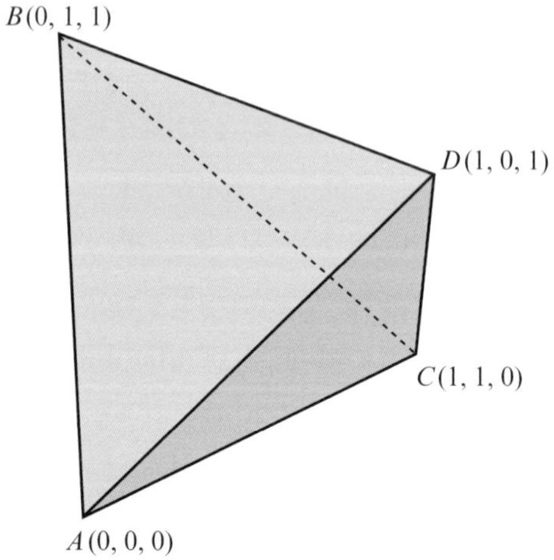

(7 marks)

(P) **14** A flagpole is supported by 3 guide ropes which are attached at a point 20 m above the base of the pole. The ends of the ropes are secured at points with position vectors (0, 8, 2), (12, −5, 3) and (−2, 6, 5) relative to the base of the pole, where the units are metres. The flagpole will be stable if the angles between adjacent guide ropes are all greater than 15°. Determine whether the flagpole will be stable, showing your working clearly. **(7 marks)**

9.5 Points of intersection

You need to be able to determine whether two lines meet and, if so, to determine their point of intersection.

$λ$ and $μ$ are the **parameters** in the vector equations of the lines.

Write the equations in column notation and set them equal to each other. → Write three linear equations involving $λ$ and $μ$ → Try to solve the first two equations simultaneously. → No solutions → Lines **do not** intersect.

Do these solutions satisfy the third equation? No → Lines **do not** intersect.

Solutions exist

Yes ↓

Lines **do** intersect. Substitute your values for $λ$ and $μ$ into the equation of one of the lines to find the point of intersection. You can use the other equation to check.

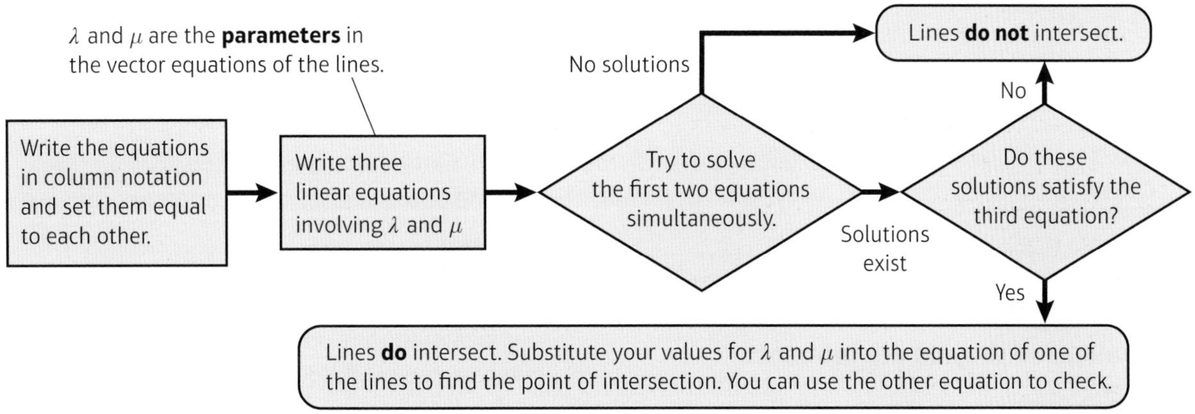

Example 22

The lines l_1 and l_2 have vector equations

$\mathbf{r} = 3\mathbf{i} + \mathbf{j} + \mathbf{k} + λ(\mathbf{i} - 2\mathbf{j} - \mathbf{k})$ and $\mathbf{r} = -2\mathbf{j} + 3\mathbf{k} + μ(-5\mathbf{i} + \mathbf{j} + 4\mathbf{k})$ respectively.

Show that the two lines intersect, and find the position vector of the point of intersection.

$$\begin{pmatrix} 3 + λ \\ 1 - 2λ \\ 1 - λ \end{pmatrix} = \begin{pmatrix} -5μ \\ -2 + μ \\ 3 + 4μ \end{pmatrix}$$

Use column vector notation for clarity, and to help to avoid errors.

Solve the simultaneous equations

$3 + λ = -5μ$ (1)

and $1 - λ = 3 + 4μ$ (2)

Adding gives $4 = 3 - μ$

and so $μ = -1$.

Substituting back into equation (1) gives $λ = 2$.

Check $μ = -1$, $λ = 2$ also satisfy the third equation.

$1 - 2λ = -2 + μ$ gives $-3 = -3$

So the lines do intersect.

Substituting $λ = 2$ into $\begin{pmatrix} 3 + λ \\ 1 - 2λ \\ 1 - λ \end{pmatrix}$ gives $\begin{pmatrix} 5 \\ -3 \\ -1 \end{pmatrix}$.

The point where the lines meet is $(5, -3, -1)$.

Choose two of the three equations obtained by equating x-, y- and z-components and solve the resulting simultaneous equations.

If the lines intersect there is a pair of values of $λ$ and $μ$ that satisfy the 3 equations simultaneously.

Check that the point which you obtain after substitution lies on both straight lines.

You also need to be able to find the coordinates of the point of intersection of a line with a plane.

Example 23

Find the coordinates of the point of intersection of the line l and the plane Π where l has equation $\mathbf{r} = -\mathbf{i} + \mathbf{j} - 5\mathbf{k} + \lambda(\mathbf{i} + \mathbf{j} + 2\mathbf{k})$ and Π has equation $\mathbf{r}.(\mathbf{i} + 2\mathbf{j} + 3\mathbf{k}) = 4$.

The line meets the plane when

$$\begin{pmatrix} -1 + \lambda \\ 1 + \lambda \\ -5 + 2\lambda \end{pmatrix} . \begin{pmatrix} 1 \\ 2 \\ 3 \end{pmatrix} = 4$$

$$-1 + \lambda + 2(1 + \lambda) + 3(-5 + 2\lambda) = 4$$

$$9\lambda - 14 = 4$$

$$9\lambda = 18$$

$$\lambda = 2$$

So the line meets the plane when $\lambda = 2$, at the point (1, 3, −1).

Write the equation of the line in column vector form as $\begin{pmatrix} x \\ y \\ z \end{pmatrix} = \begin{pmatrix} -1 + \lambda \\ 1 + \lambda \\ -5 + 2\lambda \end{pmatrix}$ and substitute into the equation of the plane $\begin{pmatrix} x \\ y \\ z \end{pmatrix} \begin{pmatrix} 1 \\ 2 \\ 3 \end{pmatrix} = 4$.

Solve to find λ and substitute its value into the equation of the line.

Watch out If the line were parallel to the plane then this equation would produce either no solutions (if the line does not lie in the plane), or infinitely many (if it does).

- **Two straight lines are skew if they are not parallel and they do not intersect.**

Example 24

The lines l_1 and l_2 have equations $\frac{x-2}{4} = \frac{y+3}{2} = z - 1$ and $\frac{x+1}{5} = \frac{y}{4} = \frac{z-4}{-2}$ respectively.

Prove that l_1 and l_2 are skew.

$$\begin{pmatrix} 2 + 4\lambda \\ -3 + 2\lambda \\ 1 + \lambda \end{pmatrix} = \begin{pmatrix} -1 + 5\mu \\ 4\mu \\ 4 - 2\mu \end{pmatrix}$$

$2 + 4\lambda = -1 + 5\mu \quad (1)$

$-3 + 2\lambda = 4\mu \quad (2)$

$(1) - 2 \times (2)$: $8 = -1 - 3\mu \Rightarrow \mu = -3$

Substituting into (2): $-3 + 2\lambda = -12 \Rightarrow \lambda = -\frac{9}{2}$

Check for consistency:

$1 + \lambda = -\frac{7}{2}$ and $4 - 2\mu = 10$.

$1 + \lambda \neq 4 - 2\mu$ so equations not consistent and lines do not intersect.

Direction of l_1 is $\begin{pmatrix} 4 \\ 2 \\ 1 \end{pmatrix}$ or $\begin{pmatrix} 8 \\ 4 \\ 2 \end{pmatrix}$.

Direction of l_2 is $\begin{pmatrix} 5 \\ 4 \\ -2 \end{pmatrix}$.

Direction vectors are not scalar multiples of each other so lines are not parallel.

Hence l_1 and l_2 are skew.

Problem-solving

To show that two lines are skew you need to show that they do not intersect **and** that they are not parallel. Start by writing the general point on each line. Equate these general points and attempt to solve the three equations simultaneously.

Solve the first two equations simultaneously, then check to see whether the answer is consistent with the third equation.

If the lines are parallel the direction vectors will be scalar multiples of each other. Multiply the direction vector of l_1 by a scalar to make one component match the direction vector of l_2, then compare the other components.

Exercise 9E

1 In each case establish whether lines l_1 and l_2 meet and, if they meet, find the coordinates of their point of intersection:

a l_1 has equation $\mathbf{r} = \mathbf{i} + 3\mathbf{j} + \lambda(\mathbf{i} - \mathbf{j} + 5\mathbf{k})$ and l_2 has equation $\mathbf{r} = -\mathbf{i} - 3\mathbf{j} + 2\mathbf{k} + \mu(\mathbf{i} + \mathbf{j} + 2\mathbf{k})$

b l_1 has equation $\mathbf{r} = 3\mathbf{i} + 2\mathbf{j} + \mathbf{k} + \lambda(\mathbf{i} + \mathbf{j} + 2\mathbf{k})$ and l_2 has equation $\mathbf{r} = 4\mathbf{i} + 3\mathbf{j} + \mu(-\mathbf{i} + \mathbf{j} - \mathbf{k})$

c l_1 has equation $\mathbf{r} = \mathbf{i} + 3\mathbf{j} + 5\mathbf{k} + \lambda(2\mathbf{i} + 3\mathbf{j} + \mathbf{k})$ and l_2 has equation $\mathbf{r} = \mathbf{i} + \frac{5}{2}\mathbf{j} + \frac{5}{2}\mathbf{k} + \mu(\mathbf{i} + \mathbf{j} - 2\mathbf{k})$

(In each of the above cases λ and μ are scalars.)

E **2** With respect to a fixed origin O, the lines l_1 and l_2 are given by the equations

l_1: $\mathbf{r} = (-6\mathbf{i} + 11\mathbf{k}) + \lambda(\mathbf{i} - \mathbf{j} + \mathbf{k})$

l_2: $\mathbf{r} = (2\mathbf{i} - 2\mathbf{j} + 9\mathbf{k}) + \mu(2\mathbf{i} + \mathbf{j} - 3\mathbf{k})$

Show that l_1 and l_2 meet and find the coordinates of their point of intersection. **(6 marks)**

E **3** The line l_1 has equation $\mathbf{r} = \begin{pmatrix} 3 \\ 1 \\ -2 \end{pmatrix} + \lambda \begin{pmatrix} 2 \\ 2 \\ 3 \end{pmatrix}$ and the line l_2 has equation $\mathbf{r} = \begin{pmatrix} 5 \\ 4 \\ 0 \end{pmatrix} + \mu \begin{pmatrix} 2 \\ 1 \\ -1 \end{pmatrix}$.

Show that l_1 and l_2 do not meet. **(4 marks)**

4 In each case, find the coordinates of the point of intersection of the line l with the plane Π.

a l: $\mathbf{r} = \mathbf{i} + \mathbf{j} + \mathbf{k} + \lambda(-2\mathbf{i} + \mathbf{j} - 4\mathbf{k})$
Π: $\mathbf{r}.(3\mathbf{i} - 4\mathbf{j} + 2\mathbf{k}) = 16$

b l: $\mathbf{r} = \mathbf{i} + \mathbf{j} + \mathbf{k} + \lambda(2\mathbf{j} - 2\mathbf{k})$
Π: $\mathbf{r}.(3\mathbf{i} - \mathbf{j} - 6\mathbf{k}) = 1$

E/P **5** The line l has equation $\mathbf{r} = \begin{pmatrix} 2 \\ 3 \\ -2 \end{pmatrix} + \lambda \begin{pmatrix} 1 \\ 1 \\ 1 \end{pmatrix}$.

a Show that l does not meet the plane with equation $\mathbf{r}.\begin{pmatrix} 1 \\ 1 \\ -2 \end{pmatrix} = 1$. **(4 marks)**

b Give a geometrical interpretation to your answer to part **a**. **(1 mark)**

E/P **6** The line with vector equation $\mathbf{r} = \begin{pmatrix} 5 \\ 4 \\ -1 \end{pmatrix} + \lambda \begin{pmatrix} 3 \\ -1 \\ 2 \end{pmatrix}$ is perpendicular to the line with vector equation $\mathbf{r} = \begin{pmatrix} 0 \\ 11 \\ 3 \end{pmatrix} + \mu \begin{pmatrix} -1 \\ p \\ p \end{pmatrix}$.

a Find the value of p. **(2 marks)**

b Show that the two lines meet, and find the coordinates of the point of intersection. **(4 marks)**

E **7** The line l_1 has vector equation $\mathbf{r} = \begin{pmatrix} 5 \\ 2 \\ 1 \end{pmatrix} + \lambda \begin{pmatrix} -1 \\ 1 \\ 2 \end{pmatrix}$ and the line l_2 has vector equation

$\mathbf{r} = \begin{pmatrix} 4 \\ 1 \\ 1 \end{pmatrix} + \mu \begin{pmatrix} 1 \\ 0 \\ -1 \end{pmatrix}$, where λ and μ are parameters.

The lines l_1 and l_2 intersect at the point A and the acute angle between l_1 and l_2 is θ.

a Find the coordinates of A. **(4 marks)**

b Find the value of $\cos\theta$ giving your answer as a simplified fraction. **(4 marks)**

E/P **8** The lines l_1 and l_2 have equations $\frac{x}{-3} = \frac{y+1}{5} = \frac{z-2}{4}$ and $x = \frac{y-1}{-2} = \frac{z+5}{2}$ respectively.

Prove that l_1 and l_2 are skew.

E/P **9** With respect to a fixed origin O the lines l_1 and l_2 are given by the equations

$$l_1: \mathbf{r} = \begin{pmatrix} 8 \\ 2 \\ -12 \end{pmatrix} + \lambda\begin{pmatrix} -1 \\ 3 \\ 2 \end{pmatrix} \qquad l_2: \mathbf{r} = \begin{pmatrix} -4 \\ 10 \\ p \end{pmatrix} + \mu\begin{pmatrix} q \\ 2 \\ -1 \end{pmatrix}$$

where λ and μ are parameters and p and q are constants. Given that l_1 and l_2 are perpendicular,

a show that $q = 4$. **(2 marks)**

Given further that l_1 and l_2 intersect, find:

b the value of p **(6 marks)**

c the coordinates of the point of intersection. **(2 marks)**

The point A lies on l_1 and has position vector $\begin{pmatrix} 9 \\ -1 \\ -14 \end{pmatrix}$. The point C lies on l_2.

Given that a circle, with centre C, cuts the line l_1 at the points A and B,

d find the position vector of B. **(3 marks)**

Problem-solving

Draw a diagram showing the lines l_1 and l_2 and the circle, and use circle properties.

E/P **10** The plane Π has equation $\mathbf{r}.\begin{pmatrix} 2 \\ 3 \\ -1 \end{pmatrix} = k$ where k is a constant.

Given the point with position vector $\begin{pmatrix} 6 \\ -2 \\ 4 \end{pmatrix}$ lies on Π,

a find the value of k **(3 marks)**

b find a Cartesian equation for Π. **(2 marks)**

The point P has coordinates (6, 4, 8). The line l passes through P and is perpendicular to Π. The line l intersects Π at the point N.

c Find the coordinates of N. **(4 marks)**

E/P **11** The line l has a Cartesian equation $\frac{x-3}{5} = \frac{y+2}{3} = \frac{4-z}{1}$

The plane Π has Cartesian equation $4x + 3y - 2z = -10$.
The line intersects the plane at the point P.

a Find the position vector of P. **(5 marks)**

b Find the acute angle between the line and the plane at the point of intersection. **(5 marks)**

9.6 Finding perpendiculars

You need to be able to calculate the **perpendicular distance** between:

- two lines
- a point and a line
- a point and a plane

In each case, the perpendicular distance is the **shortest distance** between them.

- **For any two non-intersecting lines l_1 and l_2 there is a unique line segment AB such that A lies on l_1, B lies on l_2 and AB is perpendicular to both lines.**

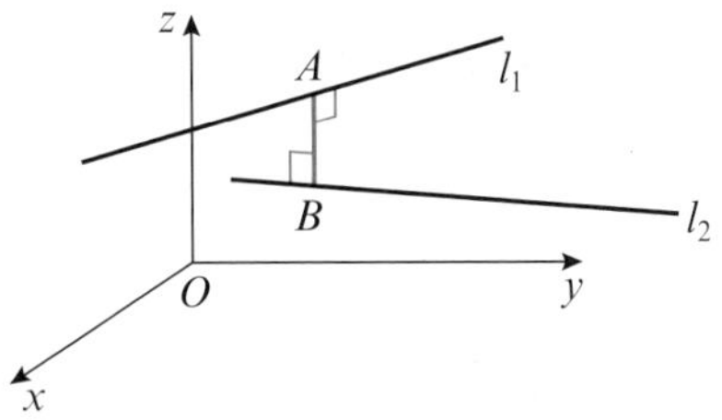

- **The perpendicular from a point P to a line l is a line through P which meets l at right angles.**

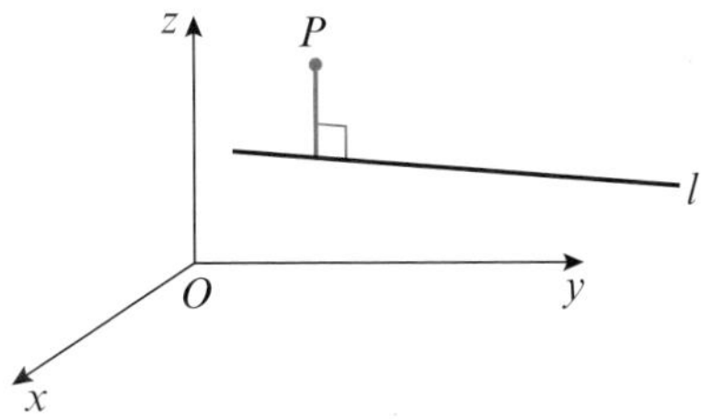

- **The perpendicular from a point P to a plane Π is a line through P which is parallel to the normal vector of the plane, n.**

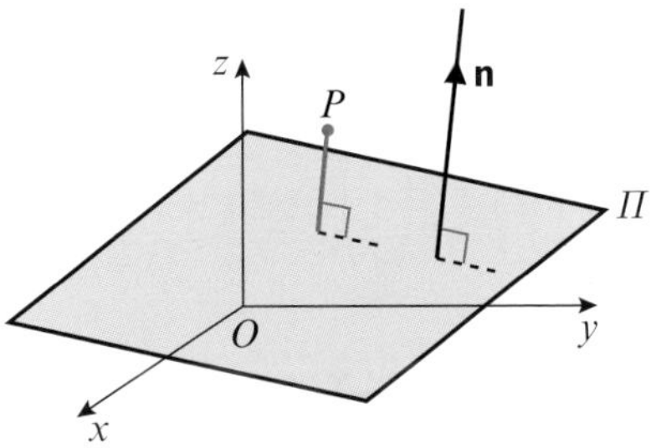

Example 24

Show that the shortest distance between the parallel lines with equations $\mathbf{r} = \mathbf{i} + 2\mathbf{j} - \mathbf{k} + \lambda(5\mathbf{i} + 4\mathbf{j} + 3\mathbf{k})$ and $\mathbf{r} = 2\mathbf{i} + \mathbf{k} + \mu(5\mathbf{i} + 4\mathbf{j} + 3\mathbf{k})$,
where λ and μ are scalars, is $\frac{21\sqrt{2}}{10}$

Let A be a general point on the first line and B be a general point on the second line, then

$$\overrightarrow{AB} = \begin{pmatrix}1\\-2\\2\end{pmatrix} + t\begin{pmatrix}5\\4\\3\end{pmatrix} \text{ where } t = \mu - \lambda.$$

As $\begin{pmatrix}1\\-2\\2\end{pmatrix} = \begin{pmatrix}2\\0\\1\end{pmatrix} - \begin{pmatrix}1\\2\\-1\end{pmatrix}$

You can set $t = \mu - \lambda$ so that there is only one independent variable.

$$\begin{pmatrix}1+5t\\-2+4t\\2+3t\end{pmatrix}\cdot\begin{pmatrix}5\\4\\3\end{pmatrix} = 0$$

As the direction of $\overrightarrow{AB}$ is perpendicular to the direction vector for each line, the scalar product is zero.

$$5 + 25t - 8 + 16t + 6 + 9t = 0$$
$$50t = -3$$
$$t = -\frac{3}{50}$$

$$\begin{pmatrix}1-5t\\-2+4t\\2+3t\end{pmatrix} = \begin{pmatrix}1-\frac{15}{50}\\-2-\frac{12}{50}\\2-\frac{9}{50}\end{pmatrix} = \begin{pmatrix}\frac{35}{50}\\-\frac{112}{50}\\\frac{91}{50}\end{pmatrix}$$

Substitute $t = -\frac{3}{50}$ into the general form of $\overrightarrow{AB}$.

$$|\overrightarrow{AB}| = \sqrt{\frac{35^2 + 112^2 + 91^2}{50^2}}$$
$$= \frac{21\sqrt{2}}{10}$$

The shortest distance between two lines is the length of the line segment that is perpendicular to both lines.

So the shortest distance between the two lines is $\frac{21\sqrt{2}}{10}$

Watch out Because the two lines are parallel, the line segment AB is not unique. There are infinitely many line segments that are perpendicular to both lines, but they will all have the same length.

Example 25

The lines l_1 and l_2 have equations $\mathbf{r} = \begin{pmatrix}1\\0\\0\end{pmatrix} + \lambda\begin{pmatrix}0\\1\\1\end{pmatrix}$ and $\mathbf{r} = \begin{pmatrix}-1\\3\\-1\end{pmatrix} + \mu\begin{pmatrix}2\\-1\\-1\end{pmatrix}$ respectively,
where λ and μ are scalars.
Find the shortest distance between these two lines.

Let A be the general point on l_1 with position vector $\begin{pmatrix}1\\\lambda\\\lambda\end{pmatrix}$ and let B be the general point on l_2 with position vector $\begin{pmatrix}-1+2\mu\\3-\mu\\-1-\mu\end{pmatrix}$

Online Explore the perpendicular distance between two lines using GeoGebra.

$$\overrightarrow{AB} = \begin{pmatrix} -1 + 2\mu \\ 3 - \mu \\ -1 - \mu \end{pmatrix} - \begin{pmatrix} 1 \\ \lambda \\ \lambda \end{pmatrix} = \begin{pmatrix} -2 + 2\mu \\ 3 - \mu - \lambda \\ -1 - \mu - \lambda \end{pmatrix}$$

Find position vectors of general points on l_1 and l_2 and use these to find $\overrightarrow{AB}$ in terms of μ and λ.

$\overrightarrow{AB}$ is perpendicular to l_1, so:

$$\begin{pmatrix} -2 + 2\mu \\ 3 - \mu - \lambda \\ -1 - \mu - \lambda \end{pmatrix} . \begin{pmatrix} 0 \\ 1 \\ 1 \end{pmatrix} = 0$$

$3 - \mu - \lambda - 1 - \mu - \lambda = 0$

$2 - 2\mu - 2\lambda = 0$ (1)

$\overrightarrow{AB}$ is perpendicular to l_2 so:

$$\begin{pmatrix} -2 + 2\mu \\ 3 - \mu - \lambda \\ -1 - \mu - \lambda \end{pmatrix} . \begin{pmatrix} 2 \\ -1 \\ -1 \end{pmatrix} = 0$$

Problem-solving

As $\overrightarrow{AB}$ is perpendicular to l_1 and l_2, the scalar product of the direction vectors of the lines is zero. You can use this fact to generate two linear equations in λ and μ.

$-4 + 4\mu - 3 + \mu + \lambda + 1 + \mu + \lambda = 0$

$-6 + 6\mu + 2\lambda = 0$ (2)

From (1), $\lambda = 1 - \mu$

From (2), $-3 + 3\mu + \lambda = 0$

So $-3 + 3\mu + 1 - \mu = 0 \Rightarrow \mu = 1$

Substituting into (1) gives

$2 - 2 - 2\lambda = 0 \Rightarrow \lambda = 0$

The equations can be solved simultaneously to find λ and μ.

$$\overrightarrow{AB} = \begin{pmatrix} -2 + 2\mu \\ 3 - \mu - \lambda \\ -1 - \mu - \lambda \end{pmatrix} = \begin{pmatrix} -2 + 2 \\ 3 - 1 - 0 \\ -1 - 1 - 0 \end{pmatrix}$$

$$= \begin{pmatrix} 0 \\ 2 \\ -2 \end{pmatrix}$$

$\left| \overrightarrow{AB} \right| = \sqrt{0^2 + 2^2 + (-2)^2} = \sqrt{8} = 2\sqrt{2}$

So the shortest distance between the two lines is $2\sqrt{2}$.

These two lines are **skew**, so in this case the line segment AB is unique.

Example 26

The line l has equation $\dfrac{x - 1}{2} = \dfrac{y - 1}{-2} = \dfrac{z + 3}{-1}$, and the point A has coordinates $(1, 2, -1)$.

a Find the shortest distance between A and l.

b Find a Cartesian equation of the line that is perpendicular to l and passes through A.

a Vector equation of l is

$$\mathbf{r} = \begin{pmatrix} 1 \\ 1 \\ -3 \end{pmatrix} + \lambda\begin{pmatrix} 2 \\ -2 \\ -1 \end{pmatrix}$$

$\frac{x-1}{2} = \lambda,\ x = 1 + 2\lambda$

$\frac{y-1}{-2} = \lambda,\ y = 1 - 2\lambda$

$\frac{z+3}{-1} = \lambda,\ z = -3 - \lambda$

So a general point B, on the line has position vector $\begin{pmatrix} 1+2\lambda \\ 1-2\lambda \\ -3-\lambda \end{pmatrix}$.

Let B be the position vector of a general point on l.

Then $\overrightarrow{AB} = \begin{pmatrix} 1+2\lambda \\ 1-2\lambda \\ -3-\lambda \end{pmatrix} - \begin{pmatrix} 1 \\ 2 \\ -1 \end{pmatrix} = \begin{pmatrix} 2\lambda \\ -1-2\lambda \\ -2-\lambda \end{pmatrix}$

Find $\overrightarrow{AB}$ in terms of λ.

$$\begin{pmatrix} 2\lambda \\ -1-2\lambda \\ -2-\lambda \end{pmatrix} \cdot \begin{pmatrix} 2 \\ -2 \\ -1 \end{pmatrix} = 0$$

$$4\lambda + 2 + 4\lambda + 2 + \lambda = 0$$

$$4 + 9\lambda = 0$$

$$\lambda = -\tfrac{4}{9}$$

Since $\overrightarrow{AB}$ is perpendicular to l the scalar product of $\overrightarrow{AB}$ with the direction vector of the line is zero. This gives you an equation which you can solve to find λ.

$$\overrightarrow{AB} = \begin{pmatrix} 2\lambda \\ -1-2\lambda \\ -2-\lambda \end{pmatrix} = \begin{pmatrix} -\frac{8}{9} \\ -\frac{1}{9} \\ -\frac{14}{9} \end{pmatrix}$$

Substitute the value of λ into your general expression for $\overrightarrow{AB}$.

$$|\overrightarrow{AB}| = \sqrt{\frac{(-8)^2 + (-1)^2 + (-14)^2}{9^2}}$$

$$= \frac{\sqrt{29}}{3} = 1.80 \text{ (3 s.f.)}$$

The shortest distance is given by $|\overrightarrow{AB}|$.

So the shortest distance between A and l is $\frac{\sqrt{29}}{3}$ or 1.80 (3 s.f.).

b $\overrightarrow{AB}$ is perpendicular to l, so direction of perpendicular is $\begin{pmatrix} -\frac{8}{9} \\ -\frac{1}{9} \\ -\frac{14}{9} \end{pmatrix}$ or $\begin{pmatrix} 8 \\ 1 \\ 14 \end{pmatrix}$.

Remember that you can multiply the direction vector by a scalar to find a simpler parallel vector.

A vector equation of the line through A perpendicular to l is $\mathbf{r} = \begin{pmatrix} 1 \\ 2 \\ -1 \end{pmatrix} + \mu\begin{pmatrix} 8 \\ 1 \\ 14 \end{pmatrix}$.

A vector equation of the line through the point with position vector **a** with direction **b** is $\mathbf{r} = \mathbf{a} + \mu\mathbf{b}$ where μ is a scalar constant.

So the Cartesian equation of the line is

$$\frac{x-1}{8} = \frac{y-2}{1} = \frac{z+1}{14}$$

You can use the principles covered above to give meaning to the constant, k, in the scalar product form of the vector equation of a plane.

- **k is the length of the perpendicular from the origin to a plane Π, where the equation of plane Π is written in the form $\mathbf{r}.\hat{\mathbf{n}} = k$, where $\hat{\mathbf{n}}$ is a unit vector perpendicular to Π.**
- **The perpendicular distance from the point with coordinates (α, β, γ) to the plane with equation $ax + by + cz = d$ is**
$$\frac{|a\alpha + b\beta + c\gamma - d|}{\sqrt{a^2 + b^2 + c^2}}$$

Note This formula is given in the formulae booklet and you can use it without proof in your exam.

Example 27

Find the perpendicular distance from the point with coordinates (3, 2, −1) to the plane with equation $2x - 3y + z = 5$.

$$\text{Distance} = \frac{|2 \times 3 - 3 \times 2 + 1 \times (-1) - 5|}{\sqrt{2^2 + (-3)^2 + 1^2}}$$

Substitute into the formula.

$$= \frac{|-6|}{\sqrt{14}}$$

Remember to use the modulus of the numerator as distance is always positive.

$$= \frac{6}{\sqrt{14}}$$

Example 28

The plane Π has equation $\mathbf{r}.(\mathbf{i} + 2\mathbf{j} + 2\mathbf{k}) = 5$. The point P has coordinates (1, 3, −2).

a Find the shortest distance between P and Π.

The point Q is the reflection of the point P in Π.

b Find the coordinates of point Q.

a Cartesian equation of Π is $x + 2y + 2z = 5$.

Use $\mathbf{r} = x\mathbf{i} + y\mathbf{j} + z\mathbf{k}$ and $\begin{pmatrix} x \\ y \\ z \end{pmatrix} . \begin{pmatrix} 1 \\ 2 \\ 3 \end{pmatrix} = 5$.

$$\text{Distance} = \frac{|1 \times 1 + 2 \times 3 + 2 \times (-2) - 5|}{\sqrt{1^2 + 2^2 + 2^2}}$$

$$= \frac{|-2|}{\sqrt{9}} = \frac{2}{3}$$

Using the formula $\frac{|a\alpha + b\beta + c\gamma - d|}{\sqrt{a^2 + b^2 + c^2}}$ for the distance between a point with coordinates (α, β, γ) and the plane with Cartesian equation $ax + by + cz = d$.

b A perpendicular vector to Π is

$\mathbf{n} = \mathbf{i} + 2\mathbf{j} + 2\mathbf{k}$

A normal vector to the plane $ax + by + cz = d$ is $a\mathbf{i} + b\mathbf{j} + c\mathbf{k}$.

Let Q have coordinates (x_1, y_1, z_1).

Let M be the midpoint of PQ.

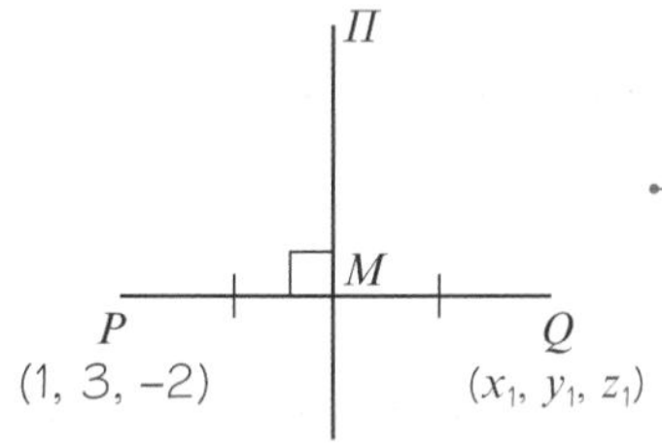

The line joining P to its reflection Q will be perpendicular to the plane, and P and Q will be the same distance from the plane. Draw a diagram showing P, Q, and the midpoint of PQ. Represent the plane Π using a vertical line.

Online Explore reflections in a plane using GeoGebra.

A vector equation of the line through P, M and Q is $\mathbf{r} = \begin{pmatrix} 1 \\ 3 \\ -2 \end{pmatrix} + \lambda \begin{pmatrix} 1 \\ 2 \\ 2 \end{pmatrix}$

M lies on this line so has position vector $\begin{pmatrix} 1 + \lambda \\ 3 + 2\lambda \\ -2 + 2\lambda \end{pmatrix}$

M also lies on Π, so $\begin{pmatrix} 1 + \lambda \\ 3 + 2\lambda \\ -2 + 2\lambda \end{pmatrix} \cdot \begin{pmatrix} 1 \\ 2 \\ 2 \end{pmatrix} = 5$

$1 + \lambda + 2(3 + 2\lambda) + 2(-2 + 2\lambda) = 5$

$3 + 9\lambda = 5$

$\lambda = \frac{2}{9}$

M has position vector $\begin{pmatrix} 1 + \frac{2}{9} \\ 3 + 2 \times \frac{2}{9} \\ -2 + 2 \times \frac{2}{9} \end{pmatrix} = \begin{pmatrix} \frac{11}{9} \\ \frac{31}{9} \\ -\frac{14}{9} \end{pmatrix}$

Use the fact that M lies on the line joining P and Q and on the plane to find M.

P is the initial point in the equation of l,

so if M has position vector $\begin{pmatrix} 1 \\ 3 \\ -2 \end{pmatrix} + \frac{2}{9} \begin{pmatrix} 1 \\ 2 \\ 2 \end{pmatrix}$

then P has position vector

$\begin{pmatrix} 1 \\ 3 \\ -2 \end{pmatrix} + 2 \times \frac{2}{9} \begin{pmatrix} 1 \\ 2 \\ 2 \end{pmatrix} = \begin{pmatrix} 1 + \frac{4}{9} \\ 3 + 2 \times \frac{4}{9} \\ -2 + 2 \times \frac{4}{9} \end{pmatrix} = \begin{pmatrix} \frac{13}{9} \\ \frac{35}{9} \\ -\frac{10}{9} \end{pmatrix}$

Point Q has coordinates $\left(\frac{13}{9}, \frac{35}{9}, -\frac{10}{9}\right)$.

Problem-solving

$\begin{pmatrix} 1 \\ 3 \\ -2 \end{pmatrix}$ is the position vector of point P, so if the point on the line $\mathbf{r} = \begin{pmatrix} 1 \\ 3 \\ -2 \end{pmatrix} + \lambda \begin{pmatrix} 1 \\ 2 \\ 2 \end{pmatrix}$ with $\lambda = k$ is a distance x away from P, then the point with $\lambda = 2k$ will be a distance $2x$ away from P.

You could also use the fact that the midpoint of the line segment joining (x_1, y_1, z_1) to (x_2, y_2, z_2) is $\left(\frac{x_1 + x_2}{2}, \frac{y_1 + y_2}{2}, \frac{z_1 + z_2}{2}\right)$

Example 29

The line l_1 has equation $\frac{x-2}{2} = \frac{y-4}{-2} = \frac{z+6}{1}$. The plane Π has equation $2x - 3y + z = 8$.

The line l_2 is the reflection of line l_1 in the plane Π. Find a vector equation of the line l_2.

A vector equation of l_1 is $\mathbf{r} = \begin{pmatrix} 2 \\ 4 \\ -6 \end{pmatrix} + \lambda \begin{pmatrix} 2 \\ -2 \\ 1 \end{pmatrix}$.

So $P(2, 4, -6)$ is a point on line l_1.

Let A be the point of intersection of l_1 and Π.

Problem-solving

You need to find **two points** on the reflected line, l_2. One is the point of intersection of l_1 and Π.
To find another point on l_2, choose any point on l_1 and reflect it in the plane.

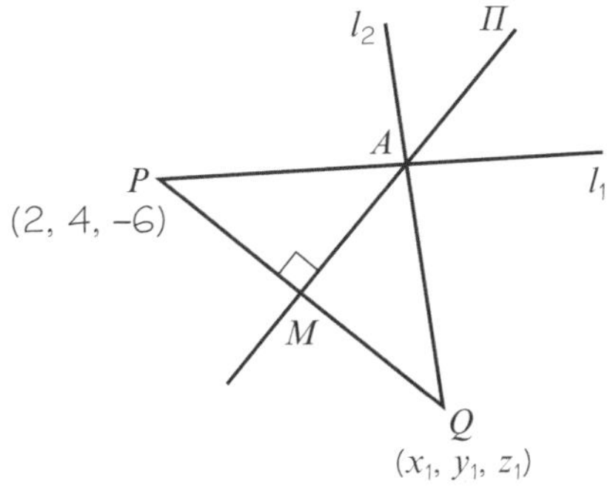

A lies on l_1 and on $2x - 3y + z = 8$.

A has position vector $\begin{pmatrix} 2 + 2\lambda \\ 4 - 2\lambda \\ -6 + \lambda \end{pmatrix}$ and satisfies

$$2(2 + 2\lambda) - 3(4 - 2\lambda) - 6 + \lambda = 8$$
$$4 + 4\lambda - 12 + 6\lambda - 6 + \lambda = 8$$
$$11\lambda = 22$$
$$\lambda = 2$$

Substitute the general position vector of a point on l_1 into the equation for Π to find the value of λ at A.

So A has coordinates $(6, 0, -4)$.

Substitute $\lambda = 2$ into the general position vector to find the coordinates of A.

A perpendicular direction vector to Π is $\mathbf{n} = 2\mathbf{i} - 3\mathbf{j} + \mathbf{k}$.

A vector equation of the line through P perpendicular to Π is $\mathbf{r} = \begin{pmatrix} 2 \\ 4 \\ -6 \end{pmatrix} + \mu \begin{pmatrix} 2 \\ -3 \\ 1 \end{pmatrix}$.

Let $Q\ (x_1, y_1, z_1)$ be the point of intersection of this line and l_2.

Q is the reflection of the point (6, 0, −4) in the plane.

Let M be midpoint of PQ.

M lies on line and on $2x - 3y + z = 8$ so satisfies

$$2(2 + 2\mu) - 3(4 - 3\mu) - 6 + \mu = 8$$
$$4 + 4\mu - 12 + 9\mu - 6 + \mu = 8$$
$$14\mu = 22$$
$$\mu = \tfrac{11}{7}$$

The general point on the line PQ is $\begin{pmatrix} 2 + 2\mu \\ 4 - 3\mu \\ -6 + \mu \end{pmatrix}$.

So M has position vector $\begin{pmatrix} 2 \\ 4 \\ -6 \end{pmatrix} + \frac{11}{7} \begin{pmatrix} 2 \\ -3 \\ 1 \end{pmatrix}$

and Q has position vector

$$\begin{pmatrix} 2 \\ 4 \\ -6 \end{pmatrix} + 2 \times \frac{11}{7}\begin{pmatrix} 2 \\ -3 \\ 1 \end{pmatrix} = \begin{pmatrix} \frac{58}{7} \\ -\frac{38}{7} \\ -\frac{20}{7} \end{pmatrix}$$

Q is twice as far along the line from P as M.

So Q has coordinates $\left(\frac{58}{7}, -\frac{38}{7}, -\frac{20}{7}\right)$.

l_2 is the line through Q and A, so has direction:

$$\overrightarrow{AQ} = \begin{pmatrix} \frac{16}{7} \\ -\frac{38}{7} \\ \frac{8}{7} \end{pmatrix}$$

The direction of l_2 can be simplified to $\begin{pmatrix} 8 \\ -19 \\ 4 \end{pmatrix}$ by multiplying the expression for $\overrightarrow{AQ}$ by $\frac{7}{2}$

A vector equation of l_2 is $\mathbf{r} = \begin{pmatrix} 6 \\ 0 \\ -4 \end{pmatrix} + t\begin{pmatrix} 8 \\ -19 \\ 4 \end{pmatrix}$.

Exercise 9F

1 Find the shortest distance between the parallel lines with equations
$\mathbf{r} = 2\mathbf{i} - \mathbf{j} + \mathbf{k} + \lambda(-3\mathbf{i} - 4\mathbf{j} + 5\mathbf{k})$ and $\mathbf{r} = \mathbf{j} + \mathbf{k} + \mu(-3\mathbf{i} - 4\mathbf{j} + 5\mathbf{k})$
where λ and μ are scalars.

2 Find the shortest distance between the two skew lines with equations
$\mathbf{r} = \mathbf{j} + \mathbf{k} + \lambda(-\mathbf{i} + \mathbf{j} - \mathbf{k})$ and $\mathbf{r} = 5\mathbf{i} - 2\mathbf{j} + 2\mathbf{k} + \mu(4\mathbf{i} - 2\mathbf{j} + 3\mathbf{k})$,
where λ and μ are scalars.

3 Determine whether the lines l_1 and l_2 meet. If they do, find their point of intersection. If they do not, find the shortest distance between them. (In each of the following cases λ and μ are scalars.)

a l_1 has equation $\mathbf{r} = 7\mathbf{i} + 3\mathbf{j} + \mathbf{k} + \lambda(6\mathbf{i} + 2\mathbf{j} - 4\mathbf{k})$ and l_2 has equation $\mathbf{r} = -\mathbf{i} + \mathbf{j} + 2\mathbf{k} - 2\mu\mathbf{j}$

b l_1 has equation $\mathbf{r} = 2\mathbf{i} + \mathbf{j} - 2\mathbf{k} + \lambda(2\mathbf{i} - 2\mathbf{j} + 2\mathbf{k})$ and l_2 has equation $\mathbf{r} = \mathbf{i} - \mathbf{j} + 3\mathbf{k} + \mu(\mathbf{i} - \mathbf{j} + \mathbf{k})$

c l_1 has equation $\mathbf{r} = \mathbf{i} + \mathbf{j} + 5\mathbf{k} + \lambda(2\mathbf{i} + \mathbf{j} - 2\mathbf{k})$ and l_2 has equation $\mathbf{r} = -\mathbf{i} - \mathbf{j} + 2\mathbf{k} + \mu(\mathbf{i} + \mathbf{j} + \mathbf{k})$

4 Find the shortest distance between the point with coordinates (4, 1, −1) and the line with equation $\mathbf{r} = 3\mathbf{i} - \mathbf{j} + 2\mathbf{k} + \mu(2\mathbf{i} - \mathbf{j} - \mathbf{k})$, where μ is a scalar.

5 Find the shortest distance between the parallel planes.

a $\mathbf{r}.(6\mathbf{i} + 6\mathbf{j} - 7\mathbf{k}) = 55$ and $\mathbf{r}.(6\mathbf{i} + 6\mathbf{j} - 7\mathbf{k}) = 22$

b $\mathbf{r} = 3\mathbf{i} + 4\mathbf{j} + \mathbf{k} + \lambda(4\mathbf{i} + \mathbf{k}) + \mu(8\mathbf{i} + 3\mathbf{j} + 3\mathbf{k})$ and $\mathbf{r} = 14\mathbf{i} + 2\mathbf{j} + 2\mathbf{k} + \lambda(3\mathbf{j} + \mathbf{k}) + \mu(8\mathbf{i} - 9\mathbf{j} - \mathbf{k})$

6 The plane Π has equation $\mathbf{r}.(10\mathbf{i} + 10\mathbf{j} + 23\mathbf{k}) = 81$.

a Find the perpendicular distance from the origin to plane Π.

b Find the perpendicular distance from the point $(-1, -1, 4)$ to the plane Π.

c Find the perpendicular distance from the point $(2, 1, 3)$ to the plane Π.

d Find the perpendicular distance from the point $(6, 12, -9)$ to the plane Π.

P **7** The line l has vector equation $\mathbf{r} = \begin{pmatrix} 2 \\ -1 \\ 3 \end{pmatrix} + \lambda\begin{pmatrix} 1 \\ 2 \\ -1 \end{pmatrix}$.

The point P has coordinates $(3, 0, 2)$.

Find the coordinates of the reflection of the point P in the line l. **(5 marks)**

Problem-solving

Let M be the midpoint of the line segment joining P to its reflection in l. This segment must be perpendicular to l and pass through it.

P **8** The plane Π has equation $-2x + y + z = 5$. The point P has coordinates $(1, 0, 3)$.

a Find the shortest distance between P and Π. **(3 marks)**

The point Q is the reflection of the point P in Π.

b Find the coordinates of point Q. **(5 marks)**

P **9** A birdwatcher is located on a hilltop. Relative to a fixed origin O, the position vector of the birdwatcher is $\begin{pmatrix} 5 \\ 4 \\ 0.7 \end{pmatrix}$ km. The birdwatcher is able to spot any bird that flies within 500 m of her position. A kestrel flies from point A to point B, where points A and B have position vectors $\begin{pmatrix} 3 \\ 5 \\ 0 \end{pmatrix}$ km and $\begin{pmatrix} 12 \\ 0 \\ 1.2 \end{pmatrix}$ km respectively. The kestrel is modelled as flying in a straight line.

a Use the model to determine whether the birdwatcher is able to spot the kestrel. **(7 marks)**

b Give one criticism of the model. **(1 mark)**

P **10** The plane Π_1 has equation $3x - 2y + 4z = 6$.

a Find the perpendicular distance from the point $(4, -1, 8)$ to Π_1. **(3 marks)**

The plane Π_2 has vector equation $\mathbf{r} = \lambda(2\mathbf{i} - 2\mathbf{j} + 3\mathbf{k}) + \mu(-3\mathbf{i} + 3\mathbf{k})$ where λ and μ are scalar parameters.

b Show that the vector $2\mathbf{i} + 5\mathbf{j} + 2\mathbf{k}$ is perpendicular to Π_2. **(2 marks)**

c Find the acute angle between Π_1 and Π_2. **(3 marks)**

P **11** The line l_1 has equation $\dfrac{x+2}{2} = \dfrac{y-2}{-1} = \dfrac{z+1}{-2}$, and the point A has coordinates $(3, -1, 2)$.

a Find the shortest distance between A and l_1. **(5 marks)**

b Find a Cartesian equation of the line that is perpendicular to l_1 and passes through A. **(3 marks)**

P **12** The line l_1 has vector equation $\mathbf{r} = \begin{pmatrix} 2 \\ -1 \\ 2 \end{pmatrix} + \lambda\begin{pmatrix} 4 \\ -3 \\ 4 \end{pmatrix}$. The plane Π has equation $\mathbf{r}.\begin{pmatrix} 1 \\ -1 \\ -1 \end{pmatrix} = 4$.

The line l_2 is the reflection of line l_1 in the plane Π.
Find a vector equation of the line l_2. **(7 marks)**

Mixed exercise 9

(E) **1** The line l passes through the points A and B with position vectors $\mathbf{i} - \mathbf{j} + 3\mathbf{k}$ and $\mathbf{i} + 2\mathbf{j} + 2\mathbf{k}$ respectively, relative to a fixed origin O.

a Find a vector equation of the line l. **(4 marks)**

b Find the position vector of the point C which lies on the line segment AB such that $AC = 2CB$. **(3 marks)**

(E) **2** Find a Cartesian equation of the straight line that passes through the points with coordinates $(7, -1, 2)$ and $(-1, 3, 8)$. **(4 marks)**

(E) **3** Find a vector equation of the straight line which passes through the point A with position vector $2\mathbf{i} + 3\mathbf{j} - 4\mathbf{k}$, and is parallel to the vector $2\mathbf{j} + 3\mathbf{k}$. **(3 marks)**

(E/P) **4** A straight line l has vector equation $\mathbf{r} = \begin{pmatrix} 3 \\ -2 \\ 1 \end{pmatrix} + \lambda\begin{pmatrix} 2 \\ 1 \\ -1 \end{pmatrix}$.

a Write down a Cartesian equation for l. **(2 marks)**

b Given that the point $(0, a, b)$ lies on l, find the value of a and the value of b. **(3 marks)**

5 A straight line l has vector equation $\mathbf{r} = (\mathbf{i} + 2\mathbf{j} - \mathbf{k}) + \lambda(3\mathbf{i} + \mathbf{j} - 2\mathbf{k})$.
Show that another vector equation of l is $\mathbf{r} = (7\mathbf{i} + 4\mathbf{j} - 5\mathbf{k}) + \lambda(9\mathbf{i} + 3\mathbf{j} - 6\mathbf{k})$.

6 A straight line l has Cartesian equation $\frac{x+2}{1} = \frac{y-2}{3} = \frac{z+3}{4}$.

a Find a vector form of the equation of l. **b** Verify that the point $(0, 8, 5)$ lies on l.

7 A plane passes through the points $A(2, -1, 2)$, $B(1, 3, -1)$ and $C(4, 2, 5)$.

a Find a vector form of the equation of the plane.

b Find a Cartesian form of the equation of the plane.

8 A Cartesian form of the equation of a plane is $3x + 2y - 4z = 18$. Find a vector form of the equation of the plane.

(E/P) **9** With respect to an origin O, the position vectors of the points L, M and N are $\begin{pmatrix} 4 \\ 7 \\ 7 \end{pmatrix}$, $\begin{pmatrix} 1 \\ 3 \\ 2 \end{pmatrix}$ and $\begin{pmatrix} 2 \\ 4 \\ 6 \end{pmatrix}$ respectively.

a Find the vectors $\overrightarrow{ML}$ and $\overrightarrow{MN}$. **(3 marks)**

b Prove that $\cos \angle LMN = \frac{9}{10}$ **(3 marks)**

(E/P) **10** Referred to a fixed origin O, the points A, B and C have position vectors $9\mathbf{i} - 2\mathbf{j} + \mathbf{k}$, $6\mathbf{i} + 2\mathbf{j} + 6\mathbf{k}$ and $3\mathbf{i} + p\mathbf{j} + q\mathbf{k}$ respectively, where p and q are constants.

a Find, in vector form, an equation of the line l which passes through A and B. **(2 marks)**

Given that C lies on l,

b find the value of p and the value of q **(2 marks)**

c calculate, in degrees, the acute angle between OC and AB. **(3 marks)**

The point D lies on AB and is such that OD is perpendicular to AB.

d Find the position vector of D. **(5 marks)**

E **11** Referred to a fixed origin O, the points A and B have position vectors $\begin{pmatrix}1\\2\\-3\end{pmatrix}$ and $\begin{pmatrix}5\\0\\-3\end{pmatrix}$ respectively.

a Find, in vector form, an equation of the line l_1 which passes through A and B. **(3 marks)**

The line l_2 has equation $\mathbf{r} = \begin{pmatrix}4\\-4\\3\end{pmatrix} + \mu\begin{pmatrix}1\\-2\\2\end{pmatrix}$, where μ is a scalar parameter.

b Show that A lies on l_2. **(2 marks)**

c Find, in degrees, the acute angle between the lines l_1 and l_2. **(4 marks)**

The point C with position vector $\begin{pmatrix}0\\4\\-5\end{pmatrix}$ lies on l_2.

d Find the shortest distance from C to the line l_1. **(4 marks)**

E **12** Two submarines are travelling in straight lines through the ocean. Relative to a fixed origin, the vector equations of the two lines, l_1 and l_2, along which they travel are

l_1: $\mathbf{r} = 3\mathbf{i} + 4\mathbf{j} - 5\mathbf{k} + \lambda(\mathbf{i} - 2\mathbf{j} + 2\mathbf{k})$
l_2: $\mathbf{r} = 9\mathbf{i} + \mathbf{j} - 2\mathbf{k} + \mu(4\mathbf{i} + \mathbf{j} - \mathbf{k})$

where λ and μ are scalars.

a Show that the submarines are moving in perpendicular directions. **(2 marks)**

b Given that l_1 and l_2 intersect at the point A, find the position vector of A. **(4 marks)**

The point B has position vector $10\mathbf{j} - 11\mathbf{k}$.

c Show that only one of the submarines passes through the point B. **(3 marks)**

d Given that 1 unit on each coordinate axis represents 100 m, find, in km, the distance AB. **(2 marks)**

E **13** Find the shortest distance between the lines with vector equations

$$\mathbf{r} = 3\mathbf{i} + s\mathbf{j} - \mathbf{k} \quad \text{and} \quad \mathbf{r} = 9\mathbf{i} - 2\mathbf{j} - \mathbf{k} + t(\mathbf{i} - 2\mathbf{j} + \mathbf{k})$$

where s and t are scalars. **(4 marks)**

E **14** Obtain the shortest distance between the lines with equations

$$\mathbf{r} = (3s - 3)\mathbf{i} - s\mathbf{j} + (s + 1)\mathbf{k} \quad \text{and} \quad \mathbf{r} = (3 + t)\mathbf{i} + (2t - 2)\mathbf{j} + \mathbf{k}$$

where s and t are parameters. **(4 marks)**

15 Find, in the form $\mathbf{r}.\mathbf{n} = p$, an equation of the plane which contains the line l and the point with position vector $\mathbf{a}$ where:

a l has equation $\mathbf{r} = \mathbf{i} + \mathbf{j} - 2\mathbf{k} + \lambda(2\mathbf{i} - \mathbf{k})$ and $\mathbf{a} = 4\mathbf{i} + 3\mathbf{j} + \mathbf{k}$

b l has equation $\mathbf{r} = \mathbf{i} + 2\mathbf{j} + 2\mathbf{k} + \lambda(2\mathbf{i} + \mathbf{j} - 3\mathbf{k})$ and $\mathbf{a} = 3\mathbf{i} + 5\mathbf{j} + \mathbf{k}$

c l has equation $\mathbf{r} = 2\mathbf{i} - \mathbf{j} + \mathbf{k} + \lambda(\mathbf{i} + 2\mathbf{j} + 2\mathbf{k})$ and $\mathbf{a} = 7\mathbf{i} + 8\mathbf{j} + 6\mathbf{k}$

16 Find a Cartesian equation of the plane which passes through the point (1, 1, 1) and contains the line with equation $\dfrac{x-2}{3} = \dfrac{y+4}{1} = \dfrac{z-1}{2}$.

(E) **17** A plane passes through the three points A, B and C, whose position vectors, referred to an origin O, are $\mathbf{i} + 3\mathbf{j} + 3\mathbf{k}$ and $3\mathbf{i} + \mathbf{j} + 4\mathbf{k}$, $2\mathbf{i} + 4\mathbf{j} + \mathbf{k}$ respectively.

a Find, in the form $l\mathbf{i} + m\mathbf{j} + n\mathbf{k}$, a unit vector normal to this plane. **(4 marks)**

b Find also a Cartesian equation of the plane. **(2 marks)**

c Find the perpendicular distance from the origin to this plane. **(4 marks)**

(E) **18** **a** Show that the vector $\mathbf{i} + \mathbf{k}$ is perpendicular to the plane with vector equation

$\mathbf{r} = \mathbf{i} + s\mathbf{j} + t(\mathbf{i} - \mathbf{k})$ **(3 marks)**

b Find the perpendicular distance from the origin to this plane. **(4 marks)**

c Hence or otherwise obtain a Cartesian equation of the plane. **(2 marks)**

(E) **19** The points A, B and C have position vectors $\mathbf{i} + \mathbf{j} + \mathbf{k}$, $5\mathbf{i} - 2\mathbf{j} + \mathbf{k}$ and $3\mathbf{i} + 2\mathbf{j} + 6\mathbf{k}$ respectively, referred to an origin O.

a Find a vector perpendicular to the plane containing the points A, B and C. **(4 marks)**

b Hence, or otherwise, find an equation for the plane which contains the points A, B and C, in the form $ax + by + cz + d = 0$. **(2 marks)**

(E/P) **20** Planes Π_1 and Π_2 have equations given by

Π_1: $\mathbf{r}.(2\mathbf{i} - \mathbf{j} + \mathbf{k}) = 0$,

Π_2: $\mathbf{r}.(\mathbf{i} + 5\mathbf{j} + 3\mathbf{k}) = 1$

a Show that the point $A(2, -2, 3)$ lies in Π_2. **(2 marks)**

b Show that Π_1 is perpendicular to Π_2. **(4 marks)**

c Find, in vector form, an equation of the straight line through A which is perpendicular to Π_1. **(2 marks)**

d Determine the coordinates of the point where this line meets Π_1. **(4 marks)**

e Find the perpendicular distance of A from Π_1. **(4 marks)**

(E/P) **21** With respect to a fixed origin O, the straight lines l_1 and l_2 are given by

$$l_1: \mathbf{r} = \begin{pmatrix} 1 \\ 1 \\ 0 \end{pmatrix} + \lambda \begin{pmatrix} 2 \\ 1 \\ -2 \end{pmatrix},$$

$$l_2: \mathbf{r} = \begin{pmatrix} 1 \\ 4 \\ -4 \end{pmatrix} + \mu \begin{pmatrix} -3 \\ 0 \\ 1 \end{pmatrix}$$

where λ and μ are scalar parameters.

a Show that the lines intersect. **(3 marks)**

b Find the position vector of their point of intersection. **(1 mark)**

c Find the cosine of the acute angle between the lines. **(4 marks)**

(E/P) **22** The line l_1 has vector equation $\mathbf{r} = 6\mathbf{i} + 8\mathbf{j} + 5\mathbf{k} + \lambda(\mathbf{i} - \mathbf{j} + \mathbf{k})$, where λ is a scalar parameter. The point A has coordinates $(3, a, 2)$, where a is a constant. The point B has coordinates $(8, 6, b)$, where b is a constant. Points A and B lie on the line l_1.

a Find the values of a and b. **(3 marks)**

Given that the point O is the origin, and that the point P lies on l_1 such that OP is perpendicular to l_1,

b find the coordinates of P. **(5 marks)**

c Hence find the distance OP, giving your answer in surd form. **(2 marks)**

P **23** Relative to a fixed origin O, the point A has position vector $6\mathbf{i} + 3\mathbf{j} + 4\mathbf{k}$, and the point B has position vector $5\mathbf{i} + 2\mathbf{j} + 6\mathbf{k}$. The line l passes through the points A and B.

a Find the vector $\overrightarrow{AB}$. **(2 marks)**

b Find a vector equation for the line l. **(2 marks)**

The point C has position vector $4\mathbf{i} + 10\mathbf{j} + 2\mathbf{k}$.

The point P lies on l. Given that the vector $\overrightarrow{CP}$ is perpendicular to l,

c find the position vector of the point P. **(6 marks)**

P **24** With respect to a fixed origin O, the lines l_1 and l_2 are given by the equations

$$l_1: \mathbf{r} = \begin{pmatrix} 3 \\ -2 \\ 4 \end{pmatrix} + \lambda \begin{pmatrix} 2 \\ 1 \\ -1 \end{pmatrix} \qquad l_2: \mathbf{r} = \begin{pmatrix} 1 \\ 12 \\ 8 \end{pmatrix} + \mu \begin{pmatrix} 1 \\ -2 \\ -1 \end{pmatrix}$$

where λ and μ are scalar parameters.

a Show that l_1 and l_2 meet and find the position vector of their point of intersection, A. **(6 marks)**

b Find, to the nearest 0.1°, the acute angle between l_1 and l_2. **(3 marks)**

The point B has position vector $\begin{pmatrix} 5 \\ -1 \\ 3 \end{pmatrix}$.

c Show that B lies on l_1. **(1 mark)**

d Find the shortest distance from B to the line l_2, giving your answer to 3 significant figures. **(4 marks)**

E/P **25** The plane P has equation $\mathbf{r}.(2\mathbf{i} - 2\mathbf{j} + 3\mathbf{k}) = 1$.
The line l passes through the point A (1, 2, 2) and meets P at (4, 2, −1).
The acute angle between the plane P and the line l is α.

a Find α to the nearest degree. **(4 marks)**

b Find the perpendicular distance from A to the plane P. **(4 marks)**

E/P **26** Two aeroplanes are modelled as travelling in straight lines. Aeroplane A travels from a point with position vector $\begin{pmatrix} 120 \\ -80 \\ 13 \end{pmatrix}$ km to a point with position vector $\begin{pmatrix} 200 \\ 20 \\ 5 \end{pmatrix}$ km, relative to a fixed origin O. Aeroplane B starts at a point with position vector $\begin{pmatrix} -20 \\ 35 \\ 5 \end{pmatrix}$ km relative to O, and flies in the direction of $\begin{pmatrix} 10 \\ -2 \\ 0.1 \end{pmatrix}$.

a Show that the flight paths of the two aeroplanes will intersect, and determine the position vector of the point of intersection. **(7 marks)**

An air traffic controller states that this means that the planes will collide.

b Explain why this conclusion is not necessarily correct. **(2 marks)**

Challenge

1 **a** Show that the equations $4x - 2y + 6z = 10$ and $-2x + y - 3z = -5$ represent the same plane.

b Hence explain why the matrix $\begin{pmatrix} 4 & -2 & 6 \\ -2 & 1 & -3 \\ a & b & c \end{pmatrix}$ is singular for all possible values of a, b and c.

c Find values of a, b and c such that the matrix equation

$\begin{pmatrix} 4 & -2 & 6 \\ -2 & 1 & -3 \\ a & b & c \end{pmatrix}\begin{pmatrix} x \\ y \\ z \end{pmatrix} = \begin{pmatrix} 10 \\ -5 \\ 15 \end{pmatrix}$ has

i no solutions

ii infinitely many solutions.

2 The points A, B and C have coordinates (4, −4, 5), (0, 4, 1) and (0, 0, 5) respectively. Find the centre and radius of the circle that passes through all three points.

Summary of key points

1 A vector equation of a straight line passing through the point A with position vector **a**, and parallel to the vector **b**, is

$$\mathbf{r} = \mathbf{a} + \lambda\mathbf{b}$$

where λ is a scalar parameter.

2 A vector equation of a straight line passing through the points C and D, with position vectors **c** and **d** respectively, is

$$\mathbf{r} = \mathbf{c} + \lambda(\mathbf{d} - \mathbf{c})$$

where λ is a scalar parameter.

3 If $\mathbf{a} = \begin{pmatrix} a_1 \\ a_2 \\ a_3 \end{pmatrix}$ and $\mathbf{b} = \begin{pmatrix} b_1 \\ b_2 \\ b_3 \end{pmatrix}$, the equation of the line with vector equation $\mathbf{r} = \mathbf{a} + \lambda\mathbf{b}$ can be given in Cartesian form as:

$$\frac{x - a_1}{b_1} = \frac{y - a_2}{b_2} = \frac{z - a_3}{b_3}$$

Each of these three expressions is equal to λ.

4 The vector equation of a plane is

$\mathbf{r} = \mathbf{a} + \lambda\mathbf{b} + \mu\mathbf{c}$, where:

- **r** is the position vector of a general point in the plane
- **a** is the position vector of a point in the plane
- **b** and **c** are non-parallel, non-zero vectors in the plane
- λ and μ are scalars

5 A Cartesian equation of a plane in three dimensions can be written in the form $ax + by + cz = d$ where a, b, c and d are constants, and $\begin{pmatrix} a \\ b \\ c \end{pmatrix}$ is the normal vector to the plane.

6 The **scalar product** of two vectors **a** and **b** is written as **a.b** (say '**a** dot **b**'), and defined as

$$\mathbf{a.b} = |\mathbf{a}||\mathbf{b}|\cos\theta$$

where θ is the angle between **a** and **b**.

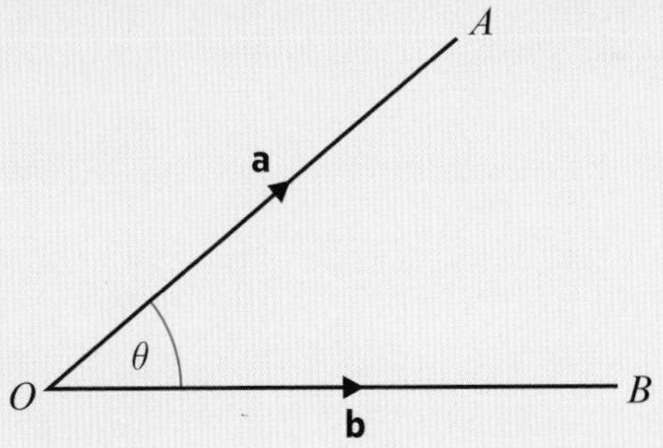

7 If **a** and **b** are the position vectors of the points A and B, then $\cos(\angle AOB) = \dfrac{\mathbf{a.b}}{|\mathbf{a}||\mathbf{b}|}$

8 The non-zero vectors **a** and **b** are perpendicular if and only if $\mathbf{a.b} = 0$.

9 If **a** and **b** are parallel, $\mathbf{a.b} = |\mathbf{a}||\mathbf{b}|$. In particular, $\mathbf{a.a} = |\mathbf{a}|^2$.

10 If $\mathbf{a} = a_1\mathbf{i} + a_2\mathbf{j} + a_3\mathbf{k}$ and $\mathbf{b} = b_1\mathbf{i} + b_2\mathbf{j} + b_3\mathbf{k}$

$$\mathbf{a.b} = \begin{pmatrix} a_1 \\ a_2 \\ a_3 \end{pmatrix} . \begin{pmatrix} b_1 \\ b_2 \\ b_3 \end{pmatrix} = a_1b_1 + a_2b_2 + a_3b_3$$

11 The acute angle θ between two intersecting straight lines is given by

$$\cos\theta = \left|\frac{\mathbf{a.b}}{|\mathbf{a}||\mathbf{b}|}\right|$$

where **a** and **b** are direction vectors of the lines.

12 The scalar product form of the equation of a plane is $\mathbf{r.n} = k$ where $k = \mathbf{a.n}$ for any point in the plane with position vector **a**.

13 The acute angle θ between the line with equation $\mathbf{r} = \mathbf{a} + \lambda\mathbf{b}$ and the plane with equation $\mathbf{r.n} = k$ is given by the formula

$$\sin\theta = \left|\frac{\mathbf{b.n}}{|\mathbf{b}||\mathbf{n}|}\right|$$

14 The acute angle θ between the plane with equation $\mathbf{r.n}_1 = k_1$ and the plane with equation $\mathbf{r.n}_2 = k_2$ is given by the formula

$$\cos\theta = \left|\frac{\mathbf{n}_1.\mathbf{n}_2}{|\mathbf{n}_1||\mathbf{n}_2|}\right|$$

15 Two lines are **skew** if they are not parallel and they do not intersect.

16 For any two non-intersecting lines l_1 and l_2 there is a unique line segment AB such that A lies on l_1, B lies on l_2 and AB is perpendicular to both lines.

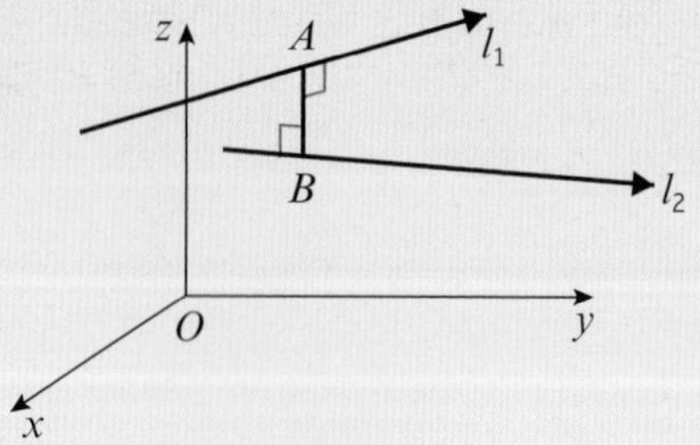

17 The perpendicular from a point P to a line l is a line drawn from P at right angles to l.

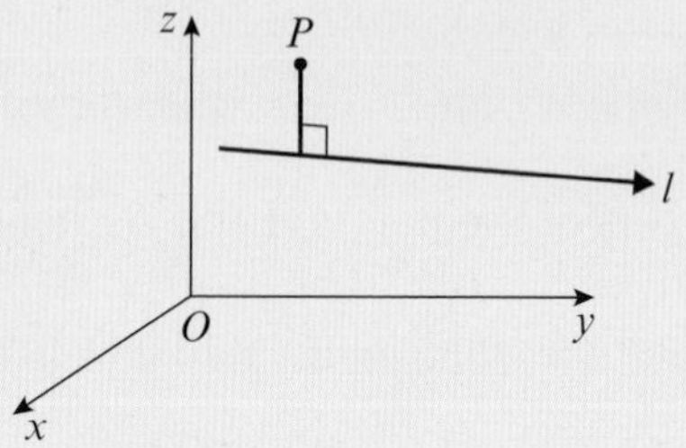

18 The perpendicular from a point P to a plane Π is a line drawn from P parallel to the normal vector **n**.

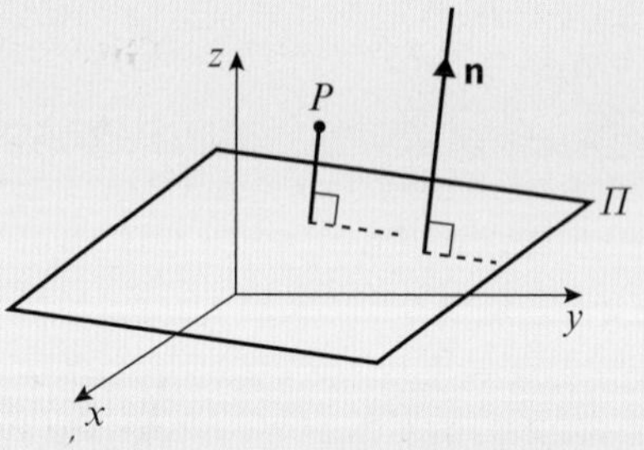

19 k is the length of the perpendicular from the origin to a plane Π, where the equation of plane Π is written in the form $\mathbf{r}.\hat{\mathbf{n}} = k$, where $\hat{\mathbf{n}}$ is a unit vector perpendicular to Π.

The perpendicular distance from the point with coordinates (α, β, γ) and the plane with equation $ax + by + cz = d$ is

$$\frac{|a\alpha + b\beta + c\gamma - d|}{\sqrt{a^2 + b^2 + c^2}}$$

Review exercise 2

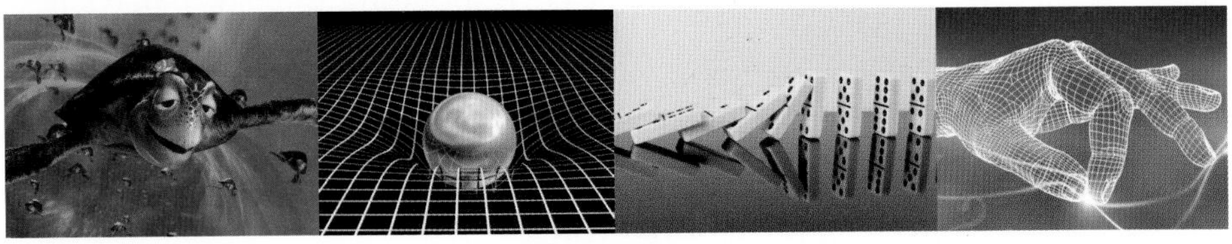

E 1 $\mathbf{A} = \begin{pmatrix} 3 & 2 & p \\ 0 & 2 & -1 \end{pmatrix}$, $\mathbf{B} = \begin{pmatrix} q & 0 \\ 3 & -1 \end{pmatrix}$, $\mathbf{C} = \begin{pmatrix} 4 \\ -3 \\ 1 \end{pmatrix}$

where p and q are integers.

Determine whether or not the following products exist. Where the product exists, evaluate the product in terms of p and q. Where the product does not exist, give a reason.

a **AB** **(1)**

b **BA** **(1)**

c **BAC** **(1)**

d **CBA** **(1)**

← Section 6.2

P 2 $\mathbf{M} = \begin{pmatrix} 0 & 3 \\ -1 & 2 \end{pmatrix}$, $\mathbf{I} = \begin{pmatrix} 1 & 0 \\ 0 & 1 \end{pmatrix}$, $\mathbf{O} = \begin{pmatrix} 0 & 0 \\ 0 & 0 \end{pmatrix}$

Find the values of the constants, a and b, such that $\mathbf{M}^2 + a\mathbf{M} + b\mathbf{I} = \mathbf{O}$. **(3)**

← Sections 6.1, 6.2

P 3 $\mathbf{A} = \begin{pmatrix} a & b \\ c & d \end{pmatrix}$

Find an expression for λ, in terms of a, b, c and d so that $\mathbf{A}^2 - (a + d)\mathbf{A} = \lambda\mathbf{I}$, where $\mathbf{I}$ is the 2×2 identity matrix. **(3)**

← Sections 6.1, 6.2

P 4 A matrix **A** is given as $\mathbf{A} = \begin{pmatrix} 1 & 2 & b \\ 3 & 0 & 1 \\ a & -1 & 2 \end{pmatrix}$

Given that $\mathbf{A}^2 = \begin{pmatrix} 3 & 3 & -1 \\ 7 & 5 & -1 \\ 9 & 6 & -1 \end{pmatrix}$, find the values of a and b. **(3)**

← Section 6.2

E/P 5 $\mathbf{A} = \begin{pmatrix} 2 & 3 \\ p & -1 \end{pmatrix}$, where p is a real constant.

Given that **A** is singular,

a find the value of p. **(2)**

Given instead that det $\mathbf{A} = 4$,

b find the value of p. **(2)**

Using the value of p found in part **b**,

c show that $\mathbf{A}^2 - \mathbf{A} = k\mathbf{I}$, stating the value of the constant k. **(2)**

← Section 6.3

E/P 6 $\mathbf{A} = \begin{pmatrix} 4 & -1 \\ -6 & 2 \end{pmatrix}$, $\mathbf{B}^{-1} = \begin{pmatrix} 2 & 0 \\ 3 & p \end{pmatrix}$

a Find $\mathbf{A}^{-1}$. **(1)**

b Find $(\mathbf{AB})^{-1}$, in terms of p. **(3)**

Given also that $\mathbf{AB} = \begin{pmatrix} -1 & 2 \\ 3 & -4 \end{pmatrix}$,

c find the value of p. **(2)**

← Section 6.4

E/P 7 $\mathbf{A} = \begin{pmatrix} k & 1 & -2 \\ 0 & -1 & k \\ 9 & 1 & 0 \end{pmatrix}$

where k is a real constant.

a Find the values of k for which **A** is singular. **(3)**

Given that **A** is non-singular,

b find $\mathbf{A}^{-1}$ in terms of k. **(4)**

← Section 6.3, 6.5

E/P 8 $\mathbf{A} = \begin{pmatrix} 2p & p & 2 \\ 3 & 0 & 0 \\ -1 & 1 & -1 \end{pmatrix}$ where p is a real constant.

Given that **A** is non-singular, find $\mathbf{A}^{-1}$ in terms of p. **(4)**

← Sections 6.3, 6.5

E/P 9 $\mathbf{A} = \begin{pmatrix} 4p & -q \\ -3p & q \end{pmatrix}$, where p and q are non-zero constants.

a Find $\mathbf{A}^{-1}$, in terms of p and q. (3)

Given that $\mathbf{AX} = \begin{pmatrix} 2p & 3q \\ -p & q \end{pmatrix}$,

b find $\mathbf{X}$, in terms of p and q. (3)

← Section 6.4

E/P 10 Three planes A, B and C are defined by the following equations:

$$\begin{aligned} A:&\quad x + y + z = 3 \\ B:&\quad 2x - y - 2z = 0 \\ C:&\quad 3x - 2y + z = -1 \end{aligned}$$

By constructing and solving a suitable matrix equation, show that these three planes intersect at a single point and find the coordinates of that point. (5)

← Section 6.6

E/P 11 A llama farmer has three types of llama: woolly, classic and Suri. Initially his flock had 2810 llamas in it. There were 160 more woolly llamas than classic.
After one year:

- the number of woolly llamas had increased by 5%
- the number of classic llamas had increased by 3%
- the number of Suri llamas had decreased by 4%
- overall the flock size had increased by 46

Form and solve a matrix equation to find out how many of each type of llama there were in the initial flock. (7)

← Section 6.6

E 12 Three planes have equations given by

$$\begin{aligned} x - 2y - pz &= -2 \\ 2x + py + 5z &= p \\ x + 3y - 2z &= -p \end{aligned}$$

Where p is a real constant.

Given that the planes do not meet at a single point

a Find the possible values of p (3)

b Identify two possible geometric configurations of the planes, stating the corresponding value of p in each case. (5)

← Section 6.6

E 13 The matrix $\mathbf{A}$ represents a reflection in the x-axis.

The matrix $\mathbf{B}$ represents a rotation of 135°, in the anti-clockwise direction, about (0, 0).

Given that $\mathbf{C} = \mathbf{AB}$,

a find the matrix $\mathbf{C}$ (2)

b show that $\mathbf{C}^2 = \mathbf{I}$. (2)

← Sections 7.2, 7.4

E/P 14 The linear transformation T is represented by the matrix $\mathbf{M}$, where

$$\mathbf{M} = \begin{pmatrix} a & b \\ c & d \end{pmatrix}$$

The transformation T maps (1, 0) to (3, 2) and maps (2, 1) to (2, 1).

a Find the values of a, b, c and d. (4)

b Show that $\mathbf{M}^2 = \mathbf{I}$. (2)

The transformation T maps (p, q) to $(8, -3)$.

c Find the value of p and the value of q. (3)

← Sections 7.1, 7.6

E 15 The linear transformation T is defined by

$$T: \begin{pmatrix} x \\ y \end{pmatrix} \mapsto \begin{pmatrix} 2y - x \\ 3y \end{pmatrix}$$

The linear transformation T is represented by the matrix $\mathbf{C}$.

a Find $\mathbf{C}$. (1)

The quadrilateral $OABC$ is mapped by T to the quadrilateral $OA'B'C'$, where the coordinates of A', B' and C' are (0, 3), (10, 15) and (10, 12) respectively.

b Show that the line $y = 2x$ is invariant under this transformation. (3)

c Find the coordinates of A, B and C. (2)

d Sketch the quadrilateral $OABC$ and verify that $OABC$ is a rectangle. (3)

← Sections 7.1, 7.6

E 16 $\mathbf{A} = \begin{pmatrix} 3 & 1 & -1 \\ 1 & 1 & 1 \\ 5 & 3 & u \end{pmatrix}, u \neq 1$

a Show that $\det \mathbf{A} = 2(u - 1)$. **(2)**

b Find the inverse of **A**. **(4)**

The image of the vector $\begin{pmatrix} a \\ b \\ c \end{pmatrix}$ when transformed by the matrix $\begin{pmatrix} 3 & 1 & -1 \\ 1 & 1 & 1 \\ 5 & 3 & 6 \end{pmatrix}$ is $\begin{pmatrix} 3 \\ 1 \\ 6 \end{pmatrix}$.

c Find the values of a, b and c. **(3)**

← Sections 6.3, 6.5, 7.6

E/P 17 The transformation R is represented by the matrix **M**, where

$$\mathbf{M} = \begin{pmatrix} 3 & a & 0 \\ 2 & b & 0 \\ c & 0 & 1 \end{pmatrix}$$

and where a, b and c are constants.

Given that $\mathbf{M} = \mathbf{M}^{-1}$,

a find the values of a, b and c **(5)**

b evaluate the determinant of **M** **(2)**

c find an equation satisfied by all the points which remain invariant under R. **(2)**

← Sections 7.5, 7.6

E/P 18 A triangle T, of area $18\,\text{cm}^2$, is transformed into a triangle T' by the matrix **A**, where

$$\mathbf{A} = \begin{pmatrix} k & k-1 \\ -3 & 2k \end{pmatrix}, k \in \mathbb{R}.$$

a Find $\det \mathbf{A}$ in terms of k. **(3)**

Given that the area of T' is $198\,\text{cm}^2$,

b find the possible values of k. **(3)**

← Sections 6.3, 7.3

E/P 19 The matrix $\mathbf{M} = \begin{pmatrix} \frac{3}{\sqrt{2}} & -\frac{3}{\sqrt{2}} \\ \frac{3}{\sqrt{2}} & \frac{3}{\sqrt{2}} \end{pmatrix}$ represents a rotation followed by an enlargement.

a Find the scale factor of the enlargement. **(2)**

b Find the angle of rotation.

A point P is mapped onto a point P' under **M**. Given that the coordinates of P' are (p, q), **(3)**

c find, in terms of p and q, the coordinates of P. **(4)**

← Sections 7.4, 7.6

E/P 20 The linear transformation T is represented by the matrix $\begin{pmatrix} 0 & 2 \\ \frac{1}{2} & 0 \end{pmatrix}$.

Show that the line $y = k - \frac{1}{2}x$, where k is a real constant, is invariant under T. **(4)**

E/P 21 Use the method of mathematical induction to prove, for $n \in \mathbb{Z}^+$, that

$$\sum_{r=1}^{n} r(r+3) = \frac{1}{3}n(n+1)(n+5) \quad \textbf{(6)}$$

E/P 22 Prove by induction that, for all $n \in \mathbb{Z}^+$,

$$\sum_{r=1}^{n} (2r-1)^2 = \frac{1}{3}n(2n-1)(2n+1) \quad \textbf{(6)}$$

E/P 23 The rth term, a_r, in a series is given by $a_r = r(r+1)(2r+1)$

Prove, by mathematical induction, that the sum of the first n terms of the series is $\frac{1}{2}n(n+1)^2(n+2)$ **(6)**

E/P 24 Prove, by induction, that for all $n \in \mathbb{Z}^+$,

$$\sum_{r=1}^{n} r^2(r-1) = \frac{1}{12}n(n-1)(n+1)(3n+2)$$

(6)

← Section 8.1

E/P 25 Given that $f(n) = 3^{4n} + 2^{4n+2}$,

a show that, for $k \in \mathbb{Z}^+$, $f(k+1) - f(k)$ is divisible by 15, **(3)**

b prove that, for all $n \in \mathbb{Z}^+$, $f(n)$ is divisible by 5. **(4)**

E/P 26 $f(n) = 24 \times 2^{4n} + 3^{4n}$, where n is a non-negative integer.

a Write down $f(n+1) - f(n)$. **(3)**

b Prove, by induction, that $f(n)$ is divisible by 5 for $n \in \mathbb{Z}^+$. **(4)**

E/P **27** Prove that the expression $7^n + 4^n + 1$ is divisible by 6 for all positive integers n. **(6)**

E/P **28** Prove by induction that $4^n + 6n - 1$ is divisible by 9 for all $n \in \mathbb{Z}^+$. **(6)**

E/P **29** Prove that the expression $3^{4n-1} + 2^{4n-1} + 5$ is divisible by 10 for all positive integers n. **(6)**

← Section 8.2

E/P **30** $\mathbf{A} = \begin{pmatrix} 1 & c \\ 0 & 2 \end{pmatrix}$, where c is a constant.

Prove by induction that, for all positive integers n,

$$\mathbf{A}^n = \begin{pmatrix} 1 & (2^n - 1)c \\ 0 & 2^n \end{pmatrix}$$ **(6)**

← Section 8.3

E/P **31** $\mathbf{A} = \begin{pmatrix} 3 & 1 \\ -4 & -1 \end{pmatrix}$

Prove by induction that, for all positive integers n,

$$\mathbf{A}^n = \begin{pmatrix} 2n + 1 & n \\ -4n & -2n + 1 \end{pmatrix}$$ **(6)**

← Section 8.3

E/P **32** Derek is attempting to prove that $2^n + 3$ is divisible by 3 for all positive integers n. He writes the following working:

> Assume true for $n = k$, so $2^k + 3$ is divisible by 3.
> Consider $n = k + 1$:
> $2^{k+1} + 3 = 2 \times 2^k + 3$
> $= 2(2^k + 3) - 3$
> By induction hypothesis $2^k + 3$ is divisible by 3, and 3 is divisible by 3, hence $2^{k+1} + 3$ is divisible by 3.
> Hence by induction $2^n + 3$ is divisible by 3 for all positive integers n.

a Explain the mistake that Derek has made. **(2)**

b Prove that $2^{2n} - 1$ is divisible by 3 for all positive integers n. **(4)**

← Section 8.2

E **33** The line l has equation

$$\mathbf{r} = \begin{pmatrix} 2 \\ -1 \\ 3 \end{pmatrix} + \lambda \begin{pmatrix} -1 \\ -2 \\ 3 \end{pmatrix}$$

A and B are the points on l with $\lambda = 4$ and $\lambda = -1$ respectively.

Find the distance AB. **(4)**

← Section 9.1

E/P **34** The points $P(1, -1, 3)$, $Q(a, 3, 8)$ and $R(5, 7, b)$, where a and b are constants, are collinear. Find the values of a and b and the vector equation of the line through the three points. **(5)**

← Section 9.1

E/P **35** $A(3, -1, 4)$, $B(1, 1, 2)$ and $C(4, -2, 0)$ lie in a plane.

a Find the Cartesian equation of the plane. **(4)**

b Write down a normal vector to the plane. **(1)**

D is a point in the plane with coordinates $(2, k, -2)$.

c Find the value of k. **(2)**

← Section 9.2

E **36** The line l_1 has vector equation

$$\mathbf{r} = 11\mathbf{i} + 5\mathbf{j} + 6\mathbf{k} + \lambda(4\mathbf{i} + 2\mathbf{j} + 4\mathbf{k})$$

and the line l_2 has vector equation

$$\mathbf{r} = 24\mathbf{i} + 4\mathbf{j} + 13\mathbf{k} + \mu(7\mathbf{i} + \mathbf{j} + 5\mathbf{k})$$

where λ and μ are parameters.

a Show that the lines l_1 and l_2 intersect. **(3)**

b Find the coordinates of their point of intersection. **(2)**

Given that θ is the acute angle between l_1 and l_2,

c find the value of $\cos\theta$. Give your answer in the form $k\sqrt{3}$, where k is a simplified fraction. **(3)**

← Sections 9.3, 9.4, 9.5

37 The line l_1 has vector equation
$\mathbf{r} = 8\mathbf{i} + 12\mathbf{j} + 14\mathbf{k} + \lambda(\mathbf{i} + \mathbf{j} - \mathbf{k})$
The points A, with coordinates $(4, 8, a)$ and B, with coordinates $(b, 13, 13)$, lie on this line.

a Find the values of a and b. **(2)**

Given that the point O is the origin, and that the point P lies on l_1 such that OP is perpendicular to l_1,

b find the coordinates of P. **(3)**

c Hence find the distance OP, giving your answer as a simplified surd. **(3)**

← Sections 9.3, 9.4, 9.5

38 Referred to a fixed origin O, the point A has position vector $a(4\mathbf{i} + \mathbf{j} + 2\mathbf{k})$ and the plane Π has equation

$\mathbf{r}.(\mathbf{i} - 5\mathbf{j} + 3\mathbf{k}) = 5a$

where a is a scalar constant.

a Show that A lies in the plane Π. **(2)**

The point B has position vector

$a(2\mathbf{i} + 11\mathbf{j} - 4\mathbf{k})$

b Show that $\overrightarrow{BA}$ is perpendicular to the plane Π. **(3)**

c Calculate $\angle OBA$ to the nearest one tenth of a degree. **(3)**

← Sections 9.3, 9.4

39 The plane Π has equation $\mathbf{r}.\begin{pmatrix}-1\\2\\1\end{pmatrix} = k$ where k is a constant.

Given the point with position vector $\begin{pmatrix}3\\-4\\1\end{pmatrix}$ lies on Π,

a find the value of k **(3)**

b find a Cartesian equation for Π. **(2)**

The point P has coordinates $(4, -3, 2)$. The line l passes through P and is perpendicular to Π. The line l intersects Π at the point N.

c Find the coordinates of N. **(4)**

← Sections 9.2, 9.5

(E) **40** The line l has Cartesian equation $\frac{x-1}{2} = \frac{y+3}{1} = \frac{2-z}{3}$. The plane Π has Cartesian equation $2x + y - z = 5$.
l intersects the Π at the point P.

a Find the position vector of P. **(5)**

b Find the acute angle between the line and the plane at the point of intersection. Give your answer in radians correct to three decimal places. **(5)**

← Sections 9.4, 9.5

(E) **41** The points A and B have position vectors $\mathbf{i} - \mathbf{j} + 3\mathbf{k}$ and $4\mathbf{i} + 3\mathbf{j} - 2\mathbf{k}$ respectively.

a Find $|\overrightarrow{AB}|$. **(2)**

b Find a vector equation for the line l_1 which passes through the points A and B. **(2)**

A second line l_2 has vector equation
$\mathbf{r} = 6\mathbf{i} + 4\mathbf{j} - 3\mathbf{k} + \mu(2\mathbf{i} + \mathbf{j} - \mathbf{k})$

c Show that the line l_2 also passes through B. **(2)**

d Find the size of the acute angle between l_1 and l_2. **(3)**

e Hence, or otherwise, find the shortest distance from A to l_2. **(3)**

← Sections 9.4, 9.5, 9.6

(E/P) **42** The plane Π has equation $-3x + 2y + 3z = -7$ and the point P has coordinates $(-1, 2, -3)$.

a Find, correct to three decimal places, the shortest distance between P and Π. **(3)**

The point Q is the reflection of the point P in Π.

b Find the exact coordinates of Q. **(5)**

← Section 9.6

E/P **43** Two fish are modelled as travelling in straight lines. A shark, swims from a point with position vector $\begin{pmatrix} 2 \\ 3 \\ -1 \end{pmatrix}$ to a point with position vector $\begin{pmatrix} -2 \\ 11 \\ 11 \end{pmatrix}$, both relative to a fixed origin O and with units given in metres.

A flounder, starts at a point with position vector $\begin{pmatrix} 2 \\ 0 \\ 1 \end{pmatrix}$ relative to O, and travels in the direction of $\begin{pmatrix} -2 \\ -1 \\ 3 \end{pmatrix}$.

a Show that, no matter how fast either fish swims, the shark will never catch the flounder. **(7)**

b Give one criticism of this model. **(1)**

← Section 9.5

Challenge

1 Find the 3 × 3 matrix representing the single transformation that is equivalent to a reflection in the plane $x = 0$, followed by a rotation of 270° about the y-axis, followed by a reflection in the plane $y = 0$. ← Section 7.5

2 The points A, B and C have coordinates $(-2, -3, 0)$, $(-1, -1, 3)$ and $(1, 1, 1)$ respectively. Find the centre and radius of the circle that passes through all three points. ← Chapter 9

Hint For question **2**, consider angle ABC.

3 The diagram below shows two different ways in which four non-parallel lines can divide the plane into regions:

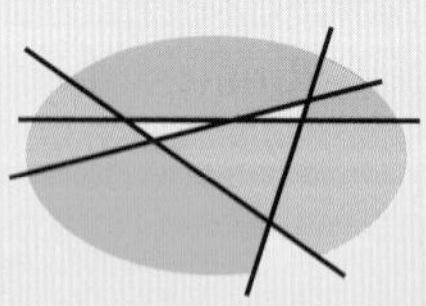

11 regions

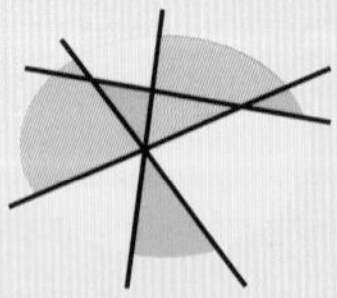

10 regions

Prove that if n non-parallel lines divide the plane into r regions, then

$2n \leqslant r \leqslant \frac{1}{2}(n^2 + n + 2)$ ← Chapter 8

Exam-style practice

Further Mathematics
AS Level
Paper 1: Core Pure Mathematics

Time: 1 hour 40 minutes
You must have: Mathematical Formulae and Statistical Tables, Calculator

---

1 With respect to a fixed origin O, the lines l_1 and l_2 are given by the equations:

l_1: $\mathbf{r} = (-3\mathbf{i} + 5\mathbf{k}) + \lambda(5\mathbf{i} - \mathbf{j} + \mathbf{k})$

l_2: $\mathbf{r} = (10\mathbf{i} - \mathbf{j} + 15\mathbf{k}) + \mu(6\mathbf{i} - 2\mathbf{j} + 4\mathbf{k})$

Show that lines l_1 and l_2 do not meet. **(4)**

2 $\mathbf{M} = \begin{pmatrix} 2 & k & 3 \\ 1 & -3 & 1 \\ 3 & -1 & 2 \end{pmatrix}$ where k is an integer.

a Find det **M**, giving your answer in terms of k. **(3)**

Three planes A, B and C are defined by the following Cartesian equations:

A: $2x + ky + 3z = 1$

B: $x - 3y + z = -2$

C: $3x - y + 2z = 3$

b Given that the planes do not meet at a single point, determine whether the three equations form a consistent system, and give a geometric interpretation of your answer. You must show sufficient working to justify your conclusions. **(5)**

3 The cubic equation

$2x^3 - 3x^2 - 7x - 1 = 0$

has roots α, β and γ.

Without solving the equation, find the cubic equation whose roots are $(2\alpha - 1)$, $(2\beta - 1)$ and $(2\gamma - 1)$, giving your answer in the form $pw^3 + qw^2 + rw + s = 0$ where p, q, r and s are integers to be found. **(6)**

4 **a** Prove by induction that for all positive integers n,

$$\sum_{r=1}^{n} r^3 = \frac{1}{4}n^2(n+1)^2$$ **(6)**

b Use the standard results for $\sum_{r=1}^{n} r^2$ and $\sum_{r=1}^{n} r$ to show that for all positive integers n,

$$\sum_{r=1}^{n} 2r(r+1) = \frac{2}{3}n(n+1)(n+2)$$ **(4)**

c Hence show that $n = 1$ is the only value of n that satisfies

$$\sum_{r=1}^{n} 2r(r+1) = 4\sum_{r=1}^{n} r^3$$ **(5)**

5 $f(z) = z^4 - 14z^3 - 78z^2 + kz + 221$, where k is a real constant.
Given that $z = 3 - 2i$ is a root of the equation $f(z) = 0$,

a show that $z^2 - 6z + 13$ is a factor of $f(z)$ **(4)**

b find the value of k **(1)**

c solve completely the equation $f(z) = 0$ and show the roots on an Argand diagram. **(4)**

6 The plane Π has Cartesian equation $x - y + 2z = 3$.

a Find a unit vector $\hat{\mathbf{n}}$ normal to Π. **(2)**

A line l has vector equation $\mathbf{r} = \begin{pmatrix} 0 \\ 3 \\ -1 \end{pmatrix} + \lambda \begin{pmatrix} 2 \\ 4 \\ -3 \end{pmatrix}$.

The line intersects Π at point P.

b Find the coordinates of P and the acute angle between l and Π, giving your answer in radians correct to two decimal places. **(7)**

7 Frances makes silver jewellery beads.
Each bead is formed by rotating the curve shown through 2π radians about the x-axis.
Each unit on the axes represents 1 mm.

Silver costs £0.05 per cubic millimetre.

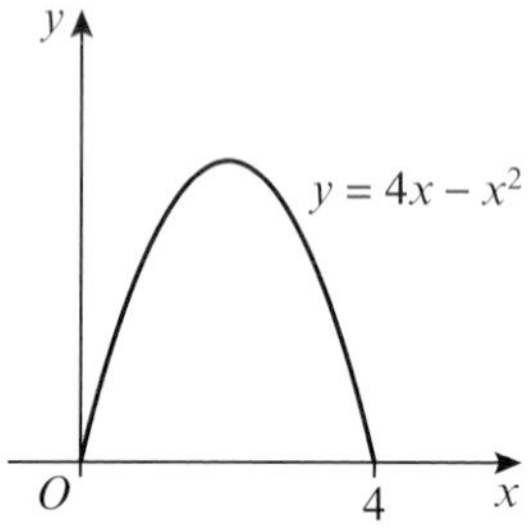

a Find the cost of the silver needed to make 500 beads. **(6)**

b State one limitation of this model. **(1)**

8 The matrix $\mathbf{M} = \begin{pmatrix} -\frac{5}{\sqrt{2}} & -\frac{5}{\sqrt{2}} \\ \frac{5}{\sqrt{2}} & -\frac{5}{\sqrt{2}} \end{pmatrix}$ represents an enlargement followed by a rotation.

a Find the scale factor of the enlargement. **(2)**

b Find the angle of rotation. **(2)**

A point P is mapped onto a point P' under $\mathbf{M}$. Given that the coordinates of P' are (a, b),

c find, in terms of a and b, the coordinates of P. **(4)**

9 Shade on an Argand diagram the set of points

$\{z \in \mathbb{C} : |z - 2i| \leqslant 3\} \cap \left\{z \in \mathbb{C} : \frac{\pi}{2} < \arg(z - 2 + i) \leqslant \frac{3\pi}{4}\right\}$ **(6)**

10 A stolen car is modelled as travelling in a straight line from a point $A(5, 3, -1)$ to a point $B(7, -5, 8)$. The police are waiting for the car to pass at a point modelled as the origin, O. The police have a 'stinger' that they can deploy to stop the car. The 'stinger' is 6 metres in length.

a Given that each unit in the model represents one metre, determine whether or not the police can successfully stop the car. **(7)**

b State one limitation of this model. **(1)**

Answers

CHAPTER 1

Prior knowledge check

1 a $5\sqrt{2}$ b $6\sqrt{3}$ c $6\sqrt{5}$
2 a 0 b 2 c 1
3 $4 \pm \sqrt{10}$
4 $\frac{28}{13} + \frac{7\sqrt{3}}{13}$

Exercise 1A

1 a 3i b 7i c 11i d 100i
e 15i f $i\sqrt{5}$ g $2i\sqrt{3}$ h $3i\sqrt{5}$
i $10i\sqrt{2}$ j $7i\sqrt{3}$
2 a 13 + 11i b 5 + 2i c 4 + i d 3 + 2i
e 9 + 9i f 7 − 4i g 4 + 2i h $2\sqrt{2} + 2i$
i 11 − 5i j 0
3 a 14 + 4i b 24 − 12i c 12 + 5i d 24 + 7i
e 3 − 2i f 3 + 5i g $3 + \frac{11}{3}i$ h $-\frac{11}{2} + \frac{7}{4}i$
4 a $2\sqrt{2} - i\sqrt{2}$ b $(-1 + \sqrt{3}) + (3 - 3\sqrt{3})i$
5 a −12i b 14
6 $a = -10, b = 11$
7 a −3 + 4i b 28 − 12i c 43 − 13i
8 a $z + w = (a + bi) + (a - bi) = 2a$
b $z - w = (a + bi) - (a - bi) = 2bi$

Exercise 1B

1 a $z = \pm 11i$ b $z = \pm 2i\sqrt{10}$ c $z = \pm 2i\sqrt{15}$
d $z = \pm 2i\sqrt{7}$ e $z = \pm 2i\sqrt{6}$ f $z = \pm\frac{1}{2}i$
2 a $z = 3 \pm i\sqrt{7}$ b $z = 7 \pm 2i\sqrt{3}$ c $z = -1 \pm \frac{3}{4}i$
3 a $z = -1 \pm 2i$ b $z = 1 \pm 3i$ c $z = -2 \pm 5i$
d $z = -5 \pm i$ e $z = -\frac{5}{2} \pm \frac{5i\sqrt{3}}{2}$ f $z = \frac{-3 \pm i\sqrt{11}}{2}$
4 a $z = -\frac{5}{4} \pm \frac{\sqrt{7}}{4}i$ b $z = \frac{3}{14} \pm \frac{5\sqrt{3}}{14}i$ c $z = \frac{1}{10} \pm \frac{\sqrt{59}}{10}i$
5 $z_1 = 4 + \sqrt{5}i$ and $z_2 = 4 - \sqrt{5}i$
6 $-\sqrt{44} < b < \sqrt{44}$ or $-2\sqrt{11} < b < 2\sqrt{11}$

Exercise 1C

1 a 11 + 23i b 36 + 33i c 15 + 23i d 2 − 110i
e −5 − 25i f 39 + 80i g −77 − 36i h 10i
i 54 − 62i j −46 + 9i
2 a 41 b 53 c They are both real
d $(a + bi)(a - bi) = a^2 + b^2$, which is real.
3 $a = 7, b = -6$ or $a = 18, b = -\frac{7}{3}$
4 a −1 b 81 c 2i d −60i
5 −8i, $a = 0, b = -8$
6 −119 − 120i, so real part is −119
7 a −2i b −49 − 66i
8 Substitute $z = 1 - 4i$ into f(z) to get f(z) = 0.
9 a $i^3 = -i, i^4 = 1$ b $i^5 = i, i^6 = -1, i^7 = -i, i^8 = 1$
c i 1 ii i iii i

Challenge

a $(a + bi)^2 = a^2 - b^2 + 2abi$
b $a^2 - b^2 = 40, 2ab = -42 \Rightarrow a = \frac{-21}{b}$
$b^4 + 40b^2 - 441 = 0 \Rightarrow (b^2 - 9)(b^2 + 49) = 0$
$b = -3\ (b < 0) \Rightarrow a = 7, b^2 \neq -49 \Rightarrow 7 - 3i$

Exercise 1D

1 a 8 − 2i b 6 + 5i c $\frac{2}{3} + \frac{1}{2}i$ d $\sqrt{5} - i\sqrt{10}$
2 a $z + z^* = 12, zz^* = 45$ b $z + z^* = 20, zz^* = 125$
c $z + z^* = \frac{3}{2}, zz^* = \frac{5}{8}$ d $z + z^* = 2\sqrt{5}, zz^* = 50$
3 a $-\frac{6}{5} - \frac{7}{5}i$ b $-\frac{11}{50} + \frac{27}{50}i$ c $\frac{31}{2} + \frac{25}{2}i$ d $\frac{6}{17} - \frac{7}{17}i$
4 $-\frac{31}{2} - \frac{17}{2}i$
5 a $\frac{3}{5} + \frac{4}{5}i$ b $\frac{7}{2} + \frac{1}{2}i$ c $\frac{41}{5} - \frac{3}{5}i$
6 $\frac{8}{5} + \frac{9}{5}i$
7 6 + 8i
8 $\frac{4}{8 - \sqrt{2}i} \times \frac{8 + \sqrt{2}i}{8 + \sqrt{2}i} = \frac{32 + 4i\sqrt{2}}{66} = \frac{16}{33} + \frac{2\sqrt{2}}{33}i$
9 $\frac{1}{1 - 9i} \times \frac{1 + 9i}{1 + 9i} = \frac{1}{82} + \frac{9}{82}i$
10 $\frac{z + 4}{z - 3} = \frac{(8 - i\sqrt{2})(1 + i\sqrt{2})}{(1 - i\sqrt{2})(1 + i\sqrt{2})} = \frac{10}{3} + \frac{7\sqrt{2}}{3}i$
11 $z - 2i = \frac{(6 - 4i)(4 - 2i)}{(4 + 2i)(4 - 2i)} = \frac{16 - 28i}{20}$
$z - 2i = \frac{4}{5} - \frac{7}{5}i \Rightarrow z = \frac{4}{5} + \frac{3}{5}i$
12 $\frac{(p - 7i)(2 - 5i)}{(2 + 5i)(2 - 5i)} = \frac{2p - 35}{29} + \frac{-5p - 14}{29}i$
13 $\frac{z}{z^*} = \frac{\sqrt{5} + 4i}{\sqrt{5} - 4i} \times \frac{\sqrt{5} + 4i}{\sqrt{5} + 4i} = -\frac{11}{21} + \frac{8\sqrt{5}}{21}i$
14 a $\frac{p + 5i}{p - 2i} \times \frac{p + 2i}{p + 2i} = \frac{p^2 - 10 + 7pi}{p^2 + 4}$
$2p^2 - 20 = p^2 + 4$
$p^2 = 24 \Rightarrow p = 2\sqrt{6}$
b $\frac{1}{2} + \frac{\sqrt{6}}{2}i$

Exercise 1E

1 a −1 + 5i, −1 − 5i b −2 c 26
2 a 4 + 3i, 4 − 3i b 8 c 25
3 a 2 − 3i b $z^2 - 4z + 13 = 0$
4 a 5 + i
b $(z - (5 - i))(z - (5 + i)) = 0$
$z^2 - (5 + i)z - (5 - i)z + (5 - i)(5 + i) = 0$
$z^2 - 10z + 26 = 0 \Rightarrow p = -10, q = 26$
5 $z^2 + 10z + 41 = 0$
6 $z^2 - 2z + 5 = 0$
7 $z^2 - 6z + 34 = 0$
8 a $z = \frac{3}{2} + \frac{1}{2}i$
b $\left(z - \left(\frac{3}{2} + \frac{1}{2}i\right)\right)\left(z - \left(\frac{3}{2} - \frac{1}{2}i\right)\right)$
$= z^2 - z\left(\frac{3}{2} - \frac{1}{2}i\right) - z\left(\frac{3}{2} + \frac{1}{2}i\right) + \left(\frac{3}{2} - \frac{1}{2}i\right)\left(\frac{3}{2} + \frac{1}{2}i\right)$
$= z^2 - 3z + \frac{5}{2}$ so $p = -3, q = \frac{5}{2}$
9 $(z - (5 + qi))(z - (5 - qi)) = z^2 - 10z + 25 + q^2$
$\Rightarrow 4p = 10 \Rightarrow p = \frac{5}{2} \Rightarrow 25 + q^2 = 34 \Rightarrow q = 3$

Exercise 1F

1 a f(2) = 8 − 24 + 42 − 26 = 0
b $z = 2, z = 2 + 3i$ or $z = 2 - 3i$
2 a Substitute $z = \frac{1}{2}$ into f(z).
b $b = 3, c = 6$
c $z = \frac{1}{2}$, or $z = -\frac{3}{2} \pm \frac{\sqrt{15}}{2}i$
3 3, $-\frac{1}{2} + \frac{1}{2}i$ and $-\frac{1}{2} - \frac{1}{2}i$
4 a $(z - (-4 + i))(z - (-4 - i)) = z^2 + 8z + 16 + 1$
$= z^2 + 8z + 17$
b $z = 4, z = -4 + i$ or $z = -4 - i$
5 a $a = 8, b = 25$ b −1, −4 + 3i, −4 − 3i c −9
6 a 3 − i b $c = 46, d = -60$
7 $-\frac{1}{2}, -\frac{1}{2} + \frac{\sqrt{3}}{2}i$ and $-\frac{1}{2} - \frac{\sqrt{3}}{2}i$
8 a $k = -40$ b 2 − 4i, 2 + 4i
9 2, −2, 2i and − 2i

10 a $(z^2 - 9)(z^2 - 12z + 40)$ **b** $z = \pm 3, 6 \pm 2i$

11 $-3 + i, -3 - i, 2 + 3i$ and $2 - 3i$
$(z - (2 - 3i))(z - (2 + 3i)) = z^2 - 4z + 13$

12 a $(z^2 - 4z + 13)(z^2 + bz + c)$
$= z^4 - 10z^3 + 71z^2 + Qz + 442$
$b = -6, c = 34$
b $Q = -214$ **c** $z = 2 + 3i, 2 - 3i, 3 + 5i$ or $3 - 5i$

Challenge

$b = 0, c = 2, d = 4, e = -8, f = 16$

Mixed exercise

1 a $6 + i$ **b** $-6 + 12i$ **c** $50 - 22i$

2 $-2\sqrt{14} < b < 2\sqrt{14}$

3 $3 + i\sqrt{3}, 3 - i\sqrt{3}$

4 $(1 + 2i)^5$
$= 1^5 + 5(1)^4(2i) + 10(1)^3(2i)^2 + 10(1)^2(2i)^3 + 5(1)(2i)^4 + (2i)^5$
$= 1 + 10i + 40i^2 + 80i^3 + 80i^4 + 32i^5$
$= 1 + 10i - 40 - 80i + 80 + 32i$
$= 41 - 38i$

5 Substitute $z = 3 + i$ into f(z) to get f(z) = 0.

6 a $4 - 2i$ **b** $-14 - 2i$ **c** $-1 - i$

7 $\dfrac{(45 - 28i)(1 - i\sqrt{3})}{(1 + \sqrt{3}i)(1 - \sqrt{3}i)} = \dfrac{45 - 28\sqrt{3}}{4} + \left(\dfrac{-45\sqrt{3} - 28}{4}\right)i$

8 $\dfrac{4 - 7i}{3 + i} = \dfrac{(4 - 7i)(3 - i)}{(3 + i)(3 - i)} = \dfrac{12 - 25i + 7i^2}{10} = \frac{1}{2} - \frac{5}{2}i$

9 a $\frac{3}{25} - \frac{4}{25}i$ **b** $\frac{-8}{5} - \frac{6}{5}i$

10 $\dfrac{z}{z^*} = \dfrac{(a + bi)(a + bi)}{(a - bi)(a + bi)} = \dfrac{a^2 + 2abi + b^2i^2}{a^2 - b^2i^2}$
$= \dfrac{a^2 - b^2}{a^2 + b^2} + \left(\dfrac{2ab}{a^2 + b^2}\right)i$

11 a $\dfrac{3 + qi}{q - 5i} \times \dfrac{q + 5i}{q + 5i} = \dfrac{3q - 5q}{q^2 + 25} + \dfrac{q^2 + 15}{q^2 + 25}i$
$\dfrac{-2q}{q^2 + 25} = \frac{1}{13} \Rightarrow q^2 + 26q + 25 = 0 \Rightarrow q = -1, q = -25$
b $\frac{1}{13} + \frac{8}{13}i, \frac{1}{13} + \frac{64}{65}i$

12 $x + yi + 4i(x - yi) = -3 + 18i$
$(x + 4y) + (4x + y)i = -3 + 18i$
$x + 4y = -3, 4x + y = 18 \Rightarrow x = 5, y = -2$

13 $\dfrac{(9 + 6i)(2 + 3i)}{(2 - 3i)(2 + 3i)} = \dfrac{18 + 39i + 18i^2}{4 - 9i^2} = 3i$

14 $\dfrac{(q + 3i)(4 - qi)}{(4 + qi)(4 - qi)} = \dfrac{7q}{q^2 + 16} + \dfrac{12 - q^2}{q^2 + 16}i$

15 a $6 + 2i$ **b** $z^2 - 12z + 40$

16 $k = 6, m = 4$

17 $z = 2 + i, 2 - i$ or -4

18 $z = -2, 1 + 3i$ or $1 - 3i$

19 a $k = 16$ **b** $-4i$ and -3

20 a $b = 4, c = 10$ **b** $z = 6, -1, -2 + \sqrt{6}i$ or $-2 - \sqrt{6}i$

21 $3 - 2i, 3 + 2i, i\sqrt{6}$ and $-i\sqrt{6}$

22 a $p = -18$ **b** $1, 4, -\frac{3}{2} + \dfrac{\sqrt{15}}{2}i$ and $-\frac{3}{2} - \dfrac{\sqrt{15}}{2}i$

Challenge

a If a root is not real, the other root must be its complex conjugate, but only real numbers are equal to their conjugate.

b $(z + i)^2(z - i)^2 = z^4 + 2z^2 + 1$

CHAPTER 2

Prior knowledge 2

1 $(x + 3)^2 + (y - 6)^2 = 25$

2 a $6 - 3i$ **b** $21 + 3i$ **c** $\frac{3}{2} + \frac{3}{2}i$

3 a 13 cm **b** 67.4°

4 $z = 4 \pm 2\sqrt{2}i$

Exercise 2A

1

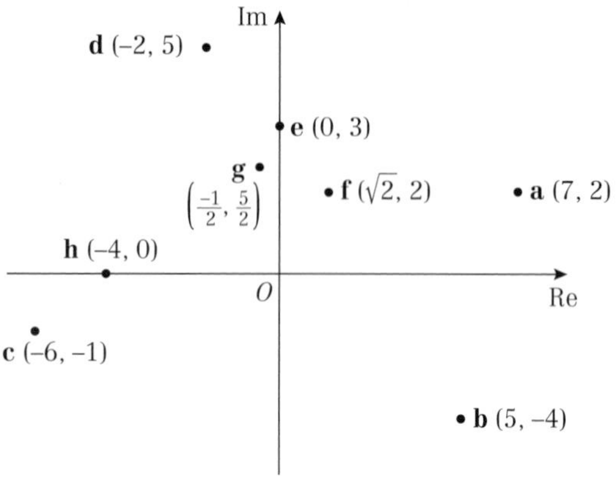

2

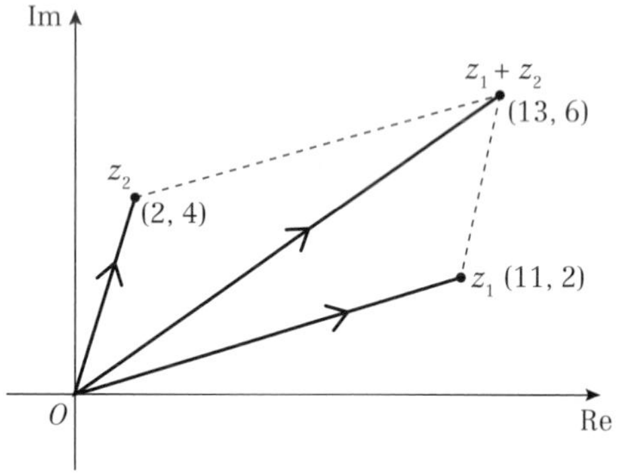

3

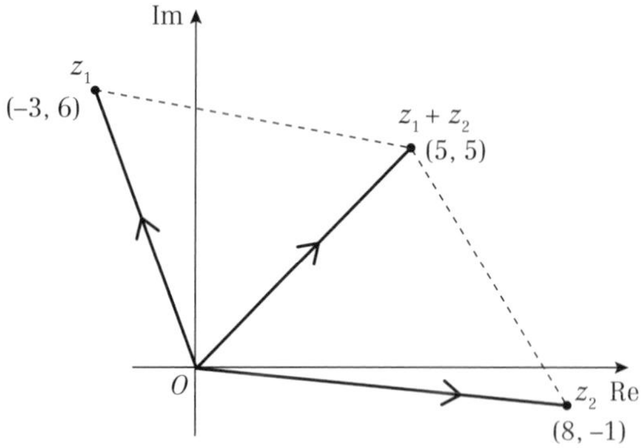

4

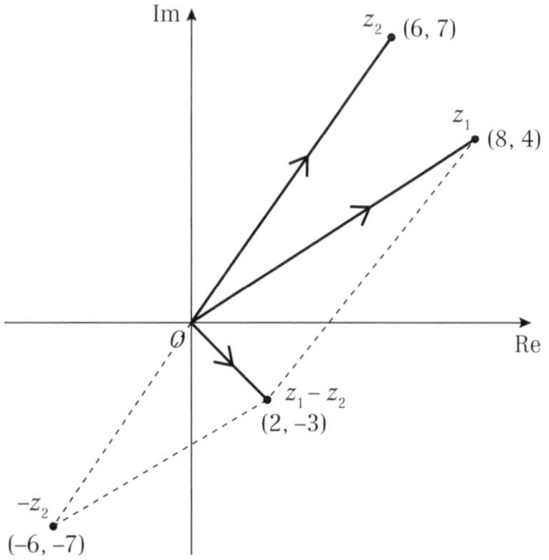

Online Full worked solutions are available in SolutionBank.

5

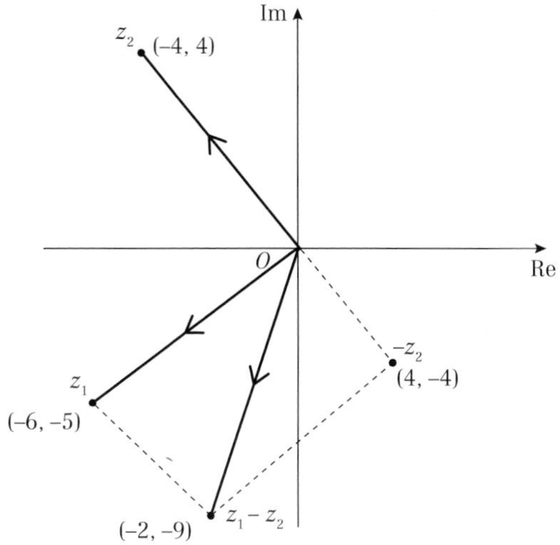

6 **a** $a = -10, b = 7$

b

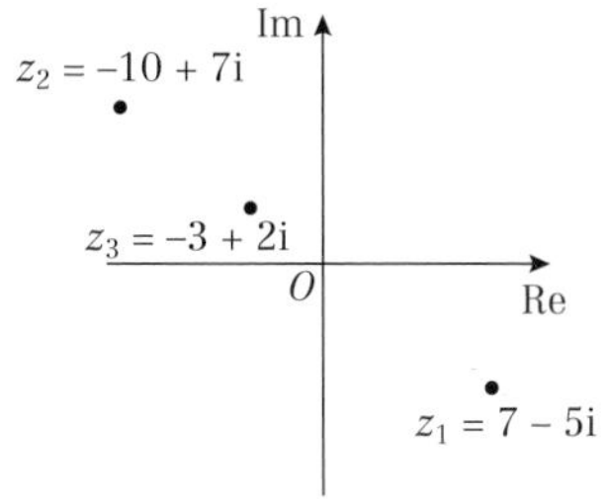

7 **a** $p = -17, q = 10$

b

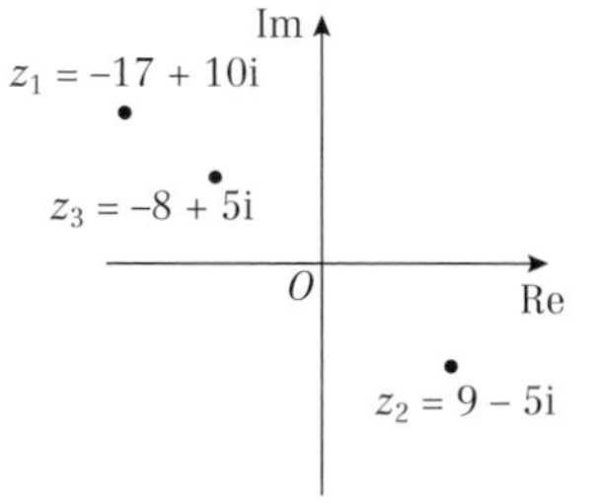

8 **a** $z_1 = 3 + i$ and $z_2 = 3 - i$. Other way round acceptable.

b

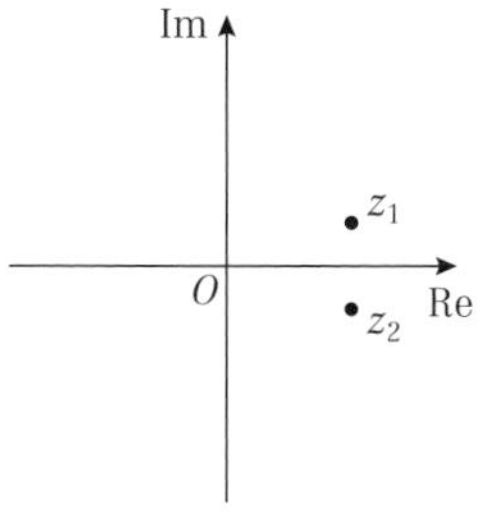

9 **a** $2\left(\frac{3}{2}\right)^3 - 19\left(\frac{3}{2}\right)^2 + 64\left(\frac{3}{2}\right) - 60 = 0$

b $\left(\frac{3}{2}\right)$, 4 + 2i, 4 – 2i.

c

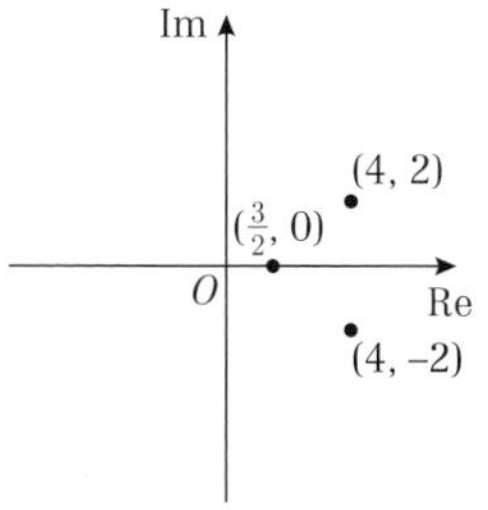

Challenge

a $1, -1, -\frac{1}{2} + \frac{\sqrt{3}}{2}i, -\frac{1}{2} - \frac{\sqrt{3}}{2}i, \frac{1}{2} + \frac{\sqrt{3}}{2}i$ and $\frac{1}{2} - \frac{\sqrt{3}}{2}i$

b

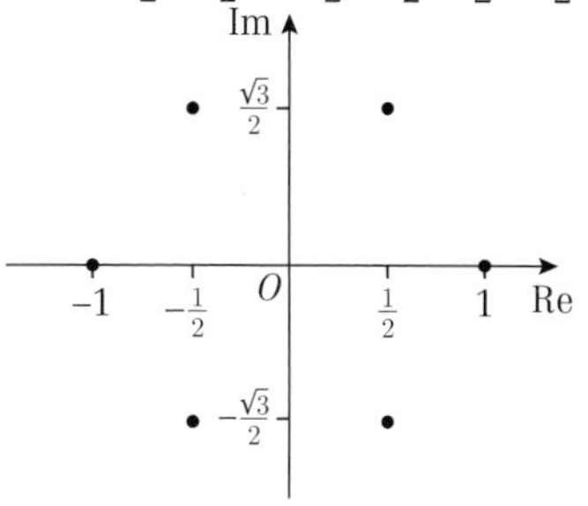

c (0, 1) and (0, –1) are on the unit circle.
$\left(-\frac{1}{2}\right)^2 + \left(\frac{\sqrt{3}}{2}\right)^2 = 1$, so other 4 points also lie on the unit circle.

Exercise 2B

1 **a** Modulus = 13, argument = 0.39
b Modulus = 2, argument = $\frac{\pi}{6}$
c Modulus = $3\sqrt{5}$, argument = 2.03
d Modulus = $2\sqrt{2}$, argument = $-\frac{\pi}{4}$
e Modulus = $\sqrt{113}$, argument = –2.42
f Modulus = $\sqrt{137}$, argument = 1.92
g Modulus = $\sqrt{15}$, argument = –0.46
h Modulus = 17, argument = –2.06

2 **a** **i** $\sqrt{8} = 2\sqrt{2}$ **ii** $\frac{\pi}{4}$
b **i** $\sqrt{50} = 5\sqrt{2}$ **ii** $\frac{\pi}{4}$
c **i** $\sqrt{72} = 6\sqrt{2}$ **ii** $\frac{3\pi}{4}$
d **i** $\sqrt{2a^2} = a\sqrt{2}$ **ii** $-\frac{3\pi}{4}$

3 **a**

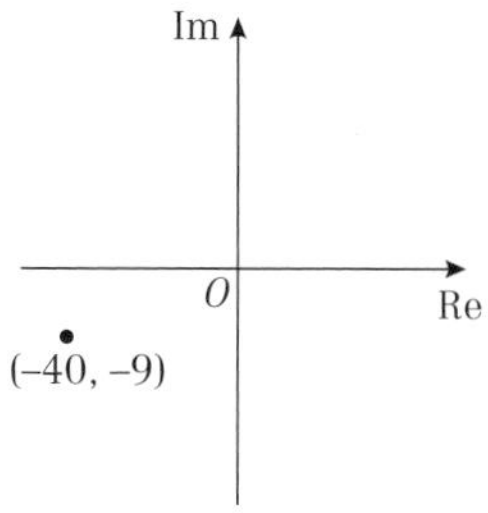

b –2.92

4 **a** $z^2 = (3 + 4i)(3 + 4i) = 9 + 16i^2 + 24i = -7 + 24i$
b 25 **c** 1.85
d

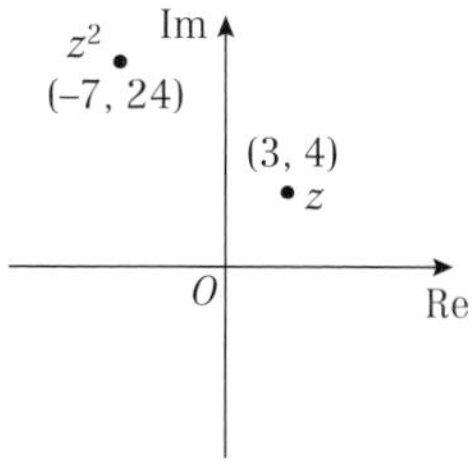

5 **a** $\frac{z_1}{z_2} = \frac{(4 + 6i)(1 - i)}{(1 + i)(1 - i)} = \frac{10 + 2i}{2} = 5 + i$
b $\sqrt{26}$ **c** 0.20

6 **a** $\left(\frac{3 - 2p}{2}\right) + \left(\frac{3 + 2p}{2}\right)i$ **b** $p = 1$
c $\frac{\sqrt{26}}{2}$
d Argand diagram showing $z_1 = 3 + i$, $z_2 = \frac{1}{2} + \frac{5}{2}i$ and $\frac{z_1}{z_2} = 1 - i$

7 a $4 + 6i$ b $-20 + 48i$
c $\sqrt{52} = 2\sqrt{13}$ d 1.97

8 a $|6 + 6i| = \sqrt{72} = 6\sqrt{2}$
b $\frac{(4 + 2i)(a + bi)(2 - 4i)}{(2 + 4i)(2 - 4i)} = \frac{(16 - 12i)(a + bi)}{20}$
$= \left(\frac{4a + 3b}{5}\right) + \left(\frac{4b - 3a}{5}\right)i$
c $a = 6, b = -1$ d -0.81

9 a $\sqrt{45} = 3\sqrt{5}$ b 0.46 c $\lambda = 2$

10 a $|-1 - i\sqrt{3}| = \sqrt{(-1)^2 + (-\sqrt{3})^2} = \sqrt{4} = 2$
b $\frac{z}{z^*} = \frac{(-1 - i\sqrt{3})(-1 - i\sqrt{3})}{(-1 + i\sqrt{3})(-1 - i\sqrt{3})} = -\frac{1}{2} + \frac{\sqrt{3}}{2}i$
$\left|\frac{z}{z^*}\right| = \left|-\frac{1}{2} + \frac{\sqrt{3}}{2}i\right| = \sqrt{\left(-\frac{1}{2}\right)^2 + \left(\frac{\sqrt{3}}{2}\right)^2} = 1$
c $\arg z = -\frac{2\pi}{3}$, $\arg z^* = \frac{2\pi}{3}$ and $\arg\frac{z}{z^*} = \frac{2\pi}{3}$,

11 $k = \frac{4\sqrt{3} - 1}{5 + \sqrt{3}} = \frac{-17 + 21\sqrt{3}}{22}$

12 Using sine rule, $|z| = \frac{5\sin\left(\frac{\pi}{10}\right)}{\sin\left(\frac{\pi}{5}\right)} \approx 2.63$

Exercise 2C

1 a $2\sqrt{2}\left(\cos\frac{\pi}{4} + i\sin\frac{\pi}{4}\right)$ b $3\left(\cos\frac{\pi}{2} + i\sin\frac{\pi}{2}\right)$
c $5(\cos 2.21 + i\sin 2.21)$ d $2\left(\cos -\frac{\pi}{3} + i\sin -\frac{\pi}{3}\right)$
e $\sqrt{29}(\cos(-1.95) + i\sin(-1.95))$
f $20(\cos\pi + i\sin\pi)$
g $25(\cos(-1.29) + i\sin(-1.29))$
h $5\sqrt{2}\left(\cos\frac{3\pi}{4} + i\sin\frac{3\pi}{4}\right)$

2 a $\frac{3}{2}\left(\cos\left(-\frac{\pi}{3}\right) + i\sin\left(-\frac{\pi}{3}\right)\right)$ b $\frac{\sqrt{5}}{5}(\cos 0.46 + i\sin 0.46)$
c $1\left(\cos\frac{\pi}{2} + i\sin\frac{\pi}{2}\right)$

3 a $5i$ b $\frac{\sqrt{3}}{4} + \frac{1}{4}i$ c $-3\sqrt{3} + 3i$
d $-\frac{3}{2} - \frac{3\sqrt{3}}{2}i$ e $2 - 2i$ f $2\sqrt{3} + 2i$

4 a $-2 + 2i\sqrt{3}$
b $-2 + 2i\sqrt{3}$ shown on an Argand diagram.

5 $p = \frac{7\sqrt{3}}{2}$, $q = -\frac{7}{2}$

6 $a = -\frac{5}{2}$, $b = \frac{5\sqrt{3}}{2}$

Exercise 2D

1 a i $|z_1 z_2| = 30$ ii $\arg(z_1 z_2) = \frac{5\pi}{4}$
iii $30\left(\cos\frac{5\pi}{4} + i\sin\frac{5\pi}{4}\right)$
b i $|z_1 z_2| = 8$ ii $\arg(z_1 z_2) = \frac{13\pi}{12}$
iii $8\left(\cos\frac{13\pi}{12} + i\sin\frac{13\pi}{12}\right)$

2 a $|z_1 z_2| = 32$, $\arg(z_1 z_2) = \frac{4\pi}{15}$
b $\left|\frac{z_1}{z_2}\right| = 2$, $\arg\left(\frac{z_1}{z_2}\right) = \frac{14\pi}{15}$
c $|z_1^2| = 64$, $\arg(z_1^2) = -\frac{4\pi}{5}$

3 a $\cos 5\theta + i\sin 5\theta$ b -1 c $-\frac{3}{4}i$
d $3\sqrt{2}$ e $-\sqrt{3} - i$ f $-5\sqrt{3} + 5i$
g $\cos 3\theta + i\sin 3\theta$ h $3 - 3i$

4 a $\cos 3\theta + i\sin 3\theta$ b $2 + 2i$ c $-\frac{11}{4}i$
d $\cos(-5\theta) + i\sin(-5\theta)$ or $\cos 5\theta - i\sin 5\theta$

5 a $z = 6\sqrt{3}\left(\cos\frac{5\pi}{6} + i\sin\frac{5\pi}{6}\right)$
b i $w = \sqrt{3}\left(\cos\frac{7\pi}{12} + i\sin\frac{7\pi}{12}\right)$
ii $18\cos\left(-\frac{7\pi}{12}\right) + i\sin\left(-\frac{7\pi}{12}\right)$
iii $6\left(\cos\frac{\pi}{4} + i\sin\frac{\pi}{4}\right)$

Challenge

a $|z| = 2$, $\arg z = \frac{\pi}{3}$, $z = 2\left(\cos\frac{\pi}{3} + i\sin\frac{\pi}{3}\right)$
$z^7 = 128\left(\cos\frac{\pi}{3} + i\sin\frac{\pi}{3}\right) = 64 + 64i\sqrt{3}$
$k = 64$

b $z^4 = 16\left(\cos\frac{4\pi}{3} + i\sin\frac{4\pi}{3}\right) = -8 - 8i\sqrt{3}$
$p = -8$

c All points lie on the same line as shown. The values of k and p are the values of the scale factor of the enlargement.

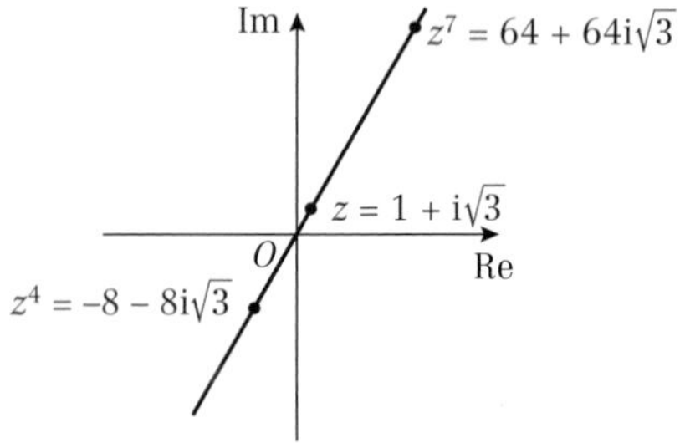

Exercise 2E

1 a
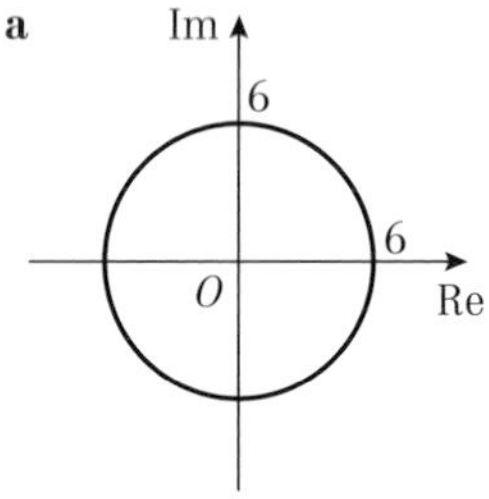

b
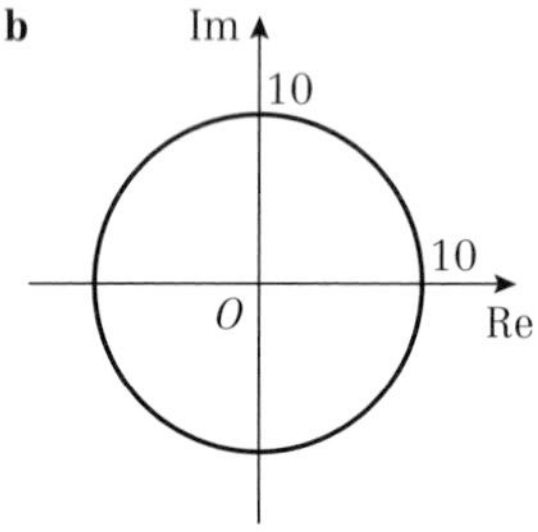

c
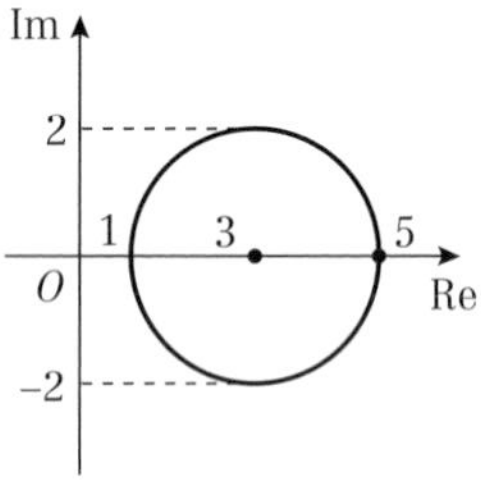

d
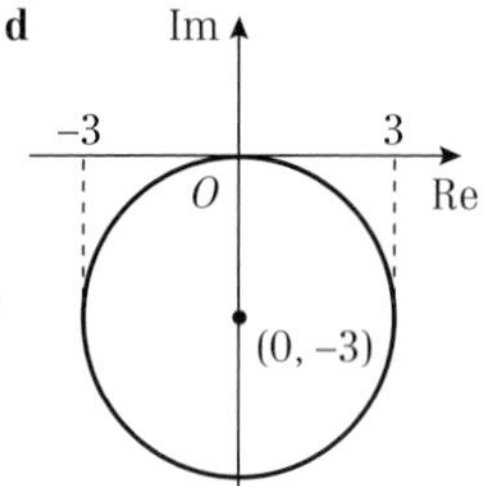

e
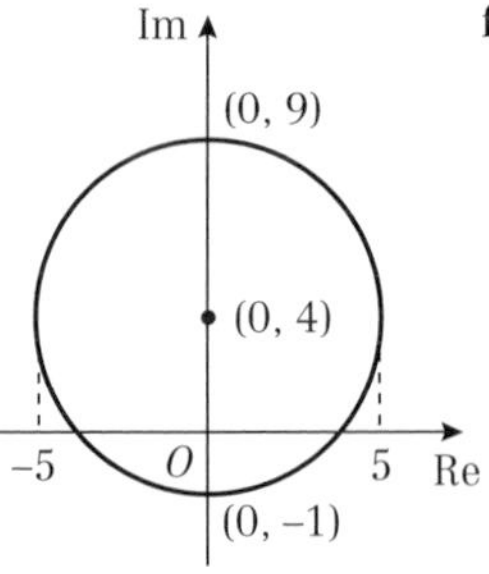

f
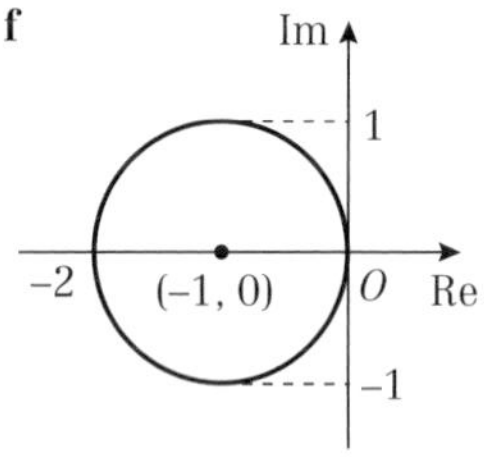

g
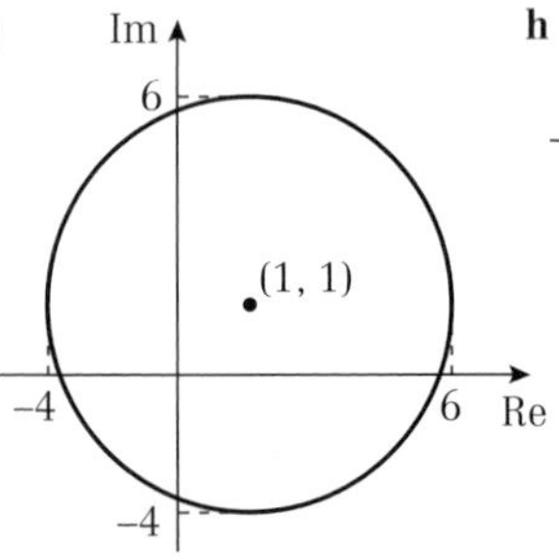

h
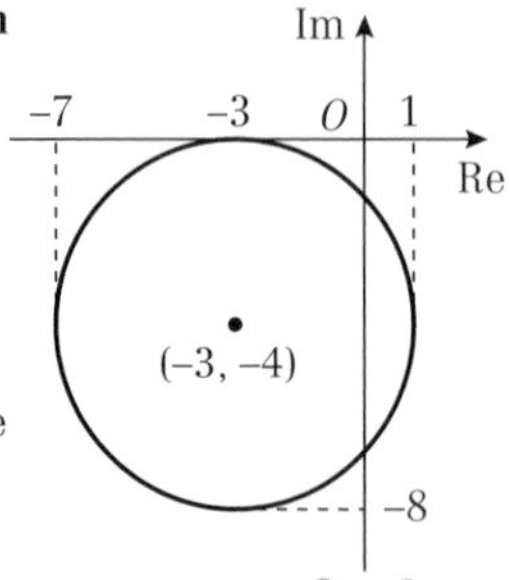

Online Full worked solutions are available in SolutionBank.

i

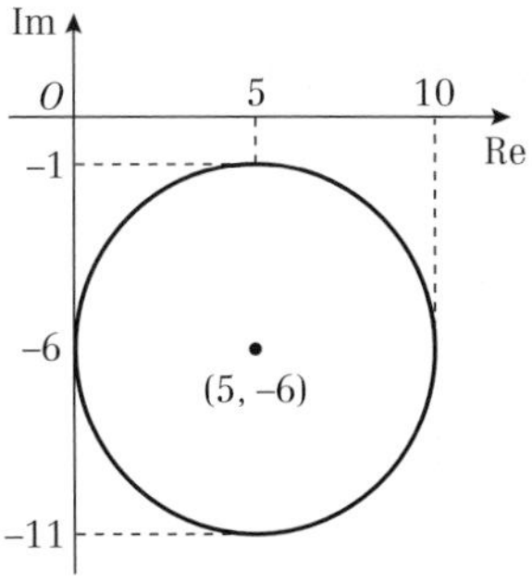

2 a

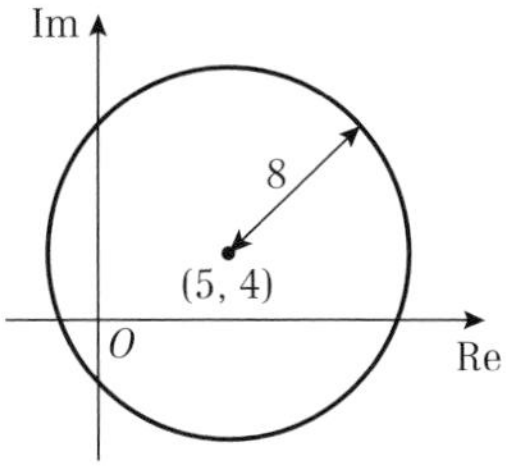

b i $(4 + \sqrt{39})\text{i}$ and $(4 - \sqrt{39})\text{i}$

ii $5 + 4\sqrt{3}$ and $5 - 4\sqrt{3}$

3 a

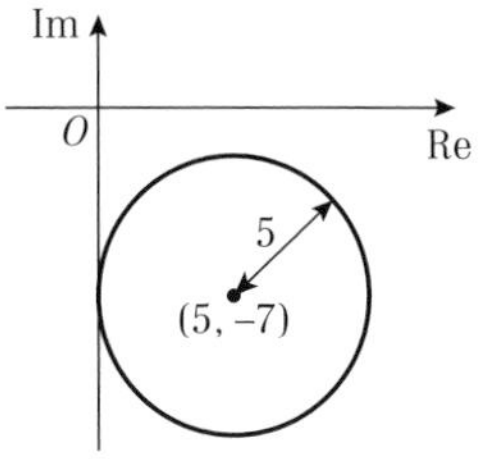

b $(x - 5)^2 + (y + 7)^2 = 25$

c $2\arctan\left(\frac{5}{7}\right) - \frac{\pi}{2} = -0.330$ rad (3 s.f.)

4 a $(x - 4)^2 + (y - 3)^2 = 8^2$

b

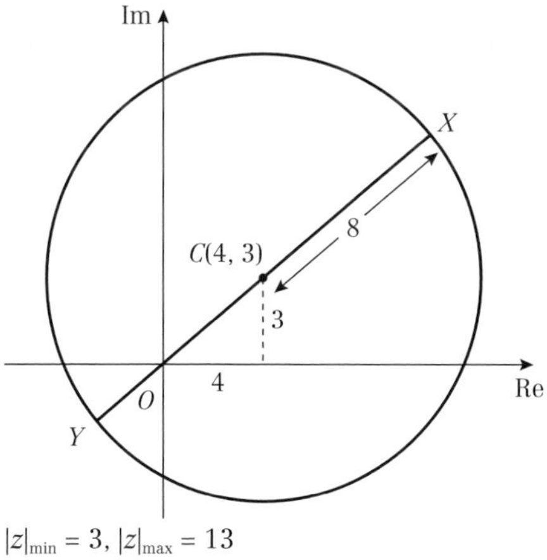

c $|z|_{min} = 3$, $|z|_{max} = 13$

5 a

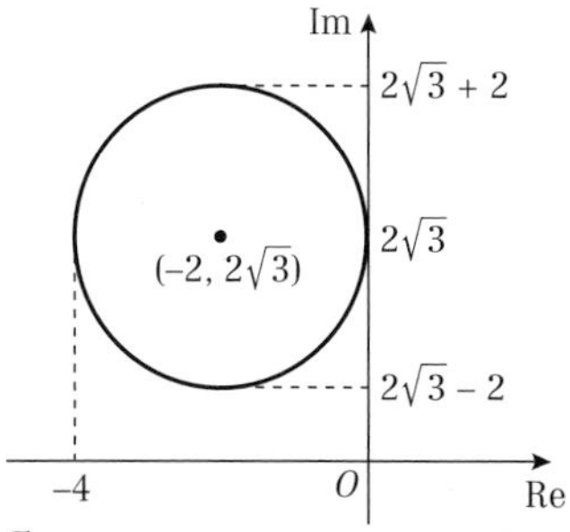

b $\frac{\pi}{2}$

c 2.51 rad

6 a

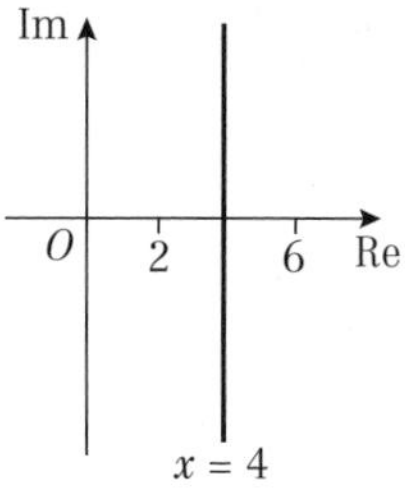

b

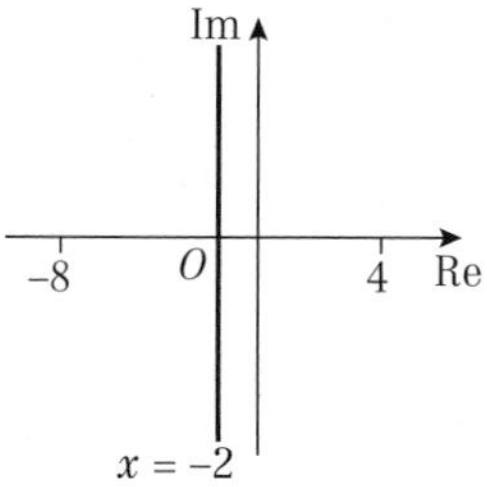

c

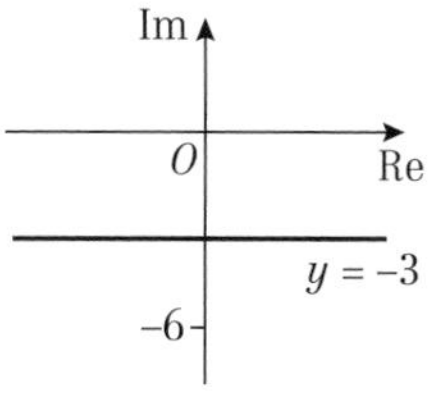

d

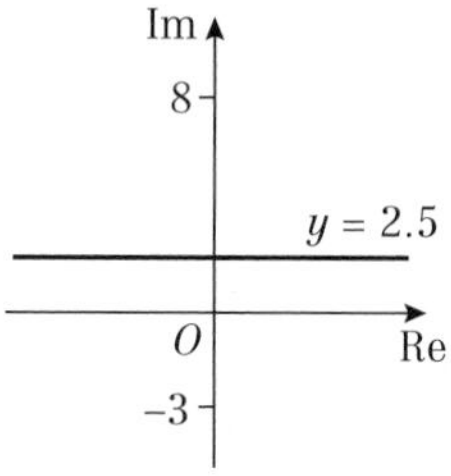

e

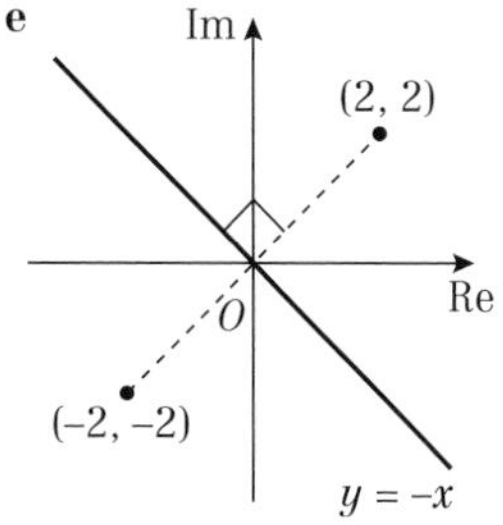

f

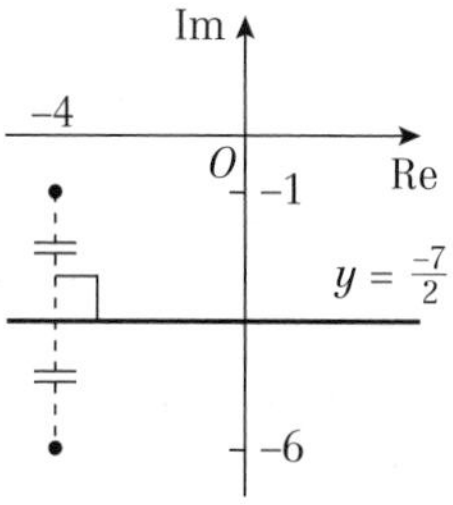

g

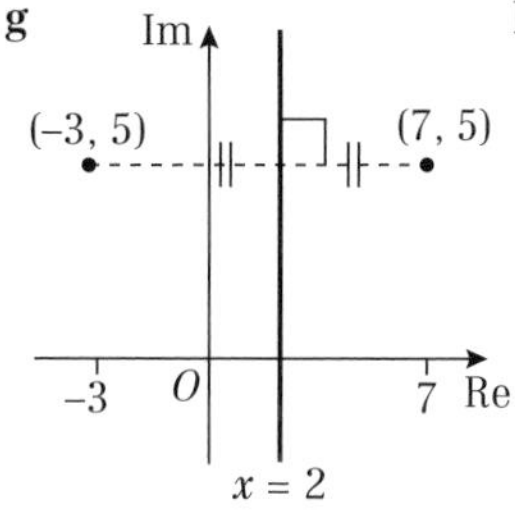

h

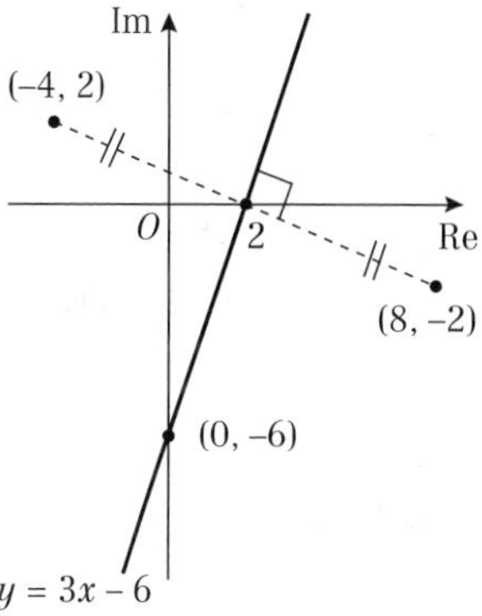

i

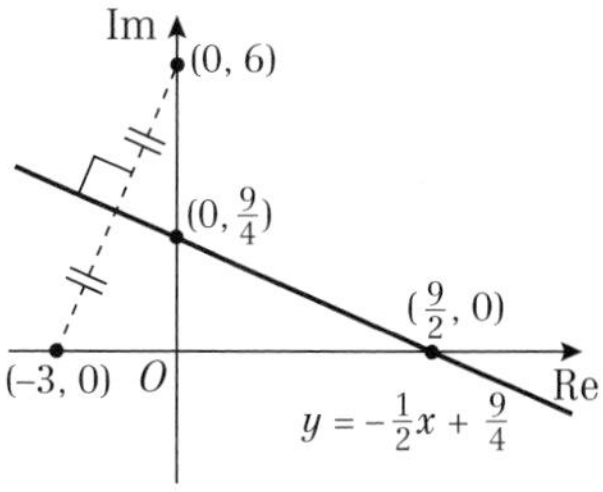

j

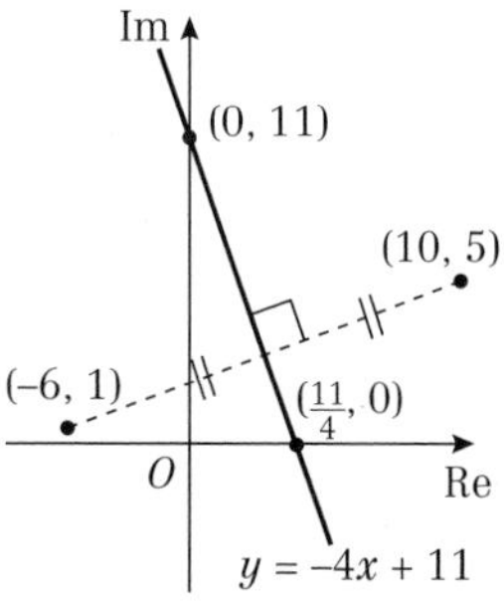

7 a

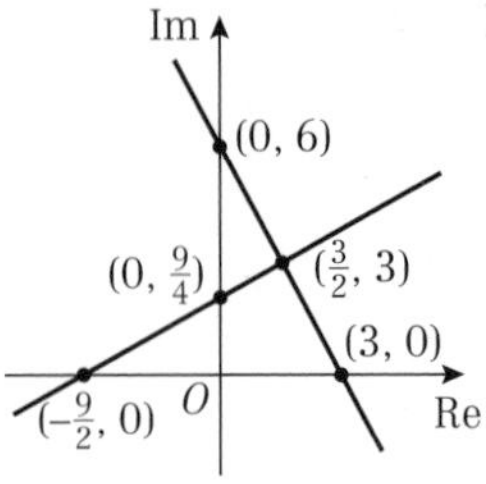

b $\frac{9\sqrt{5}}{10}$

8 a

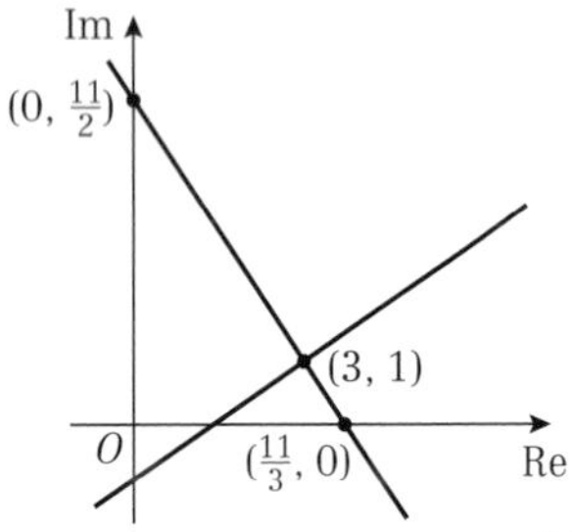

b $y = -\frac{3}{2}x + \frac{11}{2}$

c $\frac{11\sqrt{13}}{13}$

9 a

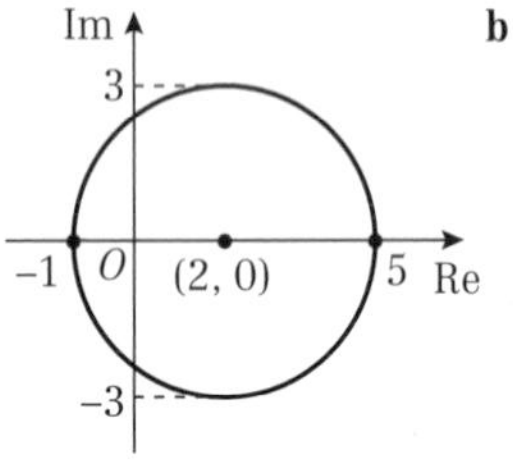

b

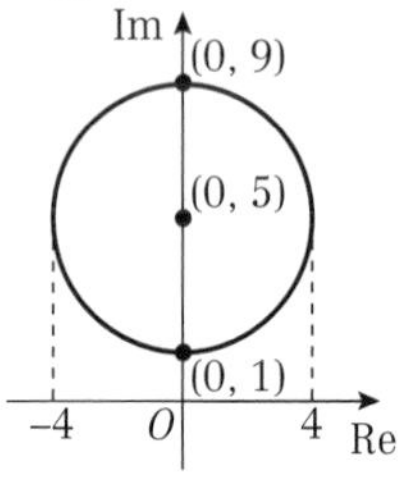

c

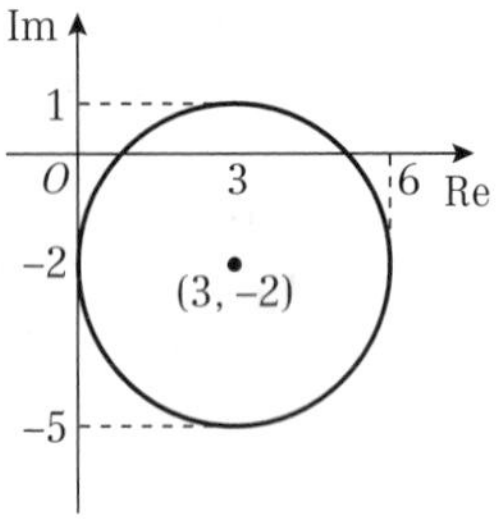

10 a

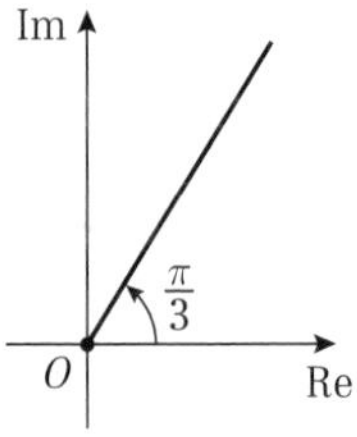

b

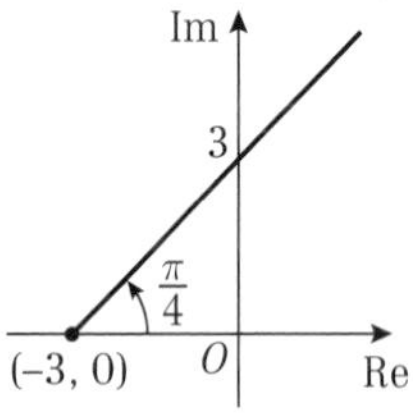

c

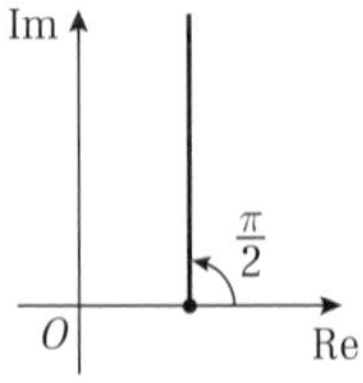

d

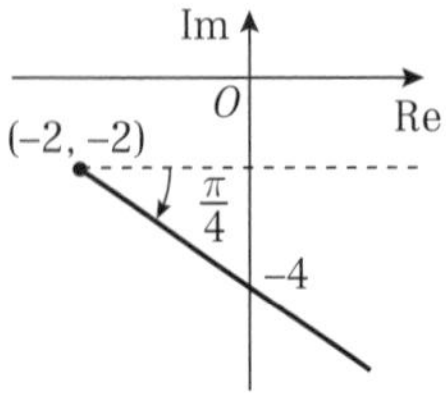

e

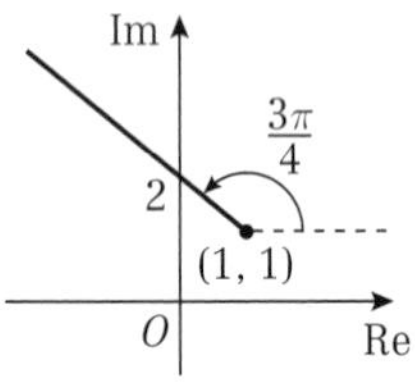

f

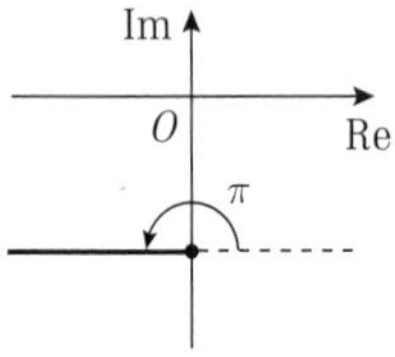

g

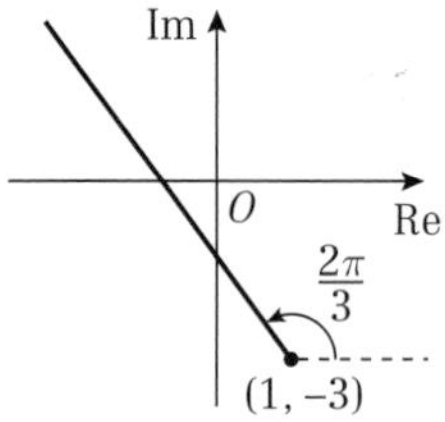

h

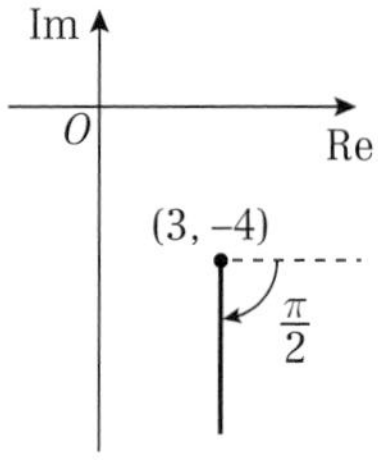

i

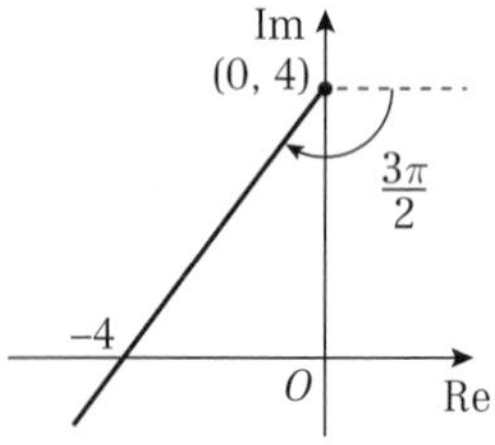

11 a

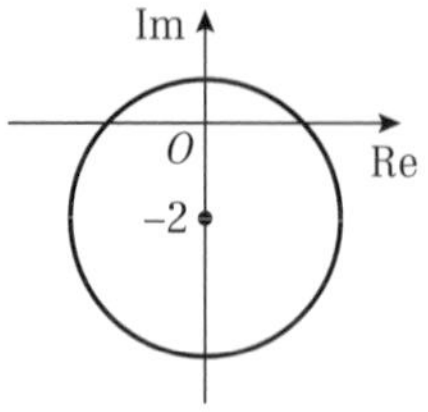

b Use the cosine rule to find $|z|^2 + 2|z| - 5 = 0$, solve to get $|z| = -1 + \sqrt{6}$

12 a $|z_{max}| = 6\sqrt{2} + 4$ and $|z_{min}| = 6\sqrt{2} - 4$

b $(-2.38, \pi)$

13 a

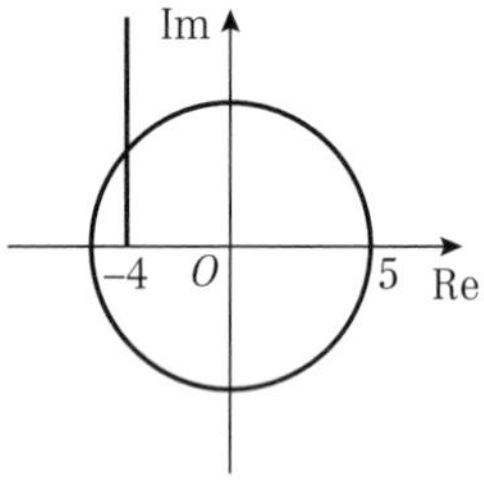

b $z = -4 + 3\text{i}$

14 a

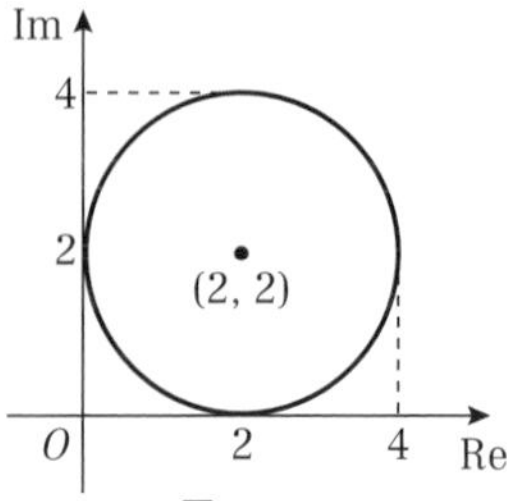

b $z = (2 + \sqrt{3}) + 3\text{i}$

15 a,b

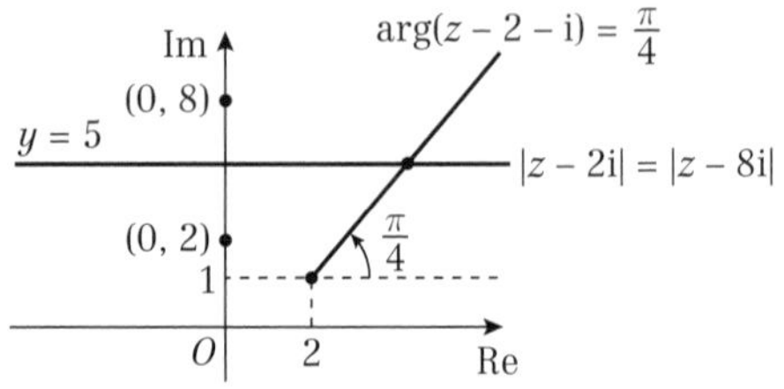

c $z = 6 + 5\text{i}$

Online Full worked solutions are available in SolutionBank.

16 a,b

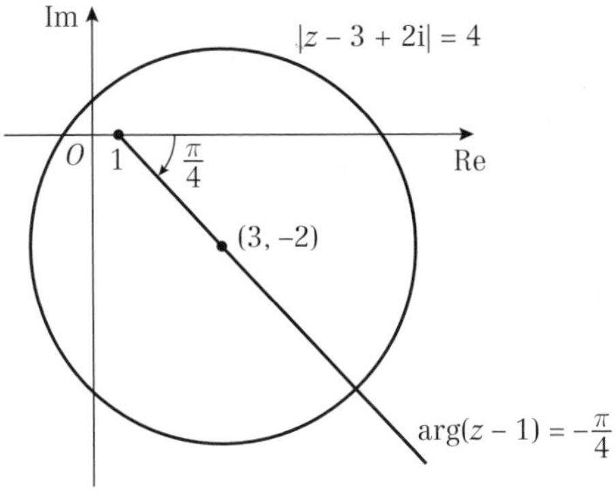

c $a = 3 + 2\sqrt{2}$, $b = -2 - 2\sqrt{2}$

17 a $z = 4 + 4i\sqrt{3}$ **b** $\arg(z - 8) = \frac{2\pi}{3}$

18 a

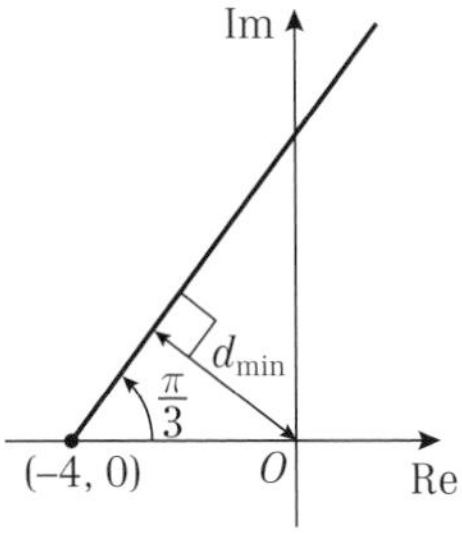

b Hence the minimum value of $|z|$ is $|z|_{min} = 2\sqrt{3}$

19 a

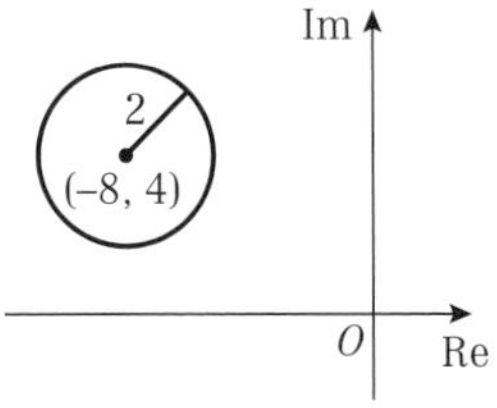

b

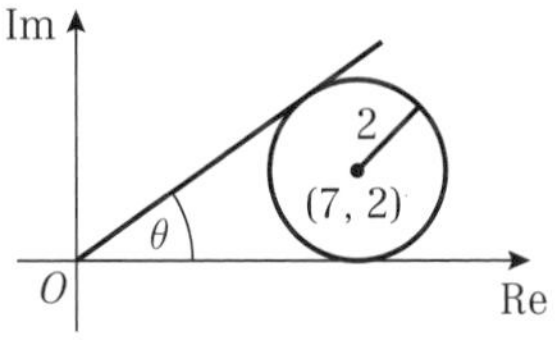

Maximum value of $\arg(z + 15 - 2i) = \theta$

$\sin\left(\frac{\theta}{2}\right) = \frac{2}{\sqrt{2^2 + 7^2}} = \frac{2}{\sqrt{53}}$

$\Rightarrow \theta = 2\arcsin\left(\frac{2}{\sqrt{53}}\right)$

c $\left(-8 + \sqrt{2}, 4 - \sqrt{2}\right)$ and $\left(-8 - \sqrt{2}, 4 + \sqrt{2}\right)$

Challenge

$0.37 < \theta < 2.77$

Exercise 2F

1 a

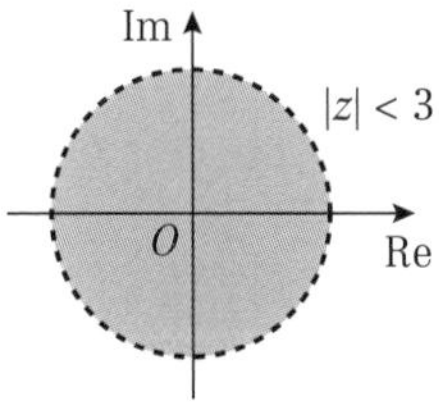

b

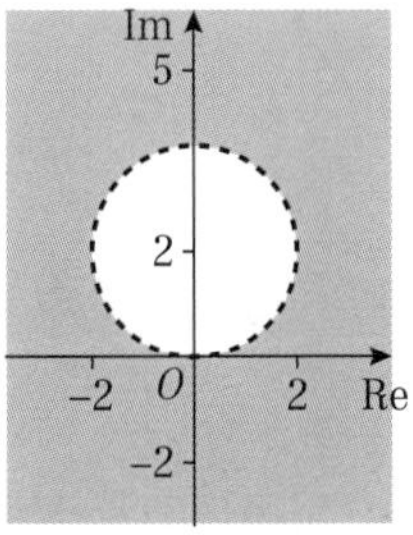

c

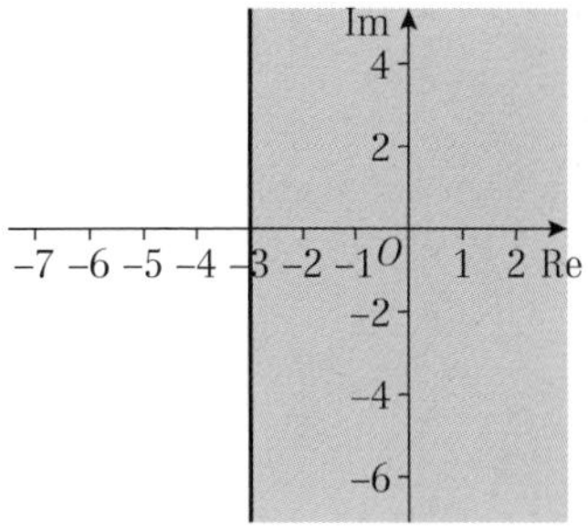

d

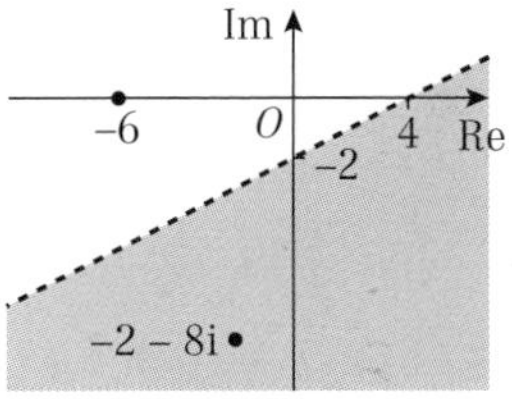

e

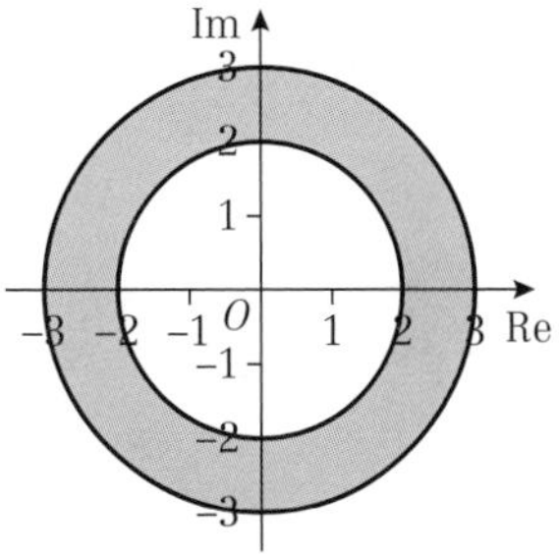

f

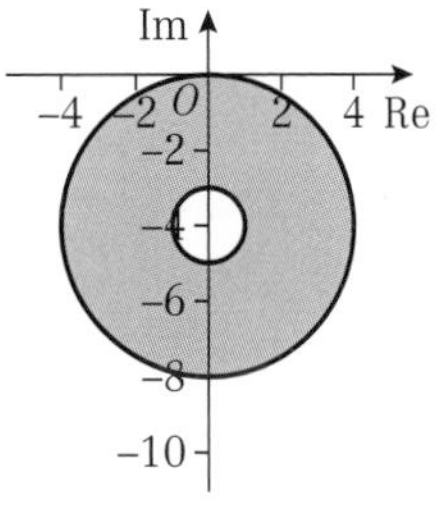

g

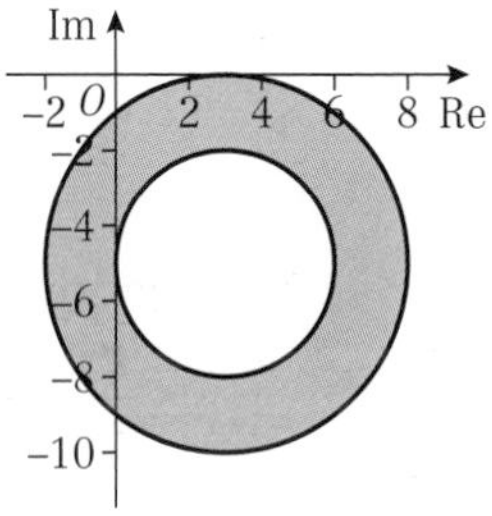

2

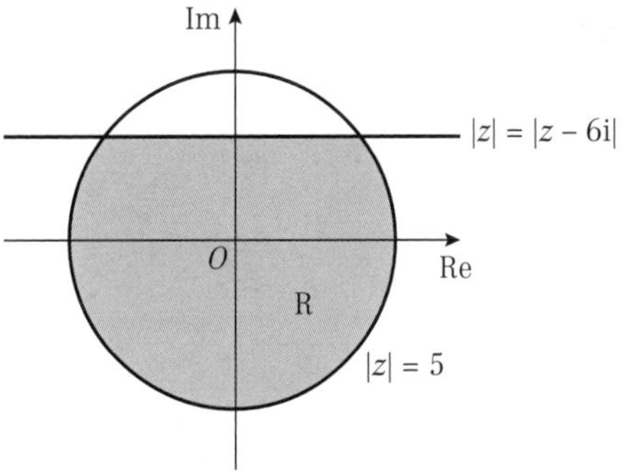

3

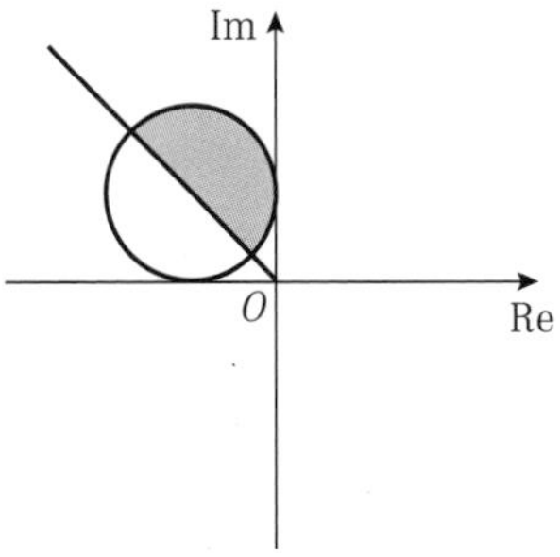

4 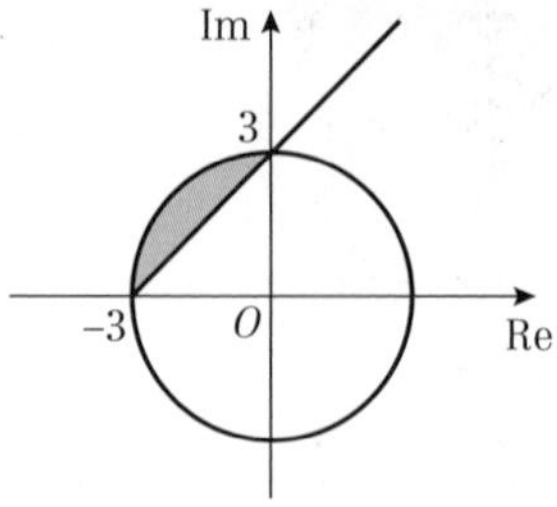

5 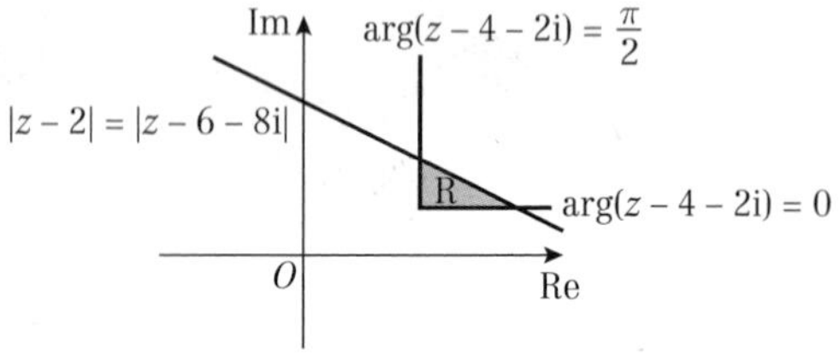

6 a i $y = -2\sqrt{2}x - 2\sqrt{2}$ ii $(x + 1)^2 + y^2 = 9$
b $z = -\sqrt{2} + 2i\sqrt{2}$ or $z = -2i$
c 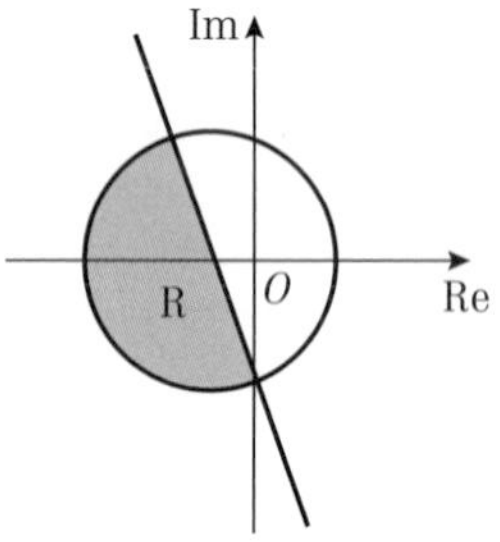

Challenge

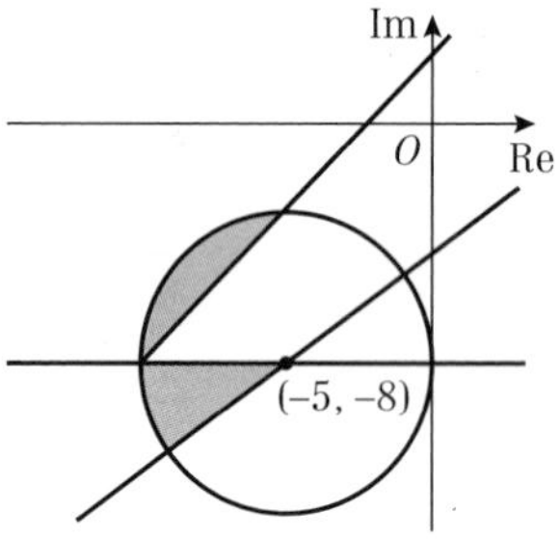

Mixed exercise 2

1 a $z = -\frac{5}{2} + \frac{\sqrt{15}}{2}i$ and $z = -\frac{5}{2} - \frac{\sqrt{15}}{2}i$
b 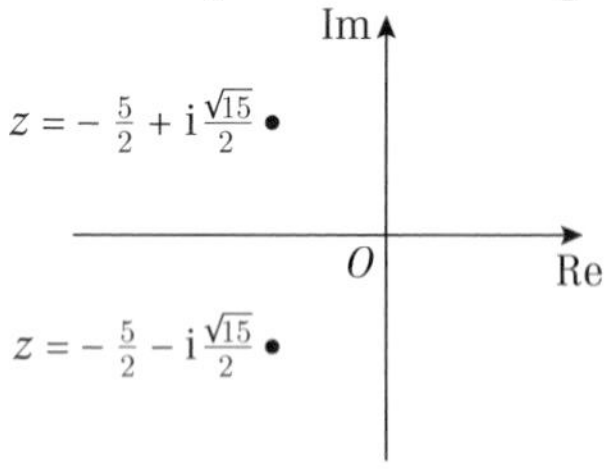

2 a –1 + 2i, –1 – 2i are two of the roots. These roots can be used to form the quadratic $z^2 + 2z + 5$.
$(z - 1)(z^2 + 2z + 5) = f(z)$, so third root is 1.
b Argand diagram showing –1 + 2i, –1 – 2i and 1.
c Sides of triangle are $\sqrt{8}$, $\sqrt{8}$ and 4. $(\sqrt{8})^2 + (\sqrt{8})^2 = 4^2$.

3 a –1 + 4i, –1 – 4i, 2, 1
b Argand diagram showing above roots.

4 a $4x - y = 3$
$-3x - 6y = 0 \Rightarrow x = -2y$
$-9y = 3 \Rightarrow y = -\frac{1}{3} \Rightarrow x = \frac{2}{3}$
b Argand diagram showing the point $z = \frac{2}{3} - \frac{1}{3}i$
c $\frac{\sqrt{5}}{3}$
d –0.46 rad

5 a 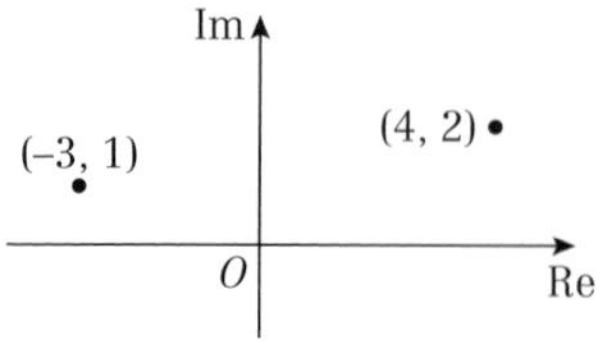

b $5\sqrt{2}$ c $-1 - i$ d $-\frac{3\pi}{4}$

6 a $z^2 = (a^2 - 16) + 8ai$
$2z = 2a + 8i$
$z^2 + 2z = (a^2 + 2a - 16) + (8 + 8a)i$
b $a = -1$
c $z = -1 + 4i$
$|z| = \sqrt{17} \approx 4.12$
$\arg z \approx 1.82$
d Show $z = -1 + 4i$, $z^2 = -15 - 8i$ and $z^2 + 2z = -17$ on a single Argand diagram.

7 a $z = \frac{(3 + 5i)(2 + i)}{(2 - i)(2 + i)} = \frac{1}{5} + \frac{13}{5}i$
$|z| = \frac{1}{5}\sqrt{170}$
b $\arg z = 1.49$

8 a $z^2 = -3 + 4i$
$z^2 - z = -4 + 2i$
$|-4 + 2i| = \sqrt{(-4)^2 + (2)^2} = \sqrt{20} = 2\sqrt{5}$
b 2.68
c 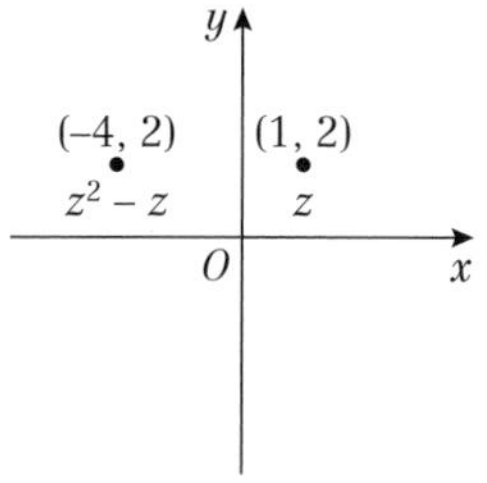

9 a i $\frac{3}{25} - \frac{4}{25}i$ ii $\frac{-8}{5} - \frac{6}{5}i$ b $\frac{1}{5}$ c –2.50

10 a $\frac{\sqrt{5}}{2}$
b $\frac{a + 3i}{2 + ai} = \frac{5a}{4 + a^2} + \frac{-a^2 + 6}{4 + a^2}i$,
for $\arg z = \frac{\pi}{4}$ real and imaginary parts must be equal
$\Rightarrow a^2 + 5a - 6 = 0$
$\Rightarrow a = -6$ or 1
a cannot be negative otherwise $\arg z$ is negative
$\therefore a = 1$

11 a $z_1 = \sqrt{2}\left(\cos\left(-\frac{3\pi}{4}\right) + i\sin\left(-\frac{3\pi}{4}\right)\right)$ and
$z_2 = 2\left(\cos\left(\frac{\pi}{3}\right) + i\sin\left(\frac{\pi}{3}\right)\right)$
b i $2\sqrt{2}$ ii $\frac{\sqrt{2}}{2}$
c i $-\frac{5\pi}{12}$ ii $\frac{11\pi}{12}$

12 a $|z| = |2 - 2i\sqrt{3}| = \sqrt{2^2 + (2\sqrt{3})^2} = \sqrt{16} = 4$
b $\arg z = -\frac{\pi}{3}$
c $\left|\frac{w}{z}\right| = 1$
d $\arg\left(\frac{w}{z}\right) = \frac{\pi}{12}$

Online Full worked solutions are available in SolutionBank.

13 $4\sqrt{2}\left(\cos\left(-\frac{\pi}{4}\right) + \mathrm{i}\sin\left(-\frac{\pi}{4}\right)\right)$

14 **a** $(x+1)^2 + (y-1)^2 = 1$

b

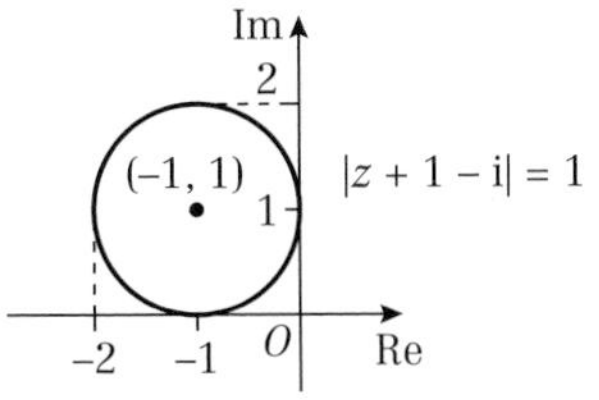

c $|z|_{min} = \sqrt{2} - 1$

$|z|_{max} = \sqrt{2} + 1$

d $|z-1|_{min} = \sqrt{5} - 1$

$|z-1|_{max} = \sqrt{5} + 1$

15 **a**

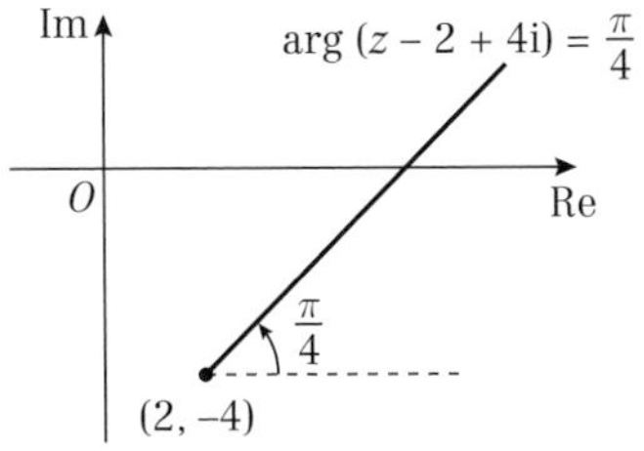

b $3\sqrt{2}$

16

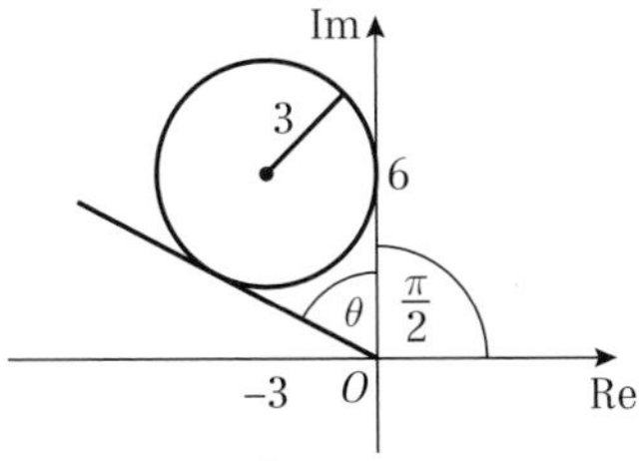

Max value $= \frac{\pi}{2} + \theta$

$\sin\left(\frac{\theta}{2}\right) = \frac{3}{\sqrt{3^2+6^2}} = \frac{3}{\sqrt{45}} = \frac{1}{\sqrt{5}}$

$\Rightarrow \frac{\pi}{2} + \theta = \frac{\pi}{2} + 2\arcsin\left(\frac{1}{\sqrt{5}}\right)$

17 **a**

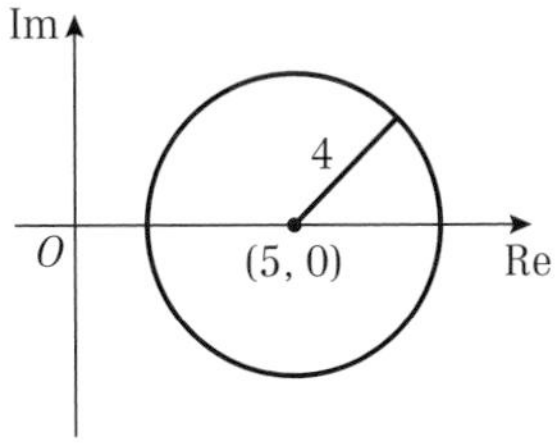

b (3.96, 3.86) and (1.14, −1.03)

c $-\pi < \theta < -0.41,\ 0.41 < \theta < \pi$

18 **a**

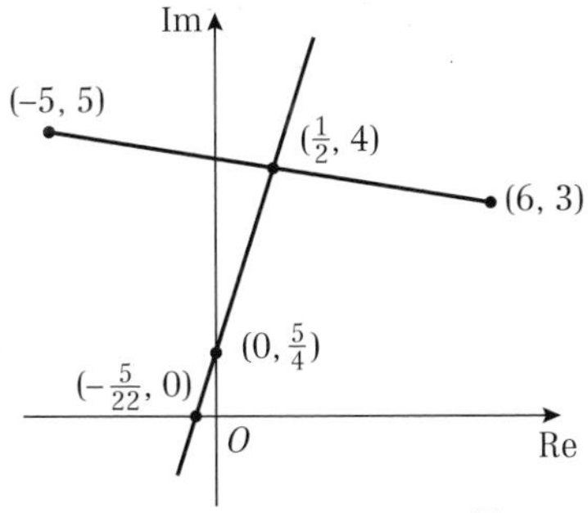

b $y = \frac{11}{2}x + \frac{5}{4}$

c $\frac{\sqrt{5}}{10}$

19 **a** $y = \frac{1}{2}x + 3$

b $6 + 6\mathrm{i}$

c

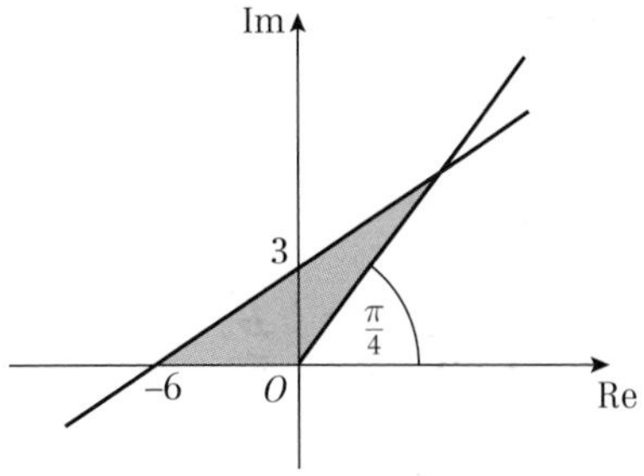

20 **a** **i** $y = x - 2$

ii $(x-2)^2 + y^2 = 8$

b $-2\mathrm{i},\ 4 + 2\mathrm{i}$

c

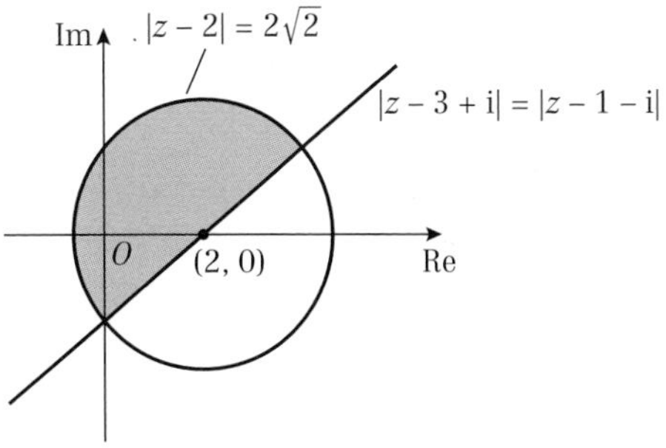

Challenge

a

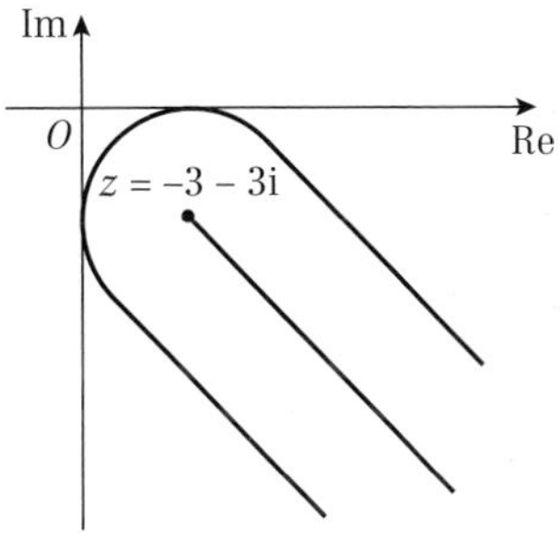

b $3\sqrt{2} - 3$

CHAPTER 3

Prior knowledge 3

1 **a** $(x+3)(x+2)$

b $(x-1)(x+4)$

c $(2x+3)(x+2)$

2 **a** $(k+1)(1+k+2) = (k+1)(k+3)$

b $\frac{1}{2}(k+1)^2(1+2k^2)$

c $(2k-1)(k^2+5)$

Exercise 3A

1 **a** 16 **b** 820 **c** 210 **d** 4950

e 775 **f** 15 150 **g** 610 **h** 3240

2 $n = 32$

3 $k = 14$

4 **a** $n(2n-1)$

b $\sum_{r=n+1}^{2n-1} r = \sum_{r=1}^{2n-1} r - \sum_{r=1}^{n} r = n(2n-1) - \frac{1}{2}n(n+1)$

$= \frac{1}{2}n(4n - 2 - n - 1) = \frac{1}{2}n(3n-3) = \frac{3}{2}n(n-1)$

5 $\sum_{r=n-1}^{2n} r = \sum_{r=1}^{2n} r - \sum_{r=1}^{n-2} r = \frac{1}{2}(2n)(2n+1) - \frac{1}{2}(n-2)(n-1)$

$= \frac{1}{2}(2n(2n+1) - (n-1)(n-2)) = \frac{1}{2}(3n^2+5n-2)$

$= \frac{1}{2}(n+2)(3n-1)$

6 a $\sum_{r=1}^{n^2} r - \sum_{r=1}^{n} r = \frac{1}{2}n^2(n^2+1) - \frac{1}{2}n(n+1)$

$= \frac{1}{2}n(n(n^2+1) - (n+1)) = \frac{1}{2}n(n^3+n-n-1)$

$= \frac{1}{2}n(n^3-1)$

b 3276

7 a 4565 **b** −28 485 **c** 2576

8 a $\sum_{r=1}^{n}(3r+2) = 3\sum_{r=1}^{n} r + 2\sum_{r=1}^{n} 1 = \frac{3}{2}n(n+1) + 2n$

$= \frac{1}{2}n(3(n+1)+4) = \frac{1}{2}n(3n+7)$

b $\sum_{r=1}^{2n}(5r-4) = 5\sum_{r=1}^{2n} r - 4\sum_{r=1}^{2n} 1 = \frac{5}{2}(2n)(2n+1) - 4(2n)$

$= 5n(2n+1) - 8n = 10n^2 - 3n = n(10n-3)$

c $\sum_{r=1}^{n+2}(2r+3) = 2\sum_{r=1}^{n+2} r + 3\sum_{r=1}^{n+2} 1 = (n+2)(n+3) + 3(n+2)$

$= (n+2)((n+3)+3) = (n+2)(n+6)$

d $\sum_{r=3}^{n}(4r+5) = 4\sum_{r=3}^{n} r + 5\sum_{r=3}^{n} 1$

$= 4\left(\sum_{r=1}^{n} r - \sum_{r=1}^{2} r\right) + 5\left(\sum_{r=1}^{n} 1 - \sum_{r=1}^{2} 1\right)$

$= 4(\frac{1}{2}n(n+1) - 3) + 5(n-2) = 2n^2 + 2n - 12 + 5n - 10$

$= 2n^2 + 7n - 22 = (2n+11)(n-2)$

9 a $\sum_{r=1}^{k}(4r-5) = 4\sum_{r=1}^{k} r - 5\sum_{r=1}^{k} 1 = \frac{4}{2}k(k+1) - 5k$

$= 2k^2 + 2k - 5k = 2k^2 - 3k$

b 51

10 $a = 7$, $b = -3$

11 a $\sum_{r=1}^{4n-1}(3r+1) = 3\sum_{r=1}^{4n-1} r + \sum_{r=1}^{4n-1} 1 = \frac{3}{2}(4n-1)(4n) + (4n-1)$

$= 6n(4n-1) + (4n-1) = (4n-1)(6n+1)$

$= 24n^2 - 2n - 1$

b 14 949

12 a $\sum_{r=1}^{2k+1}(4-5r) = -5\sum_{r=1}^{2k+1} r + 4\sum_{r=1}^{2k+1} 1$

$= -\frac{5}{2}(2k+1)(2k+2) + 4(2k+1)$

$= -5(2k+1)(k+1) + 4(2k+1)$

$= (2k+1)(-5(k+1)+4) = (2k+1)(-5k-1)$

$= -(2k+1)(5k+1)$

b −1525

c $\sum_{r=1}^{15}(5r-4) = -\sum_{r=1}^{15}(4-5r) = -(-540) = 540$

13 If $g(r) = 2r$, then $\sum_{r=1}^{n} g(r) = n^2 + n$. Hence $f(r) = 2r + 3$.

14 a $f(r) = 4r - 1$ **b** 210

Challenge

$n = 4$

Exercise 3B

1 a 30 **b** 22 140 **c** 19 270
d 24 502 500 **e** 25 502 500 **f** 379 507 500
g 173 880

2 a $\sum_{r=1}^{2n} r^2 = \frac{1}{6}(2n)(2n+1)(4n+1) = \frac{1}{3}n(2n+1)(4n+1)$

b $\sum_{r=1}^{2n-1} r^2 = \frac{1}{6}(2n-1)(2n)(2(2n-1)+1)$

$= \frac{1}{6}2n(2n-1)(4n-1) = \frac{1}{3}n(2n-1)(4n-1)$

c $\sum_{r=n}^{2n} r^2 = \sum_{r=1}^{2n} r^2 - \sum_{r=1}^{n-1} r^2$

$= \frac{1}{3}n(2n+1)(4n+1) - \frac{1}{6}(n-1)(n)(2(n-1)+1)$

$= \frac{1}{6}n(2(2n+1)(4n+1) - (n-1)(2n-1))$

$= \frac{1}{6}n(16n^2 + 12n + 2 - 2n^2 + 3n - 1)$

$= \frac{1}{6}n(n+1)(14n+1)$

3 $\sum_{r=1}^{m} r^3 = \frac{1}{4}m^2(m+1)^2$ and so $\sum_{r=1}^{n+k} r^3 = \frac{1}{4}(n+k)^2(n+k+1)^2$

4 a $\sum_{r=n+1}^{3n} r^3 = \sum_{r=1}^{3n} r^3 - \sum_{r=1}^{n} r^3 = \frac{1}{4}(3n)^2(3n+1)^2 - \frac{1}{4}n^2(n+1)^2$

$= \frac{1}{4}n^2(9(3n+1)^2 - (n+1)^2)$

$= \frac{1}{4}n^2(81n^2 - n^2 + 54n - 2n + 9 - 1)$

$= n^2(20n^2 + 13n + 2) = n^2(4n+1)(5n+2)$

b 213 200

5 a $\sum_{r=n}^{2n} r^3 = \sum_{r=1}^{2n} r^3 - \sum_{r=1}^{n-1} r^3 = \frac{1}{4}(2n)^2(2n+1)^2 - \frac{1}{4}(n-1)^2n^2$

$= \frac{1}{4}n^2(4(2n+1)^2 - (n-1)^2)$

$= \frac{1}{4}n^2(16n^2 + 16n + 4 - n^2 + 2n - 1)$

$= \frac{3}{4}n^2(5n^2 + 6n + 1) = \frac{3}{4}n^2(n+1)(5n+1)$

b 3 159 675

6 a 9425 **b** 25 420
c 10 507 320 **d** 393 825

7 a $\sum_{r=1}^{n}(r+2)(r+5) = \sum_{r=1}^{n} r^2 + 7\sum_{r=1}^{n} r + 10\sum_{r=1}^{n} 1$

$= \frac{1}{6}n(n+1)(2n+1) + \frac{7}{2}n(n+1) + 10n$

$= \frac{1}{6}n(2n^2 + 3n + 1 + 21n + 21 + 60)$

$= \frac{1}{3}n(n^2 + 12n + 41)$

b 51 660

8 a $\sum_{r=1}^{n}(r^2 + 3r + 1) = \sum_{r=1}^{n} r^2 + 3\sum_{r=1}^{n} r + \sum_{r=1}^{n} 1$

$= \frac{1}{6}n(n+1)(2n+1) + \frac{3}{2}n(n+1) + n$

$= \frac{1}{6}n(2n^2 + 3n + 1 + 9n + 9 + 6)$

$= \frac{1}{3}n(n^2 + 6n + 8) = \frac{1}{3}n(n+2)(n+4)$

$a = 2$, $b = 4$

b 22 000

9 a $\sum_{r=1}^{n} r^2(r-1) = \sum_{r=1}^{n} r^3 - \sum_{r=1}^{n} r^2$

$= \frac{1}{4}n^2(n+1)^2 - \frac{1}{6}n(n+1)(2n+1)$

$= \frac{1}{12}n(n+1)(3n(n+1) - 2(2n+1))$

$= \frac{1}{12}n(n+1)(3n^2 - n - 2)$

b $\sum_{r=1}^{2n-1} r^2(r-1)$

$= \frac{1}{12}(2n-1)((2n-1)+1)(3(2n-1)^2 - (2n-1) - 2)$

$= \frac{1}{12}2n(2n-1)(12n^2 - 14n + 2)$

$= \frac{1}{3}n(2n-1)(6n^2 - 7n + 1)$

Online Full worked solutions are available in SolutionBank.

10 a $\sum_{r=1}^{n}(r+1)(r+3)=\sum_{r=1}^{n}r^2+4\sum_{r=1}^{n}r+3\sum_{r=1}^{n}1$
$=\frac{1}{6}n(n+1)(2n+1)+2n(n+1)+3n$
$=\frac{1}{6}n(2n^2+3n+1+12n+12+18)$
$=\frac{1}{6}n(2n^2+15n+31)$

b $\frac{1}{6}n(14n^2+45n+31)$

11 a $\sum_{r=1}^{n}(r+3)(r+4)=\sum_{r=1}^{n}r^2+7\sum_{r=1}^{n}r+12\sum_{r=1}^{n}1$
$=\frac{1}{6}n(n+1)(2n+1)+\frac{7}{2}n(n+1)+12n$
$=\frac{1}{6}n(2n^2+3n+1+21n+21+72)$
$=\frac{1}{3}n(n^2+12n+47)$

b $\frac{2}{3}n(13n^2+48n+47)$

12 a $\sum_{r=1}^{n}r(r+3)^2=\sum_{r=1}^{n}r^3+6\sum_{r=1}^{n}r^2+9\sum_{r=1}^{n}r$
$=\frac{1}{4}n^2(n+1)^2+n(n+1)(2n+1)+\frac{9}{2}n(n+1)$
$=\frac{1}{4}n(n+1)(n(n+1)+4(2n+1)+18)$
$=\frac{1}{4}n(n+1)(n^2+9n+22)$

b 59070

13 a $\sum_{r=1}^{kn}(2r-1)=2\sum_{r=1}^{kn}r-\sum_{r=1}^{kn}1=kn(kn+1)-kn=k^2n^2$

b $n=9$

14 a $\sum_{r=1}^{n}(r^3-r^2)=\sum_{r=1}^{n}r^3-\sum_{r=1}^{n}r^2$
$=\frac{1}{4}n^2(n+1)^2-\frac{1}{6}n(n+1)(2n+1)$
$=\frac{1}{12}n(n+1)(3n(n+1)-2(2n+1))$
$=\frac{1}{12}n(n+1)(3n^2-n-2)=\frac{1}{12}n(n+1)(n-1)(3n+2)$

b $n=4$

Challenge

a $f_1(x)=1$, $f_2(x)\ 2x-1$, $f_3(x)=3x^2-3x+1$,
$f_4(x)=4x^3-6x^2+4x-1$

b Given $h(x)=ax^3+bx^2+cx+d$,
$nh(n)=an^4+bn^3+cn^2+dn$
$=\sum_{r=1}^{n}(af_4(r)+bf_3(r)+cf_2(r)+df_1(r))=\sum_{r=1}^{n}g(r)$
for $g(r)=af_4(r)+bf_3(r)+cf_2(r)+df_1(r)$

Mixed exercise 3

1 **a** 55 **b** 1230 **c** 385
d 3025 **e** 37400 **f** 24001875
g 75640

2 **a** $\frac{3}{2}n^2-\frac{7}{2}n$ **b** $\frac{1}{3}n(n+1)(n+2)$
c $n(n+1)(n+4)$ **d** $n(n+1)(n^2+3n+1)$
e $\frac{1}{6}n(n+1)(2n-5)$ **f** $\frac{1}{3}n(n+1)(n-4)$
g $\frac{1}{6}n(2n^2+3n-29)$ **h** $\frac{1}{2}n(n^3+4n^2+5n+10)$

3 27 900

4 a $\sum_{r=1}^{n}r^2(r-3)=\sum_{r=1}^{n}r^3-3\sum_{r=1}^{n}r^2$
$=\frac{1}{4}n^2(n+1)^2-\frac{1}{2}n(n+1)(2n+1)$
$=\frac{1}{4}n(n+1)(n(n+1)-2(2n+1))$
$=\frac{1}{4}n(n+1)(n^2-3n-2)$
so $a=-3$, $b=-2$.

b 35 490

5 a $\sum_{r=1}^{n}(2r-1)^2=4\sum_{r=1}^{n}r^2-4\sum_{r=1}^{n}r+\sum_{r=1}^{n}1$
$=\frac{2}{3}n(n+1)(2n+1)-2n(n+1)+n$
$=\frac{1}{3}n(n+1)(2(2n+1)-6)+n=\frac{1}{3}n(n+1)(4n-4)+n$
$=\frac{1}{3}n(4n^2-4+3)=\frac{1}{3}n(2n-1)(2n+1)$

b $\frac{2}{3}n(4n-1)(4n+1)$

6 a $\sum_{r=1}^{n}r(r+2)=\sum_{r=1}^{n}r^2+2\sum_{r=1}^{n}r$
$=\frac{1}{6}n(n+1)(2n+1)+n(n+1)=\frac{1}{6}n(n+1)(2n+7)$

b 9160

7 a $\sum_{r=n+1}^{2n}r^2=\sum_{r=1}^{2n}r^2-\sum_{r=1}^{n}r^2$
$=\frac{1}{6}(2n)((2n)+1)(2(2n)+1)-\frac{1}{6}n(n+1)(2n+1)$
$=\frac{1}{6}n(2n+1)(2(4n+1)-n-1)=\frac{1}{6}n(2n+1)(7n+1)$

b 8215

8 a $\sum_{r=1}^{n}(r^2-r-1)=\sum_{r=1}^{n}r^2-\sum_{r=1}^{n}r-\sum_{r=1}^{n}1$
$=\frac{1}{6}n(n+1)(2n+1)-\frac{1}{2}n(n+1)-n=\frac{1}{3}n(n^2-4)$

b 21 049 **c** $n=7$

9 a $\sum_{r=1}^{n}r(2r^2+1)=2\sum_{r=1}^{n}r^3+\sum_{r=1}^{n}r=\frac{1}{2}n^2(n+1)^2+\frac{1}{2}n(n+1)$
$=\frac{1}{2}n(n+1)(n(n+1)+1)=\frac{1}{2}n(n+1)(n^2+n+1)$

b $100\sum_{r=1}^{n}r^2-\sum_{r=1}^{n}r=\frac{1}{6}n(n+1)(200n+97)$

Now if $\sum_{r=1}^{n}r(2r^2+1)=\sum_{r=1}^{n}(100r^2-r)$, then
$=\frac{1}{2}n(n+1)(n^2+n+1)=\frac{1}{6}n(n+1)(200n+97)$
$=\frac{1}{6}n(n+1)(3(n^2+n+1)-(200n+97))=0$
$=\frac{1}{6}n(n+1)(3n^2-197n-94)=0.$
But $n\neq 0$, $n\neq -1$ and $3n^2-197n-94$ has discriminant 39 937 which is not square.

10 a $\sum_{r=1}^{n}r(r+1)^2=\sum_{r=1}^{n}r^3+2\sum_{r=1}^{n}r^2+\sum_{r=1}^{n}r$
$=\frac{1}{4}n^2(n+1)^2+\frac{1}{3}n(n+1)(2n+1)+\frac{1}{2}n(n+1)$
$=\frac{1}{12}n(n+1)(n+2)(3n+5)$

b $n=10$

11 $n=15$

Challenge

a $\sum_{i=1}^{n}\frac{1}{6}i(i+1)(2i+1)=\sum_{i=1}^{n}(\frac{1}{3}i^3+\frac{1}{2}i^2+\frac{1}{6}i)$

Then use the formulae for sums of i, i^2, i^3 to obtain the result.

b $\sum_{j=1}^{n}\left(\sum_{i=1}^{j}\left(\sum_{r=1}^{i}r\right)\right)=\sum_{j=1}^{n}\left(\frac{1}{2}\sum_{i=1}^{j}i(i+1)\right)$
$=\frac{1}{2}\sum_{j=1}^{n}\left(\sum_{i=1}^{j}i^2\right)+\frac{1}{2}\sum_{j=1}^{n}\left(\sum_{i=1}^{j}i\right)$
$=\frac{1}{24}n(n+1)^2(n+2)+\frac{1}{4}\sum_{j=1}^{n}j^2+\frac{1}{4}\sum_{j=1}^{n}j$
$=\frac{1}{24}n(n+1)^2(n+2)+\frac{1}{24}n(n+1)(2n+1)+\frac{1}{8}n(n+1)$
$=\frac{1}{24}n(n+1)(n^2+5n+6)$
$=\frac{1}{24}n(n+1)(n+2)(n+3)$

CHAPTER 4

Prior knowledge 4

1 a $x = -2 \pm \mathrm{i}$ b $x = \frac{7 \pm \mathrm{i}\sqrt{15}}{4}$

2 -2 and $1 + \mathrm{i}$

3 a -1 and 3 b 4 and 8 c $-\frac{1}{2}$ and $\frac{3}{2}$

Exercise 4A

1 a $-\frac{7}{3}$ b $-\frac{4}{3}$ c $\frac{7}{4}$ d $\frac{73}{9}$

2 a $\frac{3}{7}$ b $\frac{1}{7}$ c 3 d $-\frac{5}{49}$

3 a $\frac{3}{2}$ b $\frac{1}{9}$ c $\frac{9}{2}$ d $\frac{15}{8}$

4 $a = 1, b = 1, c = -6$

5 $a = 6, b = 5, c = 1$

6 $a = 2, b = 2, c = 1$

7 a $-1 + 4\mathrm{i}$ b $b = 2, c = 17$

8 $\frac{3}{5}$

9 -16

10 6

11 $k = \sqrt{2}$ and $m = -\frac{2\sqrt{2}}{3}$ or $k = -\sqrt{2}$ and $m = \frac{2\sqrt{2}}{3}$

12 a -2 b -26

13 a 24 b $q > 36$

Exercise 4B

1 a $-\frac{5}{2}$ b $-\frac{3}{2}$ c -1 d $\frac{2}{3}$

2 a -5 b -13 c 17 d 169

3 a $\frac{4}{7}$ b $-\frac{6}{7}$ c $-\frac{216}{343}$ d $\frac{1}{6}$

4 $a = 4, b = -12, c = 11, d = -3$

5 $a = 2, b = -5, c = 22, d = -10$

6 $a = 16, b = -4, c = -32, d = 15$

7 a $\alpha\beta + \beta\gamma + \gamma\alpha = 0, \alpha\beta\gamma = -\frac{1}{16}$

b i $\alpha = \frac{1}{2}, \beta = \frac{1}{2}, \gamma = -\frac{1}{4}$ ii 12

8 a $\alpha\beta + \beta\gamma + \gamma\alpha = 15, \alpha\beta\gamma = \frac{13}{2}, k = 2(\alpha + \beta + \gamma)$ b 9

9 a Yes – there are two other real roots, so α^* couldn't also be a root.

b $m = 13, n = 12, \alpha = 3$

10 a -1 and $3 + \mathrm{i}$ b -20

11 $\alpha = 2$ and $k = 2$ or $\alpha = 8$ and $k = \frac{1}{2}$

12 $\alpha = 1, c = 32, d = 12$

13 $\alpha = -4, c = 352, d = 768$

Challenge

If all three roots are real, then the three expressions will clearly all be real, so assume there are two complex conjugate roots, $a \pm b\mathrm{i}$, and one real root, r.

a $\alpha + \beta + \gamma = r + (a + b\mathrm{i}) + (a - b\mathrm{i}) = r + 2a \in \mathbb{R}$

b $\alpha\beta + \beta\gamma + \gamma\alpha = r(a + b\mathrm{i}) + (a + b\mathrm{i})(a - b\mathrm{i}) + (a - b\mathrm{i})r$
$= 2ar + a^2 + b^2 \in \mathbb{R}$

c $\alpha\beta\gamma = r(a + b\mathrm{i})(a - b\mathrm{i}) = r(a^2 + b^2) \in \mathbb{R}$

Exercise 4C

1 a $-\frac{3}{4}$ b $\frac{1}{2}$ c $\frac{5}{4}$ d $\frac{5}{4}$

2 a -2 b $-\frac{3}{2}$ c $\frac{1}{2}$
d 1 e $\frac{1}{2}$

3 a -3 b 2 c 1
d $\frac{1}{4}$ e 16

4 a $-\frac{6}{7}$ b $-\frac{5}{7}$ c $-\frac{4}{7}$
d $-\frac{4}{3}$ e $\frac{27}{343}$

5 $a = 12, b = 40, c = 25, d = -20, e = -12$

6 $a = 6, b = -11, c = 9, d = 4, e = -2$

7 $a = 72, b = -102, c = -25, d = 53, e = -12$

8 $x = 1, 3, 5$ or 7

9 $x = \frac{1}{2}, \frac{1}{4}, \frac{1}{8}$ or $\frac{1}{16}$

10 a $-\frac{3}{4}$ b $m = -60, n = 45$

11 a 2 b $d = -494, e = 420$

12 a $\alpha + \beta + (3 + \mathrm{i}) + (3 - \mathrm{i}) = \frac{19}{4} \Rightarrow 4\alpha + 4\beta + 5 = 0$
$\alpha\beta(3 + \mathrm{i})(3 - \mathrm{i}) = 10\alpha\beta = \frac{10}{4} \Rightarrow 4\alpha\beta - 1 = 0$

b $-1, -\frac{1}{4}, 3 + \mathrm{i}, 3 - \mathrm{i}, p = 11, q = 44$

c
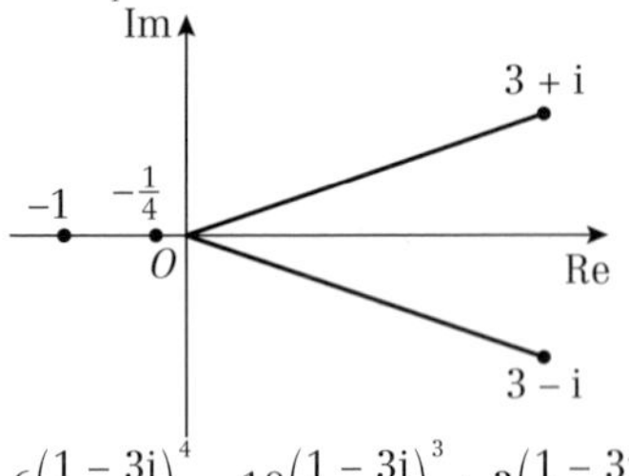

13 a $6\left(\frac{1 - 3\mathrm{i}}{2}\right)^4 - 10\left(\frac{1 - 3\mathrm{i}}{2}\right)^3 + 3\left(\frac{1 - 3\mathrm{i}}{2}\right)^2 + 6\left(\frac{1 - 3\mathrm{i}}{2}\right) - 40 = 0$

b $\frac{1 + 3\mathrm{i}}{2}, 2, -\frac{4}{3}$

c
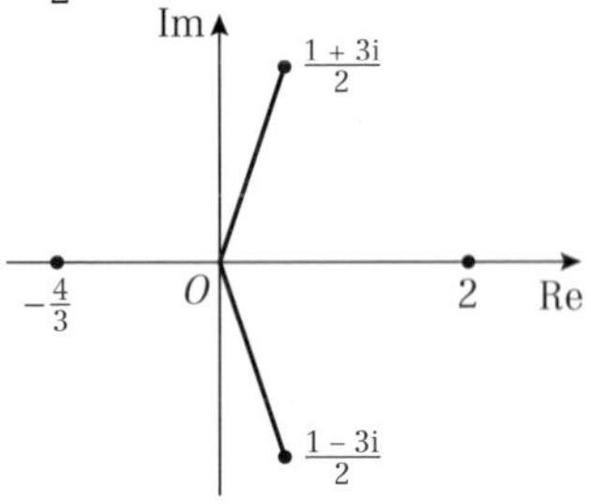

Exercise 4D

1 a $\frac{4}{3}$ b 9 c 10 d 28

2 a $-\frac{8}{9}$ b $\frac{9}{16}$ c $-\frac{19}{18}$ d $\frac{65}{54}$

3 a $\frac{37}{6}$ b $\frac{32}{3}$ c $\frac{481}{144}$

4 a $-\frac{3}{4}$ b 10 c 38 d -7

5 a $-\frac{8}{3}$ b $\frac{59}{12}$ c $\frac{87}{8}$ d $\frac{1}{8}$

6 a $\frac{71}{10}$ b $-\frac{683}{20}$ c $\frac{53}{20}$ d $\frac{13}{80}$ e $\frac{723}{1600}$

7 a 2 b -1 c 16

8 a $-\frac{3}{20}$ b $\frac{7}{4}$ c $\frac{64}{27}$ d $\frac{823}{240}$ e $\frac{51}{25}$

9 a $\frac{23}{12}$ b $\frac{59}{3}$

10 a $\alpha + \beta + \gamma = 6, \alpha\beta + \beta\gamma + \gamma\alpha = 9, \alpha\beta\gamma = 15$

b i $\frac{3}{5}$ ii 18 iii 11

11 a $\alpha + \beta + \gamma = -2, \alpha\beta + \beta\gamma + \gamma\alpha = 0, \alpha\beta\gamma = -\frac{7}{2}$

b i 4 ii $-\frac{343}{8}$ iii $-\frac{7}{2}$

12 $(\alpha + \beta + \gamma)^3 \equiv (\alpha + \beta + \gamma)(\alpha^2 + \beta^2 + \gamma^2 + 2(\alpha\beta + \beta\gamma + \gamma\alpha))$
$\equiv \alpha^3 + \beta^3 + \gamma^3 + \alpha(\beta^2 + \gamma^2) + \beta(\alpha^2 + \gamma^2)$
$+ \gamma(\alpha^2 + \beta^2) + 2(\alpha + \beta + \gamma)(\alpha\beta + \beta\gamma + \gamma\alpha)$

$(\alpha + \beta + \gamma)(\alpha\beta + \beta\gamma + \gamma\alpha)$
$\equiv \alpha^2\beta + \beta^2\alpha + \alpha^2\gamma + \gamma^2\alpha + \beta^2\gamma + \gamma^2\beta + 3\alpha\beta\gamma$
$\equiv \alpha(\beta^2 + \gamma^2) + \beta(\alpha^2 + \gamma^2) + \gamma(\alpha^2 + \beta^2) + 3\alpha\beta\gamma$

$(\alpha + \beta + \gamma)^3 \equiv \alpha^3 + \beta^3 + \gamma^3 + 3(\alpha + \beta + \gamma)(\alpha\beta + \beta\gamma + \gamma\alpha) - 3\alpha\beta\gamma$

$\alpha^3 + \beta^3 + \gamma^3 \equiv (\alpha + \beta + \gamma)^3 - 3(\alpha + \beta + \gamma)(\alpha\beta + \beta\gamma + \gamma\alpha) + 3\alpha\beta\gamma$

13 a -12 b $\alpha + \beta + \gamma = 0, \alpha\beta\gamma = -\frac{11}{3}$ c $\frac{128}{3}$

14 a $\sum\alpha = 0, \sum\alpha\beta = 2, \sum\alpha\beta\gamma = 1, \alpha\beta\gamma\delta = 3$

b i $\frac{1}{3}$ ii -4 iii 7

15 a 2 b $\sum\alpha = -\frac{3}{2}, \sum\alpha\beta = 1, \sum\alpha\beta\gamma = -\frac{1}{2}$ c $\frac{1}{6}$

16 $\left(\sum\alpha\right)^2 \equiv (\alpha + \beta + \gamma + \delta)^2$
$\equiv \alpha^2 + \beta^2 + \gamma^2 + \delta^2 + 2(\alpha\beta + \beta\gamma + \gamma\delta + \alpha\gamma + \beta\delta + \alpha\delta)$
$\Rightarrow \alpha^2 + \beta^2 + \gamma^2 + \delta^2 \equiv \left(\sum\alpha\right)^2 - 2(\alpha\beta + \beta\gamma + \gamma\delta + \alpha\gamma + \beta\delta + \alpha\delta)$

Online Full worked solutions are available in SolutionBank.

Exercise 4E

1 a $w^3 - 10w^2 + 23w - 9 = 0$
  b $w^3 - 14w^2 + 24w + 40 = 0$
2 a $3w^3 + 23w^2 + 52w + 31 = 0$
  b $24w^3 - 16w^2 - 10w + 1 = 0$
3 $w^3 - 9w^2 + 31w - 79 = 0$
4 $w^3 + 11w^2 + 3w + 9 = 0$
5 $w^3 - 4w^2 + 11w - 53 = 0$
6 a $2w^4 + 12w^3 - 45w^2 + 54w - 81 = 0$
  b $2w^4 + 12w^3 + 19w^2 + 12w + 2 = 0$
7 a $w^4 + 4w^3 - 12w^2 + 32w + 80 = 0$
  b $w^4 + 10w^3 + 33w^2 + 48w + 33 = 0$
8 a $w^4 + 5w^3 - 12w^2 - 27w + 27 = 0$
  b $3w^4 - 7w^3 - w^2 + 8w - 2 = 0$

Challenge

a $w^4 - 7w^3 + 17w^2 - 25w - 34 = 0$
b
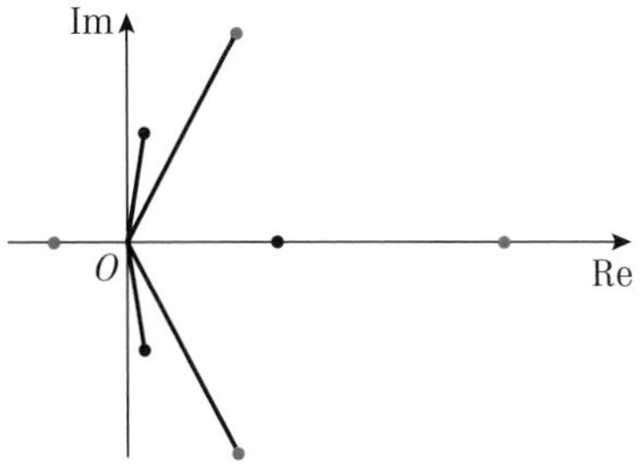

Mixed exercise 4

1 $a = 250, b = 325, c = 110, d = -7, e = -6$
2 a $\alpha\beta + \beta\gamma + \gamma\alpha = 37, \alpha\beta\gamma = 52, p = -\alpha - \beta - \gamma$
  b -10
3 a $-2 - i, \frac{3}{2}$ b -15
4 $x = 1, 7, 13$ or 19
5 a $\frac{1}{4}$
  b $d = 7, e = -2$
6 a $(-2 + i) + (-2 - i) + \gamma + \delta = -2 \Rightarrow \gamma + \delta - 2 = 0$
  $(-2 + i)(-2 - i)\gamma\delta = 85 \Rightarrow 5\gamma\delta = 85 \Rightarrow \gamma\delta - 17 = 0$
  b Roots: $-2 \pm i$, $1 \pm 4i$; $m = 14$, $n = 58$
  c
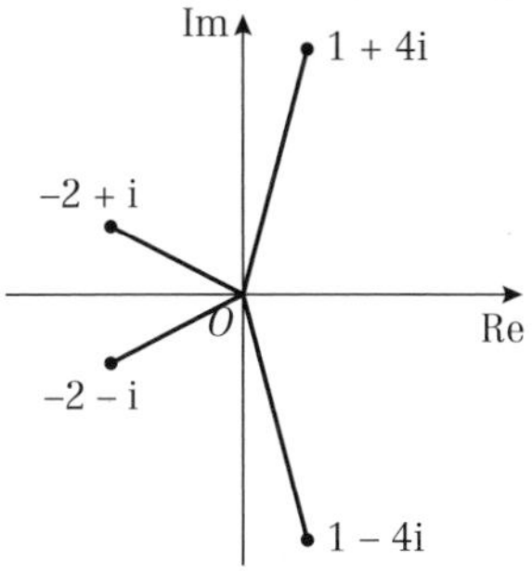

7 a $4(2 - 5i)^4 - 16(2 - 5i)^3 + 115(2 - 5i)^2 + 4(2 - 5i) - 29 = 0$
  b $2 + 5i, \pm\frac{1}{2}$
  c
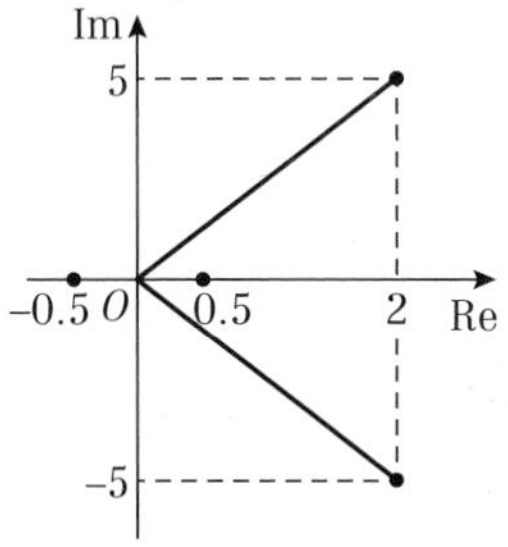

8 a $\alpha + \beta + \gamma = \frac{5}{2}, \alpha\beta + \beta\gamma + \gamma\alpha = \frac{11}{2}, \alpha\beta\gamma = \frac{9}{2}$
  b i $\frac{11}{9}$ ii $-\frac{19}{4}$ iii $\frac{1}{2}$
9 a 7 b $\sum\alpha = -\frac{12}{7}, \sum\alpha\beta = \frac{6}{7}, \sum\alpha\beta\gamma = -\frac{5}{7}$ c $\frac{60}{49}$
10 a -30 b $\alpha + \beta + \gamma = 0, \alpha\beta\gamma = -\frac{21}{5}$ c $-\frac{4}{5}$
11 $2w^3 + 9w^2 + 39w - 104 = 0$
12 a $3w^4 - 2w^3 - 10w^2 + 28w + 64 = 0$
  b $2w^4 + 14w^3 + 21w^2 + 43w + 298 = 0$

Challenge

1 $w^3 - 10w^2 + w + 1 = 0$
2 $w^3 + 4w^2 + w - 1 = 0$
3 $w^2 - 5\sqrt{w - 1} + 1 = 0$

CHAPTER 5

Prior knowledge 5

1 a 64 b $\frac{472}{5}$ c $\frac{7}{2}$
2 $\frac{64}{3}$
3 $\int_1^6 -x^2 + 7x - 6\,dx = \frac{125}{6}$

Exercise 5A

1 a 640π b $\frac{8}{3}\pi$ c 48π d $\frac{55}{24}\pi$
2 250π
3 $\frac{141\pi}{10}$
4 8π
5 a $(3, 0)$ b $\frac{2187}{20}\pi$
6 5.97
7 $\frac{279\pi}{20}$
8 $a = \frac{1}{2}$
9 $y = r, \pi\int_0^h r^2 dx = \pi[r^2x]_0^h = \pi r^2 h$

Challenge

$\frac{35}{2}\pi$

Exercise 5B

1 a $\frac{93}{4}\pi$ b $\frac{1}{80}\pi$ c $\frac{2}{3}\pi$ d 14016π
2 $\pi\int_1^4 \left(\frac{1}{2}y^2 + 1\right)^2 dy = \pi\left[\frac{y^5}{20} + \frac{y^3}{3} + y\right]_1^4 = \frac{1503\pi}{20}$
3 a $\frac{461}{36}$ b 104.21
4 a 6 b $\frac{78\pi}{5}$
5 $\frac{25\pi}{4}$
6 a $x^2 - 2x + 1 = y \Rightarrow y = (x - 1)^2$
  $\Rightarrow x = \sqrt{y} + 1\ (y > 0) \Rightarrow x^2 = y + 2\sqrt{y} + 1$
  b $\frac{248\pi}{3}$
7 8π
8 1.44
  $y = \frac{h}{r}x \Rightarrow x = \frac{ry}{h} \Rightarrow x^2 = \frac{r^2y^2}{h^2}$
9 $\pi\int_0^h x^2\,dy = \frac{r^2\pi}{h^2}\int_0^h y^2\,dy = \frac{r^2\pi}{h^2}\left[\frac{y^3}{3}\right]_0^h = \frac{1}{3}\pi r^2 h$

Exercise 5C

1 a $\frac{2187}{4}\pi$ b $\frac{729}{2}\pi$
  c Rotation about x-axis: $r = 13.5, h = 9$
  $V = \frac{1}{3}\pi(13.5)^2 \times 9 = \frac{2187}{4}\pi$
  Rotation about y-axis: $r = 9, h = 13.5$
  $V = \frac{1}{3}\pi \times 9^2 \times 13.5 = \frac{729}{2}\pi$
2 a Substitute (2, 8) into both equations.
  b $\frac{2272}{105}\pi + \frac{138}{3}\pi = \frac{6752}{105}\pi$
3 a Solve the equations simultaneously to get $x = 2$, $y = 4$
  B (4, 0)
  b $\frac{2048}{105}\pi$

4 **a** $A(-2, 2)$ and $B(2, 2)$ **b** $\frac{16\pi}{3} + \frac{12\pi}{3} = \frac{28\pi}{3}$

5 $\frac{81}{3}\pi - \frac{16}{3}\pi = \frac{65}{3}\pi$

6 $\frac{247}{6}\pi$

7 $16\pi - 6\pi = 10\pi$

8 $\frac{194\pi}{15}$

9 **a** $x = -\sqrt{2}$ and $x = \sqrt{2}$ **b** 63.98

Challenge

$\frac{1000}{9}\pi - \frac{64}{9}\pi + \frac{432}{9}\pi = 56\pi$

Exercise 5D

1 **a** $5 \leqslant k \leqslant 15$ **b** $\frac{40}{3}\pi k^3\,\text{m}^3$

c The actual tent would probably not follow the exact same shape as the model.

2 $\frac{32\,768}{15}\pi\,\text{cm}^3$

3 **a** 16π **b** 24π

c The solid formed by rotating about the x axis, as the solid formed by rotating about the y-axis will have a flatter, disc-shape which does not closely match the shape of an egg

4 $\pi\int_0^b y^{\frac{2}{3}}\,dy = 5 \times 8 \Rightarrow b = 6.25\,\text{cm}$

5 **a** 19.3 cm **b** $3162.8\,\text{cm}^3$ **c** $75.9\,\text{cm}^2$

d 50.9% or 0.509

6 **a** $325.8\,\text{cm}^3$

b Yes. Volume of water at a height of 10 cm is $280.2\,\text{cm}^3$. With the extra $50\,\text{cm}^3$, this is greater than $325.8\,\text{cm}^3$, so the water will overflow.

7 $486\pi + 259.2\pi = 745.2\pi$

Challenge

$\frac{159}{2}\pi$

Mixed exercise 5

1 $\frac{4374}{35}\pi$

2 **a** $\frac{1}{4}$ **b** $\frac{81}{8}\pi$

3 **a** $x^2 + 4x + 4 = y \Rightarrow y = (x + 2)^2$

$\Rightarrow x = \sqrt{y} - 2 \Rightarrow x^2 = 4 - 4\sqrt{y} + y$

b $\frac{11}{6}\pi$

4 **a** $\frac{56\pi}{5}$ **b** $\frac{7\pi}{2}$

5 **a** $\frac{2}{4}(2 + 1)^2 = 4.5$

$3 \times 2 + 4 \times 4.5 = 24$

b $\frac{1549}{210}\pi + \frac{81}{2}\pi = \frac{5027}{105}\pi$

6 **a** $17.6\pi - 0.8\pi = 16.8\pi\,\text{cm}^3$

b e.g. the shape of the holder is unlikely to exactly follow the curve

7 $\frac{128}{3}\pi + 4\pi = \frac{140}{3}\pi$

8 $\frac{343\pi}{16}$

Challenge

a $R = \pi y^2 = \pi(r^2 - x^2)$

b $V = \int_{-r}^{r}\pi(r^2 - x^2)\,dx = 2\pi\int_0^r\pi(r^2 - x^2)\,dx$

$= 2\pi\left[r^2x - \frac{x^3}{3}\right]_0^r = 2\pi\left(r^3 - \frac{r^3}{3}\right) = \frac{4}{3}\pi r^3$

Review exercise 1

1 **a** $4 - (5 + p)\text{i}$ **b** $5p + 4p\text{i}$ **c** $-\frac{5}{p} - \frac{4}{p}\text{i}$

2 **a** $-2\sqrt{3} < k < 2\sqrt{3}$ **b** $0, 1 \pm \sqrt{2}\text{i}$

3 $\frac{5}{2} + \text{i}\frac{3\sqrt{3}}{2}, \frac{5}{2} - \text{i}\frac{3\sqrt{3}}{2}$

4 $x = 3, y = -1$

5 **a** $\frac{1}{2}$ **b** $-\frac{1}{4}$

6 $-1 + \text{i}, -1 - \text{i}, -3$

7 **a** 21 **b** $2 + 3\text{i}, 2$

8 **a** $(z^2 - 3z - 4)(z^2 + 2z + 4)$

b $-1, 4, -1 + \text{i}\sqrt{3}, -1 - \text{i}\sqrt{3}$

9 $3 + \text{i}, 3 - \text{i}, 1 - 2\text{i}, 1 + 2\text{i}$

10 **a** $1 + \text{i}\sqrt{3}, 1 - \text{i}\sqrt{3}, 3$ **b** $-5, 10$

11 **a** $\frac{1}{3} - \frac{4}{3}\text{i}$

b

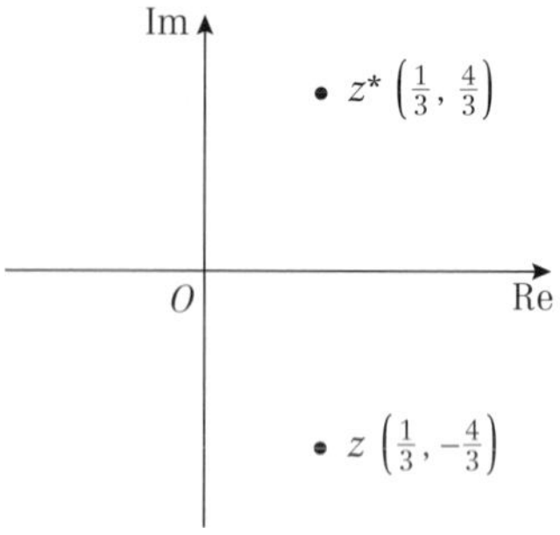

c $z = \frac{\sqrt{17}}{3}(\cos(-76°) + \text{i}\sin(-76°))$

$z^* = \frac{\sqrt{17}}{3}(\cos 76° + \text{i}\sin 76°)$

12 1.318 radians, 1.823 radians

13 **a**

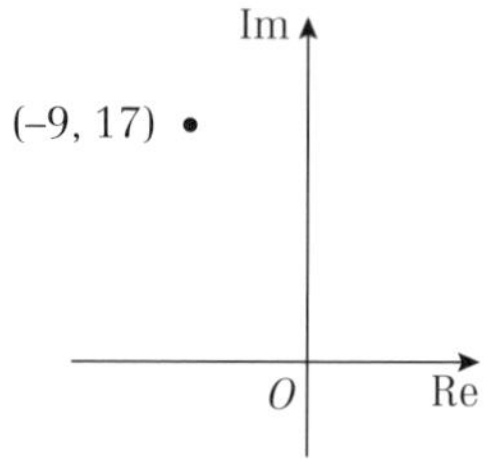

b 2.06 **c** $1 - 2\text{i}$

14 **a** $|z_1|^2 = |5 + \text{i}|^2 = 25 + 1 = 26$

$|z_2|^2 = |-2 + 3\text{i}|^2 = 4 + 9 = 13 = \frac{|z_1|^2}{2}$

b $\frac{3\pi}{4}$

15 **a** $z^2 = (2 - \text{i})^2 = 4 - 4\text{i} + \text{i}^2 = 4 - 4\text{i} - 1 = 3 - 4\text{i}$

b $2 - 2\text{i}$ and -2

c

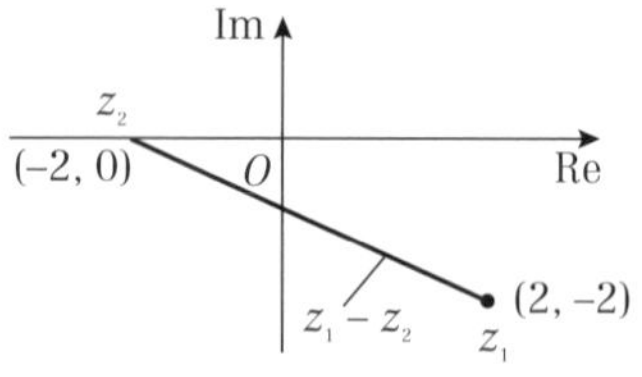

d $|z_1 - z_2|^2 = \sqrt{(-2 - 2)^2 + (-2)^2} = 2\sqrt{5}$ **e** $-\frac{\pi}{2}$

16 **a**

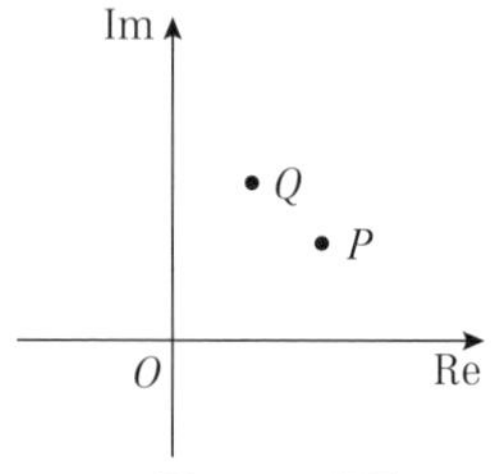

b $|z_1| = 2\sqrt{2}, |z_2| = \sqrt{10}, PQ = \sqrt{2}$

Online Full worked solutions are available in SolutionBank.

c $|z_1|^2 + PQ^2 = 8 + 2 = 10 = |z_2|^2$, so $\angle OPQ$ is a right angle.

d $-1 + i$

17 $\dfrac{(\cos 2x + i\sin 2x)(\cos 9x + i\sin 9x)}{(\cos 9x - i\sin 9x)(\cos 9x + i\sin 9x)}$

$= \dfrac{(\cos 2x\cos 9x - \sin 2x\sin 9x) + i(\cos 2x\sin 9x + \sin 2x\cos 9x)}{\cos^2 9x + \sin^2 9x}$

$= \dfrac{\cos 11x + i\sin 11x}{1}$

$n = 11$

18 a

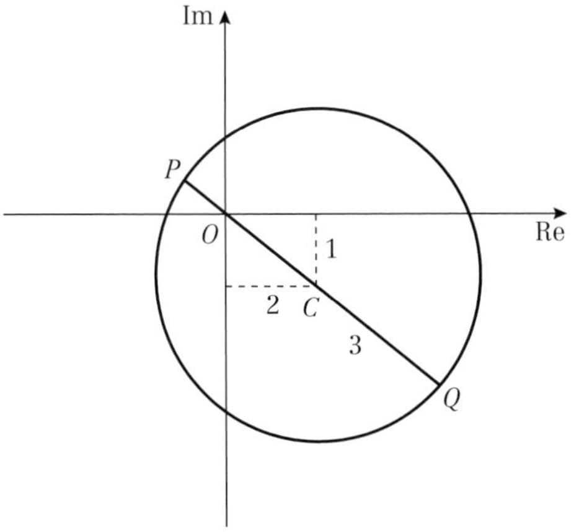

b maximum value of $|z|$ is $3 + \sqrt{5}$
minimum value of $|z|$ is $3 - \sqrt{5}$

19 a

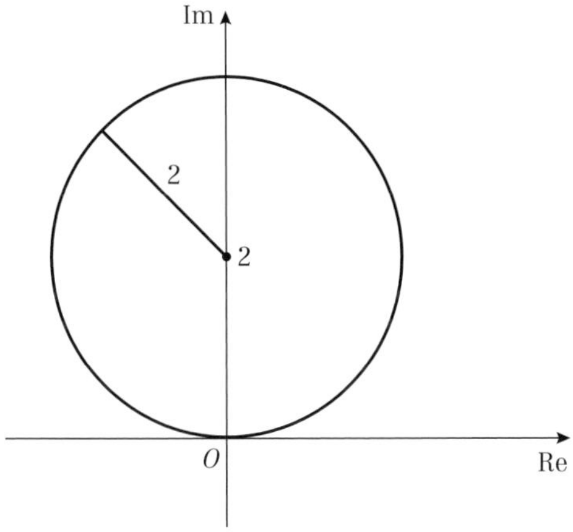

b 4

20 a

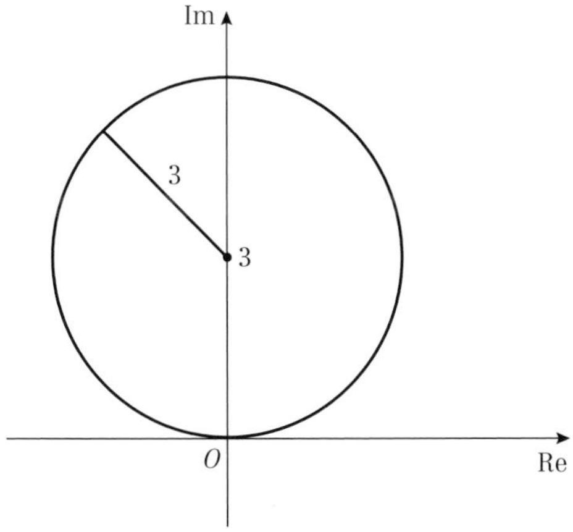

b $z = -\dfrac{3\sqrt{2}}{2} + \left(3 + \dfrac{3\sqrt{2}}{2}\right)i$

21

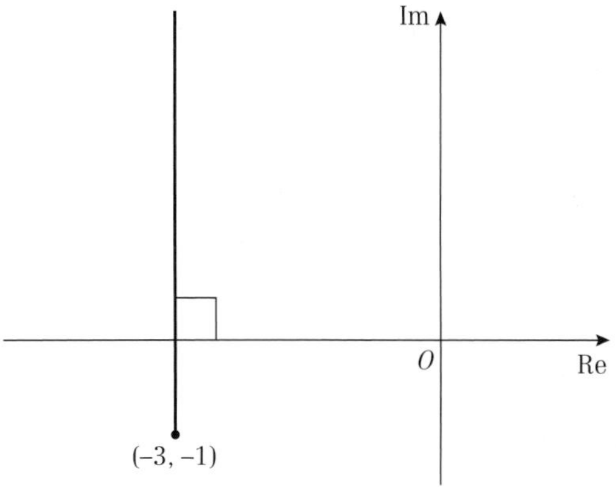

22 a

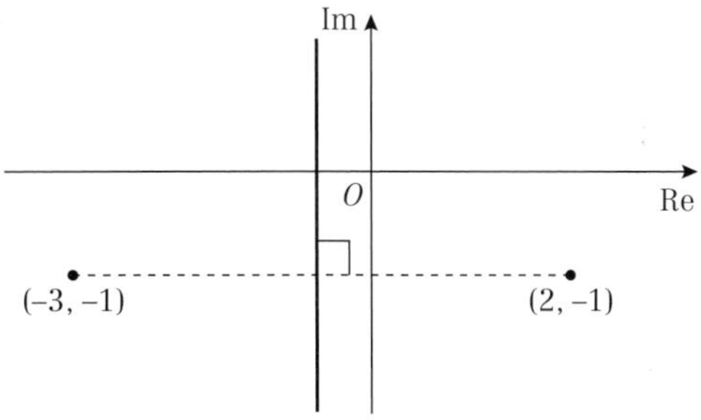

b $\frac{1}{2}$

c $-\frac{1}{2} - \frac{1}{2}i$

23

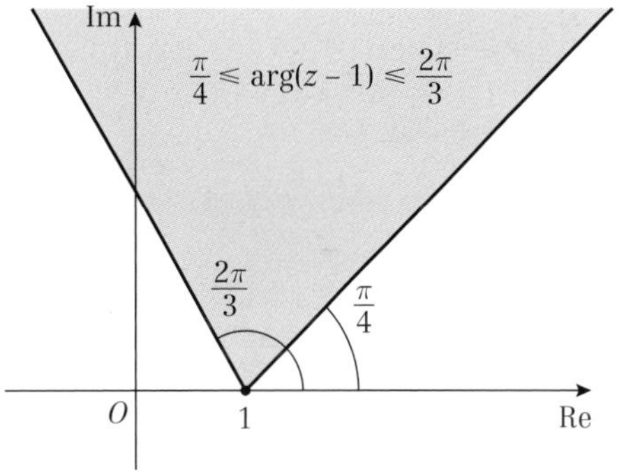

24

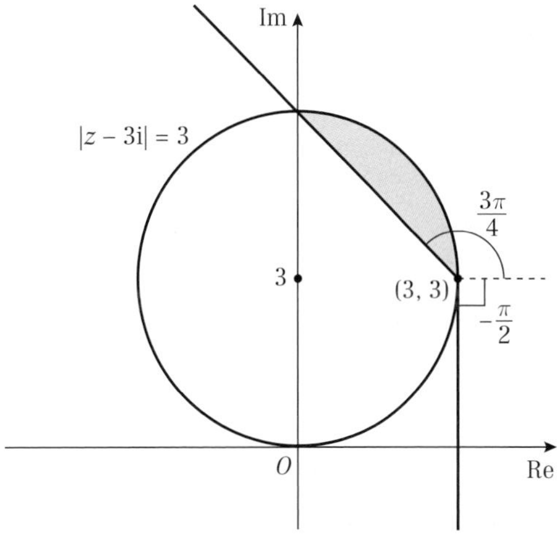

25 $\sum_{r=1}^{n}(2r-1)^2 = \sum_{r=1}^{n}(4r^2 - 4r + 1) = 4\sum_{r=1}^{n}r^2 - 4\sum_{r=1}^{n}r + n$

$= \frac{1}{3}n(n+1)(2n+1) - 2n(n+1) + n = \frac{1}{3}n(4n^2 - 1)$

26 $\sum_{r=1}^{n} r(r^2 - 3) = \sum_{r=1}^{n} r^3 - 3\sum_{r=1}^{n} r = \frac{1}{4}n^2(n + 1)^2 - \frac{3}{2}n(n + 1)$
$= \frac{1}{4}n(n + 1)(n - 2)(n + 3)$

27 **a** $\sum_{r=1}^{n} r(2r - 1) = 2\sum_{r=1}^{n} r^2 - \sum_{r=1}^{n} r$
$= \frac{1}{3}n(n + 1)(2n + 1) - \frac{1}{2}n(n + 1) = \frac{1}{6}n(n + 1)(4n - 1)$
b 17 730

28 **a** $\sum_{r=1}^{n} (6r^2 + 4r - 5) = 6\sum_{r=1}^{n} r^2 + 4\sum_{r=1}^{n} r - 5n$
$= n(n + 1)(2n + 1) + 2n(n + 1) - 5n = n(2n^2 + 5n - 2)$
b 32 480

29 **a** $\sum_{r=1}^{n} r(r + 1) = \sum_{r=1}^{n} r^2 + \sum_{r=1}^{n} r$
$= \frac{1}{6}n(n + 1)(2n + 1) + \frac{1}{2}n(n + 1) = \frac{1}{3}n(n + 1)(n + 2)$
b $\sum_{r=n}^{3n} r(r + 1) = \sum_{r=1}^{3n} r(r + 1) - \sum_{r=1}^{n-1} r(r + 1)$
$= \frac{1}{3}(3n)(3n + 1)(3n + 2) - \frac{1}{3}(n - 1)n(n + 1)$
$= \frac{1}{3}n(26n^2 + 27n + 7) = \frac{1}{3}n(2n + 1)(13n + 7)$
$p = 13, q = 7$

30 **a** $p = 3, q = -1, r = -2$
b 23 703 950

31 **a** -12 **b** $0, -\frac{11}{3}$ **c** $\frac{2}{3}$

32 **a** 4 **b** $-\frac{7}{4}, \frac{5}{4}, -\frac{3}{4}$ **c** $\frac{9}{16}$

33 $w^3 + 3w^2 + 11w - 23 = 0$

34 **a** $w^4 - 3w^3 - 18w^2 + 81w + 324 = 0$
b $w^4 + 2w^3 - 8w^2 + 6w + 79 = 0$

35 **a** 1 **b** $\frac{2\pi}{15}$

36 5

37 24.7

38 **a** 20
b $V = \pi\int_a^b 100 - (y - 20)^2 \, dy = \pi\int_a^b 40y - y^2 - 300 \, dy$
$= \pi\left[20y^2 - \frac{y^3}{3} - 300y\right]_a^b$
$= \frac{\pi}{3}(60(b^2 - a^2) - (b^3 - a^3) - 900(b - a))$
c £104.72

39 **a** 24 mm **b** 13.6 grams
c Any valid reason, e.g. the gold may have voids or impurities, the actual dimensions may differ from those modelled, answer is given to too great a degree of accuracy.

Challenge

1 **a**

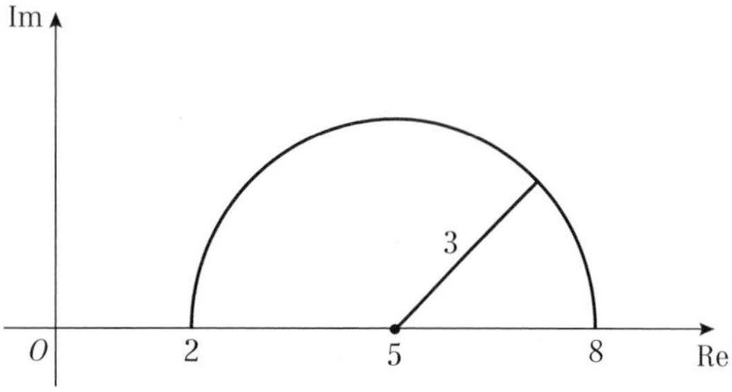

b 3

2 **a** $2r + 4$
b $\sum_{r=n}^{2n} u_r = \sum_{r=1}^{2n} u_r - \sum_{r=1}^{n-1} u_r$
$= ((2n)^2 + 5(2n)) - ((n - 1)^2 - 5(n - 1))$
$= 3n^2 + 7n + 4 = (n + 1)(3n + 4)$

3 $w^3 - 6w^2 - 20w - 200 = 0$

CHAPTER 6

Prior knowledge 6

1 **a** $\begin{pmatrix} 6 \\ 2 \end{pmatrix}$ **b** $\begin{pmatrix} -2 \\ 11 \end{pmatrix}$ **c** $\begin{pmatrix} 8 \\ -16 \end{pmatrix}$

2 **a** $x = 7, y = 3$ **b** $x = -\frac{40}{7}, y = -\frac{23}{7}$

Exercise 6A

1 **a** 2×2 **b** 2×1 **c** 2×3
d 1×3 **e** 1×2 **f** 3×3

2 $\begin{pmatrix} 1 & 0 & 0 & 0 \\ 0 & 1 & 0 & 0 \\ 0 & 0 & 1 & 0 \\ 0 & 0 & 0 & 1 \end{pmatrix}$

3 $a = 6, b = 2$

4 **a** $\begin{pmatrix} 8 & -1 \\ 1 & 4 \end{pmatrix}$ **b** $\begin{pmatrix} 2 & 2 \\ -2 & -5 \end{pmatrix}$ **c** $\begin{pmatrix} 0 & 0 \\ 0 & 0 \end{pmatrix}$

5 **a** Not possible **b** $\begin{pmatrix} -2 \\ 3 \end{pmatrix}$ **c** $(1 \quad 1 \quad 4)$
d Not possible **e** $(3 \quad -1 \quad 4)$
f Not possible **g** $(-3 \quad 1 \quad -4)$

6 $a = 6, b = 3, c = 2, d = -1$

7 $a = 4, b = 3, c = 5$

8 $a = 2, b = -2, c = 2, d = 1, e = -1, f = 3$

9 **a** $\begin{pmatrix} 7 & 2 & -1 \\ 3 & 1 & 4 \\ 1 & 4 & -3 \end{pmatrix}$ **b** $\begin{pmatrix} 8 & 3 & -7 \\ -1 & -7 & 8 \\ -1 & -1 & -4 \end{pmatrix}$ **c** $\begin{pmatrix} -1 & -1 & 6 \\ 4 & 8 & -4 \\ 2 & 5 & 1 \end{pmatrix}$
d **i** $a = 3, b = -5, c = 6$ **ii** $a = 8, b = -9, c = -1$

10 **a** $\begin{pmatrix} 6 & 0 \\ 12 & -18 \end{pmatrix}$ **b** $\begin{pmatrix} 1 & 0 \\ 2 & -3 \end{pmatrix}$ **c** $\begin{pmatrix} 2 \\ -2 \end{pmatrix}$
d **A** and **B** are not the same size, so you can't subtract them.

11 **a** $\begin{pmatrix} 13 & -4 \\ -1 & 6 \end{pmatrix}$ **b** $\begin{pmatrix} -2 & -8 \\ 10 & -12 \end{pmatrix}$
c $\begin{pmatrix} 11 & -12 \\ 9 & -6 \end{pmatrix}$ **d** $\begin{pmatrix} \frac{9}{2} & \frac{1}{2} \\ -\frac{5}{2} & \frac{9}{2} \end{pmatrix}$

12 **a** $\begin{pmatrix} 14 & 0 & 9 \\ 7 & -9 & 1 \\ 2 & 0 & 2 \end{pmatrix}$ **b** $\begin{pmatrix} 0 & 14 & -8 \\ 0 & -6 & -4 \\ -1 & 7 & 6 \end{pmatrix}$
c $\begin{pmatrix} 38 & 6 & 21 \\ 19 & -27 & 1 \\ 5 & 3 & 8 \end{pmatrix}$ **d** $\begin{pmatrix} -\frac{5}{3} & \frac{11}{3} & -\frac{19}{6} \\ -\frac{5}{6} & -\frac{1}{2} & -\frac{7}{6} \\ -\frac{1}{2} & \frac{11}{6} & \frac{4}{3} \end{pmatrix}$

13 $k = 3, x = -1$

14 $a = 3, b = -3.5, c = -1, d = 2$

15 $a = 5, b = 5, c = -2, d = 2$

16 $k = \frac{3}{2}$

17 **a** $k = 2$ **b** $p = 13, q = 3, r = 12$

18 $a = -6, b = -4, c = 2, d = 0, e = \frac{1}{2}, k = -2$

Exercise 6B

1 **a** 1×2 **b** 3×3 **c** 1×2
d 2×2 **e** 2×3 **f** 3×2

2 **a** $\begin{pmatrix} 3 \\ 6 \end{pmatrix}$ **b** $\begin{pmatrix} -2 & 1 \\ -4 & 7 \end{pmatrix}$

3 **a** $\begin{pmatrix} -3 & -2 & -1 \\ 3 & 3 & 0 \end{pmatrix}$ **b** $\begin{pmatrix} 1 & -4 \\ 0 & 9 \end{pmatrix}$

Online Full worked solutions are available in SolutionBank.

4 **a** Not possible **b** $\begin{pmatrix}-6 & -4\\ -3 & -2\end{pmatrix}$

c Not possible **d** $\begin{pmatrix}7\\ 0\end{pmatrix}$

e (-8) **f** $(-7 \quad -7)$

5 $\begin{pmatrix}2 & 6-a & 2a\\ 1 & 4 & -2\end{pmatrix}$

6 $\begin{pmatrix}3x+2 & 0\\ 0 & 3x+2\end{pmatrix}$

7 **a** $\begin{pmatrix}8 & -3\\ -10 & 6\end{pmatrix}$ **b** $\begin{pmatrix}3 & -2\\ 20 & 13\end{pmatrix}$ **c** $\begin{pmatrix}13 & -21\\ -36 & -13\end{pmatrix}$

8 **a** $\begin{pmatrix}3+k^2 & 2k\\ 2k & k^2-1\end{pmatrix}$ **b** $\begin{pmatrix}3+k^2 & 2k\\ 2k & k^2-1\end{pmatrix}$

c $\begin{pmatrix}7 & k\\ k & 5\end{pmatrix}$ **d** $\begin{pmatrix}9+2k^2 & 7k\\ 7k & 2k^2-5\end{pmatrix}$

9 **a** $\begin{pmatrix}1 & 4\\ 0 & 1\end{pmatrix}$ **b** $\begin{pmatrix}1 & 6\\ 0 & 1\end{pmatrix}$ **c** $\begin{pmatrix}1 & 2\times k\\ 0 & 1\end{pmatrix}$

10 **a** $\begin{pmatrix}a^2 & 0\\ ab & 0\end{pmatrix}$ **b** 3

11 **a** $\begin{pmatrix}-8 & -14\\ -4 & -7\\ 0 & 0\end{pmatrix}$ **b** $(-16 \quad 29)$

12 **a** $\begin{pmatrix}-1\\ 1\\ -2\end{pmatrix}$ **b** $(-3 \quad 2 \quad 3)$

13 **a** $\begin{pmatrix}1 & 0\\ 0 & 1\end{pmatrix}$

b $\mathbf{AI} = \begin{pmatrix}2 & -2\\ 1 & 3\end{pmatrix}\begin{pmatrix}1 & 0\\ 0 & 1\end{pmatrix} = \begin{pmatrix}2+0 & 0-2\\ 1+0 & 0+3\end{pmatrix} = \begin{pmatrix}2 & -2\\ 1 & 3\end{pmatrix} = \mathbf{A}$

$\mathbf{IA} = \begin{pmatrix}1 & 0\\ 0 & 1\end{pmatrix}\begin{pmatrix}2 & -2\\ 1 & 3\end{pmatrix} = \begin{pmatrix}2+0 & -2+0\\ 0+1 & 0+3\end{pmatrix} = \begin{pmatrix}2 & -2\\ 1 & 3\end{pmatrix} = \mathbf{A}$

14 $\mathbf{AB} = \begin{pmatrix}9 & 4\\ 10 & 6\end{pmatrix}$ and $\mathbf{AC} = \begin{pmatrix}2 & 5\\ 3 & 4\end{pmatrix}$, so $\mathbf{AB} + \mathbf{AC} = \begin{pmatrix}11 & 9\\ 13 & 10\end{pmatrix}$

$\mathbf{B} + \mathbf{C} = \begin{pmatrix}5 & 4\\ -1 & -1\end{pmatrix}$

so $\mathbf{A}(\mathbf{B} + \mathbf{C}) = \begin{pmatrix}2 & -1\\ 3 & 2\end{pmatrix}\begin{pmatrix}5 & 4\\ -1 & -1\end{pmatrix} = \begin{pmatrix}11 & 9\\ 13 & 10\end{pmatrix} = \mathbf{AB} + \mathbf{AC}$

15 $\mathbf{A}^2 = \begin{pmatrix}1 & 2\\ 3 & 1\end{pmatrix}\begin{pmatrix}1 & 2\\ 3 & 1\end{pmatrix} = \begin{pmatrix}7 & 4\\ 6 & 7\end{pmatrix}$

$2\mathbf{A} + 5\mathbf{I} = \begin{pmatrix}2 & 4\\ 6 & 2\end{pmatrix} + \begin{pmatrix}5 & 0\\ 0 & 5\end{pmatrix} = \begin{pmatrix}7 & 4\\ 6 & 7\end{pmatrix}$

16 $\begin{pmatrix}2a+c+1 & bc & 2+c\\ 1 & 2a+b+1 & ac-1\\ ab+1 & 2-b & b+c\end{pmatrix}$

17 $a = 3, b = -2$

18 $\mathbf{AB} = \begin{pmatrix}p & 3\\ 6 & p\end{pmatrix}\begin{pmatrix}q & 2\\ 4 & q\end{pmatrix} = \begin{pmatrix}pq+12 & 2p+3q\\ 6q+4p & 12+pq\end{pmatrix}$

$\mathbf{BA} = \begin{pmatrix}q & 2\\ 4 & q\end{pmatrix}\begin{pmatrix}p & 3\\ 6 & p\end{pmatrix} = \begin{pmatrix}pq+12 & 2p+3q\\ 6q+4p & 12+pq\end{pmatrix}$

19 $p = 2, q = -3$

Challenge

a e.g. $\begin{pmatrix}0 & 1\\ 0 & 0\end{pmatrix}$ **b** $\begin{pmatrix}1 & 1\\ -1 & -1\end{pmatrix}$

Exercise 6C

1 **a** 10 **b** 6 **c** −3
d 0 **e** 21 **f** 4

2 **a** −3 **b** −5 **c** $\frac{1}{4}$

3 $k = 2 - \sqrt{3}, k = 2 + \sqrt{3}$

4 $k = -4, k = 1$

5 **a** $\det\mathbf{A} = 0, \det\mathbf{B} = 0$ **b** $\begin{pmatrix}0 & 0\\ 0 & 0\end{pmatrix}$

6 **a** 6 **b** −56 **c** 1 **d** 0

7 **a** 20 **b** 17 **c** 0

8 17

9 −8, −2

10 **a** $\det\mathbf{A} = 2\begin{vmatrix}0 & 4\\ 10 & 8\end{vmatrix} - 5\begin{vmatrix}-2 & 4\\ 3 & 8\end{vmatrix} + 3\begin{vmatrix}-2 & 0\\ 3 & 10\end{vmatrix}$
$= 2(0-40) - 5(-16-12) + 3(-20-0) = 0$

b $\begin{pmatrix}7 & 6 & 7\\ -2 & -10 & -4\\ 13 & 7 & 12\end{pmatrix}$

c $\det(\mathbf{AB}) = 7\begin{vmatrix}-10 & -4\\ 7 & 12\end{vmatrix} - 6\begin{vmatrix}-2 & -4\\ 13 & 12\end{vmatrix} + 7\begin{vmatrix}-2 & -10\\ 13 & 7\end{vmatrix}$
$= 7(-120+28) - 6(-24+52) + 7(-14+130)$
$= 0$

11 $\det\begin{pmatrix}0 & a & -b\\ -a & 0 & c\\ b & -c & 0\end{pmatrix} = 0\begin{vmatrix}0 & c\\ -c & 0\end{vmatrix} - a\begin{vmatrix}-a & c\\ b & 0\end{vmatrix} - b\begin{vmatrix}-a & 0\\ b & -c\end{vmatrix}$
$= 0 + abc - abc = 0$

12 Determinant $= 2x^2 + 10x + 44 = 2\left(x + \frac{5}{2}\right)^2 + \frac{63}{2} \geqslant \frac{63}{2}$ for $x \in \mathbb{R}$.
So determinant ≠ 0, therefore matrix is non-singular.

13 −1, 0, 3

14 **a** 7 **b** $k = -2.5$ **c** $\begin{pmatrix}-13 & -11.5\\ 2 & -2\end{pmatrix}$

d $\det\mathbf{MN} = 26 + 23 = 49$, $\det\mathbf{M}\det\mathbf{N} = 7 \times 7 = 49$

15 **a** −35 **b** −4 **c** $\begin{pmatrix}2 & 4 & 6\\ 3 & 9 & 14\\ -20 & 6 & 5\end{pmatrix}$

d $\det(\mathbf{AB}) = \begin{vmatrix}2 & 4 & 6\\ 3 & 9 & 14\\ -20 & 6 & 5\end{vmatrix}$
$= -70 = (-35) \times 2 = \det\mathrm{A} \times \det\mathrm{B}$

Challenge

a $\begin{pmatrix}1 & 1\\ 1 & 1\end{pmatrix}, \begin{pmatrix}-1 & -1\\ 1 & 1\end{pmatrix}, \begin{pmatrix}1 & 1\\ -1 & -1\end{pmatrix}, \begin{pmatrix}-1 & 1\\ -1 & 1\end{pmatrix},$
$\begin{pmatrix}1 & -1\\ 1 & -1\end{pmatrix}, \begin{pmatrix}1 & -1\\ -1 & 1\end{pmatrix}, \begin{pmatrix}-1 & 1\\ 1 & -1\end{pmatrix}, \begin{pmatrix}-1 & -1\\ -1 & -1\end{pmatrix}$

b $\begin{pmatrix}0 & 0\\ 0 & 0\end{pmatrix}, \begin{pmatrix}1 & 0\\ 0 & 0\end{pmatrix}, \begin{pmatrix}0 & 1\\ 0 & 0\end{pmatrix}, \begin{pmatrix}0 & 0\\ 1 & 0\end{pmatrix}, \begin{pmatrix}0 & 0\\ 0 & 1\end{pmatrix},$
$\begin{pmatrix}1 & 1\\ 0 & 0\end{pmatrix}, \begin{pmatrix}1 & 0\\ 1 & 0\end{pmatrix}, \begin{pmatrix}0 & 1\\ 0 & 1\end{pmatrix}, \begin{pmatrix}0 & 0\\ 1 & 1\end{pmatrix}, \begin{pmatrix}1 & 1\\ 1 & 1\end{pmatrix}$

Exercise 6D

1 **a** Non-singular. Inverse $= \begin{pmatrix}1 & 0.5\\ 2 & 1.5\end{pmatrix}$

b Singular **c** Singular

d Non-singular. Inverse $= \begin{pmatrix}-5 & 2\\ 3 & -1\end{pmatrix}$ **e** Singular

f Non-singular. Inverse $= \begin{pmatrix}-0.2 & 0.3\\ 0.6 & -0.4\end{pmatrix}$

2 **a** $\begin{pmatrix}-(2+a) & 1+a\\ 1+a & -a\end{pmatrix}$

b $\begin{pmatrix}\frac{-1}{a} & \frac{-3}{a}\\ \frac{1}{b} & \frac{2}{b}\end{pmatrix}$ (provided $a \neq 0, b \neq 0$)

3 **a** $(\mathbf{A}^{-1}\mathbf{A})\mathbf{BC} = \mathbf{A}^{-1} \Rightarrow (\mathbf{B}^{-1}\mathbf{B})\mathbf{C} = \mathbf{B}^{-1}\mathbf{A}^{-1}$
$\Rightarrow \mathbf{CA} = \mathbf{B}^{-1}(\mathbf{A}^{-1}\mathbf{A}) = \mathbf{B}^{-1}$

b $\begin{pmatrix}3 & 4\\ -1 & -1\end{pmatrix}$

4 **a** $\mathbf{B} = \mathbf{A}^{-1}\mathbf{C}$ **b** $\begin{pmatrix}1 & 4\\ -1 & 2\end{pmatrix}$

5 **a** $\mathbf{A} = \mathbf{C}^{-1}$ **b** $\begin{pmatrix}2 & -3\\ -3 & 5\end{pmatrix}$

6 $\begin{pmatrix} 2 & 4 & -3 \\ 0 & 1 & 2 \end{pmatrix}$

7 $\begin{pmatrix} 1 & 3 \\ -2 & 1 \\ 0 & -1 \end{pmatrix}$

8 **a** $\frac{1}{2ab}\begin{pmatrix} 2b & -b \\ -4a & 3a \end{pmatrix}$ **b** $\begin{pmatrix} -3 & 2 \\ -1 & \frac{3}{2} \end{pmatrix}$

9 **a** $\mathbf{A}^2\mathbf{B} = \mathbf{ABA} \Rightarrow \mathbf{A}^2\mathbf{B} = \mathbf{B} \Rightarrow \mathbf{A}^2 = \mathbf{BB}^{-1} \Rightarrow \mathbf{A}^2 = \mathbf{I}$

b $\mathbf{AB} = \begin{pmatrix} c & d \\ a & b \end{pmatrix}, \mathbf{BA} = \begin{pmatrix} b & a \\ d & c \end{pmatrix} \Rightarrow \begin{pmatrix} c & d \\ a & b \end{pmatrix} = \begin{pmatrix} b & a \\ d & c \end{pmatrix}$

Hence $a = d$ and $b = c$.

10 **a** $k \neq -\frac{2}{3}$ **b** $\frac{1}{3k+2}\begin{pmatrix} 1 & 3 \\ k & -2 \end{pmatrix}$

11 **a** $\frac{1}{p-4}\begin{pmatrix} -1 & -\frac{p}{2} \\ 1 & 2 \end{pmatrix}$ **b** $p = 3$

12 **a** $k^2 + 3k + 12$

b Determinant $= \left(k + \frac{3}{2}\right)^2 + 9.75 \geqslant 9.75$, so non-singular.

c $k = -1$

13 **a** $\frac{1}{2a^2-6}\begin{pmatrix} 2a & -2 \\ -3 & a \end{pmatrix}$

b $a = \sqrt{3}, a = -\sqrt{3}$

Exercise 6E

1 **a** $\begin{pmatrix} 1 & 0 & 0 \\ 0 & \frac{2}{3} & -\frac{1}{3} \\ 0 & -\frac{1}{3} & \frac{2}{3} \end{pmatrix}$ **b** $\begin{pmatrix} 1 & 0 & 0 \\ 0 & \frac{1}{2} & 0 \\ 0 & 0 & \frac{1}{3} \end{pmatrix}$ **c** $\begin{pmatrix} 1 & 0 & 0 \\ 0 & \frac{3}{5} & \frac{4}{5} \\ 0 & -\frac{4}{5} & \frac{3}{5} \end{pmatrix}$

2 **a** $\begin{pmatrix} 4 & -6 & -1 \\ -3 & 4 & 1 \\ -6 & 9 & 2 \end{pmatrix}$ **b** $\begin{pmatrix} -\frac{3}{5} & -\frac{1}{5} & \frac{7}{5} \\ -\frac{1}{5} & -\frac{2}{5} & \frac{4}{5} \\ \frac{7}{5} & \frac{4}{5} & -\frac{13}{5} \end{pmatrix}$ **c** $\begin{pmatrix} 2 & -5 & -\frac{19}{2} \\ 1 & -3 & -5 \\ 1 & -3 & -\frac{11}{2} \end{pmatrix}$

3 **a** $\begin{pmatrix} -1 & 0 & 1 \\ 0 & 1 & 0 \\ 2 & 0 & -1 \end{pmatrix}$ **b** $\begin{pmatrix} \frac{1}{3} & \frac{1}{2} & -\frac{1}{6} \\ 0 & -\frac{1}{2} & \frac{1}{2} \\ -\frac{1}{3} & \frac{1}{2} & \frac{1}{6} \end{pmatrix}$

c $\mathbf{B}^{-1}\mathbf{A}^{-1} = \begin{pmatrix} \frac{1}{3} & \frac{1}{2} & -\frac{1}{6} \\ 0 & -\frac{1}{2} & \frac{1}{2} \\ -\frac{1}{3} & \frac{1}{2} & \frac{1}{6} \end{pmatrix}\begin{pmatrix} -1 & 0 & 1 \\ 0 & 1 & 0 \\ 2 & 0 & -1 \end{pmatrix} = \begin{pmatrix} -\frac{2}{3} & \frac{1}{2} & \frac{1}{2} \\ 1 & -\frac{1}{2} & -\frac{1}{2} \\ \frac{2}{3} & \frac{1}{2} & -\frac{1}{2} \end{pmatrix}$

$= (\mathbf{AB})^{-1}$

4 **a** $\det\mathbf{A} = 2\begin{vmatrix} 1 & 1 \\ 1 & 4 \end{vmatrix} - 0\begin{vmatrix} k & 1 \\ 1 & 4 \end{vmatrix} + 3\begin{vmatrix} k & 1 \\ 1 & 1 \end{vmatrix}$

$= 2(4 - 1) + 3(k - 1) = 3(k + 1)$

b $\frac{1}{3(k+1)}\begin{pmatrix} 3 & 3 & -3 \\ 1-4k & 5 & 3k-2 \\ k-1 & -2 & 2 \end{pmatrix}$

5 $a = -4, b = 8, c = 3$

6 **a** $\mathbf{A}^2 = \begin{pmatrix} 2 & -1 & 1 \\ 4 & -3 & 0 \\ -3 & 3 & 1 \end{pmatrix}\begin{pmatrix} 2 & -1 & 1 \\ 4 & -3 & 0 \\ -3 & 3 & 1 \end{pmatrix} = \begin{pmatrix} -3 & 4 & 3 \\ -4 & 5 & 4 \\ 3 & -3 & -2 \end{pmatrix}$,

$\mathbf{A}^3 = \begin{pmatrix} 2 & -1 & 1 \\ 4 & -3 & 0 \\ -3 & 3 & 1 \end{pmatrix}\begin{pmatrix} -3 & 4 & 3 \\ -4 & 5 & 4 \\ 3 & -3 & 2 \end{pmatrix} = \begin{pmatrix} 1 & 0 & 0 \\ 0 & 1 & 0 \\ 0 & 0 & 1 \end{pmatrix}$

b $\begin{pmatrix} -3 & 4 & 3 \\ -4 & 5 & 4 \\ 3 & -3 & -2 \end{pmatrix}$

7 **a** $\mathbf{A}^2 = \begin{pmatrix} 1 & 1 & 0 \\ 3 & -3 & 1 \\ 0 & 3 & 2 \end{pmatrix}\begin{pmatrix} 1 & 1 & 0 \\ 3 & -3 & 1 \\ - & 3 & 2 \end{pmatrix} = \begin{pmatrix} 4 & -2 & 1 \\ -6 & 15 & -1 \\ 9 & -3 & 7 \end{pmatrix}$

$\mathbf{A}^3 = \begin{pmatrix} 1 & 1 & 0 \\ 3 & -3 & 1 \\ 0 & 3 & 2 \end{pmatrix}\begin{pmatrix} 4 & -2 & 1 \\ -6 & 15 & -1 \\ 9 & -3 & 7 \end{pmatrix} = \begin{pmatrix} -2 & 13 & 0 \\ 39 & -54 & 13 \\ 0 & 39 & 11 \end{pmatrix}$

$13\mathbf{A} - 15\mathbf{I} = \begin{pmatrix} 13 & 13 & 0 \\ 39 & -39 & 13 \\ 0 & 39 & 26 \end{pmatrix} - \begin{pmatrix} 15 & 0 & 0 \\ 0 & 15 & 0 \\ 0 & 0 & 15 \end{pmatrix}$

$= \begin{pmatrix} -2 & 13 & 0 \\ 39 & -54 & 13 \\ 0 & 39 & 11 \end{pmatrix} = \mathbf{A}^3$

b $15\mathbf{I} = 13\mathbf{A} - \mathbf{A}^3 \Rightarrow 15\mathbf{A}^{-1} = 13\mathbf{AA}^{-1} - \mathbf{A}^3\mathbf{A}^{-1} = 13\mathbf{I} - \mathbf{A}^2$

c $\frac{1}{15}\begin{pmatrix} 9 & 2 & -1 \\ 6 & -2 & 1 \\ -9 & 3 & 6 \end{pmatrix}$

8 **a** $\det\mathbf{A} = 2\begin{vmatrix} 3 & -2 \\ 3 & -4 \end{vmatrix} + \begin{vmatrix} 4 & 3 \\ 0 & 3 \end{vmatrix} = 2(-12 + 6) + (12 - 0) = 0$

b $\begin{pmatrix} -6 & 16 & 12 \\ 3 & -8 & -6 \\ -3 & 8 & 6 \end{pmatrix}$

c $\mathbf{AC}^{\mathrm{T}} = \begin{pmatrix} 2 & 0 & 1 \\ 4 & 3 & -2 \\ 0 & 3 & -4 \end{pmatrix}\begin{pmatrix} -6 & 3 & -3 \\ 16 & -8 & 8 \\ 12 & -6 & 6 \end{pmatrix}$

$= \begin{pmatrix} -12+0+12 & 6+0-6 & -6+0+6 \\ -24+48-24 & 12-24+12 & -12+24-12 \\ 0+48-48 & -24+24 & 0+24-24 \end{pmatrix}$

$= \mathbf{0}$

9 **a** All values of k

b $\frac{1}{5}\begin{pmatrix} 1 & 3-k & 6-k \\ 0 & 5 & 5 \\ 1 & -k-2 & -k-4 \end{pmatrix}$

10 $\frac{1}{4p-21}\begin{pmatrix} 2 & -6 & 2p+3 \\ 1 & -3 & 12-p \\ -7 & 4p & -9p \end{pmatrix}$

Exercise 6F

1 **a** $x = 0, y = -2, z = 5$

b $x = 2, y = -3, z = -2$

c $x = 2, y = 1, z = -3$

d $x = -\frac{39}{7}, y = \frac{39}{7}, z = -\frac{11}{7}$

2 $\mathbf{A}\begin{pmatrix} x \\ y \\ z \end{pmatrix} = \begin{pmatrix} 3 \\ 30 \\ -3 \end{pmatrix}$, where $\mathbf{A} = \begin{pmatrix} 1 & -3 & -4 \\ 6 & 5 & -7 \\ 1 & 4 & 6 \end{pmatrix}$, $\det\mathbf{A} = 111$.

So **A** is invertible, there is a unique solution to the set of equations and the three planes meet at a single point, (1, 2, −2).

3 £1440, £1250, £310

4 500 brown, 250 grey, 1250 black

5 **a** $a = 2$.

b Not consistent. The three planes meet in a prism.

6 **a** $\begin{vmatrix} 1 & 4 & q \\ 2 & 3 & -3 \\ q & q & -2 \end{vmatrix} = 1\begin{vmatrix} 3 & -3 \\ q & -2 \end{vmatrix} - 4\begin{vmatrix} 2 & -3 \\ q & -2 \end{vmatrix} + q\begin{vmatrix} 2 & 3 \\ q & q \end{vmatrix}$

$= (-6 + 3q) - 4(-4 + 3q) + q(2q - 3q)$

$= -6 + 3q + 16 - 12q - q^2$

$= -q^2 - 9q + 10$

$\det\mathbf{M} = 0 \Rightarrow -q^2 - 9q + 10 = 0 \Rightarrow q^2 + 9q - 10 = 0$

b **i** Consistent, infinity of solutions, planes meet in a sheaf

ii Consistent, unique solution, planes meet in a point

iii Inconsistent, no solutions, planes meet in a prism

Online Full worked solutions are available in SolutionBank.

Mixed exercise 6

1 $\begin{pmatrix} 1 & 4 & 3 \\ -1 & 1 & -2 \end{pmatrix}$

2 **a** $\begin{pmatrix} \frac{3}{a} & -\frac{1}{a} \\ -\frac{2}{b} & \frac{1}{b} \end{pmatrix}$ **b** $\begin{pmatrix} -1 & 1 \\ 4 & -1 \end{pmatrix}$

3 **a** $\mathbf{X} = \mathbf{BAB}^{-1}$ **b** $\begin{pmatrix} 6 & 2 \\ -4 & -3 \end{pmatrix}$

4 $a = 1, b = 3$

5 8

6 $\frac{1}{2}\begin{pmatrix} 2 & 0 & 0 \\ -x & 1 & 0 \\ x-6 & -1 & 2 \end{pmatrix}$

7 **a** $k \neq \pm 2\sqrt{2}$ **b** $\frac{1}{k^2-8}\begin{pmatrix} k & 2 \\ 4 & k \end{pmatrix}$

8 **a** $k^2 - 2k + 6$
b $\det\mathbf{B} = (k-1)^2 + 5 \geqslant 5$, so $\det\mathbf{B} \neq 0$ and $\mathbf{B}$ non-singular.
c $k = -3$

9 **a** $m = \sqrt{2}, m = -\sqrt{2}$ **b** $\frac{1}{m^2-2}\begin{pmatrix} -1 & m \\ -m & 2 \end{pmatrix}$

10 **a** $a \neq -\frac{12}{13}$
b $\frac{1}{13a+12}\begin{pmatrix} a+1 & 1 & -4-5a \\ 1 & 13 & 8 \\ 1+2a & -11 & 3a-4 \end{pmatrix}$

11 $\mathbf{A}\begin{pmatrix} x \\ y \\ z \end{pmatrix} = \begin{pmatrix} 6 \\ -2 \\ 0 \end{pmatrix}$, where $\mathbf{A} = \begin{pmatrix} 1 & 1 & 1 \\ 1 & -4 & 2 \\ 2 & 1 & -3 \end{pmatrix}$, $\det\mathbf{A} = 26$.
So **A** is non-singular and has an inverse. There is a unique solution and the planes intersect at a single point, (2, 2, 2).

12 700 Hampshire, 1400 Dorset horn, 400 Wiltshire horn

13 **a** $a = -3, b = 13$
b The three planes form a sheaf.

Challenge

Let $\mathbf{A} = \begin{pmatrix} a & b \\ c & d \end{pmatrix}$, $\mathbf{B} = \begin{pmatrix} h & j \\ k & l \end{pmatrix}$. $\det\mathbf{A} = ad - bc$, $\det\mathbf{B} = hl - jk$

So $\det\mathbf{A}\det\mathbf{B} = (ad - bc)(hl - jk) = adhl - adjk - bchl + bcjk$

$\mathbf{AB} = \begin{pmatrix} ah+bk & aj+bl \\ ch+dk & cj+dl \end{pmatrix}$

$\det(\mathbf{AB}) = (ah+bk)(cj+dl) - (aj+bl)(ch+dk)$
$= adhl - adjk - bchl + bcjk = \det\mathbf{A}\det\mathbf{B}$

CHAPTER 7

Prior knowledge 7

1 **a** $\begin{pmatrix} 0 & 1 \\ 1 & 3 \end{pmatrix}$ **b** $\begin{pmatrix} -5 & -3 \\ 13 & 8 \end{pmatrix}$

2 **a** -10 **b** $\frac{-1}{10}\begin{pmatrix} -2 & -1 \\ -4 & 3 \end{pmatrix}$

3 $-\frac{1}{7}\begin{pmatrix} 5 & -4 & -2 \\ -1 & -2 & -1 \\ 1 & 5 & -1 \end{pmatrix}$

Exercise 7A

1 **a** Not linear **b** Not linear
c Not linear **d** Linear
e Not linear **f** Linear

2 **a** Linear $\begin{pmatrix} 2 & -1 \\ 3 & 0 \end{pmatrix}$
b Not linear ($2y + 1$ and $x - 1$ cannot be written as $ax + by$)
c Not linear (xy cannot be written as $ax + by$)
d Linear $\begin{pmatrix} 0 & 2 \\ -1 & 0 \end{pmatrix}$ **e** Linear $\begin{pmatrix} 0 & 1 \\ 1 & 0 \end{pmatrix}$

3 **a** Not linear (x^2 and y^2 cannot be written as $ax + by$)
b Linear $\begin{pmatrix} 0 & -1 \\ 1 & 0 \end{pmatrix}$ **c** Linear $\begin{pmatrix} 1 & -1 \\ 1 & -1 \end{pmatrix}$
d Linear $\begin{pmatrix} 0 & 0 \\ 0 & 0 \end{pmatrix}$ **e** Linear $\begin{pmatrix} 1 & 0 \\ 0 & 1 \end{pmatrix}$

4 **a** Linear $\begin{pmatrix} 2 & 1 \\ 0 & -1 \end{pmatrix}$ **b** Linear $\begin{pmatrix} 0 & -1 \\ 1 & 2 \end{pmatrix}$

5 **a** (1, 1), (−2, 3), (−5, 1)
b (3, −2), (14, −6), (9, −2)
c (−2, −2), (−6, 4), (−2, 10)

6 **a** (−2, 0), (0, 3), (2, 0), (0, −3)
b (−1, −1), (−1, 1), (1, 1), (1, −1)
c (−1, −1), (1, −1), (1, 1), (−1, 1)

7 **a** (−2, −1), (−4, −1), (−4, −2), (−2, −2)
b

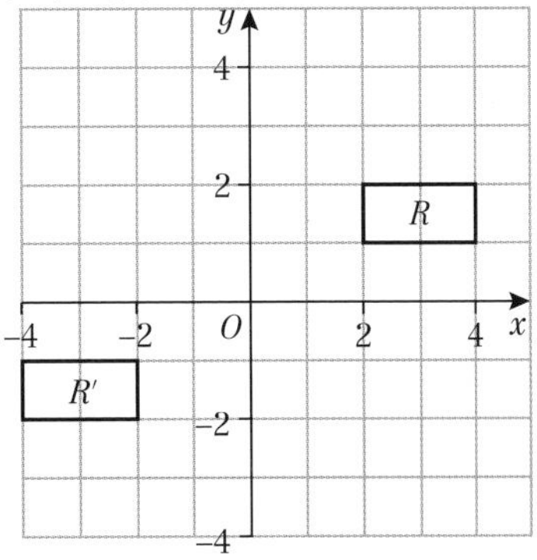

c Rotation through 180° about (0, 0)

8 **a** (2, 0), (8, 4), (6, 8), (0, 4)
b

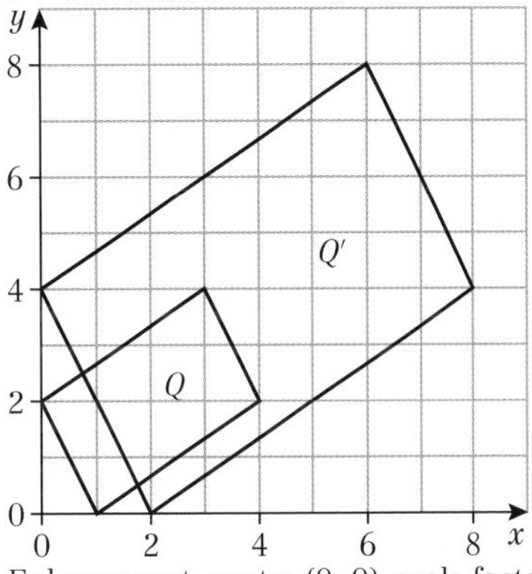

c Enlargement, centre (0, 0), scale factor 2

9 **a** (0, −2), (0, −6), (4, −6), (4, −2)
b

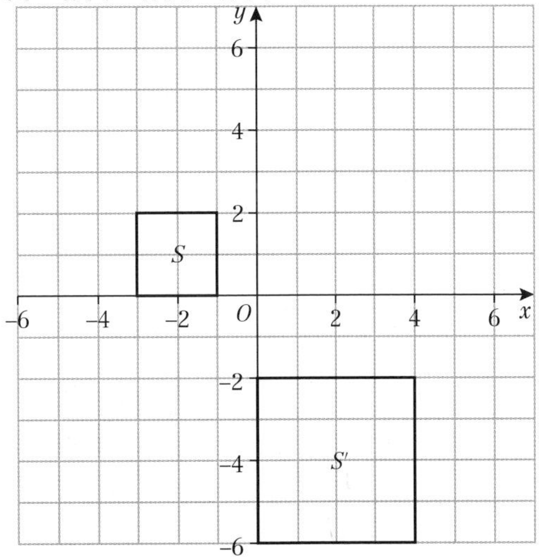

c Reflection in $y = x$ and enlargement, centre (0, 0), scale factor 2

10 a (4, 1), (4, 3), (1, 3)

b The transformation represented by the identity matrix leaves T unchanged.

Challenge

a $T = \begin{pmatrix} 2 & -3 \\ 1 & 1 \end{pmatrix}$ so $T\begin{pmatrix} kx \\ ky \end{pmatrix} = \begin{pmatrix} 2 & -3 \\ 1 & 1 \end{pmatrix}\begin{pmatrix} kx \\ ky \end{pmatrix} = \begin{pmatrix} 2kx - 3ky \\ kx + ky \end{pmatrix}$

$= k\begin{pmatrix} 2x - 3y \\ x + y \end{pmatrix} = kT\begin{pmatrix} x \\ y \end{pmatrix}$

b $T\left(\begin{pmatrix} x_1 \\ y_1 \end{pmatrix} + \begin{pmatrix} x_2 \\ y_2 \end{pmatrix}\right) = T\begin{pmatrix} x_1 + x_2 \\ y_1 + y_2 \end{pmatrix}$

$= \begin{pmatrix} 2 & -3 \\ 1 & 1 \end{pmatrix}\begin{pmatrix} x_1 + x_2 \\ y_1 + y_2 \end{pmatrix} = \begin{pmatrix} 2(x_1 + x_2) - 3(y_1 + y_2) \\ x_1 + x_2 + y_1 + y_2 \end{pmatrix}$

$= \begin{pmatrix} 2x_1 - 3y_1 \\ x_1 + y_1 \end{pmatrix} + \begin{pmatrix} 2x_2 - 3y_2 \\ x_2 + y_2 \end{pmatrix} = T\begin{pmatrix} x_1 \\ y_1 \end{pmatrix} + T\begin{pmatrix} x_2 \\ y_2 \end{pmatrix}$

Exercise 7B

1 a $\begin{pmatrix} 1 & 0 \\ 0 & -1 \end{pmatrix}$

b $\begin{pmatrix} 1 & 0 \\ 0 & -1 \end{pmatrix}\begin{pmatrix} 1 \\ 3 \end{pmatrix} = \begin{pmatrix} 1 \\ -3 \end{pmatrix}$ so $A' = (1, -3)$

$\begin{pmatrix} 1 & 0 \\ 0 & -1 \end{pmatrix}\begin{pmatrix} 3 \\ 3 \end{pmatrix} = \begin{pmatrix} 3 \\ -3 \end{pmatrix}$ so $B' = (3, -3)$

$\begin{pmatrix} 1 & 0 \\ 0 & -1 \end{pmatrix}\begin{pmatrix} 3 \\ 2 \end{pmatrix} = \begin{pmatrix} 3 \\ -2 \end{pmatrix}$ so $C' = (3, -2)$

2 a $\begin{pmatrix} 0 & -1 \\ -1 & 0 \end{pmatrix}$

b $\begin{pmatrix} 0 & -1 \\ -1 & 0 \end{pmatrix}\begin{pmatrix} 1 \\ 1 \end{pmatrix} = \begin{pmatrix} -1 \\ -1 \end{pmatrix}$ so $P' = (-1, -1)$

$\begin{pmatrix} 0 & -1 \\ -1 & 0 \end{pmatrix}\begin{pmatrix} 1 \\ 3 \end{pmatrix} = \begin{pmatrix} -3 \\ -1 \end{pmatrix}$ so $Q' = (-3, -1)$

$\begin{pmatrix} 0 & -1 \\ -1 & 0 \end{pmatrix}\begin{pmatrix} 2 \\ 3 \end{pmatrix} = \begin{pmatrix} -3 \\ -2 \end{pmatrix}$ so $R' = (-3, -2)$

$\begin{pmatrix} 0 & -1 \\ -1 & 0 \end{pmatrix}\begin{pmatrix} 2 \\ 1 \end{pmatrix} = \begin{pmatrix} -1 \\ -2 \end{pmatrix}$ so $S' = (-1, -2)$

3 a $\begin{pmatrix} 0 & -1 \\ 1 & 0 \end{pmatrix}$ **b** $\begin{pmatrix} 0 & 1 \\ -1 & 0 \end{pmatrix}$ **c** $\frac{1}{2}\begin{pmatrix} \sqrt{2} & -\sqrt{2} \\ \sqrt{2} & \sqrt{2} \end{pmatrix}$

d $\frac{1}{2}\begin{pmatrix} -\sqrt{3} & 1 \\ -1 & -\sqrt{3} \end{pmatrix}$ **e** $\frac{1}{2}\begin{pmatrix} -\sqrt{2} & \sqrt{2} \\ -\sqrt{2} & -\sqrt{2} \end{pmatrix}$

4 a $A' = (-1, 1)$, $B' = (-1, 4)$, $C' = (-2, 4)$

b $A' = \left(-\frac{\sqrt{3}}{2} - \frac{1}{2}, \frac{1}{2} - \frac{\sqrt{3}}{2}\right)$, $B' = \left(-2\sqrt{3} - \frac{1}{2}, 2 - \frac{\sqrt{3}}{2}\right)$,

$C' = (-2\sqrt{3} - 1, 2 - \sqrt{3})$

5 a $P' = (2, -2)$, $Q' = (3, -2)$, $R' = (3, -4)$, $S' = (2, -4)$

b $P' = (0, -2\sqrt{2})$, $Q' = \left(\frac{\sqrt{2}}{2}, -\frac{5\sqrt{2}}{2}\right)$,

$R' = \left(-\frac{\sqrt{2}}{2}, -\frac{7\sqrt{2}}{2}\right)$, $S' = (-\sqrt{2}, -3\sqrt{2})$

6 a **A** represents a reflection in the x-axis. **B** represents a rotation through 270° anticlockwise about (0, 0).

b (3, −2) **c** $a = 0, b = 0$

7 a Rotation through 225° anticlockwise about (0, 0)

b $p = 3, q = 1$ **c** $\begin{pmatrix} \frac{1}{\sqrt{2}} & \frac{1}{\sqrt{2}} \\ -\frac{1}{\sqrt{2}} & \frac{1}{\sqrt{2}} \end{pmatrix}$

d Rotation through 45° clockwise about (0, 0); (−3, −1)

8 a Reflection in the line $y = x$

b e.g. $y = x$, $y = -x$, $y = -x + 1$ **c** $\begin{pmatrix} 1 & 0 \\ 0 & 1 \end{pmatrix}$

9 a $a = -0.5$

b $\theta = 120°$; $\begin{pmatrix} -0.5 & -0.866 \\ 0.866 & -0.5 \end{pmatrix}$

$\theta = 240°$; $\begin{pmatrix} -0.5 & 0.866 \\ -0.866 & -0.5 \end{pmatrix}$

10 a $\begin{pmatrix} 0 & -1 \\ 1 & 0 \end{pmatrix}$ **b** $a = 0, b = 0$

Challenge

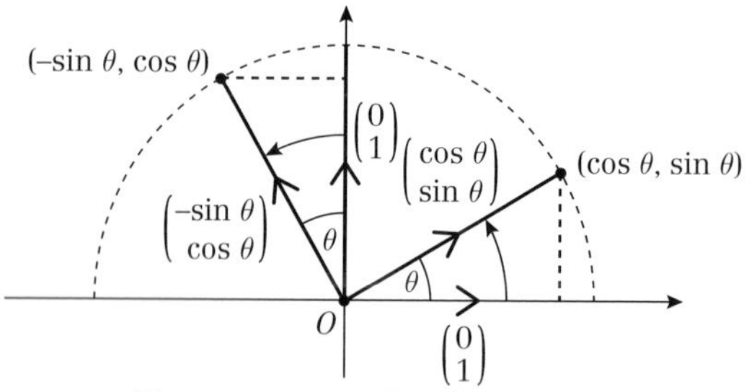

Rotating $\begin{pmatrix} 1 \\ 0 \end{pmatrix}$ by θ takes it to $\begin{pmatrix} \cos\theta \\ \sin\theta \end{pmatrix}$, and rotating $\begin{pmatrix} 0 \\ 1 \end{pmatrix}$ by θ takes it to $\begin{pmatrix} -\sin\theta \\ \cos\theta \end{pmatrix}$, so the matrix for the rotation is $\begin{pmatrix} \cos\theta & -\sin\theta \\ \sin\theta & \cos\theta \end{pmatrix}$.

Exercise 7C

1 a $\begin{pmatrix} 4 & 0 \\ 0 & 1 \end{pmatrix}$ **b** $\begin{pmatrix} 1 & 0 \\ 0 & 3 \end{pmatrix}$ **c** $\begin{pmatrix} 2 & 0 \\ 0 & 2 \end{pmatrix}$ **d** $\begin{pmatrix} 5 & 0 \\ 0 & 0.5 \end{pmatrix}$

2 a 4 **b** 3 **c** 4 **d** 2.5

3 a (0, 0) **b** 12

4 a $\begin{pmatrix} -2 & 0 \\ 0 & 1 \end{pmatrix}$ **b** $\begin{pmatrix} -3 & 0 \\ 0 & 4 \end{pmatrix}$ **c** $\begin{pmatrix} -\frac{1}{2} & 0 \\ 0 & -\frac{1}{2} \end{pmatrix}$

5 $\begin{pmatrix} a & 0 \\ 0 & a \end{pmatrix}\begin{pmatrix} x \\ kx \end{pmatrix} = \begin{pmatrix} ax \\ kax \end{pmatrix}$; $kax = k(ax)$, so this point lies on $y = kx$ and the line is invariant.

6 a Stretch parallel of the x-axis, scale factor 2 and stretch parallel to the y-axis, scale factor −3

b $k = 4$

7 a (3, 9), (15, 9), (15, 6) **b** 18

8 a (4, 0), (8, 0), (8, −15), (4, −15)

b 60

9 a Enlargement, centre (0, 0), scale factor $2\sqrt{5}$

b $a = 1$ or $a = 7$

10 a $\begin{pmatrix} p^2 + p & p + q \\ p^2 + qp & p + q^2 \end{pmatrix}$ **b** $p = -3, q = 3$

11 a $\begin{pmatrix} 8 & 0 \\ 0 & -8 \end{pmatrix}$

b Stretch parallel to the x-axis, scale factor 8; and stretch parallel to the y-axis, scale factor −8.
Or enlargement scale factor 8 and centre (0, 0) and reflection in the x-axis.

c $k = \frac{7}{2}$ or $k = -\frac{3}{2}$

12 36

13 $k = 2$

14 a $\begin{pmatrix} \frac{1}{\sqrt{2}} & -\frac{1}{\sqrt{2}} \\ \frac{1}{\sqrt{2}} & \frac{1}{\sqrt{2}} \end{pmatrix}$ **b** $(0,0)$, $(0, 7\sqrt{2})$, $\left(\frac{5}{\sqrt{2}}, \frac{1}{\sqrt{2}}\right)$

c $\det \mathbf{M} = \frac{1}{\sqrt{2}} \times \frac{1}{\sqrt{2}} - \left(-\frac{1}{\sqrt{2}} \times \frac{1}{\sqrt{2}}\right) = \frac{1}{2} - \left(-\frac{1}{2}\right) = 1$

d 17.5

Online Full worked solutions are available in SolutionBank.

Challenge

a 0

b For any x, y, $\begin{pmatrix} x' \\ y' \end{pmatrix} = \begin{pmatrix} 7 & 0 \\ 0 & 0 \end{pmatrix}\begin{pmatrix} x \\ y \end{pmatrix} = \begin{pmatrix} 7x \\ 0 \end{pmatrix}$

The y-coordinate is 0, so all points (x, y) map onto the x-axis.

Exercise 7D

1 **a** $\begin{pmatrix} 0 & 1 \\ 1 & 0 \end{pmatrix}$; Reflection in $y = x$

b $\begin{pmatrix} 0 & 1 \\ 1 & 0 \end{pmatrix}$; Reflection in $y = x$

c $\begin{pmatrix} -2 & 0 \\ 0 & -2 \end{pmatrix}$; Enlargement scale factor -2, centre (0, 0)

d $\begin{pmatrix} 1 & 0 \\ 0 & 1 \end{pmatrix}$; Identity (no transformation)

e $\begin{pmatrix} 4 & 0 \\ 0 & 4 \end{pmatrix}$; Enlargement scale factor 4, centre (0, 0)

2 **a** A: $\begin{pmatrix} 0 & -1 \\ 1 & 0 \end{pmatrix}$, B: $\begin{pmatrix} -1 & 0 \\ 0 & -1 \end{pmatrix}$ C: $\begin{pmatrix} 1 & 0 \\ 0 & -1 \end{pmatrix}$, D: $\begin{pmatrix} -1 & 0 \\ 0 & 1 \end{pmatrix}$

b **i** Reflection in y-axis
ii Reflection in y-axis
iii Rotation of 180° about (0, 0)
iv Reflection in $y = -x$
v No transformataion (Identity)
vi Rotation of 90° anticlockwise about (0, 0)
vii No transformataion (Identity)

3 **a** $\begin{pmatrix} 0 & 3 \\ 2 & 0 \end{pmatrix}$; reflection in $y = x$ with a stretch by scale factor 3 parallel to the x-axis and by scale factor 2 parallel to the y-axis.

b $\begin{pmatrix} 15 & 0 \\ 0 & -10 \end{pmatrix}$; stretch by scale factor 15 parallel to the x-axis and by scale factor -10 parallel to the y-axis

c $\begin{pmatrix} 0 & 5 \\ -5 & 0 \end{pmatrix}$; enlargement by scale factor 5 about (0, 0) and rotation through 270° anticlockwise.

d $\begin{pmatrix} 15 & 0 \\ 0 & -10 \end{pmatrix}$; stretch by scale factor 15 parallel to the x-axis and by scale factor -10 parallel to the y-axis

e $\begin{pmatrix} 0 & 5 \\ -5 & 0 \end{pmatrix}$; enlargement by scale factor 5 about (0, 0) and rotation through 270° anticlockwise.

f $\begin{pmatrix} 0 & 15 \\ 10 & 0 \end{pmatrix}$; reflection in $y = x$ with a stretch by scale factor 15 parallel to the x-axis and by scale factor 10 parallel to the y-axis.

4 **a** A: $\begin{pmatrix} 2 & 0 \\ 0 & 3 \end{pmatrix}$, B: $\begin{pmatrix} -2 & 0 \\ 0 & -2 \end{pmatrix}$, C: $\begin{pmatrix} 4 & 0 \\ 0 & 4 \end{pmatrix}$

b **i** $\begin{pmatrix} -4 & 0 \\ 0 & -6 \end{pmatrix}$ **ii** $\begin{pmatrix} 8 & 0 \\ 0 & 12 \end{pmatrix}$ **iii** $\begin{pmatrix} -8 & 0 \\ 0 & -8 \end{pmatrix}$

iv $\begin{pmatrix} 16 & 0 \\ 0 & 16 \end{pmatrix}$ **v** $\begin{pmatrix} -16 & 0 \\ 0 & -24 \end{pmatrix}$

5 Reflection in y-axis $= \mathbf{M} = \begin{pmatrix} -1 & 0 \\ 0 & 1 \end{pmatrix}$

Reflection in line $y = -x = \mathbf{N} = \begin{pmatrix} 0 & -1 \\ -1 & 0 \end{pmatrix}$

Combined transformation = **NM**

$= \begin{pmatrix} 0 & -1 \\ -1 & 0 \end{pmatrix}\begin{pmatrix} -1 & 0 \\ 0 & 1 \end{pmatrix} = \begin{pmatrix} 0 & -1 \\ 1 & 0 \end{pmatrix}$

$\begin{pmatrix} 0 & -1 \\ 1 & 0 \end{pmatrix} = \begin{pmatrix} \cos 90° & -\sin 90° \\ \sin 90° & \cos 90° \end{pmatrix}$

so it represents a rotation through 90° anticlockwise about (0, 0).

6 $\mathbf{T} = \begin{pmatrix} 1 & 0 \\ 0 & -1 \end{pmatrix}$ and $\mathbf{U} = \begin{pmatrix} 0 & -1 \\ 1 & 0 \end{pmatrix}$

$\mathbf{UT} = \begin{pmatrix} 0 & -1 \\ 1 & 0 \end{pmatrix}\begin{pmatrix} 1 & 0 \\ 0 & -1 \end{pmatrix} = \begin{pmatrix} 0 & 1 \\ 1 & 0 \end{pmatrix}$

$\mathbf{TU} = \begin{pmatrix} 1 & 0 \\ 0 & -1 \end{pmatrix}\begin{pmatrix} 0 & -1 \\ 1 & 0 \end{pmatrix} = \begin{pmatrix} 0 & -1 \\ -1 & 0 \end{pmatrix} \neq \mathbf{UT}$

7 **a** $\begin{pmatrix} -4k & 0 \\ 0 & 2k \end{pmatrix}$

b Stretch by scale factor $-4k$ parallel to the x-axis and by scale factor $2k$ parallel to the y-axis.

c $\mathbf{QP} = \begin{pmatrix} k & 0 \\ 0 & k \end{pmatrix}\begin{pmatrix} -4 & 0 \\ 0 & 2 \end{pmatrix} = \begin{pmatrix} -4k & 0 \\ 0 & 2k \end{pmatrix} = \mathbf{PQ}$ (from part a)

8 **a** $\begin{pmatrix} 9 & 0 \\ 0 & 16 \end{pmatrix}$

b Stretch by scale factor 9 parallel to the x-axis and by scale factor 16 parallel to the y-axis.

c $\begin{pmatrix} a^2 & 0 \\ 0 & b^2 \end{pmatrix}$; stretch by scale factor a^2 parallel to the x-axis and by scale factor b^2 parallel to the y-axis.

9 **a** $\begin{pmatrix} 0 & -1 \\ 1 & 0 \end{pmatrix}$

b Rotation of 90° anticlockwise about (0, 0)

c Rotation of 45° anticlockwise about (0, 0)

d $\begin{pmatrix} 1 & 0 \\ 0 & 1 \end{pmatrix}$ (Identity matrix)

10 **a** $k = -3$ **b** $\theta = 45°$

11 3

12 **a** $\begin{pmatrix} -\frac{1}{\sqrt{2}} & \frac{1}{\sqrt{2}} \\ -\frac{1}{\sqrt{2}} & -\frac{1}{\sqrt{2}} \end{pmatrix}$ **b** $\begin{pmatrix} 0 & 1 \\ 1 & 0 \end{pmatrix}$ **c** $\begin{pmatrix} \frac{1}{\sqrt{2}} & -\frac{1}{\sqrt{2}} \\ -\frac{1}{\sqrt{2}} & -\frac{1}{\sqrt{2}} \end{pmatrix}$

13 **a** $\begin{pmatrix} k^2 + 3 & 0 \\ 0 & 3 + k^2 \end{pmatrix}$

b Enlargement by scale factor $k^2 + 3$ about (0, 0).

14 $\mathbf{P}^2 = \begin{pmatrix} a & b \\ b & -a \end{pmatrix}\begin{pmatrix} a & b \\ b & -a \end{pmatrix} = \begin{pmatrix} a^2 + b^2 & ab - ba \\ ab - ba & b^2 + a^2 \end{pmatrix}$

$= \begin{pmatrix} a^2 + b^2 & 0 \\ 0 & a^2 + b^2 \end{pmatrix}$

This represents an enlargement about the origin, scale factor $a^2 + b^2$.

Challenge

a $\mathbf{P}^2 = \begin{pmatrix} \cos\theta & -\sin\theta \\ \sin\theta & \cos\theta \end{pmatrix}\begin{pmatrix} \cos\theta & -\sin\theta \\ \sin\theta & \cos\theta \end{pmatrix}$

$= \begin{pmatrix} \cos^2\theta - \sin^2\theta & -2\sin\theta\cos\theta \\ 2\sin\theta\cos\theta & \cos^2\theta - \sin^2\theta \end{pmatrix} = \begin{pmatrix} \cos 2\theta & -\sin 2\theta \\ \sin 2\theta & \cos 2\theta \end{pmatrix}$

b Two successive anticlockwise rotations about the origin by an angle θ are equivalent to a single anticlockwise rotation by an angle 2θ.

Exercise 7E

1 **a** $\begin{pmatrix} -1 & 0 & 0 \\ 0 & 1 & 0 \\ 0 & 0 & 1 \end{pmatrix}$ **b** $\begin{pmatrix} 1 & 0 & 0 \\ 0 & -1 & 0 \\ 0 & 0 & 1 \end{pmatrix}$ **c** $\begin{pmatrix} -1 & 0 & 0 \\ 0 & 1 & 0 \\ 0 & 0 & -1 \end{pmatrix}$

d $\begin{pmatrix} 0 & -1 & 0 \\ 1 & 0 & 0 \\ 0 & 0 & 1 \end{pmatrix}$ **e** $\begin{pmatrix} 0 & 0 & -1 \\ 0 & 1 & 0 \\ 1 & 0 & 0 \end{pmatrix}$ **f** $\begin{pmatrix} 1 & 0 & 0 \\ 0 & \frac{1}{2} & \frac{\sqrt{3}}{2} \\ 0 & -\frac{\sqrt{3}}{2} & \frac{1}{2} \end{pmatrix}$

2 **a** Reflection in the plane $z = 0$

b Rotation anticlockwise 90° about y-axis

c Rotation anticlockwise 135° about z-axis

3 **a** Rotation anticlockwise 90° about x-axis
b (3, −4, −1) **c** $a = 2$

4 **a** $\begin{pmatrix} -\frac{1}{2} & -\frac{\sqrt{3}}{2} & 0 \\ \frac{\sqrt{3}}{2} & -\frac{1}{2} & 0 \\ 0 & 0 & 1 \end{pmatrix}$ **b** $\left(\frac{-3+\sqrt{3}}{2}, \frac{3\sqrt{3}+1}{2}, 0\right)$
c $\left(-\frac{k}{2}, \frac{\sqrt{3}k}{2}, k\right)$

5 **a** $\mathbf{A} = \begin{pmatrix} -1 & 0 & 0 \\ 0 & 1 & 0 \\ 0 & 0 & 1 \end{pmatrix}$, $\mathbf{B} = \begin{pmatrix} 1 & 0 & 0 \\ 0 & -1 & 0 \\ 0 & 0 & 1 \end{pmatrix}$
b $(-a, b, c)$ **c** $(-a, -b, c)$

6 **a** Rotation 210° anticlockwise about y-axis.
b $\left(\frac{-k\sqrt{3}}{2}, -k, \frac{k}{2}\right)$

7 **a** $\begin{pmatrix} \frac{1}{\sqrt{2}} & 0 & -\frac{1}{\sqrt{2}} \\ 0 & 1 & 0 \\ \frac{1}{\sqrt{2}} & 0 & \frac{1}{\sqrt{2}} \end{pmatrix}$
b $(\sqrt{2}, 0, 0)$, $(\sqrt{2}, 1, 0)$, $(0, 2, 3\sqrt{2})$, $(0, 0, 0)$
c 1

Challenge

a $\begin{pmatrix} -1 & 0 & 0 \\ 0 & -1 & 0 \\ 0 & 0 & 1 \end{pmatrix}$ **b** $\begin{pmatrix} -1 & 0 & 0 \\ 0 & \frac{1}{\sqrt{2}} & -\frac{1}{\sqrt{2}} \\ 0 & \frac{1}{\sqrt{2}} & \frac{1}{\sqrt{2}} \end{pmatrix}$

Exercise 7F

1 **a** Rotation of 90° anticlockwise about (0, 0)
b $\begin{pmatrix} 0 & 1 \\ -1 & 0 \end{pmatrix}$
c Rotation of 270° anticlockwise about (0, 0)

2 **a** **i** Rotation of 180° about (0, 0)
ii $\mathbf{S}^2 = \begin{pmatrix} -1 & 0 \\ 0 & -1 \end{pmatrix}\begin{pmatrix} -1 & 0 \\ 0 & -1 \end{pmatrix} = \begin{pmatrix} 1+0 & 0+0 \\ 0+0 & 0+1 \end{pmatrix} = \begin{pmatrix} 1 & 0 \\ 0 & 1 \end{pmatrix}$
$= \mathbf{I}$
iii Rotation of 180° about (0, 0)
b **i** Reflection in $y = -x$
ii $\mathbf{T}^2 = \begin{pmatrix} 0 & -1 \\ -1 & 0 \end{pmatrix}\begin{pmatrix} 0 & -1 \\ -1 & 0 \end{pmatrix} = \begin{pmatrix} 0+1 & 0+0 \\ 0+0 & 1+0 \end{pmatrix} = \begin{pmatrix} 1 & 0 \\ 0 & 1 \end{pmatrix}$
$= \mathbf{I}$
iii Reflection in $y = -x$
c $\det \mathbf{S} = 1$; the area of a shape is unchanged by a rotation.
$\det \mathbf{T} = -1$; the area of a shape is unchanged by a reflection (the minus sign indicating that it has been reflected).

3 **a** $\begin{pmatrix} 1 & 0 \\ 0 & -1 \end{pmatrix}$; reflection in $y = 0$
b $\begin{pmatrix} 1 & 0 \\ 0 & -1 \end{pmatrix}$; reflection in $y = 0$
c $\begin{pmatrix} -1 & 0 \\ 0 & 1 \end{pmatrix}$; reflection in $x = 0$
d $\begin{pmatrix} -1 & 0 \\ 0 & 1 \end{pmatrix}$; reflection in $x = 0$

4 **a** $\left(-\frac{13}{2}, \frac{23}{4}\right)$ **b** $\begin{pmatrix} -\frac{3}{4} & -\frac{1}{4} \\ \frac{1}{2} & \frac{1}{2} \end{pmatrix}$

5 **a** Enlargement, scale factor 4, centre (0, 0)
b $\begin{pmatrix} \frac{1}{4} & 0 \\ 0 & \frac{1}{4} \end{pmatrix}$ **c** $\left(1, \frac{3}{2}\right)$, $\left(\frac{9}{4}, \frac{7}{4}\right)$, $\left(\frac{3}{4}, \frac{1}{4}\right)$

6 **a** $\begin{pmatrix} \frac{1}{a} & 0 \\ 0 & \frac{1}{b} \end{pmatrix}$ **b** $\left(\frac{-6}{a}, \frac{8}{b}\right)$

7 **a** Rotation of 330° anticlockwise about (0, 0)
b $p = -2$, $q = 1$

8 $\begin{pmatrix} \frac{1}{2} & -1 \\ -\frac{3}{2} & 2 \end{pmatrix}$

9 $\left(-\frac{a}{2} - b, -2a - 3b\right)$

10 $\begin{pmatrix} 2 & \frac{3}{2} & -1 \\ -1 & -1 & 1 \\ -1 & -\frac{3}{2} & 1 \end{pmatrix}$; $a = 10$, $b = -6$, $c = -7$

11 **a** $\begin{pmatrix} 0 & 1 & 0 \\ 1 & 0 & 0 \\ 0 & 0 & 1 \end{pmatrix}$; reflection in $y = x$
b $\begin{pmatrix} 0 & 1 & 0 \\ 1 & 0 & 0 \\ 0 & 0 & 1 \end{pmatrix}$; reflection in $y = x$
c $\begin{pmatrix} 0 & -1 & 0 \\ -1 & 0 & 0 \\ 0 & 0 & 1 \end{pmatrix}$; reflection in $y = -x$
d $\mathbf{N} = \mathbf{QP} \Rightarrow \mathbf{Q}^{-1}\mathbf{N} = \mathbf{P} \Rightarrow \mathbf{P}^{-1}\mathbf{Q}^{-1}\mathbf{N} = \mathbf{I} \Rightarrow \mathbf{P}^{-1}\mathbf{Q}^{-1} = \mathbf{N}^{-1}$
$\mathbf{N}^{-1} = \begin{pmatrix} 0 & -1 & 0 \\ -1 & 0 & 0 \\ 0 & 0 & 1 \end{pmatrix}$; reflection in $y = -x$

12 $\begin{pmatrix} -0.1 & 0.3 & -0.7 \\ 0.3 & 0.1 & 0.1 \\ -0.2 & 0.6 & -0.4 \end{pmatrix}$

Mixed exercise 7

1 **a** $\begin{pmatrix} 0 & -1 \\ 1 & 0 \end{pmatrix}$ **b** $\begin{pmatrix} -2 & 3 \\ -1 & 2 \end{pmatrix}$
c $\begin{pmatrix} 1 & 0 \\ 0 & 1 \end{pmatrix}$ (Identity matrix); four successive anticlockwise rotations of 90° about (0, 0).

2 **a** $\begin{pmatrix} 2 & 0 \\ 0 & -2 \end{pmatrix}$; reflection in x-axis and enlargement s.f. 2, centre (0, 0)
b $\begin{pmatrix} \frac{1}{2} & 0 \\ 0 & -\frac{1}{2} \end{pmatrix}$; reflection in x-axis and enlargement s.f. 2, centre (0, 0)

3 **a** $\begin{pmatrix} k & 0 \\ 0 & k \end{pmatrix}$ **b** $k = -3$ **c** $m = 1$ or $m = -1$

4 **a** 4
b 30° anticlockwise about (0, 0)
c $\left(\frac{\sqrt{3}}{8}a + \frac{b}{8}, -\frac{a}{8} + \frac{\sqrt{3}}{8}b\right)$

5 **a** **A** represents a reflection in the line $y = x$;
B represents a rotation through 270° anticlockwise about (0, 0)
b $(-p, q)$

6 **a** $k = -2.8$ or $k = 14.8$
b $\begin{pmatrix} -4 & 3 \\ 1 & -2 \end{pmatrix}\begin{pmatrix} x \\ -\frac{1}{3}x \end{pmatrix} = \begin{pmatrix} -5x \\ \frac{5}{3}x \end{pmatrix}$; $(-5x) + 3\left(\frac{5}{3}x\right) = 0$ so the point satisfies the equation of the original line.

7 **a** $\begin{pmatrix} -4 & 0 \\ 0 & 3 \end{pmatrix}$
b 5

Online Full worked solutions are available in SolutionBank.

8 a Rotation 150° anticlockwise about the z-axis

b $\left(-\frac{a\sqrt{3}}{2}-\frac{b}{2}, \frac{a}{2}-\frac{b\sqrt{3}}{2}, -a\right)$

9 a $\begin{pmatrix} \frac{1}{a} & 0 \\ 0 & \frac{1}{a} \end{pmatrix}$ b $\left(\frac{4}{a}, \frac{7}{a}\right)$

10 $\begin{pmatrix} 1 & 4 \\ \frac{1}{2} & \frac{5}{2} \end{pmatrix}$

11 a $\begin{pmatrix} -4 & 1 \\ -3 & 2 \end{pmatrix}$ b 7 c $\begin{pmatrix} -\frac{2}{5} & \frac{1}{5} \\ -\frac{3}{5} & \frac{4}{5} \end{pmatrix}$

12 $\begin{pmatrix} 1 & 0 & 0 \\ 0 & -\frac{\sqrt{2}}{2} & -\frac{\sqrt{2}}{2} \\ 0 & \frac{\sqrt{2}}{2} & -\frac{\sqrt{2}}{2} \end{pmatrix}$; $a = 0, b = -\sqrt{2}, c = 0$

13 a $\left(-\frac{1}{6}, -\frac{19}{6}, \frac{11}{6}\right)$ b $\frac{1}{6}\begin{pmatrix} -8 & 1 & 4 \\ -4 & -1 & 2 \\ -14 & 1 & 4 \end{pmatrix}$

Challenge

1 $\begin{pmatrix} 1 & 0 & 0 \\ 0 & 0 & -1 \\ 0 & 1 & 0 \end{pmatrix}$

2 a Let the point be $P(a, b)$: $\begin{pmatrix} 0 & 1 \\ 0 & 1 \end{pmatrix}\begin{pmatrix} a \\ b \end{pmatrix} = \begin{pmatrix} 0+b \\ 0+b \end{pmatrix} = \begin{pmatrix} b \\ b \end{pmatrix}$

So P′ is (b, b); its x- and y-coordinates are equal, so it is on $y = x$

b $\begin{pmatrix} 0 & 1 \\ 0 & m \end{pmatrix}$

c If $c = 0$, then the line $ax + by = c$ does not go through the origin. Hence the origin cannot be mapped to itself, and the transformation is not linear.

CHAPTER 8

Prior knowledge 8

1 a $\frac{1}{2}n(n+1)$ b $\frac{1}{6}(n+1)(n+2)(2n+3)$

2 $3^{n+2} - 3^n = 3^n(9-1) = 3^n \times 8$

3 $\begin{pmatrix} 5k & 4-8k \\ 5k+5 & -8k-14 \end{pmatrix}$

Exercise 8A

1 Basis: When $n = 1$: LHS = 1; RHS = $\frac{1}{2}(1)(1+1) = 1$

Assumption: $\sum_{r=1}^{k} r = \frac{1}{2}k(k+1)$

Induction: $\sum_{r=1}^{k+1} r = \sum_{r=1}^{k} r + (k+1) = \frac{1}{2}k(k+1) + (k+1)$
$= \frac{1}{2}(k+1)(k+2)$

So if the statement holds for $n = k$, it holds for $n = k + 1$.

Conclusion: The statement holds for all $n \in \mathbb{Z}^+$.

2 Basis step: When $n = 1$: LHS=1; RHS=$\frac{1}{4}(1)^2(1+1)^2 = 1$

Assumption: $\sum_{r=1}^{k} r^3 = \frac{1}{4}k^2(k+1)^2$

Induction: $\sum_{r=1}^{k+1} r^3 = \sum_{r=1}^{k} r^3 + (k+1)^3 = \frac{1}{4}k^2(k+1)^2 + (k+1)^3$
$= \frac{1}{4}(k+1)^2(k^2 + 4(k+1)) = \frac{1}{4}(k+1)^2(k+2)^2$

So if the statement holds for $n = k$, it holds for $n = k + 1$.

Conclusion: The statement holds for all $n \in \mathbb{Z}^+$.

3 a Basis: $n = 1$: LHS = 0; RHS = $\frac{1}{3}(1)(1+1)(1-1) = 0$

Assumption: $\sum_{r=1}^{k} r(r-1) = \frac{1}{3}k(k+1)(k-1)$

Induction: $\sum_{r=1}^{k+1} r(r-1) = \sum_{r=1}^{k} r(r-1) + (k+1)k$
$= \frac{1}{3}k(k+1)(k-1) + k(k+1)$
$= \frac{1}{3}k(k+1)(k-1+3) = \frac{1}{3}k(k+1)(k+2)$

So if the statement holds for $n = k$, it holds for $n = k + 1$.

Conclusion: The statement holds for all $n \in \mathbb{Z}^+$.

b $\sum_{r=1}^{2n+1} r(r-1) = \frac{1}{3}(2n+1)((2n+1)+1)((2n+1)-1)$
$= \frac{4}{3}n(2n+1)(n+1)$

4 a Basis: $n = 1$: LHS = 2; RHS = $(1)^2(1+1) = 2$

Assumption: $\sum_{r=1}^{k} r(3r-1) = k^2(k+1)$

Induction: $\sum_{r=1}^{k+1} r(3r-1) = \sum_{r=1}^{k} r(3r-1)$
$+ (k+1)(3k+2)$
$= k^2(k+1) + (k+1)(3k+2)$
$= (k+1)^2(k+2)$

So if the statement holds for $n = k$, it holds for $n = k + 1$.

Conclusion: The statement holds for all $n \in \mathbb{Z}^+$.

b $n = 15$

5 a Basis: $n = 1$: LHS = $\frac{1}{2}$; RHS=$1 - \frac{1}{2} = \frac{1}{2}$

Assumption: $\sum_{r=1}^{k}\left(\frac{1}{2}\right)^r = 1 - \frac{1}{2^k}$

Induction: $\sum_{r=1}^{k+1}\left(\frac{1}{2}\right)^r = \sum_{r=1}^{k}\left(\frac{1}{2}\right) + \frac{1}{2^{k+1}} = 1 - \frac{1}{2^k} + \frac{1}{2^{k+1}}$
$= 1 - \frac{2}{2^{k+1}} + \frac{1}{2^{k+1}} = 1 - \frac{1}{2^{k+1}}$

So if the statement holds for $n = k$, it holds for $n = k + 1$.

Conclusion: The statement holds for all $n \in \mathbb{Z}^+$.

b Basis: $n = 1$: LHS = $1 \times 1! = 1$; RHS = $(1+1)! - 1 = 1$

Assumption: $\sum_{r=1}^{k} r(r!) = (n+1)! - 1$

Induction: $\sum_{r=1}^{k+1} r(r!) = \sum_{r=1}^{k} r(r!) + (k+1)(k+1)!$
$= (k+1)! - 1 + (k+1)(k+1)!$
$= (k+1)!(k+2) - 1 = ((k+1)+1)! - 1$

So if the statement holds for $n = k$, it holds for $n = k + 1$.

Conclusion: The statement holds for all $n \in \mathbb{Z}^+$.

c Basis: $n = 1$: LHS = $\frac{4}{1 \times 3} = \frac{4}{3}$; RHS = $\frac{1 \times 8}{2 \times 3} = \frac{4}{3}$

Assumption: $\sum_{r=1}^{k} \frac{4}{r(r+2)} = \frac{k(3k+5)}{(k+1)(k+2)}$

Induction: $\sum_{r=1}^{k+1} \frac{4}{r(r+2)} = \sum_{r=1}^{k} \frac{4}{r(r+2)} + \frac{4}{(k+1)(k+3)}$

$= \frac{k(3k+5)}{(k+1)(k+2)} + \frac{4}{(k+1)(k+3)}$

$= \frac{k(3k+5)(k+3)}{(k+1)(k+2)(k+3)} + \frac{4(k+2)}{(k+1)(k+2)(k+3)}$

$= \frac{k(3k+5)(k+3) + 4(k+2)}{(k+1)(k+2)(k+3)} = \frac{(k+1)(3k+8)}{(k+2)(k+3)}$

$= \frac{(k+1)(3(k+1)+5)}{((k+1)+1)((k+1)+2)}$

So if the statement holds for $n = k$, it holds for $n = k + 1$.
Conclusion: The statement holds for all $n \in \mathbb{Z}^+$.

6 **a** The student has just stated and not shown that the statement is true for $n = k + 1$.
b e.g. $n = 2$: LHS $= (1 + 2)^2 = 9$; RHS $= 1^2 + 2^2 \neq 9$, so that LHS $\neq$ RHS.

7 **a** The student has not completed the basis step.
b e.g. $n = 1$: LHS = 1; RHS$=\frac{1}{2}(1^2 + 1 + 1) = \frac{3}{2} \neq 1$

Challenge

Basis: $n = 1$: LHS $= (-1)^1 \times 1^2 = -1$; RHS $= \frac{1}{2}(-1)^1(1)(1 + 1) = -1$

Assumption: $\sum_{r=1}^{k} (-1)^r r^2 = \frac{1}{2}(-1)^k k(k + 1)$

Induction: $\sum_{r=1}^{k+1} (-1)^r r^2 = \sum_{r=1}^{k} (-1)^r r^2 + (-1)^{k+1}(k + 1)^2$

$= \frac{1}{2}(-1)^k k(k + 1) + (-1)^{k+1}(k + 1)^2$

$= \frac{1}{2}(-1)^{k+1}(k + 1)(-k + 2(k + 1)) = \frac{1}{2}(-1)^{k+1}(k + 1)(k + 2)$

So if the statement holds for $n = k$, it holds for $n = k + 1$.
Conclusion: The statement holds for all $n \in \mathbb{Z}^+$.

Exercise 8B

1 **a** Let $f(n) = 8^n - 1$ where $n \in \mathbb{Z}^+$.
Basis: $n = 1$: $f(1) = 8 - 1 = 7$ is divisible by 7.
Assumption: $f(k)$ is divisible by 7.
Induction: $f(k + 1) = 8^{k+1} - 1 = 8 \times 8^k - 1 = 8f(k) + 7$
So if the statement holds for n = k, it holds for $n = k + 1$.
Conclusion: The statement holds for al $n \in \mathbb{Z}^+$.

b Let $f(n) = 3^{2n} - 1$ where $n \in \mathbb{Z}^+$.
Basis: $n = 1$: $f(1) = 3^2 - 1 = 8$ is divisible by 8.
Assumption: f(k) is divisible by 8.
Induction: $f(k + 1) = 3^{2(k+1)} - 1 = 3^{2k} \times 3^2 - 1$
$f(k + 1) - f(k) = (3^{2k} \times 3^2 - 1) - (3^{2k} - 1) = 8 \times 3^{2k}$
So if the statement holds for $n = k$, it holds for $n = k + 1$.
Conclusion: The statement holds for all $n \in \mathbb{Z}^+$.

c Let $f(n) = 5^n + 9^n + 2$ where $n \in \mathbb{Z}^+$.
Basis: $n = 1$: $f(1) = 5 + 9 + 2 = 16$ is divisible by 4.
Assumption: $f(k)$ is divisible by 4.
Induction: $f(k + 1) = 5^{k+1} + 9^{k+1} + 2 = 5 \times 5^k + 9 \times 9^k + 2$
$f(k + 1) - f(k) = (5 \times 5^k + 9 \times 9^k + 2) - (5^k + 9^k + 2)$
$= 4 \times 5^k + 8 \times 9^k$
So if the statement holds for $n = k$, it holds for $n = k + 1$.
Conclusion: The statement holds for all $n \in \mathbb{Z}^+$.

d Let $f(n) = 2^{4n} - 1$ where $n \in \mathbb{Z}^+$.
Basis: $n = 1$: $f(1) = 2^4 - 1 = 15$ is divisible by 15.
Assumption: $f(k)$ is divisible by 15.
Induction: $f(k + 1) = 2^{4(k+1)} - 1 = 16 \times 2^{4k} - 1$
$f(k + 1) - f(k) = (16 \times 2^{4k} - 1) - (2^{4k} - 1)$
$= 15 \times 2^{4k}$
So if the statement holds for $n = k$, it holds for $n = k + 1$.
Conclusion: The statement holds for all $n \in \mathbb{Z}^+$.

e Let $f(n) = 3^{2n-1} + 1$ where $n \in \mathbb{Z}^+$
Basis: $n = 1$: $f(1) = 3^{2-1} + 1 = 4$ is divisible by 4.
Assumption: $f(k)$ is divisible by 4.
Induction: $f(k + 1) = 3^{2(k+1)-1} + 1 = 3^2 \times 3^{2k-1} + 1$
$f(k + 1) - f(k) = (3^2 \times 3^{2k-1} + 1) - (3^{2k-1} + 1)$
$= 8 \times 3^{2k-1}$
So if the statement holds for $n = k$, it holds for $n = k + 1$.
Conclusion: The statement holds for all $n \in \mathbb{Z}^+$.

f Let $f(n) = n^3 + 6n^2 + 8n$ where $n \in \mathbb{Z}^+$
Basis: $n = 1$: $f(1) = 1^3 + 6 \times 1^2 + 8 = 15$ is divisible by 3.
Assumption: $f(k)$ is divisible by 3.
Induction: $f(k + 1) = (k + 1)^3 + 6(k + 1)^2 + 8(k + 1)$
$= k^3 + 9k^2 + 23k + 15$
$f(k + 1) - f(k) = 3(k^2 + 5k + 5)$
So if the statement holds for $n = k$, it holds for $n = k + 1$.
Conclusion: The statement holds for all $n \in \mathbb{Z}^+$.

g Let $f(n) = n^3 + 5n$ where $n \in \mathbb{Z}^+$.
Basis: $n = 1$: $f(1) = 1^3 + 5 = 6$ is divisible by 6.
Assumption: $f(k)$ is divisible by 6.
Induction: $f(k + 1) = (k + 1)^3 + 5(k + 1) = k^3 + 3k^2 + 8k + 6$
$f(k + 1) - f(k) = 3k^2 + 3k + 6 = 3k(k + 1) + 6$
where $3k(k + 1)$ is divisible by 3, and one of k and $k + 1$ must be even, so $3k(k + 1)$ is divisible by 6.
Therefore, if the statement holds for $n = k$, it holds for $n = k + 1$.
Conclusion: The statement holds for all $n \in \mathbb{Z}^+$.

h Let $f(n) = 2^n \times 3^{2n} - 1$ where $n \in \mathbb{Z}^+$.
Basis: $f(1) = 2 \times 3^2 - 1 = 17$ is divisible by 17.
Assumption: $f(k)$ is divisible by 17.
Induction: $f(k + 1) = 2^{k+1} \times 3^{2(k+1)} - 1 = 18 \times 2^k \times 3^{2k} - 1$
$f(k + 1) - f(k) = (18 \times 2^k \times 3^{2k} - 1) - (2^k \times 3^{2k} - 1)$
$= 17 \times 2^k \times 3^{2k}$
So if the statement holds for $n = k$, it holds for $n = k + 1$.
Conclusion: The statement holds for all $n \in \mathbb{Z}^+$.

2 **a** $f(k + 1) = 13^{k+1} - 6^{k+1} = 13 \times 13^k - 6 \times 6^k$
$= 6(13^k - 6^k) + 7 \times 13^k = 6f(k) + 7(13^k)$

b Basis: $n = 1$: $f(1) = 13 - 6 = 7$ is divisible by 7.
Assumption: $f(k)$ is divisible by 7.
Induction: $f(k + 1) = 6f(k) + 7(13^k)$ by part **a**.
So if the statement holds for $n = k$, it holds for $n = k + 1$.
Conclusion: The statement holds for all $n \in \mathbb{Z}^+$.

3 **a** $g(k + 1) = 5^{2(k+1)} - 6(k + 1) + 8 = 25 \times 5^{2k} - 6k + 2$
$= 25(5^{2k} - 6k + 8) + 144k - 198$
$= 25g(k) + 9(16k - 22)$

b Basis: $n = 1$: $g(1) = 5^2 - 6 + 8 = 27$ is divisible by 9.
Assumption: $g(k)$ is divisible by 9.
Induction: $g(k + 1) = 25\,g(k) + 9(16k - 22)$ by part **a**.
So if the statement holds for $n = k$, it holds for $n = k + 1$.
Conclusion: The statement holds for all $n \in \mathbb{Z}^+$.

4 Let $f(n) = 8^n - 3^n$ where $n \in \mathbb{Z}^+$.
Basis: $n = 1$: $f(1) = 8 - 3 = 5$ is divisible by 5.
Assumption: $f(k)$ is divisible by 5.
Induction: $f(k + 1) = 8^{k+1} - 3^{k+1} = 8 \times 8^k - 3 \times 3^k$
$f(k + 1) - 3f(k) = (8 \times 8^k - 3 \times 3^k) - 3(8^k - 3^k)$
$= 5 \times 8^k$
So if the statement holds for $n = k$, it holds for $n = k + 1$.
Conclusion: The statement holds for all $n \in \mathbb{Z}^+$.

5 Let $f(n) = 3^{2n+2} + 8n - 9$ where $n \in \mathbb{Z}^+$.
Basis: $n = 1$: $f(1) = 3^4 + 8 - 9 = 80$ is divisible by 8.

Online Full worked solutions are available in SolutionBank.

Assumption: f(k) is divisible by 8.
Induction: $f(k+1) = 3^{2(k+1)+2} + 8(k+1) - 9 = 9 \times 3^{2k+2} + 8k - 1$

$f(k+1) - f(k) = (9 \times 3^{2k+2} + 8k - 1) - (3^{2k+2} + 8k - 9)$
$= 8 \times 3^{2k+2} + 8 = 8(3^{2k+2} + 1)$

So if the statement holds for $n = k$, it holds for $n = k + 1$.
Conclusion: By induction, the statement holds for all $n \in \mathbb{Z}^+$.

6 Let $f(n) = 2^{6n} + 3^{2n-2}$ where $n \in \mathbb{Z}^+$
Basis: $n = 1$: $f(1) = 2^6 + 3^0 = 65$ is divisible by 5.
Assumption: f(k) is divisible by 5.
Induction: $f(k+1) = 2^{6(k+1)} + 3^{2(k+1)-2} = 64 \times 2^{6k} + 9 \times 3^{2k-2}$

$f(k+1) + f(k) = (64 \times 2^{6k} + 9 \times 3^{2k-2}) + (2^{6k} + 3^{2k-2})$
$= 65 \times 2^{6k} + 10 \times 3^{2k-2}$

So if the statement holds for $n = k$, it holds for $n = k + 1$.
Conclusion: The statement holds for all $n \in \mathbb{Z}^+$.

Exercise 8C

1 Basis: $n = 1$: LHS = RHS = $\begin{pmatrix} 1 & 2 \\ 0 & 1 \end{pmatrix}$

Assumption: $\begin{pmatrix} 1 & 2 \\ 0 & 1 \end{pmatrix}^k = \begin{pmatrix} 1 & 2k \\ 0 & 1 \end{pmatrix}$

Induction: $\begin{pmatrix} 1 & 2 \\ 0 & 1 \end{pmatrix}^{k+1} = \begin{pmatrix} 1 & 2 \\ 0 & 1 \end{pmatrix}^k \begin{pmatrix} 1 & 2 \\ 0 & 1 \end{pmatrix} = \begin{pmatrix} 1 & 2k \\ 0 & 1 \end{pmatrix}\begin{pmatrix} 1 & 2 \\ 0 & 1 \end{pmatrix}$

$= \begin{pmatrix} 1+0 & 2+2k \\ 0+0 & 0+1 \end{pmatrix} = \begin{pmatrix} 1 & 2(k+1) \\ 0 & 1 \end{pmatrix}$

So if the statement holds for $n = k$, it holds for $n = k + 1$.
Conclusion: The statement holds for all $n \in \mathbb{Z}^+$.

2 Basis: $n = 1$: LHS = RHS = $\begin{pmatrix} 3 & -4 \\ 1 & -1 \end{pmatrix}$

Assumption: $\begin{pmatrix} 3 & -4 \\ 1 & -1 \end{pmatrix}^k = \begin{pmatrix} 2k+1 & -4k \\ k & -2k+1 \end{pmatrix}$

Induction: $\begin{pmatrix} 3 & -4 \\ 1 & -1 \end{pmatrix}^{k+1} = \begin{pmatrix} 3 & -4 \\ 1 & -1 \end{pmatrix}^k \begin{pmatrix} 3 & -4 \\ 1 & -1 \end{pmatrix}$

$= \begin{pmatrix} 2k+1 & -4k \\ k & -2k+1 \end{pmatrix}\begin{pmatrix} 3 & -4 \\ 1 & -1 \end{pmatrix}$

$= \begin{pmatrix} 6k+3-4k & -8k-4+4k \\ 3k-2k+1 & -4k+2k-1 \end{pmatrix}$

$= \begin{pmatrix} 2(k+1)+1 & -4(k+1) \\ k+1 & -2(k+1)+1 \end{pmatrix}$

So if the statement holds for $n = k$, it holds for $n = k + 1$.
Conclusion: The statement holds for all $n \in \mathbb{Z}^+$.

3 Basis: $n = 1$: LHS= RHS=$\begin{pmatrix} 2 & 0 \\ 1 & 1 \end{pmatrix}$

Assumption: $\begin{pmatrix} 2 & 0 \\ 1 & 1 \end{pmatrix}^k = \begin{pmatrix} 2^k & 0 \\ 2^k - 1 & 1 \end{pmatrix}$

Induction: $\begin{pmatrix} 2 & 0 \\ 1 & 1 \end{pmatrix}^{k+1} = \begin{pmatrix} 2 & 0 \\ 1 & 1 \end{pmatrix}^k \begin{pmatrix} 2 & 0 \\ 1 & 1 \end{pmatrix} = \begin{pmatrix} 2^k & 0 \\ 2^k - 1 & 1 \end{pmatrix}\begin{pmatrix} 2 & 0 \\ 1 & 1 \end{pmatrix}$

$= \begin{pmatrix} 2^{k+1} + 0 & 0+0 \\ 2^{k+1} - 2 + 1 & 0+1 \end{pmatrix}$

$= \begin{pmatrix} 2^{k+1} & 0 \\ 2^{k+1} - 1 & 1 \end{pmatrix}$

So if the statement holds for $n = k$, it holds for $n = k + 1$.
Conclusion: The statement holds for all $n \in \mathbb{Z}^+$.

4 a Basis: $n = 1$: LHS= RHS=$\begin{pmatrix} 5 & -8 \\ 2 & -3 \end{pmatrix}$

Assumption: $\begin{pmatrix} 5 & -8 \\ 2 & -3 \end{pmatrix}^k = \begin{pmatrix} 4k+1 & -8k \\ 2k & 1-4k \end{pmatrix}$

Induction: $\begin{pmatrix} 5 & -8 \\ 2 & -3 \end{pmatrix}^{k+1} = \begin{pmatrix} 5 & -8 \\ 2 & -3 \end{pmatrix}^k \begin{pmatrix} 5 & -8 \\ 2 & -3 \end{pmatrix}$

$= \begin{pmatrix} 4k+1 & -8k \\ 2k & 1-4k \end{pmatrix}\begin{pmatrix} 5 & -8 \\ 2 & -3 \end{pmatrix}$

$= \begin{pmatrix} 20k+5-16k & -32k-8+24k \\ 10k+2-8k & -16k-3+12k \end{pmatrix}$

$= \begin{pmatrix} 4(k+1)+1 & -8(k+1) \\ 2(k+1) & 1-4(k+1) \end{pmatrix}$

So if the statement holds for $n = k$, it holds for $n = k + 1$.
Conclusion: The statement holds for all $n \in \mathbb{Z}^+$.

b $n = 6$

5 a Basis: $n = 1$: LHS = RHS = $\begin{pmatrix} 2 & 5 \\ 0 & 1 \end{pmatrix}$

Assumption: $\mathbf{M}^k = \begin{pmatrix} 2^k & 5(2^k - 1) \\ 0 & 1 \end{pmatrix}$

Induction:

$\mathbf{M}^{k+1} = \mathbf{M}^k \begin{pmatrix} 2 & 5 \\ 0 & 1 \end{pmatrix} = \begin{pmatrix} 2^k & 5(2^k - 1) \\ 0 & 1 \end{pmatrix}\begin{pmatrix} 2 & 5 \\ 0 & 1 \end{pmatrix}$

$= \begin{pmatrix} 2^{k+1} & 5 \times 2^k + 5(2^k - 1) \\ 0 & 1 \end{pmatrix} = \begin{pmatrix} 2^{k+1} & 5(2^{k+1} - 1) \\ 0 & 1 \end{pmatrix}$

So if the statement holds for $n = k$, it holds for $n = k + 1$.
Conclusion: The statement holds for all $n \in \mathbb{Z}^+$.

b $\begin{pmatrix} 2^{-n} & 5(2^{-n} - 1) \\ 0 & 1 \end{pmatrix}$

Challenge

Basis: $n = 1$: LHS = RHS = $\begin{pmatrix} 3 & 1 & 0 \\ 0 & 1 & 0 \\ 0 & -1 & 4 \end{pmatrix}$

Assumption: $\begin{pmatrix} 3 & 1 & 0 \\ 0 & 1 & 0 \\ 0 & -1 & 4 \end{pmatrix}^k = \begin{pmatrix} 3^k & \frac{3^k - 1}{2} & 0 \\ 0 & 1 & 0 \\ 0 & \frac{1 - 4^k}{3} & 4^k \end{pmatrix}$

Induction:

$\begin{pmatrix} 3 & 1 & 0 \\ 0 & 1 & 0 \\ 0 & -1 & 4 \end{pmatrix}^{k+1} = \begin{pmatrix} 3 & 1 & 0 \\ 0 & 1 & 0 \\ 0 & -1 & 4 \end{pmatrix}^k \begin{pmatrix} 3 & 1 & 0 \\ 0 & 1 & 0 \\ 0 & -1 & 4 \end{pmatrix}$

$= \begin{pmatrix} 3^k & \frac{3^k - 1}{2} & 0 \\ 0 & 1 & 0 \\ 0 & \frac{1 - 4^k}{3} & 4^k \end{pmatrix}\begin{pmatrix} 3 & 1 & 0 \\ 0 & 1 & 0 \\ 0 & -1 & 4 \end{pmatrix}$

$= \begin{pmatrix} 3^{k+1} & 3^k + \frac{3^k - 1}{2} & 0 \\ 0 & 1 & 0 \\ 0 & \frac{1 - 4^k}{3} - 4^k & 4^k \times 4 \end{pmatrix} = \begin{pmatrix} 3^{k+1} & \frac{3^{k+1} - 1}{2} & 0 \\ 0 & 1 & 0 \\ 0 & \frac{1 - 4^{k+1}}{3} & 4^{k+1} \end{pmatrix}$

So if the statement holds for $n = k$, it holds for $n = k + 1$.
Conclusion: The statement holds for all $n \in \mathbb{Z}^+$.

Mixed exercise 8

1 Let $f(n) = 9^n - 1$ where $n \in \mathbb{Z}^+$.
Basis: $f(1) = 9^1 - 1 = 8$ is divisible by 8.
Assumption: f(k) is divisible by 8.
Induction: $f(k+1) = 9^{k+1} - 1 = 9 \times 9^k - 1$
$f(k+1) - f(k) = (9 \times 9^k - 1) - (9^k - 1) = 8 \times 9^k$
So if the statement holds for $n = k$, it holds for $n = k + 1$.
Conclusion: The statement holds for all $n \in \mathbb{Z}^+$.

2 a $\mathbf{B}^2 = \begin{pmatrix} 1 & 0 \\ 0 & 9 \end{pmatrix}$, $\mathbf{B}^3 = \begin{pmatrix} 1 & 0 \\ 0 & 27 \end{pmatrix}$

b $\mathbf{B}^n = \begin{pmatrix} 1 & 0 \\ 0 & 3^n \end{pmatrix}$

c Basis: $n = 1$: LHS= RHS = $\begin{pmatrix} 1 & 0 \\ 0 & 3 \end{pmatrix}$

Assumption: $\mathbf{B}^k = \begin{pmatrix} 1 & 0 \\ 0 & 3^k \end{pmatrix}$

Induction: $\mathbf{B}^{k+1} = \mathbf{B}^k\begin{pmatrix} 1 & 0 \\ 0 & 3 \end{pmatrix} = \begin{pmatrix} 1 & 0 \\ 0 & 3^k \end{pmatrix}\begin{pmatrix} 1 & 0 \\ 0 & 3 \end{pmatrix}$

$= \begin{pmatrix} 1+0 & 0+0 \\ 0+0 & 0+3^{k+1} \end{pmatrix} = \begin{pmatrix} 1 & 0 \\ 0 & 3^{k+1} \end{pmatrix}$

So if the statement holds for $n = k$, it holds for $n = k + 1$.

Conclusion: The statement holds for all $n \in \mathbb{Z}^+$.

3 Basis: $n = 1$: LHS $= 3 \times 1 + 4 = 7$;
RHS $= \frac{1}{2} \times 1(3 \times 1 + 11) = 7$

Assumption: $\sum_{r=1}^{k}(3r + 4) = \frac{1}{2}k(3k + 11)$

Induction: $\sum_{r=1}^{k+1}(3r + 4) = \sum_{r=1}^{k}(3r + 4) + 3(k + 1) + 4$

$= \frac{1}{2}k(3k + 11) + 3(k + 1) + 4 = \frac{1}{2}(3k^2 + 17k + 14)$

$= \frac{1}{2}(k + 1)(3(k + 1) + 11)$

So if the statement holds for $n = k$, it holds for $n = k + 1$.
Conclusion: The statement holds for all $n \in \mathbb{Z}^+$.

4 a Basis: $n = 1$: LHS = RHS = $\begin{pmatrix} 9 & 16 \\ -4 & -7 \end{pmatrix}$

Assumption: $\mathbf{A}^k = \begin{pmatrix} 8k+1 & 16k \\ -4k & 1-8k \end{pmatrix}$

Induction:

$\mathbf{A}^{k+1} = \mathbf{A}^k\begin{pmatrix} 9 & 16 \\ -4 & -7 \end{pmatrix} = \begin{pmatrix} 8k+1 & 16k \\ -4k & 1-8k \end{pmatrix}\begin{pmatrix} 9 & 16 \\ -4 & -7 \end{pmatrix}$

$= \begin{pmatrix} 72k+9-64k & 128k+16-112k \\ -36k-4+32k & -64k-7+56k \end{pmatrix}$

$= \begin{pmatrix} 8(k+1)+1 & 16(k+1) \\ -4(k+1) & 1-8(k+1) \end{pmatrix}$

So if the statement holds for $n = k$, it holds for $n = k + 1$.

Conclusion: The statement holds for all $n \in \mathbb{Z}^+$.

b $\begin{pmatrix} 1-8n & -16n \\ 4n & 8n+1 \end{pmatrix}$

5 a $\mathrm{f}(n + 1) = 5^{2(n+1)-1} + 1 = 25 \times 5^{2n-1} + 1$

$\mathrm{f}(n + 1) - \mathrm{f}(n) = (25 \times 5^{2n-1} + 1) - (5^{2n-1} + 1)$

$= 24 \times 5^{2n-1}$; $\mu = 24$

b Let $\mathrm{f}(n) = 5^{2n-1} + 1$ where $n \in \mathbb{Z}^+$.

Basis: $n = 1$: $\mathrm{f}(1) = 5^{2-1} + 1 = 6$ is divisible by 6.

Assumption: $\mathrm{f}(k)$ is divisible by 6.

Induction: $\mathrm{f}(k + 1) - \mathrm{f}(k) = 24 \times 5^{2k-1} = 6 \times 4 \times 5^{2k-1}$

So if the statement holds for $n = k$, it holds for $n = k + 1$.

Conclusion: The statement holds for all $n \in \mathbb{Z}^+$.

6 Let $\mathrm{f}(n) = 7^n + 4^n + 1$ where $n \in \mathbb{Z}^+$.

Basis: $n = 1$: $\mathrm{f}(1) = 7^1 + 4^1 + 1 = 12$ is divisible by 6.

Assumption: $\mathrm{f}(k)$ is divisible by 6.

Induction: $\mathrm{f}(k + 1) = 7^{k+1} + 4^{k+1} + 1 = 7 \times 7^k + 4 \times 4^k + 1$

$\mathrm{f}(k + 1) - \mathrm{f}(k) = (7 \times 7^k + 4 \times 4^{k+1} + 1) - (7^k + 4^k + 1)$

$= 6 \times 7^k + 3 \times 4^k$

where both 6×7^k and 3×4^k are divisible by 6, since 4 is even.

So if the statement holds for $n = k$, it holds for $n = k + 1$.

Conclusion: The statement holds for all $n \in \mathbb{Z}^+$.

7 Basis:

$n = 1$: LHS $= 1 \times 5 = 5$; RHS $= \frac{1}{6} \times 1 \times 2 \times (2 + 13) = 5$

Assumption: $\sum_{r=1}^{k} r(r + 4) = \frac{1}{6}k(k + 1)(2k + 13)$

Induction: $\sum_{r=1}^{k+1} r(r + 4) = \sum_{r=1}^{k} r(r + 4) + (k + 1)(k + 5)$

$= \frac{1}{6}k(k + 1)(2k + 13) + (k + 1)(k + 5)$

$= \frac{1}{6}(2k^3 + 21k^2 + 49k + 30) = \frac{1}{6}(k + 1)(k + 2)(2(k + 1) + 13)$

So if the statement holds for $n = k$, it holds for $n = k + 1$.

Conclusion: The statement holds for all $n \in \mathbb{Z}^+$.

8 a Basis: $n = 1$: LHS $= 1 + 4 = 5$; RHS $= \frac{1}{3} \times 1 \times 3 \times 5 = 5$

Assumption: $\sum_{r=1}^{2k} r^2 = \frac{1}{3}k(2k + 1)(4k + 1)$

Induction: $\sum_{r=1}^{2(k+1)} r^2 = \sum_{r=1}^{2k} r^2 + (2k + 1)^2 + (2k + 2)^2$

$= \frac{1}{3}k(2k + 1)(4k + 1) + (2k + 1)^2 + (2k + 2)^2$

$= \frac{1}{3}(8k^3 + 30k^2 + 37k + 15)$

$= \frac{1}{3}(k + 1)(2(k + 1) + 1)(4(k + 1) + 1)$

So if the statement holds for $n = k$, it holds for $n = k + 1$.

Conclusion: The statement holds for all $n \in \mathbb{Z}^+$.

b Using **a** and the formula for $\sum_{r=1}^{n} r^2$,

$\frac{1}{6} \times 2n(2n + 1)(4n + 1) = \frac{1}{6}kn(n + 1)(2n + 1$

$2n(2n + 1)(4n + 1) = kn(n + 1)(2n + 1)$

$16n^3 + 12n^2 + 2n = k(2n^3 + 3n^2 + n)$

$\Rightarrow k = \frac{2n(8n^2 + 6n + 1)}{n(2n^2 + 3n + 1)} = \frac{2(2n + 1)(4n + 1)}{(2n + 1)(n + 1)} = \frac{8n + 2}{n + 1}$

$\Rightarrow kn + k = 8n + 2 \Rightarrow n(k - 8) = 2 - k \Rightarrow n = \frac{2 - k}{k - 8}$

9 a Basis: $n = 1$: LHS = RHS = $\begin{pmatrix} 2c & 1 \\ 0 & c \end{pmatrix}$

Assumption: $\mathbf{M}^k = c^k\begin{pmatrix} 2^k & \frac{2^k - 1}{c} \\ 0 & 1 \end{pmatrix}$

Induction: $\mathbf{M}^{k+1} = \mathbf{M}^k\begin{pmatrix} 2c & 1 \\ 0 & c \end{pmatrix}$

$= c^k\begin{pmatrix} 2^k & \frac{2^k - 1}{c} \\ 0 & 1 \end{pmatrix}\begin{pmatrix} 2c & 1 \\ 0 & c \end{pmatrix} = c^{k+1}\begin{pmatrix} 2^k & \frac{2^k - 1}{c} \\ 0 & 1 \end{pmatrix}\begin{pmatrix} 2 & \frac{1}{c} \\ 0 & 1 \end{pmatrix}$

$= c^{k+1}\begin{pmatrix} 2^{k+1} & \frac{2^k + 2^k - 1}{c} \\ 0 & 1 \end{pmatrix} = c^{k+1}\begin{pmatrix} 2^{k+1} & \frac{2^{k+1} - 1}{c} \\ 0 & 1 \end{pmatrix}$

So if the statement holds for $n = k$, it holds for $n = k + 1$.

Conclusion: The statement holds for all $n \in \mathbb{Z}^+$.

b Consider $n = 1$: $\det\mathbf{M} = 50 \Rightarrow 2c^2 = 50$
So $c = 5$, since c is +ve.

Challenge

a Basis: $n = 1$: LHS = RHS = $\begin{pmatrix} \cos\theta & -\sin\theta \\ \sin\theta & \cos\theta \end{pmatrix}$

Assumption: $\mathbf{M}^k = \begin{pmatrix} \cos k\theta & -\sin k\theta \\ \sin k\theta & \cos k\theta \end{pmatrix}$

Induction:

$\mathbf{M}^{k+1} = \mathbf{M}^k\begin{pmatrix} \cos\theta & -\sin\theta \\ \sin\theta & \cos\theta \end{pmatrix} = \begin{pmatrix} \cos k\theta & -\sin k\theta \\ \sin k\theta & \cos k\theta \end{pmatrix}\begin{pmatrix} \cos\theta & -\sin\theta \\ \sin\theta & \cos\theta \end{pmatrix}$

$= \begin{pmatrix} \cos k\theta\cos\theta - \sin k\theta\sin\theta & -\cos k\theta\sin\theta - \sin k\theta\cos\theta \\ \sin k\theta\cos\theta + \cos k\theta\sin\theta & -\sin k\theta\sin\theta + \cos k\theta\cos\theta \end{pmatrix}$

$= \begin{pmatrix} \cos((k+1)\theta) & -\sin((k+1)\theta) \\ \sin((k+1)\theta) & \cos((k+1)\theta) \end{pmatrix}$

So if the statement holds for $n = k$, it holds for $n = k + 1$.
Conclusion: The statement holds for all $n \in \mathbb{Z}^+$.

Online Full worked solutions are available in SolutionBank.

b The matrix $\mathbf{M}$ represents a rotation through angle θ, and so $\mathbf{M}^n$ represents a rotation through angle $n\theta$.

CHAPTER 9

Prior knowledge 9

1 **a** $\begin{pmatrix}-5\\2\end{pmatrix}$ **b** $\begin{pmatrix}7\\-5\\-7\end{pmatrix}$

2 **a** $2\sqrt{13}$ **b** $\sqrt{26}$

3 **a** $\sqrt{29}$ **b** $\frac{1}{\sqrt{29}}(4\mathbf{i} - 3\mathbf{j} + 2\mathbf{k})$

4 $(1, -1)$

Exercise 9A

1 **a** $\mathbf{r} = \begin{pmatrix}6\\5\\-1\end{pmatrix} + \lambda\begin{pmatrix}2\\-3\\-1\end{pmatrix}$ **b** $\mathbf{r} = \begin{pmatrix}2\\5\\0\end{pmatrix} + \lambda\begin{pmatrix}1\\1\\1\end{pmatrix}$

c $\mathbf{r} = \begin{pmatrix}-7\\6\\2\end{pmatrix} + \lambda\begin{pmatrix}3\\1\\2\end{pmatrix}$ **d** $\mathbf{r} = \begin{pmatrix}2\\0\\4\end{pmatrix} + \lambda\begin{pmatrix}-3\\2\\1\end{pmatrix}$

e $\mathbf{r} = \begin{pmatrix}6\\-11\\2\end{pmatrix} + \lambda\begin{pmatrix}0\\5\\-2\end{pmatrix}$

2 **a** **i** $2\mathbf{i} + 7\mathbf{j} - 3\mathbf{k}$ **ii** $\mathbf{r} = (3\mathbf{i} - 4\mathbf{j} + 2\mathbf{k}) + \lambda(2\mathbf{i} + 7\mathbf{j} - 3\mathbf{k})$

b **i** $2\mathbf{i} - 3\mathbf{j} + 4\mathbf{k}$ **ii** $\mathbf{r} = (2\mathbf{i} + \mathbf{j} - 3\mathbf{k}) + \lambda(2\mathbf{i} - 3\mathbf{j} + 4\mathbf{k})$

c **i** $-3\mathbf{i} - \mathbf{j} - 2\mathbf{k}$ **ii** $\mathbf{r} = (\mathbf{i} - 2\mathbf{j} + 4\mathbf{k}) + \lambda(-3\mathbf{i} - \mathbf{j} - 2\mathbf{k})$

d **i** $\begin{pmatrix}-5\\4\\-3\end{pmatrix}$ **ii** $\mathbf{r} = \begin{pmatrix}3\\-1\\4\end{pmatrix} + \lambda\begin{pmatrix}-5\\4\\-3\end{pmatrix}$

e **i** $\begin{pmatrix}-6\\4\\1\end{pmatrix}$ **ii** $\mathbf{r} = \begin{pmatrix}4\\-2\\3\end{pmatrix} + \lambda\begin{pmatrix}-6\\4\\1\end{pmatrix}$

3 **a** $\mathbf{r} = \begin{pmatrix}4\\-3\\8\end{pmatrix} + \lambda\begin{pmatrix}0\\0\\1\end{pmatrix}$

4 **a** **i** $\mathbf{r} = \begin{pmatrix}2\\1\\9\end{pmatrix} + \lambda\begin{pmatrix}2\\-2\\-1\end{pmatrix}$ **ii** $\mathbf{r} = \begin{pmatrix}-3\\5\\0\end{pmatrix} + \lambda\begin{pmatrix}10\\-3\\2\end{pmatrix}$

iii $\mathbf{r} = \begin{pmatrix}1\\11\\-4\end{pmatrix} + \lambda\begin{pmatrix}4\\-2\\6\end{pmatrix}$ **iv** $\mathbf{r} = \begin{pmatrix}-2\\-3\\-7\end{pmatrix} + \lambda\begin{pmatrix}14\\7\\4\end{pmatrix}$

b **i** $\frac{x-2}{2} = \frac{y-1}{-2} = \frac{z-9}{-1}$ **ii** $\frac{x+3}{10} = \frac{y-5}{-3} = \frac{z}{2}$

iii $\frac{x-1}{4} = \frac{y-11}{-2} = \frac{z+4}{6}$ **iv** $\frac{x+2}{14} = \frac{y+3}{7} = \frac{z+7}{4}$

5 **a** $p = 1, q = 10$ **b** $p = -6\frac{1}{2}, q = -21$

c $p = -19, q = -15$

6 Direction of l_1: $\begin{pmatrix}-1\\2\\4\end{pmatrix}$, direction of l_2: $\begin{pmatrix}2\\-4\\-8\end{pmatrix} = -2\begin{pmatrix}-1\\2\\4\end{pmatrix}$, so parallel

7 Direction of l_1: $\begin{pmatrix}2\\-3\\4\end{pmatrix}$, $\overrightarrow{AB} = \begin{pmatrix}-2\\3\\-4\end{pmatrix} = -\begin{pmatrix}2\\-3\\4\end{pmatrix}$, so parallel

8 $\overrightarrow{AB} = \begin{pmatrix}6\\3\\-3\end{pmatrix}$, $\overrightarrow{BC} = \begin{pmatrix}6\\3\\-3\end{pmatrix}$ same direction and a point in common

9 $\begin{pmatrix}3\\-1\\8\end{pmatrix} - \begin{pmatrix}1\\7\\-2\end{pmatrix} = \begin{pmatrix}2\\-8\\10\end{pmatrix} = 2\begin{pmatrix}1\\-4\\5\end{pmatrix}$; $\begin{pmatrix}10\\4\\0\end{pmatrix} - \begin{pmatrix}1\\7\\-2\end{pmatrix} = \begin{pmatrix}9\\-3\\2\end{pmatrix}$

so not collinear

10 $a = 2.5, b = -2$

11 $\mathbf{r} = (2\mathbf{i} - 7\mathbf{j} + 16\mathbf{k}) + \lambda(2\mathbf{i} - 4\mathbf{j} + \mathbf{k})$

12 **a** $a = 14, b = -2$ **b** $X(9, 9, -10)$

13 $AB = 9$

14 $B\begin{pmatrix}11\\3\\-2\end{pmatrix}$

15 $(4, -4, 8), (7, 5, 2)$

16 **a** $A\begin{pmatrix}-2\\4\\7\end{pmatrix}, B\begin{pmatrix}1\\1\\10\end{pmatrix}$ **b** $\mathbf{r} = \begin{pmatrix}0\\2\\3\end{pmatrix} + \lambda\begin{pmatrix}1\\-1\\1\end{pmatrix}$

c $C\begin{pmatrix}1\\1\\4\end{pmatrix}, D\begin{pmatrix}-1\\3\\2\end{pmatrix}$ so P is midpoint

17 **a** $A(10, 9, 8)$

b Tightrope will bow in the middle with acrobat's weight

Exercise 9B

1 **a** $\mathbf{r} = \mathbf{i} + 2\mathbf{j} + \lambda(2\mathbf{i} - \mathbf{j} - \mathbf{k}) + \mu(3\mathbf{i} + \mathbf{j} + 2\mathbf{k})$

b $\mathbf{r} = 3\mathbf{i} + 4\mathbf{j} + \mathbf{k} + \lambda(-4\mathbf{i} - 6\mathbf{j} - \mathbf{k}) + \mu(-\mathbf{i} - 3\mathbf{j} + 3\mathbf{k})$

c $\mathbf{r} = 2\mathbf{i} - \mathbf{j} - \mathbf{k} + \lambda(\mathbf{i} + 2\mathbf{j} + 3\mathbf{k}) + \mu(2\mathbf{i} + \mathbf{j} + 2\mathbf{k})$

d $\mathbf{r} = -\mathbf{i} + \mathbf{j} + 3\mathbf{k} + \lambda(\mathbf{j} + 2\mathbf{k}) + \mu(\mathbf{i} + 3\mathbf{j} + \mathbf{k})$

2 $-x + 3y + 2z = 2$

3 **a** 7 **b** −5 **c** 2.5 **d** 6

4 **a** **i** $2 - 6 + 5 = 1$ **ii** $4 + 12 - 15 = 1$

b $\mathbf{n} = 2\mathbf{i} - 3\mathbf{j} + 5\mathbf{k}$

5 **a** $\mathbf{r} = \begin{pmatrix}2\\3\\-2\end{pmatrix} + \lambda\begin{pmatrix}5\\-3\\-4\end{pmatrix}$

b $\frac{x-2}{5} = \frac{y-3}{-3} = \frac{z+2}{-4}$

6 $x = 0, x = 3, y = 0, y = 3, z = 0, z = 3$

7 All lie on plane with an equation

$\mathbf{r} = \begin{pmatrix}2\\2\\3\end{pmatrix} + \lambda\begin{pmatrix}-1\\3\\0\end{pmatrix} + \mu\begin{pmatrix}2\\1\\-4\end{pmatrix}$

8 A, B and C lie on plane with an equation

$\mathbf{r} = \begin{pmatrix}2\\3\\4\end{pmatrix} + \lambda\begin{pmatrix}0\\-4\\-1\end{pmatrix} + \mu\begin{pmatrix}3\\0\\-6\end{pmatrix}$

9 **a** $\begin{pmatrix}9\\5\\0\end{pmatrix}$ **b** $\lambda = -1, \mu = -1$ gives $B(1, -7, -1)$

c $\mathbf{r} = \begin{pmatrix}9\\5\\0\end{pmatrix} + \lambda\begin{pmatrix}-8\\-12\\-1\end{pmatrix}$

d $C = \begin{pmatrix}-\frac{21}{19}\\-\frac{193}{19}\\-\frac{24}{19}\end{pmatrix}$

Challenge

A: $2\mathbf{i} + 6\mathbf{j} + \mathbf{k}$ lies on plane; $\lambda = -2, \mu = 1$

$\overrightarrow{AB} = 5\mathbf{i} - 7\mathbf{j} + 6\mathbf{k}$

B: $7\mathbf{i} - \mathbf{j} + 7\mathbf{k}$ lies on plane; $\lambda = 1, \mu = 2$

So line lies entirely within plane.

Exercise 9C

1 $\frac{9}{2}$

2 **a** 2 **b** 17 **c** −6 **d** 20 **e** 0

3 **a** 55.5° **b** 94.8° **c** 87.4° **d** 79.0°

e 100.9° **f** 53.7° **g** 74.3° **h** 70.5°

4 **a** −10 **b** 5 **c** $2\frac{3}{5}$

d $-2\frac{1}{2}$ **e** −5 or 2

5 **a** 32.9° **b** 117.8°

6 **a** 20.5° **b** 109.9°

7 $\frac{2\sqrt{2}}{3}$

8 Use $\begin{pmatrix}1\\3\\0\end{pmatrix}.\begin{pmatrix}0\\1\\\lambda\end{pmatrix} = 3 = \sqrt{10}\sqrt{\lambda^2+1}\cos 60°$

9 a $\mathbf{i} + 2\mathbf{j} + \mathbf{k}$ b $3\mathbf{i} + 2\mathbf{j} + 3\mathbf{k}$ c $3\mathbf{i} + 2\mathbf{j} + 4\mathbf{k}$

10 64.7°, 64.7°, 50.6°

11 a $|\overrightarrow{AB}| = \sqrt{33}$, $|\overrightarrow{BC}| = \sqrt{173}$
b 29.1°

12 a $\cos\theta = \frac{26}{27}$ b area $= \frac{1}{2} \times 9 \times 3 \times \frac{\sqrt{53}}{27}$

13 Let $\overrightarrow{OA} = \mathbf{a}$, $\overrightarrow{OP} = \mathbf{p}$; then $\overrightarrow{OB} = \mathbf{b} = -\mathbf{a}$ and find scalar product

14 a $\overrightarrow{CA} = \begin{pmatrix}-1\\0\\-4\end{pmatrix}$, $\overrightarrow{CB} = \begin{pmatrix}-4\\5\\6\end{pmatrix}$
b $\frac{3\sqrt{101}}{2}$
c (9, −6, −6), (1, 4, 6), (3, 4, 6)
d $3\sqrt{101}$

15 a $\overrightarrow{PQ} = \begin{pmatrix}-3\\6\\-2\end{pmatrix}$, $\overrightarrow{QR} = \begin{pmatrix}2\\-2\\9\end{pmatrix}$; scalar product = 0
b centre (0.5, 1, 0.5), radius $= \frac{\sqrt{138}}{2}$

Challenge

1 $\mathbf{a.b} = |\mathbf{a}||\mathbf{b}|\cos\theta$, $\mathbf{b.a} = |\mathbf{b}||\mathbf{a}|\cos\theta$ so $\mathbf{a.b} = \mathbf{b.a}$

2 a i $\mathbf{a}.(\mathbf{b}+\mathbf{c}) = |\mathbf{a}||\mathbf{b}+\mathbf{c}|\cos\theta$, but $\cos\theta = \frac{PQ}{|\mathbf{b}+\mathbf{c}|}$
so $\mathbf{a}.(\mathbf{b}+\mathbf{c}) = |\mathbf{a}| \times PQ$
ii $\mathbf{a.b} = |\mathbf{a}||\mathbf{b}|\cos\alpha$, but $\cos\alpha = \frac{PR}{|\mathbf{b}|}$
so $\mathbf{a.b} = |\mathbf{a}| \times PR$
iii $\mathbf{a.c} = |\mathbf{a}||\mathbf{c}|\cos\beta$, but $\cos\beta = \frac{MN}{|\mathbf{c}|} = \frac{RQ}{|\mathbf{c}|}$
so $\mathbf{a.c} = |\mathbf{a}| \times RQ$
b $\mathbf{a}.(\mathbf{b}+\mathbf{c}) = |\mathbf{a}| \times PQ = |\mathbf{a}| \times (PR + RQ) = (|\mathbf{a}| \times PR) + (|\mathbf{a}| \times RQ) = \mathbf{a.b} + \mathbf{a.c}$; so $\mathbf{a}.(\mathbf{b}+\mathbf{c}) = \mathbf{a.b} + \mathbf{a.c}$

Exercise 9D

1 a 79.5° b 40.7° c 81.6° d 72.7° e 76.9°

2 a $\mathbf{r}.(2\mathbf{i} + \mathbf{j} + \mathbf{k}) = 0$ b $\mathbf{r}.(5\mathbf{i} - \mathbf{j} - 3\mathbf{k}) = 0$
c $\mathbf{r}.(\mathbf{i} + 3\mathbf{j} + 4\mathbf{k}) = -10$ d $\mathbf{r}.(4\mathbf{i} + \mathbf{j} - 5\mathbf{k}) = 9$

3 a $2x + y + z = 0$ b $5x - y - 3z = 0$
c $x + 3y + 4z = -10$ d $4x + y - 5z = 9$

4 $n_1x + n_2y + n_3z = k$

5 $\alpha = 4.25°$ (to 3 s.f.)

6 $\alpha = 43.1°$ (to 3 s.f.)

7 $\alpha = 68.3°$ (to 3 s.f.)

8 $\alpha = 40.2°$ (to 3 s.f.)

9 a Q lies on l_1 when $\lambda = 1$ b $\frac{29}{30}$
c (10, 7, 2) or (−8, 1, 2)

10 a lies on on l_1: $\frac{x-6}{-1} = \frac{y+3}{2} = \frac{z+2}{3} = 3$
lies on on l_2: $\frac{x+5}{2} = \frac{y-15}{-3} = \frac{z-3}{1} = 4$
b 69.1°

11 50.1°

12 a Points A, B and C lie on plane with equation
$\mathbf{r} = \begin{pmatrix}3\\5\\-1\end{pmatrix} + \lambda\begin{pmatrix}-1\\-7\\5\end{pmatrix} + \mu\begin{pmatrix}1\\-2\\1\end{pmatrix}$
but D does not.
b 63.0°

13 Find equation of plane for any two faces in the form $\mathbf{r.n}_1 = p_1$ and $\mathbf{r.n}_2 = p_2$ and use $\cos\theta = \left|\frac{\mathbf{n}_1.\mathbf{n}_2}{|\mathbf{n}_1||\mathbf{n}_2|}\right|$

14 Let $F(0, 0, 20)$, $A(0.\ 8.\ 2)$, $B(12, -5, 3)$, $C(-2, 6, 5)$
angle between FA and FB is 50.9°
angle between FB and FC is 54.8°
angle between FC and FA is 7.4°
No not stable.

Exercise 9E

1 a The two lines do meet at the point (3, 1, 10)
b The lines do not meet.
c The two lines do meet at the point $(0, 1\frac{1}{2}, 4\frac{1}{2})$

2 l_1 and l_2 meet when $\lambda = 4$ and $\mu = -2$
coordinates of point of intersection (−2, −4, 15)

3 No solution for λ and μ

4 a $(2\frac{2}{3}, \frac{1}{6}, 4\frac{1}{3})$ b (1, 2, 0)

5 a $\begin{pmatrix}2+\lambda\\3+\lambda\\-2+\lambda\end{pmatrix}.\begin{pmatrix}1\\1\\-2\end{pmatrix} = 1$ give $9 = 1$ i.e. no solutions for λ
b The line is parallel to, and not contained in, the plane.

6 a p = 3 b (2, 5, −3)

7 a (6, 1, −1) b $\cos\theta = \frac{\sqrt{3}}{2}$

8 $-3\lambda = \mu$ and $-1 + 5\lambda = 1 - 2\mu$ give $\lambda = -2$ and $\mu = 6$, but these are not consistent with $2 + 4\lambda = -5 + 2\mu$ so the lines do not intersect.
Direction of l_2 is $\begin{pmatrix}1\\-2\\2\end{pmatrix}$ which is not parallel to $\begin{pmatrix}-3\\5\\4\end{pmatrix}$, so lines are skew.

9 a Solve $\begin{pmatrix}-1\\3\\2\end{pmatrix}.\begin{pmatrix}q\\2\\-1\end{pmatrix} = 0$ b p = −2
c (4, 14, −4) d $\begin{pmatrix}-1\\29\\6\end{pmatrix}$

10 a $k = 2$ b $2x + 3y - z = 2$
c $N(4, 1, 9)$

11 a $P\left(\frac{53}{31}, -\frac{86}{31}, \frac{132}{31}\right)$ b 76.7°

Exercise 9F

1 $\frac{\sqrt{198}}{5}$ or 2.81 (3 s.f.)

2 $\sqrt{6}$

3 a Lines do not meet
Shortest distance = 3.61 (3 s.f)
b Lines do not meet
Shortest distance = 4.24 (3 s.f)
c Lines do not meet
Shortest distance = 3.61 (3 s.f)

4 3.54 (3 s.f.)

5 a 3 b 1

6 a 3 b $\frac{1}{3}$ c $\frac{2}{3}$ d 4

7 $\left(\frac{7}{3}, \frac{2}{3}, \frac{8}{3}\right)$

8 a $\frac{2\sqrt{6}}{3}$ b $Q\left(-\frac{5}{3}, \frac{4}{3}, \frac{13}{3}\right)$

9 a Shortest distance from line AB to birdwatcher is 0.45 km, which is less than 0.5 km so yes.
b In practice the bird is very unlikely to fly in a straight line from A to B

Online Full worked solutions are available in SolutionBank.

10 a $\frac{40\sqrt{29}}{29}$

b $\begin{pmatrix}2\\-2\\3\end{pmatrix}.\begin{pmatrix}2\\5\\2\end{pmatrix} = \begin{pmatrix}-3\\0\\3\end{pmatrix}.\begin{pmatrix}2\\5\\2\end{pmatrix} = 0$ so perpendicular

c 82.6°

11 a $\frac{13\sqrt{2}}{3}$ **b** $\frac{x-3}{-31} = \frac{y+1}{20} = \frac{z-2}{-41}$

12 $\mathbf{r} = \begin{pmatrix}4\\-3\\0\end{pmatrix} + \mu\begin{pmatrix}2\\-1\\6\end{pmatrix}$

Mixed exercise 9

1 a $\mathbf{r} = (\mathbf{i} - \mathbf{j} + 3\mathbf{k}) + \lambda(3\mathbf{j} - \mathbf{k})$ **b** $\mathbf{i} + \mathbf{j} + \frac{7}{3}\mathbf{k}$

2 $\frac{x-7}{4} = \frac{y+1}{-2} = \frac{z-2}{-3}$

3 $\mathbf{r} = (2\mathbf{i} + 3\mathbf{j} - 4\mathbf{k}) + \lambda(2\mathbf{j} + 3\mathbf{k})$

4 a $\frac{x-3}{2} = \frac{y+2}{1} = \frac{z-1}{-1}$ **b** $a = -\frac{7}{2}$, $b = \frac{5}{2}$

5 $7\mathbf{i} + 4\mathbf{j} - 5\mathbf{k}$ lies on l when $\lambda = 2$
$9\mathbf{i} + 3\mathbf{j} - 6\mathbf{k} = 3(3\mathbf{i} + \mathbf{j} - 2\mathbf{k})$ so parallel

6 a $\mathbf{r} = \begin{pmatrix}-2\\2\\-3\end{pmatrix} + \lambda\begin{pmatrix}1\\3\\4\end{pmatrix}$

b (0, 8, 5) lies on l when $\lambda = 2$

7 a $\mathbf{r} = \begin{pmatrix}2\\-1\\2\end{pmatrix} + \lambda\begin{pmatrix}-1\\4\\-3\end{pmatrix} + \mu\begin{pmatrix}2\\3\\3\end{pmatrix}$

b $21x - 3y - 11z = 23$

8 $\mathbf{r} = \begin{pmatrix}6\\0\\0\end{pmatrix} + \lambda\begin{pmatrix}2\\-3\\0\end{pmatrix} + \mu\begin{pmatrix}4\\0\\3\end{pmatrix}$

9 a $3\mathbf{i} + 4\mathbf{j} + 5\mathbf{k}$, $\mathbf{i} + \mathbf{j} + 4\mathbf{k}$

b $\frac{\overrightarrow{ML}.\overrightarrow{MN}}{|\overrightarrow{ML}||\overrightarrow{MN}|} = \frac{27}{5\sqrt{2}\,3\sqrt{2}} = \frac{9}{10}$

10 a $\mathbf{r} = \begin{pmatrix}9\\-2\\1\end{pmatrix} + \mu\begin{pmatrix}-3\\4\\5\end{pmatrix}$ **b** $p = 6$, $q = 11$

c 39.8° **d** $\frac{36}{5}\mathbf{i} + \frac{2}{5}\mathbf{j} + 4\mathbf{k}$

11 a $\mathbf{r} = \begin{pmatrix}1\\2\\-3\end{pmatrix} + \mu\begin{pmatrix}4\\-2\\0\end{pmatrix}$ **b** $\mu = -3$

c 53.4° **d** $\frac{\sqrt{145}}{5}$

12 a $\mathbf{r}_1.\mathbf{r}_2 = 0$, therefore vectors are perpendicular
b $5\mathbf{i} - \mathbf{k}$ **c** $l_1 : \lambda = -3$ **d** 1.5 km

13 $3\sqrt{2}$ or 4.24

14 $\frac{7}{9}\sqrt{6}$ or 1.91 (3 s.f.)

15 a $\mathbf{r}.(2\mathbf{i} - 9\mathbf{j} + 4\mathbf{k}) = -15$
b $\mathbf{r}.(2\mathbf{i} - \mathbf{j} + \mathbf{k}) = 2$
c $\mathbf{r}.(8\mathbf{i} - 5\mathbf{j} + \mathbf{k}) = 22$

16 $-10x - 2y + 16z = 4$

17 a $\frac{1}{\sqrt{50}}(3\mathbf{i} + 5\mathbf{j} + 4\mathbf{k})$
b $3x + 5y + 4z = 30$ **c** $3\sqrt{2}$

18 a Vector is perpendicular to both $\mathbf{j}$ and $\mathbf{i} - \mathbf{k}$.
b $\frac{\sqrt{2}}{2}$ or 0.707 (to 3 s.f.) **c** $x + z = 1$

19 a $-15\mathbf{i} - 20\mathbf{j} + 10\mathbf{k}$ or a multiple of $(3\mathbf{i} + 4\mathbf{j} - 2\mathbf{k})$
b $3x + 4y - 2z - 5 = 0$

20 a $\begin{pmatrix}2\\-2\\3\end{pmatrix}.\begin{pmatrix}1\\5\\3\end{pmatrix} = 1$ **b** $\begin{pmatrix}2\\-1\\1\end{pmatrix}.\begin{pmatrix}1\\5\\3\end{pmatrix} = 0$
c $\mathbf{r} = 2\mathbf{i} - 2\mathbf{j} + 3\mathbf{k} + \lambda(2\mathbf{i} - \mathbf{j} + \mathbf{k})$
d $\left(-1, -\frac{1}{2}, \frac{3}{2}\right)$ **e** 3.67 (3 s.f.)

21 a intersect when $\lambda = 3$, $\mu = -2$
b $\begin{pmatrix}7\\4\\-6\end{pmatrix}$ **c** $\frac{4\sqrt{10}}{15}$

22 a $a = 11$, $b = 7$ **b** $P(5, 9, 4)$ **c** $\sqrt{122}$

23 a $\overrightarrow{AB} = \begin{pmatrix}-1\\-1\\2\end{pmatrix}$ **b** $\mathbf{r} = \begin{pmatrix}6\\3\\4\end{pmatrix} + \lambda\begin{pmatrix}-1\\-1\\2\end{pmatrix}$
c $\begin{pmatrix}7.5\\4.5\\1\end{pmatrix}$

24 a meet when $\lambda = 2$, $\mu = 6$, (7, 0, 2)
b 80.4°
c lies on l_2 when $\lambda = 1$
d 2.42

25 a 10° **b** $\frac{3\sqrt{17}}{17}$

26 a intersect at (180, −5, 7)
b pass through same point but not necessarily at the same time

Challenge

1 a $-2(-2x + y - 3z) = (-2) \times (-5)$ gives $4x - 2y + 6z = 10$
b matrix $\mathbf{A}$ is singular if $\det\mathbf{A} = 0$
$4(c + 3b) + 2(-2c + 3a) + 6(-2b - a) = 6a - 6a + 12b - 12b + 4c - 4c = 0$
c i $a = 2n$, $b = -n$, $c = 30$ where $n \in \mathbb{R}$, $n \neq 3$
ii $a = 6$, $b = -3$ and $c = 9$

2 Centre of circle (4, 0, 1), radius $4\sqrt{2}$

Review exercise 2

1 a Does not exist: $\mathbf{B}$ doesn't have 3 rows.
b $\begin{pmatrix}3q & 2q & pq\\9 & 4 & 3p+1\end{pmatrix}$
c $\begin{pmatrix}6q + pq\\3p + 25\end{pmatrix}$
d Does not exist: $\mathbf{C}$ doesn't have 2 columns.

2 a $a = -2$, $b = 3$

3 $bc - ad$

4 $a = 4$, $b = -1$

5 a $-\frac{2}{3}$ **b** −2 **c** −4

6 a $\begin{pmatrix}1 & \frac{1}{2}\\3 & 2\end{pmatrix}$ **b** $\begin{pmatrix}2 & 1\\3p+3 & 2p+\frac{3}{2}\end{pmatrix}$ **c** $-\frac{1}{2}$

7 a $k = 3, 6$
b $\frac{1}{-k^2 + 9k - 18}\begin{pmatrix}-k & -2 & k-2\\9k & 18 & -k^2\\9 & 9-k & -k\end{pmatrix}$

8 $\frac{1}{3(p+2)}\begin{pmatrix}0 & p+2 & 0\\3 & -2p+2 & 6\\3 & -3p & -3p\end{pmatrix}$

9 a $\frac{1}{pq}\begin{pmatrix}q & q\\3p & 4p\end{pmatrix}$ **b** $\frac{1}{pq}\begin{pmatrix}pq & 4q^2\\2p^2 & 13pq\end{pmatrix}$

10 (1, 2, 0)

11 1060 Woolly, 900 Classic and 850 Suri

12 a $p = -3$ or $p = 11$
b $p = -3$: planes form a sheaf
$p = 11$: planes form a prism

13 a $\begin{pmatrix} -\frac{1}{\sqrt{2}} & -\frac{1}{\sqrt{2}} \\ -\frac{1}{\sqrt{2}} & \frac{1}{\sqrt{2}} \end{pmatrix}$

b $\begin{pmatrix} -\frac{1}{\sqrt{2}} & -\frac{1}{\sqrt{2}} \\ -\frac{1}{\sqrt{2}} & \frac{1}{\sqrt{2}} \end{pmatrix}\begin{pmatrix} -\frac{1}{\sqrt{2}} & -\frac{1}{\sqrt{2}} \\ -\frac{1}{\sqrt{2}} & \frac{1}{\sqrt{2}} \end{pmatrix}$

$= \begin{pmatrix} \frac{1}{2}+\frac{1}{2} & \frac{1}{2}-\frac{1}{2} \\ \frac{1}{2}-\frac{1}{2} & \frac{1}{2}+\frac{1}{2} \end{pmatrix} = \begin{pmatrix} 1 & 0 \\ 0 & 1 \end{pmatrix}$

14 a $a = 3, b = -4, c = 2, d = -3$

b $\begin{pmatrix} 3 & -4 \\ 2 & -3 \end{pmatrix}\begin{pmatrix} 3 & -4 \\ 2 & -3 \end{pmatrix} = \begin{pmatrix} 9-8 & -12+12 \\ 6-6 & -8+9 \end{pmatrix} = \begin{pmatrix} 1 & 0 \\ 0 & 1 \end{pmatrix}$

c $p = 36, q = 25$

15 a $\begin{pmatrix} -1 & 2 \\ 0 & 3 \end{pmatrix}$

b $\begin{pmatrix} -1 & 2 \\ 0 & 3 \end{pmatrix}\begin{pmatrix} x \\ 2x \end{pmatrix} = \begin{pmatrix} 3x \\ 6x \end{pmatrix}$; $6x = 2(3x)$ so the point satisfies the equation of the original line.

c $A(2, 1)$, $B(0, 5)$, $C(-2, 4)$

d

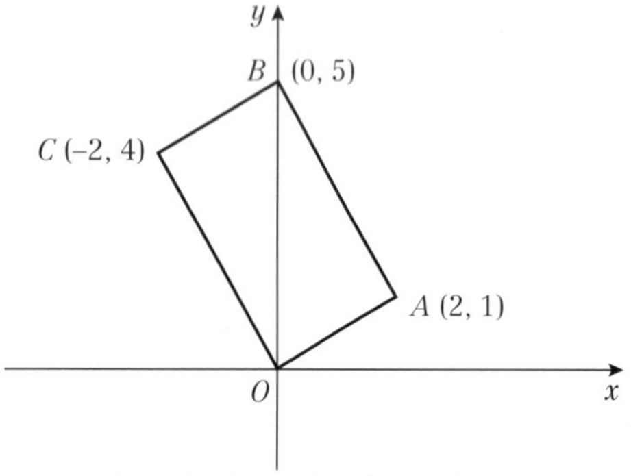

16 a $\det \mathbf{A} = 3\begin{vmatrix} 1 & 1 \\ 3 & u \end{vmatrix} - \begin{vmatrix} 1 & -1 \\ 3 & u \end{vmatrix} + 5\begin{vmatrix} 1 & -1 \\ 1 & 1 \end{vmatrix}$

$= 3(u - 3) - (u + 3) + 5(1 + 1) = 2(u - 1)$

b $\frac{1}{2(u-1)}\begin{pmatrix} u-3 & -u-3 & 2 \\ -u+5 & 3u+5 & -4 \\ -2 & -4 & 2 \end{pmatrix}$

c $a = 1.2, b = -0.4, c = 0.2$

17 a $a = -4$
$b = -3$
$c = 0$

b -1

c $x = 2y$

18 a $2k^2 + 3k - 3$

b $-\frac{7}{2}$ or 2

19 a Scale factor 3

b 45° anti-clockwise about (0, 0)

c $\left(\frac{p+q}{3\sqrt{2}}, \frac{-p+q}{3\sqrt{2}}\right)$

20 $\begin{pmatrix} 0 & 2 \\ \frac{1}{2} & 0 \end{pmatrix}\begin{pmatrix} x \\ k - \frac{1}{2}x \end{pmatrix} = \begin{pmatrix} 2k - x \\ \frac{1}{2}x \end{pmatrix}$

$k - \frac{1}{2}(2k - x) = k - k + \frac{1}{2}x = \frac{1}{2}x$ so the transformed point satisfies the equation of the original line.

21 Basis: When $n = 1$, LHS = RHS = 4

Assumption: $\sum_{r=1}^{k} r(r+3) = \frac{1}{3}k(k+1)(k+5)$

Induction: $\sum_{r=1}^{k+1} r(r+3) = \sum_{r=1}^{k} r(r+3) + (k+1)(k+4)$

$= \frac{1}{3}k(k+1)(k+5) + (k+1)(k+4)$

$= (k+1)\left(\frac{1}{3}k(k+5) + (k+4)\right)$

$= \frac{1}{3}(k+1)(k+2)(k+6)$

So if the statement holds for $n = k$, it holds for $n = k + 1$.

Conclusion: The statement holds for all $n \geqslant 1$.

22 Basis: When $n = 1$, LHS = RHS = 1

Assumption: $\sum_{r=1}^{k} (2r-1)^2 = \frac{1}{3}k(2k-1)(2k+1)$

Induction: $\sum_{r=1}^{k+1} (2r-1)^2 = \sum_{r=1}^{k} (2r-1)^2 + (2k+2-1)^2$

$= \frac{1}{3}k(2k-1)(2k+1) + (2k+1)^2$

$= (2k+1)(\frac{1}{3}k(2k-1) + 2k + 1)$

$= \frac{1}{3}(k+1)(2(k+1) - 1)(2(k+1) + 1)$

So if the statement holds for $n = k$, it holds for $n = k + 1$.

Conclusion: The statement holds for all $n \in \mathbb{Z}^+$.

23 Basis: $S_1 = a_1 = 6 = \frac{1}{2} \times 1 \times (1+1)^2 \times (1+2)$

Assumption: $S_k = \frac{1}{2}k(k+1)^2(k+2)$

Induction:

$S_{k+1} = S_k + a_{k+1} = \frac{1}{2}k(k+1)^2(k+2) + (k+1)(k+2)(2k+3)$

$= (k+1)(k+2)(\frac{1}{2}k(k+1) + (2k+3))$

$= \frac{1}{2}(k+1)(k+2)^2(k+3)$

So if the statement holds for $n = k$, it holds for $n = k + 1$.

Conclusion: The statement holds for all $n \geq 1$.

24 Basis: When $n = 1$, LHS = RHS = 0

Assumption: $\sum_{r=1}^{k} r^2(r-1) = \frac{1}{12}k(k-1)(k+1)(3k+2)$

Induction: $\sum_{r=1}^{k+1} r^2(r-1) = \sum_{r=1}^{k} r^2(r-1) + (k+1)^2 k$

$= \frac{1}{12}k(k-1)(k+1)(3k+2) + (k+1)^2 k$

$= \frac{1}{12}(k+1)k(k+2)(3(k+1)+2)$

So if the statement holds for $n = k$, it holds for $n = k + 1$.

Conclusion: The statement holds for all $n \geqslant 1$.

25 a $f(k+1) - f(k) = 3^{4k+4} + 2^{4k+6} - 3^{4k} - 2^{4k+2}$

$= 3^{4k}(3^4 - 1) + 2^{4k+2}(2^4 - 1) = 80 \times 3^{4k} + 15 \times 2^{4k+2}$

The first term is divisible by 15 since it is clearly divisible by 3, and 5 divides 80. Therefore $f(k+1) - f(k)$ is divisible by 15.

b Basis: When $n = 1$, $f(n) = 3^4 + 2^6 = 145 = 5 \times 29$

Assumption: $f(k)$ is divisible by 5.

Induction: From part **a**, $f(k+1) - f(k)$ is divisible by 15 and hence also by 5. Since $f(k)$ is divisible by 5, $f(k+1)$ is also divisible by 5.

So if the statement holds for $n = k$, it holds for $n = k + 1$.

Conclusion: The statement holds for all $n \in \mathbb{Z}^+$.

26 a $24 \times 2^{4(n+1)} + 3^{4(n+1)} - 24 \times 2^{4n} - 3^{4n}$

b Basis: When $n = 1$, $f(n) = 24 \times 2^4 + 3^4 = 465 = 5 \times 93$

Assumption: $f(k)$ is divisible by 5.

Induction: From part **a**,

$f(k+1) - f(k) = 24 \times 2^{4k+4} + 3^{4k+4} - 24 \times 2^{4k} - 3^{4k}$

$= 24 \times 2^{4k}(2^4 - 1) + 3^{4k}(3^4 - 1)$

$= 5(72 \times 2^{4k} + 16 \times 3^{4k})$

So if $f(k)$ is divisible by 5, $f(k+1)$ is divisible by 5.

Conclusion: $f(n)$ is divisible by 5 for all $n \in \mathbb{Z}^+$.

Online Full worked solutions are available in SolutionBank.

27 Let $f(n) = 7^n + 4^n + 1$.
Basis: $f(1) = 7 + 4 + 1 = 12$, which is divisible by 6.
Assumption: $f(k)$ is divisible by 6.
Induction: $f(k+1) - f(k) = 7^{k+1} + 4^{k+1} + 1 - 7^k - 4^k - 1$
$= 7^k(7-1) + 4^k(4-1) = 6 \times 7^k + 3 \times 4^k$
The first term is divisible by 6, and since 4^k is even, the second term is divisible by 6. So if $f(k)$ is divisible by 6, then $f(k+1)$ is also divisible by 6.
Conclusion: $f(n)$ is divisible by 6 for all $n \in \mathbb{Z}^+$.

28 Let $f(n) = 4^n + 6n - 1$.
Basis: When $n = 1$, $f(n) = 4^1 + 6(1) - 1 = 9$, which is divisible by 9.
Assumption: $f(k)$ is divisible by 9.
Induction: $f(k+1) - f(k) = 4^{k+1} + 6(k+1) - 1 - 4^k - 6k + 1$
$= 4^k(4-1) + 6 = 3(4^k - 1) + 9$
$4^k - 1$ is divisible by $4 - 1 = 3$.
First term has two factors of 3 so is divisible by 9 and the second term is divisible by 9. So if $f(k)$ is divisible by 9, then $f(k+1)$ is also divisible by 9
Conclusion: $f(n)$ is divisible by 9 for all $n \in \mathbb{Z}^+$.

29 Let $f(n) = 3^{4n-1} + 2^{4n-1} + 5$
Basis: $f(1) = 3^3 + 2^3 + 5 = 27 + 8 + 5 = 40 = 10 \times 4$
Assumption: $f(k)$ is divisible by 10.
Induction: $f(k+1) - f(k) = 3^{4k+3} + 2^{4k+3} - 3^{4k-1} + 2^{4k-1}$
$= 3^{4k-1}(3^4 - 1) + 2^{4k-1}(2^4 - 1) = 80 \times 3^{4k-1} + 15 \times 2^{4k-1}$
This is divisible by 10: 15 is divisible by 5 and 2^{4k-1} is even.
So if $f(k)$ is divisible by 10, then $f(k+1)$ is divisible by 10.
Conclusion: $f(n)$ is divisible by 10 for all positive integers, n.

30 Basis: When $n = 1$, $\mathbf{A}^1 = \begin{pmatrix} 1 & c \\ 0 & 2 \end{pmatrix} = \begin{pmatrix} 1 & (2^1-1)c \\ 0 & 2^1 \end{pmatrix}$

Assumption: $\mathbf{A}^k = \begin{pmatrix} 1 & (2^k-1)c \\ 0 & 2^k \end{pmatrix}$

Induction: $\mathbf{A}^{k+1} = \mathbf{A}^k\begin{pmatrix} 1 & c \\ 0 & 2 \end{pmatrix} = \begin{pmatrix} 1 & (2^k-1)c \\ 0 & 2^k \end{pmatrix}\begin{pmatrix} 1 & c \\ 0 & 2 \end{pmatrix}$
$= \begin{pmatrix} 1 & c + 2c(2^k-1) \\ 0 & 2^{k+1} \end{pmatrix} = \begin{pmatrix} 1 & (2^{k+1}-1)c \\ 0 & 2^{k+1} \end{pmatrix}$
So if the statement holds for $n = k$, it holds for $n = k + 1$.
Conclusion: The statement holds for all positive integers, n.

31 Basis:
When $n = 1$, $\mathbf{A}^1 = \begin{pmatrix} 3 & 1 \\ -4 & -1 \end{pmatrix} = \begin{pmatrix} 2 \times 1 + 1 & 1 \\ -4 \times 1 & -2 \times 1 + 1 \end{pmatrix}$

Assumption: $\mathbf{A}^k = \begin{pmatrix} 2k+1 & k \\ -4k & -2k+1 \end{pmatrix}$
Induction:
$\mathbf{A}^{k+1} = \mathbf{A}^k\begin{pmatrix} 3 & 1 \\ -4 & -1 \end{pmatrix} = \begin{pmatrix} 2k+1 & k \\ -4k & -2k+1 \end{pmatrix}\begin{pmatrix} 3 & 1 \\ -4 & -1 \end{pmatrix}$
$= \begin{pmatrix} 6k+3-4k & 2k+1-k \\ -12k+8k-4 & -4k+2k-1 \end{pmatrix}$
$= \begin{pmatrix} 2(k+1)+1 & k+1 \\ -4(k+1) & -2(k+1)+1 \end{pmatrix}$
So if the statement holds for $n = k$, it holds for $n = k + 1$.
Conclusion: The statement holds for all positive integers, n.

32 **a** He has not shown it true for $k = 1$
b Let $f(n) = 2^{2n} - 1$
Basis: When $n = 1$, $f(n) = 2^2 - 1 = 3$
Assumption: $f(k) = 2^{2k} - 1$ is divisible by 3.
Induction: $f(k+1) = 2^{2(k+1)} - 1 = 4f(k) + 3$
So if $f(k)$ is divisible by 3, then $f(k+1)$ is divisible by 3.
Conclusion: $f(n)$ is divisible by 3 for all positive integers, n.

33 $5\sqrt{14}$

34 $a = 3$, $b = 13$, $\mathbf{r} = \begin{pmatrix} 1 \\ -1 \\ 3 \end{pmatrix} + \lambda\begin{pmatrix} 2 \\ 4 \\ 5 \end{pmatrix}$ or any equivalent

35 **a** $\frac{x}{2} + \frac{y}{2} = 1$ **b** $\begin{pmatrix} 1 \\ 1 \\ 0 \end{pmatrix}$ or any equivalent **c** 0

36 **a** As the solution $\lambda = -2$, $\mu = -3$ satisfies all three equations, the lines *do* meet.
b $(3, 1, -2)$ is point of intersection.
c $\frac{5}{9}\sqrt{3}$

37 **a** $a = 18$ $b = 9$
b $(6, 10, 16)$
c $14\sqrt{2}$

38 **a** $a(4\mathbf{i} + \mathbf{j} + 2\mathbf{k}).(\mathbf{i} - 5\mathbf{j} + 3\mathbf{k}) = a(4 - 5 + 6) = 5a$
b $\overrightarrow{BA} = a(2\mathbf{i} - 10\mathbf{j} + 6\mathbf{k})$
$$\sin\theta = \left|\frac{(2 + 50 + 18)}{\sqrt{2^2 + (-10)^2 + 6^2} \times \sqrt{1^2 + (-5^2) + 3^2}}\right| = 1$$
Therefore $\overrightarrow{BA}$ is perpendicular to the plane.
c 22.3° (nearest one tenth of a degree)

39 **a** $k = -10$
b $-x + 2y + z = -10$
c $\left(\frac{13}{3}, -\frac{11}{3}, \frac{5}{3}\right)$

40 **a** $(3, -2, -1)$ **b** 1.061

41 **a** $5\sqrt{2}$ **b** $\mathbf{r} = \mathbf{i} - \mathbf{j} + 3\mathbf{k} + \lambda(3\mathbf{i} + 4\mathbf{j} - 5\mathbf{k})$
c $\mu = -1$ **d** $\frac{\pi}{6}$ **e** $\frac{5\sqrt{2}}{2}$

42 **a** 1.066 **b** $\left(\frac{4}{11}, \frac{12}{11}, -\frac{48}{11}\right)$

43 **a** Lines do not intersect.
b Unlikely that the shark will not adjust course to intercept flounder.

Challenge

1 $\begin{pmatrix} 0 & 0 & -1 \\ 0 & -1 & 0 \\ -1 & 0 & 0 \end{pmatrix}$

2 $\left(-\frac{1}{2}, -1, \frac{1}{2}\right)$, radius $= \sqrt{\frac{13}{2}}$

3 Basis: When $n = 1$, $r = 2 \Rightarrow 2(1) \leqslant 2 \leqslant \frac{1}{2}(1^2 + 1 + 2)$

Assumption: $2k \leqslant r \leqslant \frac{1}{2}(k^2 + k + 2)$
Induction:
Lower bound – all lines pass through a single point, $r_k = 2k$
One more line added ⇒ two more regions.
$r_{k+1} = 2k + 2 = 2(k+1)$
Upper bound – lines do not pass through the intersection of any other pairs of lines, $r_k = \frac{1}{2}(k^2 + k + 2)$.
One more line added ⇒ $k + 1$ more regions.
$r_{k+1} = \frac{1}{2}(k^2 + k + 2) + k + 1 = \frac{1}{2}(k^2 + 3k + 4)$
$= \frac{1}{2}((k+1)^2 + (k+1) + 1)$
So if the statement holds for $n = k$, it holds for $n = k + 1$.
Conclusion: The statement holds for all positive integers, n.

Exam-style practice

1 General point A on l_1 is $\begin{pmatrix} -3 + 5\lambda \\ -\lambda \\ 5 + \lambda \end{pmatrix}$.

General point B on l_2 is $\begin{pmatrix} 10 + 6\mu \\ -1 - 2\mu \\ 15 + 4\mu \end{pmatrix}$.

$\overrightarrow{AB} = \begin{pmatrix} 13 - 5\lambda + 6\mu \\ -1 + \lambda - 2\mu \\ 10 - \lambda + 4\mu \end{pmatrix}$

Shortest distance between l_1 and l_2 is when $\overrightarrow{AB}$ is perpendicular to l_1 and l_2, so the scalar product with their direction vectors is 0, which gives equations:

$76 - 27\lambda + 36\mu = 0$ and $30 - 9\lambda + 14\mu = 0$

Solve to get $\lambda = -\frac{8}{27}$ and $\mu = -\frac{7}{3}$.

Then $\overrightarrow{AB} = \frac{13}{27}\begin{pmatrix} 1 \\ 7 \\ 2 \end{pmatrix} \neq \begin{pmatrix} 0 \\ 0 \\ 0 \end{pmatrix}$, so the lines do not meet.

2 **a** $\det \mathbf{M} = k + 14$

b Equations do not form consistent system. Planes form a prism.

3 $w^3 - 17w - 20 = 0$. $p = 1$, $q = 0$, $r = -17$, $s = -20$

4 **a** Basis: When $n = 1$, LHS = RHS = 1.

Assumption: $\sum_{r=1}^{k} r^3 = \frac{1}{4}k^2(k+1)^2$

Induction:

$\sum_{r=1}^{k+1} r^3 = \sum_{r=1}^{k} r^3 + (k+1)^3 = \frac{1}{4}k^2(k+1)^2 + (k+1)^3$

$= (k+1)^2(\frac{1}{4}k^2 + k + 1) = \frac{1}{4}(k+1)^2(k+2)^2$

So if the result holds for $n = k$, it holds for $n = k + 1$.

Conclusion: The statement holds for all positive integers n.

b $\sum_{r=1}^{n} 2r(r+1) = 2\sum_{r=1}^{n} r^2 + 2\sum_{r=1}^{n} r$

$= 2 \times \frac{1}{6}n(n+1)(n+2) + 2 \times \frac{1}{2}n(n+1)$

$= \frac{2}{3}n(n+1)(n+2)$

c $n^2(n+1)^2 = \frac{2}{3}n(n+1)(n+2)$

$\Rightarrow 3n^4 + 4n^3 - 3n^2 - 4n = 0$

$\Rightarrow n(n-1)(n+1)(3n+4) = 0$

So $n = 0, 1, -1$ or $-\frac{4}{3}$

The only positive integer that the result holds for is $n = 1$.

5 **a** $3 + 2i$ and $3 - 2i$ are roots, so $(z - (3 + 2i))(z - (3 - 2i))$ is a factor of f(z).

$(z - (3 + 2i))(z - (3 - 2i))$

$= z^2 - (3 + 2i + 3 - 2i)z + (3 + 2i)(3 - 2i)$

$= z^2 - 6z + (3^2 + 6i - 6i - (2i)^2) = z^2 - 6z + 3^2 - (-4)$

$= z^2 - 6z + 13$

So $z^2 - 6z + 13$ is a factor of f(z).

b -206

c $z = 3 - 2i, 3 + 2i, 4 - i$ or $4 + i$

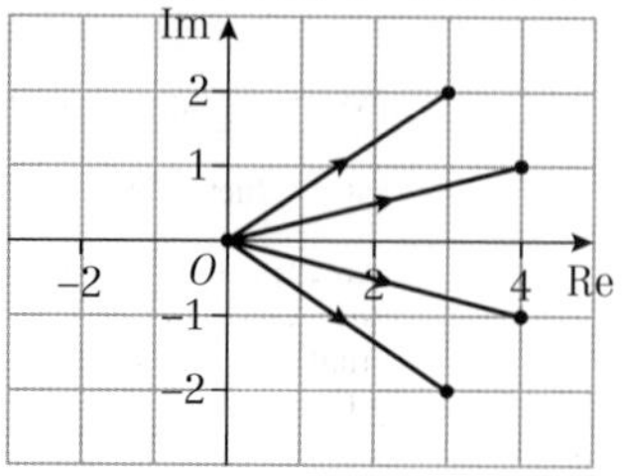

6 **a** $\begin{pmatrix} \frac{1}{\sqrt{6}} \\ \frac{-1}{\sqrt{6}} \\ \frac{2}{\sqrt{6}} \end{pmatrix}$ **b** $(-2, -1, 2)$; 0.92 radians

7 **a** £2680.83

b e.g. Doesn't account for a hole through the middle of the bead.

8 **a** 5 **b** Rotation 135° anticlockwise

c $\left(\frac{-a+b}{5\sqrt{2}}, \frac{-a-b}{5\sqrt{2}}\right)$

9

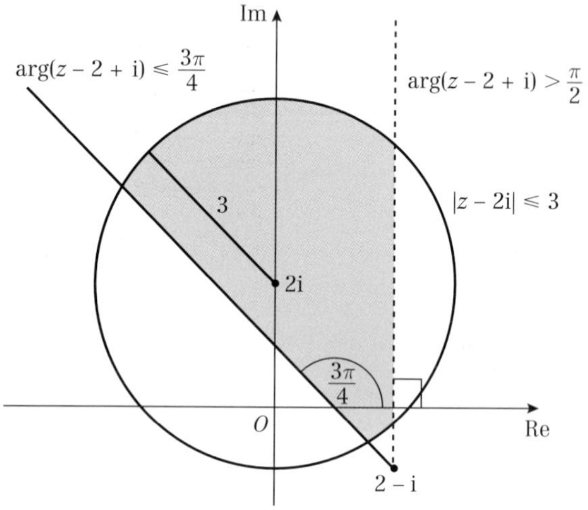

10 **a** Yes: closest point is 5.6…m from the origin.

b e.g. The car would be unlikely to drive in a perfectly straight line.

Online Full worked solutions are available in SolutionBank.

Index